SCHAUM'S OUTLINE OF

THEORY AND PROBLEMS

OF

DIFFERENTIAL EQUATIONS

Second Edition

•

RICHARD BRONSON, Ph.D.

Professor of Mathematics and Computer Science
Fairleigh Dickinson University

SCHAUM'S OUTLINE SERIES
McGraw-Hill

New York San Francisco Washington, D.C. Auckland Bogotá Caracas
Lisbon London Madrid Mexico City Milan Montreal
New Delhi San Juan Singapore Sydney Tokyo Toronto

RICHARD BRONSON, Ph.D. is Professor of Mathematics at Fairleigh Dickinson University. He received his Ph.D. in applied mathematics from Stevens Institute of Technology in 1968. Dr. Bronson has served as an associate editor of the journal *Simulation,* as a contributing editor to *SIAM News,* and as a consultant to Bell Laboratories. He has conducted joint research in mathematical modeling and computer simulation at the Technion – Israel Institute of Technology and the Wharton School of Business at the University of Pennsylvania. Dr. Bronson has published over 30 technical articles and books, the latter including *Schaum's Outline of Matrix Operations* and *Schaum's Outline of Operations Research.*

Schaum's Outline of Theory and Problems of
DIFFERENTIAL EQUATIONS

16 17 18 19 VFM VFM 0 5 4

ISBN 0-07-008019-4

Sponsoring Editor: David Beckwith
Production Supervisor: Fred Schulte
Editing Supervisor: Maureen Walker

Library of Congress Cataloging-in-Publication Data

Bronson, Richard.
Schaum's outline of theory and problems of differential equations / Richard Bronson.—2nd ed.
p. cm.
Updated ed. of Schaum's outline of modern introductory differential equations. c1973.
Includes index.
ISBN 0-07-008019-4
1. Differential equations. I. Bronson, Richard. Schaum's outline of modern introductory differential equations. II. Title.
QA372.B856 1993
515'.35—dc20 92-41809
CIP

McGraw-Hill

A Division of The ***McGraw-Hill*** *Companies*

To Ignace and Gwendolyn Bronson
Samuel and Rose Feldschuh

Preface

Differential equations is one of the linchpins of modern mathematics which, along with matrices, is essential for analyzing and solving complex problems in engineering, the natural sciences, economics, and even business. The emergence of low-cost, high-speed computers has spawned new techniques for solving differential equations, which allows problem solvers to model and solve complex problems based on systems of differential equations.

As with the first edition, this book outlines both the classical theory of differential equations and the more recently developed solution procedures favored by practitioners. Included are the systems approach, which lends itself to both matrix methods and Laplace transforms. Numerical methods for implementation on computers are covered as are eigenvalue problems. Many solved problems clarify each of the methods presented in this book, and a variety of applications illustrate the types of problems that are modeled and solved with differential equations.

There are significant differences between this edition and the first. Perhaps most importantly, the number of solved problems was increased and now numbers over 500; the number of supplementary problems was more than doubled and totals over 800. The Adams–Bashforth–Moulton method was added to the section on numerical techniques, and greater emphasis is now placed on Runge–Kutta methods. A variety of older, and now antiquated, numerical procedures were deleted. Interest rate problems, buoyancy problems, and graphical methods, including direction fields, were added.

This new edition organizes the material differently to provide more cohesion to the subject matter. As before, the bulk of the book addresses both initial-value problems and differential equations without subsidiary conditions. Now, however, all the solution techniques for linear differential equations are grouped together in consecutive chapters. Thus, solutions by characteristic equations are followed directly by undetermined coefficients, variation of parameters, Laplace transforms, and matrix methods. This material is followed by chapters on nonlinear differential equations and then by numerical methods, which are applicable to both linear and nonlinear equations. The last two chapters are an introduction to boundary-value problems.

Each chapter of the book is divided into three parts. The first outlines salient points of the theory and concisely summarizes solution procedures, drawing attention to potential difficulties and subtleties that too easily can be overlooked. The second part consists of worked-out problems to clarify and, on occasion, to augment the material presented in the first part. Finally, there is a section of problems with answers that readers can use to test their understanding of the material.

Contents

Chapter *1* **BASIC CONCEPTS** **1**
Differential equations. Notation. Solutions. Initial-value and boundary-value problems.

Chapter *2* **CLASSIFICATIONS OF FIRST-ORDER DIFFERENTIAL EQUATIONS** **8**
Standard form and differential form. Linear equations. Bernoulli equations. Homogeneous equations. Separable equations. Exact equations.

Chapter *3* **SEPARABLE FIRST-ORDER DIFFERENTIAL EQUATIONS** **14**
General solution. Solutions to the initial-value problem. Reduction of homogeneous equations.

Chapter *4* **EXACT FIRST-ORDER DIFFERENTIAL EQUATIONS** **24**
Defining properties. Method of solution. Integrating factors.

Chapter *5* **LINEAR FIRST-ORDER DIFFERENTIAL EQUATIONS** **35**
Method of solution. Reduction of Bernoulli equations.

Chapter *6* **APPLICATIONS OF FIRST-ORDER DIFFERENTIAL EQUATIONS** **43**
Growth and decay problems. Temperature problems. Falling body problems. Dilution problems. Electrical circuits. Orthogonal trajectories.

Chapter *7* **LINEAR DIFFERENTIAL EQUATIONS: THEORY OF SOLUTIONS** **67**
Linear differential equations. Linearly independent solutions. The Wronskian. Nonhomogeneous equations.

Chapter *8* **SECOND-ORDER LINEAR HOMOGENEOUS DIFFERENTIAL EQUATIONS WITH CONSTANT COEFFICIENTS** **77**
The characteristic equation. The general solution.

Chapter *9* ***n*TH-ORDER LINEAR HOMOGENEOUS DIFFERENTIAL EQUATIONS WITH CONSTANT COEFFICIENTS** **83**
The characteristic equation. The general solution.

Chapter *10* **THE METHOD OF UNDETERMINED COEFFICIENTS** **88**
Simple form of the method. Generalizations. Modifications. Limitations of the method.

Chapter *11* **VARIATION OF PARAMETERS** **97**
The method. Scope of the method.

Chapter *12* **INITIAL-VALUE PROBLEMS** **104**

Chapter *13* **APPLICATIONS OF SECOND-ORDER LINEAR DIFFERENTIAL EQUATIONS** **108**
Spring problems. Electrical circuit problems. Buoyancy problems. Classifying solutions.

Chapter *14* **THE LAPLACE TRANSFORM** **125**
Definition. Properties of Laplace transforms. Functions of other independent variables.

Chapter *15* **INVERSE LAPLACE TRANSFORMS** **138**
Definition. Manipulating denominators. Manipulating numerators.

Chapter *16* **CONVOLUTIONS AND THE UNIT STEP FUNCTION** **146**
Convolutions. Unit step function. Translations.

Chapter *17* **SOLUTIONS OF LINEAR DIFFERENTIAL EQUATIONS WITH CONSTANT COEFFICIENTS BY LAPLACE TRANSFORMS** **154**
Laplace transforms of derivatives. Solutions of differential equations.

Chapter *18* **SOLUTIONS OF LINEAR SYSTEMS BY LAPLACE TRANSFORMS** **161**
The method.

Chapter *19* **MATRICES** **166**
Matrices and vectors. Matrix addition. Scalar and matrix multiplication. Powers of a square matrix. Differentiation and integration of matrices. The characteristic equation.

Chapter 20 e^{At} 175
Definition. Computation of e^{At}.

Chapter 21 REDUCTION OF LINEAR DIFFERENTIAL EQUATIONS TO A FIRST-ORDER SYSTEM 183
Reduction of one equation. Reduction of a system.

Chapter 22 SOLUTIONS OF LINEAR DIFFERENTIAL EQUATIONS WITH CONSTANT COEFFICIENTS BY MATRIX METHODS 191
Solution of the initial-value problem. Solution with no initial conditions.

Chapter 23 LINEAR DIFFERENTIAL EQUATIONS WITH VARIABLE COEFFICIENTS 199
Second-order equations. Analytic functions and ordinary points. Solutions around the origin of homogeneous equations. Solutions around the origin of nonhomogeneous equations. Initial-value problems. Solutions around other points.

Chapter 24 REGULAR SINGULAR POINTS AND THE METHOD OF FROBENIUS 213
Regular singular points. Method of Frobenius. General solution.

Chapter 25 GAMMA AND BESSEL FUNCTIONS 228
Gamma function. Bessel functions. Algebraic operations on infinite series.

Chapter 26 GRAPHICAL METHODS FOR SOLVING FIRST-ORDER DIFFERENTIAL EQUATIONS 236
Direction fields. Euler's method. Stability.

Chapter 27 NUMERICAL METHODS FOR SOLVING FIRST-ORDER DIFFERENTIAL EQUATIONS 254
General remarks. Modified Euler's method. Runge–Kutta method. Adams–Bashforth–Moulton method. Milne's method. Starting values. Order of a numerical method.

Chapter 28 NUMERICAL METHODS FOR SYSTEMS 272
First-order systems. Euler's method. Runge–Kutta method. Adams–Bashforth–Moulton method.

Chapter *29* **SECOND-ORDER BOUNDARY-VALUE PROBLEMS** **288**
Standard form. Solutions. Eigenvalue problems. Sturm–Liouville problems. Properties of Sturm–Liouville problems.

Chapter *30* **EIGENFUNCTION EXPANSIONS** **298**
Piecewise smooth functions. Fourier sine series. Fourier cosine series.

Appendix *A* **LAPLACE TRANSFORMS** **305**

ANSWERS TO SUPPLEMENTARY PROBLEMS **311**

INDEX **355**

Chapter 1

Basic Concepts

DIFFERENTIAL EQUATIONS

A *differential equation* is an equation involving an unknown function and its derivatives.

Example 1.1. The following are differential equations involving the unknown function y.

$$\frac{dy}{dx} = 5x + 3 \tag{1.1}$$

$$e^y \frac{d^2y}{dx^2} + 2\left(\frac{dy}{dx}\right)^2 = 1 \tag{1.2}$$

$$4\frac{d^3y}{dx^3} + (\sin x)\frac{d^2y}{dx^2} + 5xy = 0 \tag{1.3}$$

$$\left(\frac{d^2y}{dx^2}\right)^3 + 3y\left(\frac{dy}{dx}\right)^7 + y^3\left(\frac{dy}{dx}\right)^2 = 5x \tag{1.4}$$

$$\frac{\partial^2 y}{\partial t^2} - 4\frac{\partial^2 y}{\partial x^2} = 0 \tag{1.5}$$

A differential equation is an *ordinary differential equation* if the unknown function depends on only one independent variable. If the unknown function depends on two or more independent variables, the differential equation is a *partial differential equation. In this book we will be concerned solely with ordinary differential equations.*

Example 1.2. Equations (*1.1*) through (*1.4*) are examples of ordinary differential equations, since the unknown function y depends solely on the variable x. Equation (*1.5*) is a partial differential equation, since y depends on both the independent variables t and x.

The *order* of a differential equation is the order of the highest derivative appearing in the equation.

Example 1.3. Equation (*1.1*) is a first-order differential equation; (*1.2*), (*1.4*), and (*1.5*) are second-order differential equations. [Note in (*1.4*) that the order of the highest *derivative* appearing in the equation is two.] Equation (*1.3*) is a third-order differential equation.

NOTATION

The expressions $y', y'', y''', y^{(4)}, \ldots, y^{(n)}$ are often used to represent, respectively, the first, second, third, fourth, $\ldots$, nth derivatives of y with respect to the independent variable under consideration. Thus, y'' represents d^2y/dx^2 if the independent variable is x, but represents d^2y/dp^2 if the independent variable is p. Observe that parentheses are used in $y^{(n)}$ to distinguish it from the nth power, y^n. If the independent variable is time, usually denoted by t, primes are often replaced by dots. Thus, $\dot{y}$, $\ddot{y}$, and $\dddot{y}$ represent dy/dt, d^2y/dt^2, and d^3y/dt^3, respectively.

SOLUTIONS

A *solution* of a differential equation in the unknown function y and the independent variable x on the interval $\mathcal{I}$ is a function $y(x)$ that satisfies the differential equation identically for all x in $\mathcal{I}$.

Example 1.4. Is $y(x) = c_1 \sin 2x + c_2 \cos 2x$, where c_1 and c_2 are arbitrary constants, a solution of $y'' + 4y = 0$?

Differentiating y, we find

$$y' = 2c_1 \cos 2x - 2c_2 \sin 2x \qquad \text{and} \qquad y'' = -4c_1 \sin 2x - 4c_2 \cos 2x$$

Hence,

$$\begin{aligned} y'' + 4y &= (-4c_1 \sin 2x - 4c_2 \cos 2x) + 4(c_1 \sin 2x + c_2 \cos 2x) \\ &= (-4c_1 + 4c_1) \sin 2x + (-4c_2 + 4c_2) \cos 2x \\ &= 0 \end{aligned}$$

Thus, $y = c_1 \sin 2x + c_2 \cos 2x$ satisfies the differential equation for all values of x and is a solution on the interval $(-\infty, \infty)$.

Example 1.5. Determine whether $y = x^2 - 1$ is a solution of $(y')^4 + y^2 = -1$.

Note that the left side of the differential equation must be nonnegative for every real function $y(x)$ and any x, since it is the sum of terms raised to the second and fourth powers, while the right side of the equation is negative. Since no function $y(x)$ will satisfy this equation, the given differential equation has no solution.

We see that some differential equations have infinitely many solutions (Example 1.4), whereas other differential equations have no solutions (Example 1.5). It is also possible that a differential equation has exactly one solution. Consider $(y')^4 + y^2 = 0$, which for reasons identical to those given in Example 1.5 has only one solution $y \equiv 0$.

A *particular solution* of a differential equation is any one solution. The *general solution* of a differential equation is the set of all solutions.

Example 1.6. The general solution to the differential equation in Example 1.4 can be shown to be (see Chapters 7 and 8) $y = c_1 \sin 2x + c_2 \cos 2x$. That is, every particular solution of the differential equation has this general form. A few particular solutions are: (*a*) $y = 5 \sin 2x - 3 \cos 2x$ (choose $c_1 = 5$ and $c_2 = -3$), (*b*) $y = \sin 2x$ (choose $c_1 = 1$ and $c_2 = 0$), and (*c*) $y \equiv 0$ (choose $c_1 = c_2 = 0$).

The general solution of a differential equation cannot always be expressed by a single formula. As an example consider the differential equation $y' + y^2 = 0$, which has two particular solutions $y = 1/x$ and $y \equiv 0$.

INITIAL-VALUE AND BOUNDARY-VALUE PROBLEMS

A differential equation along with subsidiary conditions on the unknown function and its derivatives, all given at the same value of the independent variable, constitutes an *initial-value problem.* The subsidiary conditions are *initial conditions.* If the subsidiary conditions are given at more than one value of the independent variable, the problem is a *boundary-value problem* and the conditions are *boundary conditions.*

Example 1.7. The problem $y'' + 2y' = e^x$; $y(\pi) = 1$, $y'(\pi) = 2$ is an initial-value problem, because the two subsidiary conditions are both given at $x = \pi$. The problem $y'' + 2y' = e^x$; $y(0) = 1$, $y(1) = 1$ is a boundary-value problem, because the two subsidiary conditions are given at the different values $x = 0$ and $x = 1$.

A *solution* to an initial-value or boundary-value problem is a function $y(x)$ that both solves the differential equation and satisfies all given subsidiary conditions.

Solved Problems

1.1. Determine the order, unknown function, and the independent variable in each of the following differential equations:

(*a*) $y''' - 5xy' = e^x + 1$ (*b*) $t\ddot{y} + t^2\dot{y} - (\sin t)\sqrt{y} = t^2 - t + 1$

(*c*) $s^2 \dfrac{d^2t}{ds^2} + st\dfrac{dt}{ds} = s$ (*d*) $5\left(\dfrac{d^4b}{dp^4}\right)^5 + 7\left(\dfrac{db}{dp}\right)^{10} + b^7 - b^5 = p$

(*a*) Third-order, because the highest-order derivative is the third. The unknown function is y; the independent variable is x.

(*b*) Second-order, because the highest-order derivative is the second. The unknown function is y; the independent variable is t.

(*c*) Second-order, because the highest-order derivative is the second. The unknown function is t; the independent variable is s.

(*d*) Fourth-order, because the highest-order derivative is the fourth. Raising derivatives to various powers does not alter the number of derivatives involved. The unknown function is b; the independent variable is p.

1.2. Determine the order, unknown function, and the independent variable in each of the following differential equations:

(*a*) $y\dfrac{d^2x}{dy^2} = y^2 + 1$ (*b*) $y\left(\dfrac{dx}{dy}\right)^2 = x^2 + 1$

(*c*) $2\dddot{x} + 3\dot{x} - 5x = 0$ (*d*) $17y^{(4)} - t^6y^{(2)} - 4.2y^5 = 3\cos t$

(*a*) Second-order. The unknown function is x; the independent variable is y.

(*b*) First-order, because the highest-order derivative is the first even though it is raised to the second power. The unknown function is x; the independent variable is y.

(*c*) Third-order. The unknown function is x; the independent variable is t.

(*d*) Fourth-order. The unknown function is y; the independent variable is t. Note the difference in notation between the fourth derivative $y^{(4)}$, with parentheses, and the fifth power y^5, without parentheses.

1.3. Determine whether $y(x) = 2e^{-x} + xe^{-x}$ is a solution of $y'' + 2y' + y = 0$.

Differentiating $y(x)$, it follows that

$$y'(x) = -2e^{-x} + e^{-x} - xe^{-x} = -e^{-x} - xe^{-x}$$

$$y''(x) = e^{-x} - e^{-x} + xe^{-x} = xe^{-x}$$

Substituting these values into the differential equation, we obtain

$$y'' + 2y' + y = xe^{-x} + 2(-e^{-x} - xe^{-x}) + (2e^{-x} + xe^{-x}) = 0$$

Thus, $y(x)$ is a solution.

1.4. Is $y(x) \equiv 1$ a solution of $y'' + 2y' + y = x$?

From $y(x) \equiv 1$ it follows that $y'(x) \equiv 0$ and $y''(x) \equiv 0$. Substituting these values into the differential equation, we obtain

$$y'' + 2y' + y = 0 + 2(0) + 1 = 1 \neq x$$

Thus, $y(x) \equiv 1$ is not a solution.

1.5. Show that $y = \ln x$ is a solution of $xy'' + y' = 0$ on $\mathscr{I} = (0, \infty)$ but is not a solution on $\mathscr{I} = (-\infty, \infty)$.

On $(0, \infty)$ we have $y' = 1/x$ and $y'' = -1/x^2$. Substituting these values into the differential equation, we obtain

$$xy'' + y' = x\left(-\frac{1}{x^2}\right) + \frac{1}{x} = 0$$

Thus, $y = \ln x$ is a solution on $(0, \infty)$.

Note that $y = \ln x$ could not be a solution on $(-\infty, \infty)$, since the logarithm is undefined for negative numbers and zero.

1.6. Show that $y = 1/(x^2 - 1)$ is a solution of $y' + 2xy^2 = 0$ on $\mathscr{I} = (-1, 1)$ but not on any larger interval containing $\mathscr{I}$.

On $(-1, 1)$, $y = 1/(x^2 - 1)$ and its derivative $y' = -2x/(x^2 - 1)^2$ are well-defined functions. Substituting these values into the differential equation, we have

$$y' + 2xy^2 = -\frac{2x}{(x^2 - 1)^2} + 2x\left[\frac{1}{x^2 - 1}\right]^2 = 0$$

Thus, $y = 1/(x^2 - 1)$ is a solution on $\mathscr{I} = (-1, 1)$.

Note, however, that $1/(x^2 - 1)$ is not defined at $x = \pm 1$ and therefore could not be a solution on any interval containing either of these two points.

1.7. Determine whether any of the functions (*a*) $y_1 = \sin 2x$, (*b*) $y_2(x) = x$, or (*c*) $y_3(x) = \frac{1}{2}\sin 2x$ is a solution to the initial-value problem $y'' + 4y = 0$; $y(0) = 0$, $y'(0) = 1$.

(*a*) $y_1(x)$ is a solution to the differential equation and satisfies the first initial condition $y(0) = 0$. However, $y_1(x)$ does not satisfy the second initial condition ($y_1'(x) = 2\cos 2x$; $y_1'(0) = 2\cos 0 = 2 \neq 1$); hence it is not a solution to the initial-value problem. (*b*) $y_2(x)$ satisfies both initial conditions but does not satisfy the differential equation; hence $y_2(x)$ is not a solution. (*c*) $y_3(x)$ satisfies the differential equation and both initial conditions; therefore, it is a solution to the initial-value problem.

1.8. Find the solution to the initial-value problem $y' + y = 0$; $y(3) = 2$, if the general solution to the differential equation is known to be (see Chapter 7) $y(x) = c_1 e^{-x}$, where c_1 is an arbitrary constant.

Since $y(x)$ is a solution of the differential equation for every value of c_1, we seek that value of c_1 which will also satisfy the initial condition. Note that $y(3) = c_1 e^{-3}$. To satisfy the initial condition $y(3) = 2$, it is sufficient to choose c_1 so that $c_1 e^{-3} = 2$, that is, to choose $c_1 = 2e^3$. Substituting this value for c_1 into $y(x)$, we obtain $y(x) = 2e^3 e^{-x} = 2e^{3-x}$ as the solution of the initial-value problem.

1.9. Find a solution to the initial-value problem $y'' + 4y = 0$; $y(0) = 0$, $y'(0) = 1$, if the general solution to the differential equation is known to be (see Chapter 8) $y(x) = c_1 \sin 2x + c_2 \cos 2x$.

Since $y(x)$ is a solution of the differential equation for all values of c_1 and c_2 (see Example 1.4), we seek those values of c_1 and c_2 that will also satisfy the initial conditions. Note that $y(0) = c_1 \sin 0 + c_2 \cos 0 = c_2$. To satisfy the first initial condition, $y(0) = 0$, we choose $c_2 = 0$. Furthermore, $y'(x) = 2c_1 \cos 2x - 2c_2 \sin 2x$; thus, $y'(0) = 2c_1 \cos 0 - 2c_2 \sin 0 = 2c_1$. To satisfy the second initial condition, $y'(0) = 1$, we choose $2c_1 = 1$, or $c_1 = \frac{1}{2}$. Substituting these values of c_1 and c_2 into $y(x)$, we obtain $y(x) = \frac{1}{2}\sin 2x$ as the solution of the initial-value problem.

1.10. Find a solution to the boundary-value problem $y'' + 4y = 0$; $y(\pi/8) = 0$, $y(\pi/6) = 1$, if the general solution to the differential equation is $y(x) = c_1 \sin 2x + c_2 \cos 2x$.

Note that

$$y\left(\frac{\pi}{8}\right) = c_1 \sin\left(\frac{\pi}{4}\right) + c_2 \cos\left(\frac{\pi}{4}\right) = c_1\left(\frac{1}{2}\sqrt{2}\right) + c_2\left(\frac{1}{2}\sqrt{2}\right)$$

To satisfy the condition $y(\pi/8) = 0$, we require

$$c_1\left(\frac{1}{2}\sqrt{2}\right) + c_2\left(\frac{1}{2}\sqrt{2}\right) = 0 \tag{1}$$

Furthermore,
$$y\left(\frac{\pi}{6}\right) = c_1 \sin\left(\frac{\pi}{3}\right) + c_2 \cos\left(\frac{\pi}{3}\right) = c_1\left(\frac{1}{2}\sqrt{3}\right) + c_2\left(\frac{1}{2}\right)$$

To satisfy the second condition, $y(\pi/6) = 1$, we require

$$\frac{1}{2}\sqrt{3}\,c_1 + \frac{1}{2}c_2 = 1 \tag{2}$$

Solving (*1*) and (*2*) simultaneously, we find

$$c_1 = -c_2 = \frac{2}{\sqrt{3} - 1}$$

Substituting these values into $y(x)$, we obtain

$$y(x) = \frac{2}{\sqrt{3} - 1}(\sin 2x - \cos 2x)$$

as the solution of the boundary-value problem.

1.11. Find a solution to the boundary-value problem $y'' + 4y = 0$; $y(0) = 1$, $y(\pi/2) = 2$, if the general solution to the differential equation is known to be $y(x) = c_1 \sin 2x + c_2 \cos 2x$.

Since $y(0) = c_1 \sin 0 + c_2 \cos 0 = c_2$, we must choose $c_2 = 1$ to satisfy the condition $y(0) = 1$. Since $y(\pi/2) = c_1 \sin \pi + c_2 \cos \pi = -c_2$, we must choose $c_2 = -2$ to satisfy the second condition, $y(\pi/2) = 2$. Thus, to satisfy both boundary conditions simultaneously, we must require c_2 to equal both 1 and -2, which is impossible. Therefore, there does not exist a solution to this problem.

1.12. Determine c_1 and c_2 so that $y(x) = c_1 \sin 2x + c_2 \cos 2x + 1$ will satisfy the conditions $y(\pi/8) = 0$ and $y'(\pi/8) = \sqrt{2}$.

Note that

$$y\left(\frac{\pi}{8}\right) = c_1 \sin\left(\frac{\pi}{4}\right) + c_2 \cos\left(\frac{\pi}{4}\right) + 1 = c_1\left(\frac{1}{2}\sqrt{2}\right) + c_2\left(\frac{1}{2}\sqrt{2}\right) + 1$$

To satisfy the condition $y(\pi/8) = 0$, we require $c_1(\frac{1}{2}\sqrt{2}) + c_2(\frac{1}{2}\sqrt{2}) + 1 = 0$, or equivalently,

$$c_1 + c_2 = -\sqrt{2} \tag{1}$$

Since $y'(x) = 2c_1 \cos 2x - 2c_2 \sin 2x$,

$$y'\left(\frac{\pi}{8}\right) = 2c_1 \cos\left(\frac{\pi}{4}\right) - 2c_2 \sin\left(\frac{\pi}{4}\right)$$

$$= 2c_1\left(\frac{1}{2}\sqrt{2}\right) - 2c_2\left(\frac{1}{2}\sqrt{2}\right) = \sqrt{2}\,c_1 - \sqrt{2}\,c_2$$

To satisfy the condition $y'(\pi/8) = \sqrt{2}$, we require $\sqrt{2}\,c_1 - \sqrt{2}\,c_2 = \sqrt{2}$, or equivalently,

$$c_1 - c_2 = 1 \tag{2}$$

Solving (*1*) and (*2*) simultaneously, we obtain $c_1 = -\frac{1}{2}(\sqrt{2} - 1)$ and $c_2 = -\frac{1}{2}(\sqrt{2} + 1)$.

1.13. Determine c_1 and c_2 so that $y(x) = c_1 e^{2x} + c_2 e^x + 2 \sin x$ will satisfy the conditions $y(0) = 0$ and $y'(0) = 1$.

Because $\sin 0 = 0$, $y(0) = c_1 + c_2$. To satisfy the condition $y(0) = 0$, we require

$$c_1 + c_2 = 0 \tag{1}$$

From

$$y'(x) = 2c_1e^{2x} + c_2e^x + 2\cos x$$

we have $y'(0) = 2c_1 + c_2 + 2$. To satisfy the condition $y'(0) = 1$, we require $2c_1 + c_2 + 2 = 1$, or

$$2c_1 + c_2 = -1 \tag{2}$$

Solving (*1*) and (*2*) simultaneously, we obtain $c_1 = -1$ and $c_2 = 1$.

Supplementary Problems

In Problems 1.14 through 1.23, determine (*a*) the order, (*b*) the unknown function, and (*c*) the independent variable for each of the given differential equations.

1.14. $(y'')^2 - 3yy' + xy = 0$

1.15. $x^4y^{(4)} + xy''' = e^x$

1.16. $t^2\ddot{s} - t\dot{s} = 1 - \sin t$

1.17. $y^{(4)} + xy''' + x^2y'' - xy' + \sin y = 0$

1.18. $\dfrac{d^nx}{dy^n} = y^2 + 1$

1.19. $\left(\dfrac{d^2r}{dy^2}\right)^2 + \dfrac{d^2r}{dy^2} + y\dfrac{dr}{dy} = 0$

1.20. $\left(\dfrac{d^2y}{dx^2}\right)^{3/2} + y = x$

1.21. $\dfrac{d^7b}{dp^7} = 3p$

1.22. $\left(\dfrac{db}{dp}\right)^7 = 3p$

1.23. $y^{(6)} + 2y^4y^{(3)} + 5y^8 = e^x$

1.24. Which of the following functions are solutions of the differential equation $y' - 5y = 0$?

(*a*) $y = 5$, (*b*) $y = 5x$, (*c*) $y = x^5$, (*d*) $y = e^{5x}$, (*e*) $y = 2e^{5x}$, (*f*) $y = 5e^{2x}$

1.25. Which of the following functions are solutions of the differential equation $y' - 3y = 6$?

(*a*) $y = -2$, (*b*) $y = 0$, (*c*) $y = e^{3x} - 2$, (*d*) $y = e^{2x} - 3$, (*e*) $y = 4e^{3x} - 2$

1.26. Which of the following functions are solutions of the differential equation $\dot{y} - 2ty = t$?

(*a*) $y = 2$, (*b*) $y = -\frac{1}{2}$, (*c*) $y = e^{t^2}$, (*d*) $y = e^{t^2} - \frac{1}{2}$, (*e*) $y = -7e^{t^2} - \frac{1}{2}$

1.27. Which of the following functions are solutions of the differential equation $dy/dt = y/t$?

(*a*) $y = 0$, (*b*) $y = 2$, (*c*) $y = 2t$, (*d*) $y = -3t$, (*e*) $y = t^2$

1.28. Which of the following functions are solutions of the differential equation

$$\frac{dy}{dx} = \frac{2y^4 + x^4}{xy^3}$$

(*a*) $y = x$, (*b*) $y = x^8 - x^4$, (*c*) $y = \sqrt{x^8 - x^4}$, (*d*) $y = (x^8 - x^4)^{1/4}$

1.29. Which of the following functions are solutions of the differential equation $y'' - y = 0$?

(a) $y = e^x$, (b) $y = \sin x$, (c) $y = 4e^{-x}$, (d) $y = 0$, (e) $y = \frac{1}{2}x^2 + 1$

1.30. Which of the following functions are solutions of the differential equation $y'' - xy' + y = 0$?

(a) $y = x^2$, (b) $y = x$, (c) $y = 1 - x^2$, (d) $y = 2x^2 - 2$, (e) $y = 0$

1.31. Which of the following functions are solutions of the differential equation $\ddot{x} - 4\dot{x} + 4x = e^t$?

(a) $x = e^t$, (b) $x = e^{2t}$, (c) $x = e^{2t} + e^t$, (d) $x = te^{2t} + e^t$, (e) $x = e^{2t} + te^t$

In Problems 1.32 through 1.35, find c so that $x(t) = ce^{2t}$ satisfies the given initial condition.

1.32. $x(0) = 0$ **1.33.** $x(0) = 1$ **1.34.** $x(1) = 1$ **1.35.** $x(2) = -3$

In Problems 1.36 through 1.39, find c so that $y(x) = c(1 - x^2)$ satisfies the given initial condition.

1.36. $y(0) = 1$ **1.37.** $y(1) = 0$ **1.38.** $y(2) = 1$ **1.39.** $y(1) = 2$

In Problems 1.40 through 1.49, find c_1 and c_2 so that $y(x) = c_1 \sin x + c_2 \cos x$ will satisfy the given conditions. Determine whether the given conditions are initial conditions or boundary conditions.

1.40. $y(0) = 1, \quad y'(0) = 2$

1.41. $y(0) = 2, \quad y'(0) = 1$

1.42. $y\left(\frac{\pi}{2}\right) = 1, \quad y'\left(\frac{\pi}{2}\right) = 2$

1.43. $y(0) = 1, \quad y\left(\frac{\pi}{2}\right) = 1$

1.44. $y'(0) = 1, \quad y'\left(\frac{\pi}{2}\right) = 1$

1.45. $y(0) = 1, \quad y'(\pi) = 1$

1.46. $y(0) = 1, \quad y(\pi) = 2$

1.47. $y(0) = 0, \quad y'(0) = 0$

1.48. $y\left(\frac{\pi}{4}\right) = 0, \quad y\left(\frac{\pi}{6}\right) = 1$

1.49. $y(0) = 0, \quad y'\left(\frac{\pi}{2}\right) = 1$

In Problems 1.50 through 1.54, find values of c_1 and c_2 so that the given functions will satisfy the prescribed initial conditions.

1.50. $y(x) = c_1 e^x + c_2 e^{-x} + 4 \sin x; \quad y(0) = 1, \quad y'(0) = -1$

1.51. $y(x) = c_1 x + c_2 + x^2 - 1; \quad y(1) = 1, \quad y'(1) = 2$

1.52. $y(x) = c_1 e^x + c_2 e^{2x} + 3e^{3x}; \quad y(0) = 0, \quad y'(0) = 0$

1.53. $y(x) = c_1 \sin x + c_2 \cos x + 1; \quad y(\pi) = 0, \quad y'(\pi) = 0$

1.54. $y(x) = c_1 e^x + c_2 xe^x + x^2 e^x; \quad y(1) = 1, \quad y'(1) = -1$

Chapter 2

Classifications of First-Order Differential Equations

STANDARD FORM AND DIFFERENTIAL FORM

Standard form for a first-order differential equation in the unknown function $y(x)$ is

$$y' = f(x, y) \tag{2.1}$$

where the derivative y' appears only on the left side of (*2.1*). Many, but not all, first-order differential equations can be written in standard form by algebraically solving for y' and then setting $f(x, y)$ equal to the right side of the resulting equation.

The right side of (*2.1*) can always be written as a quotient of two other functions $M(x, y)$ and $-N(x, y)$. Then (*2.1*) becomes $dy/dx = M(x, y)/-N(x, y)$, which is equivalent to the *differential form*

$$M(x, y)\,dx + N(x, y)\,dy = 0 \tag{2.2}$$

LINEAR EQUATIONS

Consider a differential equation in standard form (*2.1*). If $f(x, y)$ can be written as $f(x, y) = -p(x)y + q(x)$ (that is, as a function of x times y, plus another function of x), the differential equation is *linear*. First-order linear differential equations can always be expressed as

$$y' + p(x)y = q(x) \tag{2.3}$$

Linear equations are solved in Chapter 5.

BERNOULLI EQUATIONS

A *Bernoulli* differential equation is an equation of the form

$$y' + p(x)y = q(x)y^n \tag{2.4}$$

where n denotes a real number. When $n = 1$ or $n = 0$, a Bernoulli equation reduces to a linear equation. Bernoulli equations are solved in Chapter 5.

HOMOGENEOUS EQUATIONS

A differential equation in standard form (*2.1*) is *homogeneous* if

$$f(tx, ty) = f(x, y) \tag{2.5}$$

for every real number t. Homogeneous equations are solved in Chapter 3.

Note: In the general framework of differential equations, the word "homogeneous" has an entirely different meaning (see Chapter 7). Only in the context of first-order differential equations does "homogeneous" have the meaning defined above.

SEPARABLE EQUATIONS

Consider a differential equation in differential form (*2.2*). If $M(x, y) = A(x)$ (a function only of x) and $N(x, y) = B(y)$ (a function only of y), the differential equation is *separable*, or has its *variables separated.* Separable equations are solved in Chapter 3.

EXACT EQUATIONS

A differential equation in differential form (*2.2*) is *exact* if

$$\frac{\partial M(x, y)}{\partial y} = \frac{\partial N(x, y)}{\partial x} \tag{2.6}$$

Exact equations are solved in Chapter 4 (where a more precise definition of exactness is given).

Solved Problems

2.1. Write the differential equation $xy' - y^2 = 0$ in standard form.

Solving for y', we obtain $y' = y^2/x$ which has form (*2.1*) with $f(x, y) = y^2/x$.

2.2. Write the differential equation $e^x y' + e^{2x} y = \sin x$ in standard form.

Solving for y', we obtain

$$e^x y' = -e^{2x} y + \sin x$$

or

$$y' = -e^x y + e^{-x} \sin x$$

which has form (*2.1*) with $f(x, y) = -e^x y + e^{-x} \sin x$.

2.3. Write the differential equation $(y' + y)^5 = \sin (y'/x)$ in standard form.

This equation cannot be solved algebraically for y', and *cannot* be written in standard form.

2.4. Write the differential equation $y(yy' - 1) = x$ in differential form.

Solving for y', we have

$$y^2 y' - y = x$$

$$y^2 y' = x + y$$

or

$$y' = \frac{x + y}{y^2} \tag{1}$$

which is in standard form with $f(x, y) = (x + y)/y^2$. There are infinitely many different differential forms associated with (*1*). Four such forms are:

(*a*) Take $M(x, y) = x + y$, $N(x, y) = -y^2$. Then

$$\frac{M(x, y)}{-N(x, y)} = \frac{x + y}{-(-y^2)} = \frac{x + y}{y^2}$$

and (*1*) is equivalent to the differential form

$$(x + y)\, dx + (-y^2)\, dy = 0$$

(*b*) Take $M(x, y) = -1$, $N(x, y) = \dfrac{y^2}{x + y}$. Then

$$\frac{M(x, y)}{-N(x, y)} = \frac{-1}{-y^2/(x + y)} = \frac{x + y}{y^2}$$

and (1) is equivalent to the differential form

$$(-1)\,dx + \left(\frac{y^2}{x+y}\right) dy = 0$$

(c) Take $M(x, y) = \dfrac{x+y}{2}$, $N(x, y) = \dfrac{-y^2}{2}$. Then

$$\frac{M(x, y)}{-N(x, y)} = \frac{(x+y)/2}{-(-y^2/2)} = \frac{x+y}{y^2}$$

and (1) is equivalent to the differential form

$$\left(\frac{x+y}{2}\right) dx + \left(\frac{-y^2}{2}\right) dy = 0$$

(d) Take $M(x, y) = \dfrac{-x-y}{x^2}$, $N(x, y) = \dfrac{y^2}{x^2}$. Then

$$\frac{M(x, y)}{-N(x, y)} = \frac{(-x-y)/x^2}{-y^2/x^2} = \frac{x+y}{y^2}$$

and (1) is equivalent to the differential form

$$\left(\frac{-x-y}{x^2}\right) dx + \left(\frac{y^2}{x^2}\right) dy = 0$$

2.5. Write the differential equation $dy/dx = y/x$ in differential form.

This equation has infinitely many differential forms. One is

$$dy = \frac{y}{x}\,dx$$

which can be written in form (2.2) as

$$\frac{y}{x}\,dx + (-1)\,dy = 0 \tag{1}$$

Multiplying (1) through by x, we obtain

$$y\,dx + (-x)\,dy = 0 \tag{2}$$

as a second differential form. Multiplying (1) through by $1/y$, we obtain

$$\frac{1}{x}\,dx + \frac{-1}{y}\,dy = 0 \tag{3}$$

as a third differential form. Still other differential forms are derived from (1) by multiplying that equation through by any other function of x and y.

2.6. Write the differential equation $(xy+3)\,dx + (2x - y^2 + 1)\,dy = 0$ in standard form.

This equation is in differential form. We rewrite it as

$$(2x - y^2 + 1)\,dy = -(xy+3)\,dx$$

which has the standard form

$$\frac{dy}{dx} = \frac{-(xy+3)}{2x - y^2 + 1}$$

or

$$y' = \frac{xy+3}{y^2 - 2x - 1}$$

2.7. Determine if the following differential equations are linear:

(*a*) $y' = (\sin x)y + e^x$ (*b*) $y' = x \sin y + e^x$ (*c*) $y' = 5$ (*d*) $y' = y^2 + x$

(*e*) $y' + xy^5 = 0$ (*f*) $xy' + y = \sqrt{y}$ (*g*) $y' + xy = e^x y$ (*h*) $y' + \dfrac{x}{y} = 0$

(*a*) The equation is linear; here $p(x) = -\sin x$ and $q(x) = e^x$.

(*b*) The equation is not linear because of the term $\sin y$.

(*c*) The equation is linear; here $p(x) = 0$ and $q(x) = 5$.

(*d*) The equation is not linear because of the term y^2.

(*e*) The equation is not linear because of the y^5 term.

(*f*) The equation is not linear because of the $y^{1/2}$ term.

(*g*) The equation is linear. Rewrite it as $y' + (x - e^x)y = 0$ with $p(x) = x - e^x$ and $q(x) = 0$.

(*h*) The equation is not linear because of the $1/y$ term.

2.8. Determine whether any of the differential equations in Problem 2.7 are Bernoulli equations.

All of the linear equations are Bernoulli equations with $n = 0$. In addition, two of the nonlinear equations, (*e*) and (*f*), are as well. Rewrite (*e*) as $y' = -xy^5$; it has form (*2.4*) with $p(x) = 0$, $q(x) = -x$, and $n = 5$. Rewrite (*f*) as

$$y' + \frac{1}{x}y = \frac{1}{x}y^{1/2}$$

It has form (*2.4*) with $p(x) = q(x) = 1/x$ and $n = 1/2$.

2.9. Determine if the following differential equations are homogeneous:

(*a*) $y' = \dfrac{y + x}{x}$ (*b*) $y' = \dfrac{y^2}{x}$ (*c*) $y' = \dfrac{2xye^{x/y}}{x^2 + y^2 \sin \dfrac{x}{y}}$ (*d*) $y' = \dfrac{x^2 + y}{x^3}$

(*a*) The equation is homogeneous, since

$$f(tx, ty) = \frac{ty + tx}{tx} = \frac{t(y + x)}{tx} = \frac{y + x}{x} = f(x, y)$$

(*b*) The equation is not homogeneous, since

$$f(tx, ty) = \frac{(ty)^2}{tx} = \frac{t^2y^2}{tx} = t\frac{y^2}{x} \neq f(x, y)$$

(*c*) The equation is homogeneous, since

$$f(tx, ty) = \frac{2(tx)(ty)e^{tx/ty}}{(tx)^2 + (ty)^2 \sin \dfrac{tx}{ty}} = \frac{t^2 2xye^{x/y}}{t^2x^2 + t^2y^2 \sin \dfrac{x}{y}}$$

$$= \frac{2xye^{x/y}}{x^2 + y^2 \sin \dfrac{x}{y}} = f(x, y)$$

(*d*) The equation is not homogeneous, since

$$f(tx, ty) = \frac{(tx)^2 + ty}{(tx)^3} = \frac{t^2x^2 + ty}{t^3x^3} = \frac{tx^2 + y}{t^2x^3} \neq f(x, y)$$

2.10. Determine if the following differential equations are separable:

(*a*) $\sin x\,dx + y^2\,dy = 0$ (*b*) $xy^2\,dx - x^2y^2\,dy = 0$ (*c*) $(1 + xy)\,dx + y\,dy = 0$

(*a*) The differential equation is separable; here $M(x, y) = A(x) = \sin x$ and $N(x, y) = B(y) = y^2$.

(*b*) The equation is not separable in its present form, since $M(x, y) = xy^2$ is not a function of x alone. But if we divide both sides of the equation by x^2y^2, we obtain the equation $(1/x)\,dx + (-1)\,dy = 0$, which is separable. Here, $A(x) = 1/x$ and $B(y) = -1$.

(*c*) The equation is not separable, since $M(x, y) = 1 + xy$, which is not a function of x alone.

2.11. Determine whether the following differential equations are exact:

(*a*) $3x^2y\,dx + (y + x^3)\,dy = 0$ (*b*) $xy\,dx + y^2\,dy = 0$

(*a*) The equation is exact; here $M(x, y) = 3x^2y$, $N(x, y) = y + x^3$, and $\partial M/\partial y = \partial N/\partial x = 3x^2$.

(*b*) The equation is not exact. Here $M(x, y) = xy$ and $N(x, y) = y^2$; hence $\partial M/\partial y = x$, $\partial N/\partial x = 0$, and $\partial M/\partial y \neq \partial N/\partial x$.

2.12. Determine whether the differential equation $y' = y/x$ is exact.

Exactness is only defined for equations in differential form, not standard form. The given differential equation has many differential forms. One such form is given in Problem 2.5, Eq. (*1*), as

$$\frac{y}{x}\,dx + (-1)\,dy = 0$$

Here $M(x, y) = y/x$, $N(x, y) = -1$,

$$\frac{\partial M}{\partial y} = \frac{1}{x} \neq 0 = \frac{\partial N}{\partial x}$$

and the equation is not exact. A second differential form for the same differential equation is given in Eq. (*3*) of Problem 2.5 as

$$\frac{1}{x}\,dx + \frac{-1}{y}\,dy = 0$$

Here $M(x, y) = 1/x$, $N(x, y) = -1/y$,

$$\frac{\partial M}{\partial y} = 0 = \frac{\partial N}{\partial x}$$

and the equation is exact. Thus, a given differential equation has many differential forms, some of which may be exact.

2.13. Prove that a separable equation is always exact.

For a separable differential equation, $M(x, y) = A(x)$ and $N(x, y) = B(y)$. Thus,

$$\frac{\partial M(x, y)}{\partial y} = \frac{\partial A(x)}{\partial y} = 0 \quad \text{and} \quad \frac{\partial N(x, y)}{\partial x} = \frac{\partial B(y)}{\partial x} = 0$$

Since $\partial M/\partial y = \partial N/\partial x$, the differential equation is exact.

2.14. A theorem of first-order differential equations states that if $f(x, y)$ and $\partial f(x, y)/\partial y$ are continuous in a rectangle $\mathcal{R}$: $|x - x_0| \leq a$, $|y - y_0| \leq b$, then there exists an interval about x_0 in which the initial-value problem $y' = f(x, y)$; $y(x_0) = y_0$ has a unique solution. The initial-value problem $y' = 2\sqrt{|y|}$; $y(0) = 0$ has the two solutions $y = x\,|x|$ and $y \equiv 0$. Does this result violate the theorem?

No. Here, $f(x, y) = 2\sqrt{|y|}$ and, therefore, $\partial f/\partial y$ does not exist at the origin.

Supplementary Problems

In Problems 2.15 through 2.25, write the given differential equations in standard form.

2.15. $xy' + y^2 = 0$

2.16. $e^x y' - x = y'$

2.17. $(y')^3 + y^2 + y = \sin x$

2.18. $xy' + \cos(y' + y) = 1$

2.19. $e^{(y'+y)} = x$

2.20. $(y')^2 - 5y' + 6 = (x + y)(y' - 2)$

2.21. $(x - y)\,dx + y^2\,dy = 0$

2.22. $\dfrac{x + y}{x - y}dx - dy = 0$

2.23. $dx + \dfrac{x + y}{x - y}dy = 0$

2.24. $(e^{2x} - y)\,dx + e^x\,dy = 0$

2.25. $dy + dx = 0$

In Problems 2.26 through 2.35, differential equations are given in both standard and differential form. Determine whether the equations in standard form are homogeneous and/or linear, and, if not linear, whether they are Bernoulli; determine whether the equations in differential form, *as given*, are separable and/or exact.

2.26. $y' = xy;\quad xy\,dx - dy = 0$

2.27. $y' = xy;\quad x\,dx - \dfrac{1}{y}dy = 0$

2.28. $y' = xy + 1;\quad (xy + 1)\,dx - dy = 0$

2.29. $y' = \dfrac{x^2}{y^2};\quad \dfrac{x^2}{y^2}dx - dy = 0$

2.30. $y' = \dfrac{x^2}{y^2};\quad -x^2\,dx + y^2\,dy = 0$

2.31. $y' = -\dfrac{2y}{x};\quad 2xy\,dx + x^2\,dy = 0$

2.32. $y' = \dfrac{xy^2}{x^2y + y^3};\quad xy^2\,dx - (x^2y + y^3)\,dy = 0$

2.33. $y' = \dfrac{-xy^2}{x^2y + y^2};\quad xy^2\,dx + (x^2y + y^2)\,dy = 0$

2.34. $y' = x^3y + xy^3;\quad (x^2 + y^2)\,dx - \dfrac{1}{xy}dy = 0$

2.35. $y' = 2xy + x;\quad (2xye^{-x^2} + xe^{-x^2})\,dx - e^{-x^2}\,dy = 0$

Chapter 3

Separable First-Order Differential Equations

GENERAL SOLUTION

The solution to the first-order separable differential equation (see Chapter 2)

$$A(x)\,dx + B(y)\,dy = 0 \tag{3.1}$$

is

$$\int A(x)\,dx + \int B(y)\,dy = c \tag{3.2}$$

where c represents an arbitrary constant.

The integrals obtained in Eq. (*3.2*) may be, for all practical purposes, impossible to evaluate. In such case, numerical techniques (see Chapter 27) are used to obtain an approximate solution. Even if the indicated integrations in (*3.2*) can be performed, it may not be algebraically possible to solve for y explicitly in terms of x. In that case, the solution is left in implicit form.

SOLUTIONS TO THE INITIAL-VALUE PROBLEM

The solution to the initial-value problem

$$A(x)\,dx + B(y)\,dy = 0; \qquad y(x_0) = y_0 \tag{3.3}$$

can be obtained, as usual, by first using Eq. (*3.2*) to solve the differential equation and then applying the initial condition directly to evaluate c.

Alternatively, the solution to Eq. (*3.3*) can be obtained from

$$\int_{x_0}^{x} A(x)\,dx + \int_{y_0}^{y} B(y)\,dy = 0 \tag{3.4}$$

Equation (*3.4*), however, may not determine the solution of (*3.3*) *uniquely*; that is, (*3.4*) may have many solutions, of which only one will satisfy the initial-value problem.

REDUCTION OF HOMOGENEOUS EQUATIONS

The homogeneous differential equation

$$\frac{dy}{dx} = f(x, y) \tag{3.5}$$

having the property that $f(tx, ty) = f(x, y)$ (see Chapter 2) can be transformed into a separable equation by making the substitution

$$y = xv \tag{3.6}$$

along with its corresponding derivative

$$\frac{dy}{dx} = v + x\frac{dv}{dx} \tag{3.7}$$

The resulting equation in the variables v and x is solved as a separable differential equation; the required solution to Eq. (*3.5*) is obtained by back substitution.

Alternatively, the solution to (*3.5*) can be obtained by rewriting the differential equation as

$$\frac{dx}{dy} = \frac{1}{f(x, y)} \tag{3.8}$$

and then substituting

$$x = yu \tag{3.9}$$

and the corresponding derivative

$$\frac{dx}{dy} = u + y\frac{du}{dy} \tag{3.10}$$

into Eq. (*3.8*). After simplifying, the resulting differential equation will be one with variables (this time, u and y) separable.

Ordinarily, it is immaterial which method of solution is used (see Problems 3.12 and 3.13). Occasionally, however, one of the substitutions (*3.6*) or (*3.9*) is definitely superior to the other one. In such cases, the better substitution is usually apparent from the form of the differential equation itself. (See Problem 3.17.)

Solved Problems

3.1. Solve $x\,dx - y^2\,dy = 0$.

For this differential equation, $A(x) = x$ and $B(y) = -y^2$. Substituting these values into Eq. (*3.2*), we have

$$\int x\,dx + \int (-y^2)\,dy = c$$

which, after the indicated integrations are performed, becomes $x^2/2 - y^3/3 = c$. Solving for y explicitly, we obtain the solution as

$$y = \left(\frac{3}{2}x^2 + k\right)^{1/3}, \qquad k = -3c$$

3.2. Solve $y' = y^2x^3$.

We first rewrite this equation in the differential form (see Chapter 2) $x^3\,dx - (1/y^2)\,dy = 0$. Then $A(x) = x^3$ and $B(y) = -1/y^2$. Substituting these values into Eq. (*3.2*), we have

$$\int x^3\,dx + \int (-1/y^2)\,dy = c$$

or, by performing the indicated integrations, $x^4/4 + 1/y = c$. Solving explicitly for y, we obtain the solution as

$$y = \frac{-4}{x^4 + k}, \qquad k = -4c$$

3.3. Solve $\dfrac{dy}{dx} = \dfrac{x^2 + 2}{y}$.

This equation may be rewritten in the differential form

$$(x^2 + 2)\,dx - y\,dy = 0$$

which is separable with $A(x) = x^2 + 2$ and $B(y) = -y$. Its solution is

$$\int (x^2 + 2)\,dx - \int y\,dy = c$$

or

$$\frac{1}{3}x^3 + 2x - \frac{1}{2}y^2 = c$$

Solving for y, we obtain the solution in implicit form as

$$y^2 = \frac{2}{3}x^3 + 4x + k$$

with $k = -2c$. Solving for y implicitly, we obtain the two solutions

$$y = \sqrt{\frac{2}{3}x^3 + 4x + k} \qquad \text{and} \qquad y = -\sqrt{\frac{2}{3}x^3 + 4x + k}$$

3.4. Solve $y' = 5y$.

First rewrite this equation in the differential form $5\,dx - (1/y)\,dy = 0$, which is separable. Its solution is

$$\int 5\,dx + \int (-1/y)\,dy = c$$

or, by evaluating, $5x - \ln|y| = c$.

To solve for y explicitly, we first rewrite the solution as $\ln|y| = 5x - c$ and then take the exponential of both sides. Thus, $e^{\ln|y|} = e^{5x-c}$. Noting that $e^{\ln|y|} = |y|$, we obtain $|y| = e^{5x}e^{-c}$, or $y = \pm e^{-c}e^{5x}$. The solution is given explicitly by $y = ke^{5x}$, $k = \pm e^{-c}$.

Note that the presence of the term $(-1/y)$ in the differential form of the differential equation requires the restriction $y \neq 0$ in our derivation of the solution. This restriction is equivalent to the restriction $k \neq 0$, since $y = ke^{5x}$. However, by inspection, $y \equiv 0$ is a solution of the differential equation as originally given. Thus, $y = ke^{5x}$ is the solution for all k.

The differential equation as originally given is also linear. See Problem 5.9 for an alternate method of solution.

3.5. Solve $y' = \dfrac{x+1}{y^4+1}$.

This equation, in differential form, is $(x+1)\,dx + (-y^4 - 1)\,dy = 0$, which is separable. Its solution is

$$\int (x+1)\,dx + \int (-y^4 - 1)\,dy = c$$

or, by evaluating,

$$\frac{x^2}{2} + x - \frac{y^5}{5} - y = c$$

Since it is impossible algebraically to solve this equation explicitly for y, the solution must be left in its present implicit form.

3.6. Solve $dy = 2t(y^2 + 9)\,dt$.

This equation may be rewritten as

$$\frac{dy}{y^2 + 9} - 2t\,dt = 0$$

which is separable in variables y and t. Its solution is

$$\int \frac{dy}{y^2 + 9} - \int 2t\,dt = c$$

or, upon evaluating the given integrals,

$$\frac{1}{3}\arctan\left(\frac{y}{3}\right) - t^2 = c$$

Solving for y, we obtain

$$\arctan\left(\frac{y}{3}\right) = 3(t^2 + c)$$

$$\frac{y}{3} = \tan(3t^2 + 3c)$$

or

$$y = 3\tan(3t^2 + k)$$

with $k = 3c$.

3.7. Solve $\dfrac{dx}{dt} = x^2 - 2x + 2$.

This equation may be rewritten in differential form

$$\frac{dx}{x^2 - 2x + 2} - dt = 0$$

which is separable in the variables x and t. Its solution is

$$\int \frac{dx}{x^2 - 2x + 2} - \int dt = c$$

Evaluating the first integral by first completing the square, we obtain

$$\int \frac{dx}{(x-1)^2 + 1} - \int dt = c$$

or

$$\arctan(x - 1) - t = c$$

Solving for x as a function of t, we obtain

$$\arctan(x - 1) = t + c$$

$$x - 1 = \tan(t + c)$$

or

$$x = 1 + \tan(t + c)$$

3.8. Solve $e^x\,dx - y\,dy = 0;\ y(0) = 1$.

The solution to the differential equation is given by Eq. (*3.2*) as

$$\int e^x\,dx + \int (-y)\,dy = c$$

or, by evaluating, as $y^2 = 2e^x + k$, $k = -2c$. Applying the initial condition, we obtain $(1)^2 = 2e^0 + k$, $1 = 2 + k$, or $k = -1$. Thus, the solution to the initial-value problem is

$$y^2 = 2e^x - 1 \quad \text{or} \quad y = \sqrt{2e^x - 1}$$

[Note that we cannot choose the negative square root, since then $y(0) = -1$, which violates the initial condition.]

To ensure that y remains real, we must restrict x so that $2e^x - 1 \geq 0$. To guarantee that y' exists [note that $y'(x) = dy/dx = e^x/y$], we must restrict x so that $2e^x - 1 \neq 0$. Together these conditions imply that $2e^x - 1 > 0$, or $x > \ln \frac{1}{2}$.

3.9. Use Eq. (*3.4*) to solve Problem 3.8.

For this problem, $x_0 = 0$, $y_0 = 1$, $A(x) = e^x$, and $B(y) = -y$. Substituting these values into Eq. (*3.4*), we obtain

$$\int_0^x e^x \, dx + \int_1^y (-y) \, dy = 0$$

Evaluating these integrals, we have

$$e^x \Big|_0^x + \left(\frac{-y^2}{2}\right)\Big|_1^y = 0 \qquad \text{or} \qquad e^x - e^0 + \left(\frac{-y^2}{2}\right) - \left(-\frac{1}{2}\right) = 0$$

Thus, $y^2 = 2e^x - 1$, and, as in Problem 3.8, $y = \sqrt{2e^x - 1}$, $x > \ln \frac{1}{2}$.

3.10. Solve $x \cos x \, dx + (1 - 6y^5) \, dy = 0$; $y(\pi) = 0$.

Here, $x_0 = \pi$, $y_0 = 0$, $A(x) = x \cos x$, and $B(y) = 1 - 6y^5$. Substituting these values into Eq. (*3.4*), we obtain

$$\int_\pi^x x \cos x \, dx + \int_0^y (1 - 6y^5) \, dy = 0$$

Evaluating these integrals (the first one by integration by parts), we find

$$x \sin x \Big|_\pi^x + \cos x \Big|_\pi^x + (y - y^6)\Big|_0^y = 0$$

or

$$x \sin x + \cos x + 1 = y^6 - y$$

Since we cannot solve this last equation for y explicitly, we must be content with the solution in its present implicit form.

3.11. Solve $y' = \dfrac{y + x}{x}$.

This differential equation is not separable, but it is homogeneous as shown in Problem 2.9(*a*). Substituting Eqs. (*3.6*) and (*3.7*) into the equation, we obtain

$$v + x\frac{dv}{dx} = \frac{xv + x}{x}$$

which can be algebraically simplified to

$$x\frac{dv}{dx} = 1 \qquad \text{or} \qquad \frac{1}{x}dx - dv = 0$$

This last equation is separable; its solution is

$$\int \frac{1}{x} dx - \int dv = c$$

which, when evaluated, yields $v = \ln|x| - c$, or

$$v = \ln|kx| \tag{1}$$

where we have set $c = -\ln|k|$ and have noted that $\ln|x| + \ln|k| = \ln|kx|$. Finally, substituting $v = y/x$ back into (*1*), we obtain the solution to the given differential equation as $y = x \ln|kx|$.

3.12. Solve $y' = \dfrac{2y^4 + x^4}{xy^3}$.

This differential equation is not separable. Instead it has the form $y' = f(x, y)$, with

$$f(x, y) = \frac{2y^4 + x^4}{xy^3}$$

where
$$f(tx, ty) = \frac{2(ty)^4 + (tx)^4}{(tx)(ty)^3} = \frac{t^4(2y^4 + x^4)}{t^4(xy^3)} = \frac{2y^4 + x^4}{xy^3} = f(x, y)$$

so it is homogeneous. Substituting Eqs. (*3.6*) and (*3.7*) into the differential equation as originally given, we obtain

$$v + x\frac{dv}{dx} = \frac{2(xv)^4 + x^4}{x(xv)^3}$$

which can be algebraically simplified to

$$x\frac{dv}{dx} = \frac{v^4 + 1}{v^3} \qquad \text{or} \qquad \frac{1}{x}dx - \frac{v^3}{v^4 + 1}dv = 0$$

This last equation is separable; its solution is

$$\int \frac{1}{x}dx - \int \frac{v^3}{v^4 + 1}dv = c$$

Integrating, we obtain $\ln|x| - \frac{1}{4}\ln(v^4 + 1) = c$, or

$$v^4 + 1 = (kx)^4 \tag{1}$$

where we have set $c = -\ln|k|$ and then used the identities

$$\ln|x| + \ln|k| = \ln|kx| \qquad \text{and} \qquad 4\ln|kx| = \ln(kx)^4$$

Finally, substituting $v = y/x$ back into (*1*), we obtain

$$y^4 = c_1x^8 - x^4 \qquad (c_1 = k^4) \tag{2}$$

3.13. Solve the differential equation of Problem 3.12 by using Eqs. (*3.9*) and (*3.10*).

We first rewrite the differential equation as

$$\frac{dx}{dy} = \frac{xy^3}{2y^4 + x^4}$$

Then substituting (*3.9*) and (*3.10*) into this new differential equation, we obtain

$$u + y\frac{du}{dy} = \frac{(yu)y^3}{2y^4 + (yu)^4}$$

which can be algebraically simplified to

$$y\frac{du}{dy} = -\frac{u + u^5}{2 + u^4}$$

or
$$\frac{1}{y}dy + \frac{2 + u^4}{u + u^5}du = 0 \tag{1}$$

Equation (*1*) is separable; its solution is

$$\int \frac{1}{y}dy + \int \frac{2 + u^4}{u + u^5}du = c$$

The first integral is $\ln|y|$. To evaluate the second integral, we use partial fractions on the integrand to

obtain

$$\frac{2+u^4}{u+u^5} = \frac{2+u^4}{u(1+u^4)} = \frac{2}{u} - \frac{u^3}{1+u^4}$$

Therefore,

$$\int \frac{2+u^4}{u+u^5}\,du = \int \frac{2}{u}\,du - \int \frac{u^3}{1+u^4}\,du = 2\ln|u| - \frac{1}{4}\ln(1+u^4)$$

The solution to (*1*) is $\ln|y| + 2\ln|u| - \frac{1}{4}\ln(1+u^4) = c$, which can be rewritten as

$$ky^4u^8 = 1 + u^4 \tag{2}$$

where $c = -\frac{1}{4}\ln|k|$. Substituting $u = x/y$ back into (*2*), we once again have (*2*) of Problem 3.12.

3.14. Solve $y' = \dfrac{2xy}{x^2 - y^2}$.

This differential equation is not separable. Instead it has the form $y' = f(x, y)$, with

$$f(x, y) = \frac{2xy}{x^2 - y^2}$$

where

$$f(tx, ty) = \frac{2(tx)(ty)}{(tx)^2 - (ty)^2} = \frac{t^2(2xy)}{t^2(x^2 - y^2)} = \frac{2xy}{x^2 - y^2} = f(x, y)$$

so it is homogeneous. Substituting Eqs. (*3.6*) and (*3.7*) into the differential equation as originally given, we obtain

$$v + x\frac{dv}{dx} = \frac{2x(xv)}{x^2 - (xv)^2}$$

which can be algebraically simplified to

$$x\frac{dv}{dx} = -\frac{v(v^2+1)}{v^2 - 1}$$

or

$$\frac{1}{x}dx + \frac{v^2 - 1}{v(v^2+1)}dv = 0 \tag{1}$$

Using partial fractions, we can expand (*1*) to

$$\frac{1}{x}dx + \left(-\frac{1}{v} + \frac{2v}{v^2+1}\right)dv = 0 \tag{2}$$

The solution to this separable equation is found by integrating both sides of (*2*). Doing so, we obtain $\ln|x| - \ln|v| + \ln(v^2+1) = c$, which can be simplified to

$$x(v^2 + 1) = kv \qquad (c = \ln|k|) \tag{3}$$

Substituting $v = y/x$ into (*3*), we find the solution of the given differential equation is $x^2 + y^2 = ky$.

3.15. Solve $y' = \dfrac{x^2 + y^2}{xy}$.

This differential equation is homogeneous. Substituting Eqs. (*3.6*) and (*3.7*) into it, we obtain

$$v + x\frac{dv}{dx} = \frac{x^2 + (xv)^2}{x(xv)}$$

which can be algebraically simplified to

$$x\frac{dv}{dx} = \frac{1}{v} \qquad \text{or} \qquad \frac{1}{x}dx - v\,dv = 0$$

The solution to this separable equation is $\ln|x| - v^2/2 = c$, or equivalently

$$v^2 = \ln x^2 + k \qquad (k = -2c) \tag{1}$$

Substituting $v = y/x$ into (*1*), we find that the solution to the given differential equation is

$$y^2 = x^2 \ln x^2 + kx^2$$

3.16. Solve $y' = \dfrac{x^2 + y^2}{xy}$; $y(1) = -2$.

The solution to the differential equation is given in Problem 3.15 as $y^2 = x^2 \ln x^2 + kx^2$. Applying the initial condition, we obtain $(-2)^2 = (1)^2 \ln (1)^2 + k(1)^2$, or $k = 4$. (Recall that $\ln 1 = 0$.) Thus, the solution to the initial-value problem is

$$y^2 = x^2 \ln x^2 + 4x^2 \qquad \text{or} \qquad y = -\sqrt{x^2 \ln x^2 + 4x^2}$$

The negative square root is taken, to be consistent with the initial condition.

3.17. Solve $y' = \dfrac{2xye^{(x/y)^2}}{y^2 + y^2e^{(x/y)^2} + 2x^2e^{(x/y)^2}}$.

This differential equation is not separable, but it is homogeneous. Noting the (x/y)-term in the exponential, we try the substitution $u = x/y$, which is an equivalent form of (*3.9*). Rewriting the differential equation as

$$\frac{dx}{dy} = \frac{y^2 + y^2e^{(x/y)^2} + 2x^2e^{(x/y)^2}}{2xye^{(x/y)^2}}$$

we have upon using substitutions (*3.9*) and (*3.10*) and simplifying,

$$y\frac{du}{dy} = \frac{1 + e^{u^2}}{2ue^{u^2}} \qquad \text{or} \qquad \frac{1}{y}dy - \frac{2ue^{u^2}}{1 + e^{u^2}}du = 0$$

This equation is separable; its solution is

$$\ln |y| - \ln (1 + e^{u^2}) = c$$

which can be rewritten as

$$y = k(1 + e^{u^2}) \qquad (c = \ln |k|) \tag{1}$$

Substituting $u = x/y$ into (*1*), we obtain the solution of the given differential equation as

$$y = k[1 + e^{(x/y)^2}]$$

3.18. Prove that every solution of Eq. (*3.2*) satisfies Eq. (*3.1*).

Rewrite (*3.1*) as $A(x) + B(y)y' = 0$. If $y(x)$ is a solution, it must satisfy this equation identically in x; hence,

$$A(x) + B[y(x)]y'(x) = 0$$

Integrating both sides of this last equation with respect to x, we obtain

$$\int A(x)\,dx + \int B[y(x)]y'(x)\,dx = c$$

In the second integral, make the change of variables $y = y(x)$, hence $dy = y'(x)\,dx$. The result of this substitution is (*3.2*).

3.19. Prove that every solution of system (*3.3*) is a solution of (*3.4*).

Following the same reasoning as in Problem 3.18, except now integrating from $x = x_0$ to $x = x$, we obtain

$$\int_{x_0}^{x} A(x)\,dx + \int_{x_0}^{x} B[y(x)]y'(x)\,dx = 0$$

The substitution $y = y(x)$ again gives the desired result. Note that as x varies from x_0 to x, y will vary from $y(x_0) = y_0$ to $y(x) = y$.

3.20. Prove that if $y' = f(x, y)$ is homogeneous, then the differential equation can be rewritten as $y' = g(y/x)$, where $g(y/x)$ depends only on the quotient y/x.

We have that $f(x, y) = f(tx, ty)$. Since this equation is valid for all t, it must be true, in particular, for $t = 1/x$. Thus, $f(x, y) = f(1, y/x)$. If we now define $g(y/x) = f(1, y/x)$, we then have $y' = f(x, y) = f(1, y/x) = g(y/x)$ as required.

Note that this form suggests the substitution $v = y/x$ which is equivalent to (*3.6*). If, in the above, we had set $t = 1/y$, then $f(x, y) = f(x/y, 1) = h(x/y)$, which suggests the alternate substitution (*3.9*).

3.21. A function $g(x, y)$ is *homogeneous of degree n* if $g(tx, ty) = t^n g(x, y)$ for all t. Determine whether the following functions are homogeneous, and, if so, find their degree:
(*a*) $xy + y^2$,
(*b*) $x + y\sin(y/x)^2$, (*c*) $x^3 + xy^2e^{x/y}$, and (*d*) $x + xy$.

(*a*) $(tx)(ty) + (ty)^2 = t^2(xy + y^2)$; homogeneous of degree two.

(*b*) $tx + ty\sin\left(\frac{ty}{tx}\right)^2 = t\left[x + y\sin\left(\frac{y}{x}\right)^2\right]$; homogeneous of degree one.

(*c*) $(tx)^3 + (tx)(ty)^2e^{tx/ty} = t^3(x^3 + xy^2e^{x/y})$; homogeneous of degree three.

(*d*) $tx + (tx)(ty) = tx + t^2xy$; not homogeneous.

3.22. An alternate definition of a homogeneous differential equation is as follows: A differential equation $M(x, y)\,dx + N(x, y)\,dy = 0$ is *homogeneous* if both $M(x, y)$ and $N(x, y)$ are homogeneous of the same degree (see Problem 3.21). Show that this definition implies the definition given in Chapter 2.

If $M(x, y)$ and $N(x, y)$ are homogeneous of degree n, then

$$f(tx, ty) = \frac{M(tx, ty)}{-N(tx, ty)} = \frac{t^n M(x, y)}{-t^n N(x, y)} = \frac{M(x, y)}{-N(x, y)} = f(x, y)$$

Supplementary Problems

In Problems 3.23 through 3.45, solve the given differential equations or initial-value problems.

3.23. $x\,dx + y\,dy = 0$

3.24. $x\,dx - y^3\,dy = 0$

3.25. $dx + \frac{1}{y^4}\,dy = 0$

3.26. $(t + 1)\,dt - \frac{1}{y^2}\,dy = 0$

3.27. $\frac{1}{x}\,dx - \frac{1}{y}\,dy = 0$

3.28. $\frac{1}{x}\,dx + dy = 0$

3.29. $x\,dx + \frac{1}{y}\,dy = 0$

3.30. $(t^2 + 1)\,dt + (y^2 + y)\,dy = 0$

3.31. $\frac{4}{t}\,dt - \frac{y - 3}{y}\,dy = 0$

3.32. $dx - \frac{1}{1 + y^2}\,dy = 0$

3.33. $dx - \dfrac{1}{y^2 - 6y + 13}dy = 0$

3.34. $y' = \dfrac{y}{x^2}$

3.35. $y' = \dfrac{xe^x}{2y}$

3.36. $\dfrac{dy}{dx} = \dfrac{x+1}{y}$

3.37. $\dfrac{dy}{dx} = y^2$

3.38. $\dfrac{dx}{dt} = x^2t^2$

3.39. $\dfrac{dx}{dt} = \dfrac{x}{t}$

3.40. $\dfrac{dy}{dt} = 3 + 5y$

3.41. $\sin x\,dx + y\,dy = 0;\quad y(0) = -2$

3.42. $(x^2 + 1)\,dx + \dfrac{1}{y}dy = 0;\quad y(-1) = 1$

3.43. $xe^{x^2}\,dx + (y^5 - 1)\,dy = 0;\quad y(0) = 0$

3.44. $y' = \dfrac{x^2y - y}{y + 1};\quad y(3) = -1$

3.45. $\dfrac{dx}{dt} = 8 - 3x;\quad x(0) = 4$

In Problems 3.46 through 3.54, determine whether the given differential equations are homogeneous and, if so, solve them.

3.46. $y' = \dfrac{y - x}{x}$

3.47. $y' = \dfrac{2y + x}{x}$

3.48. $y' = \dfrac{x^2 + 2y^2}{xy}$

3.49. $y' = \dfrac{2x + y^2}{xy}$

3.50. $y' = \dfrac{x^2 + y^2}{2xy}$

3.51. $y' = \dfrac{2xy}{y^2 - x^2}$

3.52. $y' = \dfrac{y}{x + \sqrt{xy}}$

3.53. $y' = \dfrac{y^2}{xy + (xy^2)^{1/3}}$

3.54. $y' = \dfrac{x^4 + 3x^2y^2 + y^4}{x^3y}$

Chapter 4

Exact First-Order Differential Equations

DEFINING PROPERTIES

A differential equation

$$M(x, y)\,dx + N(x, y)\,dy = 0 \tag{4.1}$$

is *exact* if there exists a function $g(x, y)$ such that

$$dg(x, y) = M(x, y)\,dx + N(x, y)\,dy \tag{4.2}$$

Test for exactness: If $M(x, y)$ and $N(x, y)$ are continuous functions and have continuous first partial derivatives on some rectangle of the xy-plane, then (*4.1*) is exact if and only if

$$\frac{\partial M(x, y)}{\partial y} = \frac{\partial N(x, y)}{\partial x} \tag{4.3}$$

METHOD OF SOLUTION

To solve Eq. (*4.1*), assuming that it is exact, first solve the equations

$$\frac{\partial g(x, y)}{\partial x} = M(x, y) \tag{4.4}$$

$$\frac{\partial g(x, y)}{\partial y} = N(x, y) \tag{4.5}$$

for $g(x, y)$. The solution to (*4.1*) is then given implicitly by

$$g(x, y) = c \tag{4.6}$$

where c represents an arbitrary constant.

Equation (*4.6*) is immediate from Eqs. (*4.1*) and (*4.2*). If (*4.2*) is substituted into (*4.1*), we obtain $dg(x, y(x)) = 0$. Integrating this equation (note that we can write 0 as $0\,dx$), we have $\int dg(x, y(x)) = \int 0\,dx$, which, in turn, implies (*4.6*).

INTEGRATING FACTORS

In general, Eq. (*4.1*) is not exact. Occasionally, it is possible to transform (*4.1*) into an exact differential equation by a judicious multiplication. A function $I(x, y)$ is an *integrating factor* for (*4.1*) if the equation

$$I(x, y)[M(x, y)\,dx + N(x, y)\,dy] = 0 \tag{4.7}$$

is exact. A solution to (*4.1*) is obtained by solving the exact differential equation defined by (*4.7*). Some of the more common integrating factors are displayed in Table 4-1 and the conditions that follow:

If $\dfrac{1}{N}\left(\dfrac{\partial M}{\partial y} - \dfrac{\partial N}{\partial x}\right) \equiv g(x)$, a function of x alone, then

$$I(x, y) = e^{\int g(x)\,dx} \tag{4.8}$$

If $\dfrac{1}{M}\left(\dfrac{\partial M}{\partial y} - \dfrac{\partial N}{\partial x}\right) \equiv h(y)$, a function of y alone, then

$$I(x, y) = e^{-\int h(y)\,dy} \tag{4.9}$$

If $M = yf(xy)$ and $N = xg(xy)$, then

$$I(x, y) = \frac{1}{xM - yN} \tag{4.10}$$

Table 4-1

Group of terms	Integrating factor $I(x, y)$	Exact differential $dy(x, y)$
$y\,dx - x\,dy$	$-\dfrac{1}{x^2}$	$\dfrac{x\,dy - y\,dx}{x^2} = d\left(\dfrac{y}{x}\right)$
$y\,dx - x\,dy$	$\dfrac{1}{y^2}$	$\dfrac{y\,dx - x\,dy}{y^2} = d\left(\dfrac{x}{y}\right)$
$y\,dx - x\,dy$	$-\dfrac{1}{xy}$	$\dfrac{x\,dy - y\,dx}{xy} = d\left(\ln\dfrac{y}{x}\right)$
$y\,dx - x\,dy$	$-\dfrac{1}{x^2 + y^2}$	$\dfrac{x\,dy - y\,dx}{x^2 + y^2} = d\left(\arctan\dfrac{y}{x}\right)$
$y\,dx + x\,dy$	$\dfrac{1}{xy}$	$\dfrac{y\,dx + x\,dy}{xy} = d(\ln xy)$
$y\,dx + x\,dy$	$\dfrac{1}{(xy)^n}$, $n > 1$	$\dfrac{y\,dx + x\,dy}{(xy)^n} = d\left[\dfrac{-1}{(n-1)(xy)^{n-1}}\right]$
$y\,dy + x\,dx$	$\dfrac{1}{x^2 + y^2}$	$\dfrac{y\,dy + x\,dx}{x^2 + y^2} = d\left[\dfrac{1}{2}\ln(x^2 + y^2)\right]$
$y\,dy + x\,dx$	$\dfrac{1}{(x^2 + y^2)^n}$, $n > 1$	$\dfrac{y\,dy + x\,dx}{(x^2 + y^2)^n} = d\left[\dfrac{-1}{2(n-1)(x^2 + y^2)^{n-1}}\right]$
$ay\,dx + bx\,dy$ (a, b constants)	$x^{a-1}y^{b-1}$	$x^{a-1}y^{b-1}(ay\,dx + bx\,dy) = d(x^a y^b)$

In general, integrating factors are difficult to uncover. If a differential equation does not have one of the forms given above, then a search for an integrating factor likely will not be successful, and other methods of solution are recommended.

Solved Problems

4.1. Determine whether the differential equation $2xy\,dx + (1+x^2)\,dy = 0$ is exact.

This equation has the form of Eq. (*4.1*) with $M(x, y) = 2xy$ and $N(x, y) = 1 + x^2$. Since $\partial M/\partial y = \partial N/\partial x = 2x$, the differential equation is exact.

4.2. Solve the differential equation given in Problem 4.1.

This equation was shown to be exact. We now determine a function $g(x, y)$ that satisfies Eqs. (*4.4*) and (*4.5*). Substituting $M(x, y) = 2xy$ into (*4.4*), we obtain $\partial g/\partial x = 2xy$. Integrating both sides of this equation with respect to x, we find

$$\int \frac{\partial g}{\partial x}\,dx = \int 2xy\,dx$$

or

$$g(x, y) = x^2y + h(y) \tag{1}$$

Note that when integrating with respect to x, the constant (*with respect to x*) of integration can depend on y.

We now determine $h(y)$. Differentiating (*1*) with respect to y, we obtain $\partial g/\partial y = x^2 + h'(y)$. Substituting this equation along with $N(x, y) = 1 + x^2$ into (*4.5*), we have

$$x^2 + h'(y) = 1 + x^2 \qquad \text{or} \qquad h'(y) = 1$$

Integrating this last equation with respect to y, we obtain $h(y) = y + c_1$ (c_1 = constant). Substituting this expression into (*1*) yields

$$g(x, y) = x^2y + y + c_1$$

The solution to the differential equation, which is given implicitly by (*4.6*) as $g(x, y) = c$, is

$$x^2y + y = c_2 \qquad (c_2 = c - c_1)$$

Solving for y explicitly, we obtain the solution as $y = c_2/(x^2 + 1)$.

4.3. Determine whether the differential equation $y\,dx - x\,dy = 0$ is exact.

This equation has the form of Eq. (*4.1*) with $M(x, y) = y$ and $N(x, y) = -x$. Here

$$\frac{\partial M}{\partial y} = 1 \qquad \text{and} \qquad \frac{\partial N}{\partial x} = -1$$

which are not equal, so the differential equation as given is not exact.

4.4. Determine whether the differential equation

$$(x + \sin y)\,dx + (x\cos y - 2y)\,dy = 0$$

is exact.

Here $M(x, y) = x + \sin y$ and $N(x, y) = x\cos y - 2y$. Thus, $\partial M/\partial y = \partial N/\partial x = \cos y$, and the differential equation is exact.

4.5. Solve the differential equation given in Problem 4.4.

This equation was shown to be exact. We now seek a function $g(x, y)$ that satisfies (*4.4*) and (*4.5*). Substituting $M(x, y)$ into (*4.4*), we obtain $\partial g/\partial x = x + \sin y$. Integrating both sides of this equation with respect to x, we find

$$\int \frac{\partial g}{\partial x}\,dx = \int (x + \sin y)\,dx$$

or

$$g(x, y) = \frac{1}{2}x^2 + x \sin y + h(y) \tag{1}$$

To find $h(y)$, we differentiate (1) with respect to y, yielding $\partial g/\partial y = x \cos y + h'(y)$, and then substitute this result along with $N(x, y) = x \cos y - 2y$ into (4.5). Thus we find

$$x \cos y + h'(y) = x \cos y - 2y \qquad \text{or} \qquad h'(y) = -2y$$

from which it follows that $h(y) = -y^2 + c_1$. Substituting this $h(y)$ into (1), we obtain

$$g(x, y) = \frac{1}{2}x^2 + x \sin y - y^2 + c_1$$

The solution of the differential equation is given implicitly by (4.6) as

$$\frac{1}{2}x^2 + x \sin y - y^2 = c_2 \qquad (c_2 = c - c_1)$$

4.6. Solve $y' = \dfrac{2 + ye^{xy}}{2y - xe^{xy}}$.

Rewriting this equation in differential form, we obtain

$$(2 + ye^{xy})\,dx + (xe^{xy} - 2y)\,dy = 0$$

Here, $M(x, y) = 2 + ye^{xy}$ and $N(x, y) = xe^{xy} - 2y$ and, since $\partial M/\partial y = \partial N/\partial x = e^{xy} + xye^{xy}$, the differential equation is exact. Substituting $M(x, y)$ into (4.4), we find $\partial g/\partial x = 2 + ye^{xy}$; then integrating with respect to x, we obtain

$$\int \frac{\partial g}{\partial x}\,dx = \int [2 + ye^{xy}]\,dx$$

or

$$g(x, y) = 2x + e^{xy} + h(y) \tag{1}$$

To find $h(y)$, first differentiate (1) with respect to y, obtaining $\partial g/\partial y = xe^{xy} + h'(y)$; then substitute this result along with $N(x, y)$ into (4.5) to obtain

$$xe^{xy} + h'(y) = xe^{xy} - 2y \qquad \text{or} \qquad h'(y) = -2y$$

It follows that $h(y) = -y^2 + c_1$. Substituting this $h(y)$ into (1), we obtain

$$g(x, y) = 2x + e^{xy} - y^2 + c_1$$

The solution to the differential equation is given implicitly by (4.6) as

$$2x + e^{xy} - y^2 = c_2 \qquad (c_2 = c - c_1)$$

4.7. Determine whether the differential equation $y^2\,dt + (2yt + 1)\,dy = 0$ is exact.

This is an equation for the unknown function $y(t)$. In terms of the variables t and y, we have $M(t, y) = y^2$, $N(t, y) = 2yt + 1$, and

$$\frac{\partial M}{\partial y} = \frac{\partial}{\partial y}(y^2) = 2y = \frac{\partial}{\partial t}(2yt + 1) = \frac{\partial N}{\partial t}$$

so the differential equation is exact.

4.8. Solve the differential equation given in Problem 4.7.

This equation was shown to be exact, so the solution procedure given by Eqs. (*4.4*) through (*4.6*), with t replacing x, is applicable. Here

$$\frac{\partial g}{\partial t} = y^2$$

Integrating both sides with respect to t, we have

$$\int \frac{\partial g}{\partial t}\,dt = \int y^2\,dt$$

or

$$g(y, t) = y^2t + h(y) \tag{1}$$

Differentiating (*1*) with respect to y, we obtain

$$\frac{\partial g}{\partial y} = 2yt + \frac{dh}{dy}$$

Hence,

$$2yt + \frac{dh}{dy} = 2yt + 1$$

where the right side of this last equation is the coefficient of dy in the original differential equation. It follows that

$$\frac{dh}{dy} = 1$$

$h(y) = y + c_1$, and (*1*) becomes $g(t, y) = y^2t + y + c_1$. The solution to the differential equation is given implicitly by (*4.6*) as

$$y^2t + y = c_2 \qquad (c_2 = c - c_1) \tag{2}$$

We can solve for y explicitly with the quadratic formula, whence

$$y = \frac{-1 \pm \sqrt{1 + 4c_2t}}{2t}$$

4.9. Determine whether the differential equation

$$(2x^2t - 2x^3)\,dt + (4x^3 - 6x^2t + 2xt^2)\,dx = 0$$

is exact.

This is an equation for the unknown function $x(t)$. In terms of the variables t and x, we find

$$\frac{\partial}{\partial x}(2x^2t - 2x^3) = 4xt - 6x^2 = \frac{\partial}{\partial t}(4x^3 - 6x^2t + 2xt^2)$$

so the differential equation is exact.

4.10. Solve the differential equation given in Problem 4.9.

This equation was shown to be exact, so the solution procedure given by Eqs. (*4.4*) through (*4.6*), with t and x replacing x and y, respectively, is applicable. We seek a function $g(t, x)$ having the property that dg is the right side of the given differential equation. Here

$$\frac{\partial g}{\partial t} = 2x^2t - 2x^3$$

Integrating both sides with respect to t, we have

$$\int \frac{\partial g}{\partial t}\,dt = \int (2x^2t - 2x^3)\,dt$$

or

$$g(x, t) = x^2t^2 - 2x^3t + h(x) \tag{1}$$

Differentiating (*1*) with respect to x, we obtain

$$\frac{\partial g}{\partial x} = 2xt^2 - 6x^2t + \frac{dh}{dx}$$

Hence, $$2xt^2 - 6x^2t + \frac{dh}{dx} = 4x^3 - 6x^2t + 2xt^2$$

where the right side of this last equation is the coefficient of dx in the original differential equation. It follows that

$$\frac{dh}{dx} = 4x^3$$

Now $h(x) = x^4 + c_1$, and (*1*) becomes

$$g(t, x) = x^2t^2 - 2x^3t + x^4 + c_1 = (x^2 - xt)^2 + c_1$$

The solution to the differential equation is given implicitly by (*4.6*) as

$$(x^2 - xt)^2 = c_2 \qquad (c_2 = c - c_1)$$

or, by taking the square roots of both sides of this last equation, as

$$x^2 - xt = c_3 \qquad c_3 = \pm\sqrt{c_2} \tag{2}$$

We can solve for x explicitly with the quadratic formula, whence

$$x = \frac{t \pm \sqrt{t^2 + 4c_3}}{2}$$

4.11. Solve $y' = \dfrac{-2xy}{1+x^2}$; $y(2) = -5$.

The differential equation has the differential form given in Problem 4.1. Its solution is given in (2) of Problem 4.2 as $x^2y + y = c_2$. Using the initial condition, $y = -5$ when $x = 2$, we obtain $(2)^2(-5) + (-5) = c_2$, or $c_2 = -25$. The solution to the initial-value problem is therefore $x^2y + y = -25$ or $y = -25/(x^2 + 1)$.

4.12. Solve $\dot{y} = \dfrac{-y^2}{2yt+1}$; $y(1) = -2$.

This differential equation in standard form has the differential form of Problem 4.7. Its solution is given in (*2*) of Problem 4.8 as $y^2t + y = c_2$. Using the initial condition $y = -2$ when $t = 1$, we obtain $(-2)^2(1) + (-2) = c_2$, or $c_2 = 2$. The solution to the initial-value problem is $y^2t + y = 2$, in implicit form. Solving for y directly, using the quadratic formula, we have

$$y = \frac{-1 - \sqrt{1 + 8t}}{2t}$$

where the negative sign in front of the radical was chosen to be consistent with the given initial condition.

4.13. Solve $\dot{x} = \dfrac{2x^2(x-t)}{4x^3 - 6x^2t + 2xt^2}$; $x(2) = 3$.

This differential equation in standard form has the differential form of Problem 4.9. Its solution is given in (*2*) of Problem 4.10 as $x^2 - xt = c_3$. Using the initial condition $x = 3$ when $t = 2$, we obtain $(3)^2 - 3(2) = c_3$, or $c_3 = 3$. The solution to the initial-value problem is $x^2 + xt = 3$, in implicit form.

Solving for x directly, using the quadratic formula, we have

$$x = \frac{1}{2}(t + \sqrt{t^2 + 12})$$

where the positive sign in front of the radical was chosen to be consistent with the given initial condition.

4.14. Determine whether $-1/x^2$ is an integrating factor for the differential equation $y\,dx - x\,dy = 0$.

It was shown in Problem 4.3 that the differential equation is not exact. Multiplying it by $-1/x^2$, we obtain

$$\frac{-1}{x^2}(y\,dx - x\,dy) = 0 \qquad \text{or} \qquad \frac{-y}{x^2}dx + \frac{1}{x}dy = 0 \tag{1}$$

Equation (*1*) has the form of Eq. (*4.1*) with $M(x, y) = -y/x^2$ and $N(x, y) = 1/x$. Now

$$\frac{\partial M}{\partial y} = \frac{\partial}{\partial y}\left(\frac{-y}{x^2}\right) = \frac{-1}{x^2} = \frac{\partial}{\partial x}\left(\frac{1}{x}\right) = \frac{\partial N}{\partial x}$$

so (*1*) is exact, which implies that $-1/x^2$ is an integrating factor for the original differential equation.

4.15. Solve $y\,dx - x\,dy = 0$.

Using the results of Problem 4.14, we can rewrite the given differential equation as

$$\frac{x\,dy - y\,dx}{x^2} = 0$$

which is exact. Equation (*1*) can be solved using the steps described in Eqs. (*4.4*) through (*4.6*).

Alternatively, we note from Table 4-1 that (*1*) can be rewritten as $d(y/x) = 0$. Hence, by direct integration, we have $y/x = c$, or $y = cx$, as the solution.

4.16. Determine whether $-1/(xy)$ is also an integrating factor for the differential equation defined in Problem 4.14.

Multiplying the differential equation $y\,dx - x\,dy = 0$ by $-1/(xy)$, we obtain

$$\frac{-1}{xy}(y\,dx - x\,dy) = 0 \qquad \text{or} \qquad -\frac{1}{x}dx + \frac{1}{y}dy = 0 \tag{1}$$

Equation (*1*) has the form of Eq. (*4.1*) with $M(x, y) = -1/x$ and $N(x, y) = 1/y$. Now

$$\frac{\partial M}{\partial y} = \frac{\partial}{\partial y}\left(-\frac{1}{x}\right) = 0 = \frac{\partial}{\partial x}\left(\frac{1}{y}\right) = \frac{\partial N}{\partial x}$$

so (*1*) is exact, which implies that $-1/xy$ is also an integrating factor for the original differential equation.

4.17. Solve Problem 4.15 using the integrating factor given in Problem 4.16.

Using the results of Problem 4.16, we can rewrite the given differential equation as

$$\frac{x\,dy - y\,dx}{xy} = 0 \tag{1}$$

which is exact. Equation (*1*) can be solved using the steps described in Eqs. (*4.4*) through (*4.6*).

Alternatively, we note from Table 4-1 that (*1*) can be rewritten as $d[\ln(y/x)] = 0$. Then, by direct integration, $\ln(y/x) = c_1$. Taking the exponential of both sides, we find $y/x = e^{c_1}$, or finally,

$$y = cx \qquad (c = e^{c_1})$$

4.18. Solve $(y^2 - y)\,dx + x\,dy = 0$.

This differential equation is not exact, and no integrating factor is immediately apparent. Note, however, that if terms are strategically regrouped, the differential equation can be rewritten as

$$-(y\,dx - x\,dy) + y^2\,dx = 0 \tag{1}$$

The group of terms in parentheses has many integrating factors (see Table 4-1). Trying each integrating factor separately, we find that the only one that makes the entire equation exact is $I(x, y) = 1/y^2$. Using this integrating factor, we can rewrite *(1)* as

$$-\frac{y\,dx - x\,dy}{y^2} + 1\,dx = 0 \tag{2}$$

Since *(2)* is exact, it can be solved using the steps described in Eqs. *(4.4)* through *(4.6)*.

Alternatively, we note from Table 4-1 that *(2)* can be rewritten as $-d(x/y) + 1\,dx = 0$, or as $d(x/y) = 1\,dx$. Integrating, we obtain the solution

$$\frac{x}{y} = x + c \qquad \text{or} \qquad y = \frac{x}{x + c}$$

4.19. Solve $(y - xy^2)\,dx + (x + x^2y^2)\,dy = 0$.

This differential equation is not exact, and no integrating factor is immediately apparent. Note, however, that the differential equation can be rewritten as

$$(y\,dx + x\,dy) + (-xy^2\,dx + x^2y^2\,dy) = 0 \tag{1}$$

The first group of terms has many integrating factors (see Table 4-1). One of these factors, namely $I(x, y) = 1/(xy)^2$, is an integrating factor for the entire equation. Multiplying *(1)* by $1/(xy)^2$, we find

$$\frac{y\,dx + x\,dy}{(xy)^2} + \frac{-xy^2\,dx + x^2y^2\,dy}{(xy)^2} = 0$$

or equivalently,

$$\frac{y\,dx + x\,dy}{(xy)^2} = \frac{1}{x}dx - 1\,dy \tag{2}$$

Since *(2)* is exact, it can be solved using the steps described in Eqs. *(4.4)* through *(4.6)*.

Alternatively, we note from Table 4-1

$$\frac{y\,dx + x\,dy}{(xy)^2} = d\left(\frac{-1}{xy}\right)$$

so that *(2)* can be rewritten as

$$d\left(\frac{-1}{xy}\right) = \frac{1}{x}dx - 1\,dy$$

Integrating both sides of this last equation, we find

$$\frac{-1}{xy} = \ln|x| - y + c$$

which is the solution in implicit form.

4.20. Solve $y' = \dfrac{3yx^2}{x^3 + 2y^4}$.

Rewriting this equation in differential form, we have

$$(3yx^2)\,dx + (-x^3 - 2y^4)\,dy = 0$$

which is not exact. Furthermore, no integrating factor is immediately apparent. We can, however, rearrange this equation as

$$x^2(3y\,dx - x\,dy) - 2y^4\,dy = 0 \tag{1}$$

The group in parentheses is of the form $ay\,dx + bx\,dy$, where $a = 3$ and $b = -1$, which has an integrating factor x^2y^{-2}. Since the expression in parentheses is already multiplied by x^2, we try an integrating factor of the form $I(x, y) = y^{-2}$. Multiplying (*1*) by y^{-2}, we have

$$x^2y^{-2}(3y\,dx - x\,dy) - 2y^2\,dy = 0$$

which can be simplified (see Table 4-1) to

$$d(x^3y^{-1}) = 2y^2\,dy \tag{2}$$

Integrating both sides of (*2*), we obtain

$$x^3y^{-1} = \frac{2}{3}y^3 + c$$

as the solution in implicit form.

4.21. Convert $y' = 2xy - x$ into an exact differential equation.

Rewriting this equation in differential form, we have

$$(-2xy + x)\,dx + dy = 0 \tag{1}$$

Here $M(x, y) = -2xy + x$ and $N(x, y) = 1$. Since

$$\frac{\partial M}{\partial y} = -2x \qquad \text{and} \qquad \frac{\partial N}{\partial x} = 0$$

are not equal, (*1*) is not exact. But

$$\frac{1}{N}\left(\frac{\partial M}{\partial y} - \frac{\partial N}{\partial x}\right) = \frac{(-2x) - (0)}{1} = -2x$$

is a function of x alone. Using Eq. (*4.8*), we have $I(x, y) = e^{\int -2x\,dx} = e^{-x^2}$ as an integrating factor. Multiplying (*1*) by e^{-x^2}, we obtain

$$(-2xye^{-x^2} + xe^{-x^2})\,dx + e^{-x^2}\,dy = 0 \tag{2}$$

which is exact.

4.22. Convert $y^2\,dx + xy\,dy = 0$ into an exact differential equation.

Here $M(x, y) = y^2$ and $N(x, y) = xy$. Since

$$\frac{\partial M}{\partial y} = 2y \qquad \text{and} \qquad \frac{\partial N}{\partial x} = y$$

are not equal, (*1*) is not exact. But

$$\frac{1}{M}\left(\frac{\partial M}{\partial y} - \frac{\partial N}{\partial x}\right) = \frac{2y - y}{y^2} = \frac{1}{y}$$

is a function of y alone. Using Eq. (*4.9*), we have as an integrating factor $I(x, y) = e^{-\int (1/y)\,dy} = e^{-\ln y} = 1/y$. Multiplying the given differential equation by $I(x, y) = 1/y$, we obtain the exact equation $y\,dx + x\,dy = 0$.

4.23 Convert $y' = \dfrac{xy^2 - y}{x}$ into an exact differential equation.

Rewriting this equation in differential form, we have

$$y(1 - xy)\,dx + x\,dy = 0 \tag{1}$$

Here $M(x, y) = y(1 - xy)$ and $N(x, y) = x$. Since

$$\frac{\partial M}{\partial y} = 1 - 2xy \quad \text{and} \quad \frac{\partial N}{\partial x} = 1$$

are not equal, (*1*) is not exact. Equation (*4.10*), however, is applicable and provides the integrating factor

$$I(x, y) = \frac{1}{x[y(1 - xy)] - yx} = \frac{-1}{(xy)^2}$$

Multiplying (*1*) by $I(x, y)$, we obtain

$$\frac{xy - 1}{x^2y}\,dx - \frac{1}{xy^2}\,dy = 0$$

which is exact.

Supplementary Problems

In Problems 4.24 through 4.40, test whether the differential equations are exact and solve those that are.

4.24. $(y + 2xy^3)\,dx + (1 + 3x^2y^2 + x)\,dy = 0$

4.25. $(xy + 1)\,dx + (xy - 1)\,dy = 0$

4.26. $e^{x^3}(3x^2y - x^2)\,dx + e^{x^3}\,dy = 0$

4.27. $3x^2y^2\,dx + (2x^3y + 4y^3)\,dy = 0$

4.28. $y\,dx + x\,dy = 0$

4.29. $(x - y)\,dx + (x + y)\,dy = 0$

4.30. $(y \sin x + xy \cos x)\,dx + (x \sin x + 1)\,dy = 0$

4.31. $-\dfrac{y^2}{t^2}\,dt + \dfrac{2y}{t}\,dy = 0$

4.32. $-\dfrac{2y}{t^3}\,dt + \dfrac{1}{t^2}\,dy = 0$

4.33. $y^2\,dt + t^2\,dy = 0$

4.34. $(4t^3y^3 - 2ty)\,dt + (3t^4y^2 - t^2)\,dy = 0$

4.35. $\dfrac{ty - 1}{t^2y}\,dt - \dfrac{1}{ty^2}\,dy = 0$

4.36. $(t^2 - x)\,dt - t\,dx = 0$

4.37. $(t^2 + x^2)\,dt + (2tx - x)\,dx = 0$

4.38. $2xe^{2t}\,dt + (1 + e^{2t})\,dx = 0$

4.39. $\sin t \cos x\,dt - \sin x \cos t\,dx = 0$

4.40. $(\cos x + x \cos t)\,dt + (\sin t - t \sin x)\,dx = 0$

In Problems 4.41 through 4.55, find an appropriate integrating factor for each differential equation and solve.

4.41. $(y + 1)\,dx - x\,dy = 0$

4.42. $y\,dx + (1 - x)\,dy = 0$

4.43. $(x^2 + y + y^2)\,dx - x\,dy = 0$

4.44. $(y + x^3y^3)\,dx + x\,dy = 0$

4.45. $(y + x^4y^2)\,dx + x\,dy = 0$

4.46. $(3x^2y - x^2)\,dx + dy = 0$

4.47. $dx - 2xy\,dy = 0$

4.48. $2xy\,dx + y^2\,dy = 0$

4.49. $y\,dx + 3x\,dy = 0$

4.50. $\left(2xy^2 + \dfrac{x}{y^2}\right) dx + 4x^2y\,dy = 0$

4.51. $xy^2\,dx + (x^2y^2 + x^2y)\,dy = 0$

4.52. $xy^2\,dx + x^2y\,dy = 0$

4.53. $(y + x^3 + xy^2)\,dx - x\,dy = 0$

4.54. $(x^3y^2 - y)\,dx + (x^2y^4 - x)\,dy = 0$

4.55. $3x^2y^2\,dx + (2x^3y + x^3y^4)\,dy = 0$

In Problems 4.56 through 4.65, solve the initial-value problems.

4.56. Problem 4.10 with $x(0) = 2$

4.57. Problem 4.10 with $x(2) = 0$

4.58. Problem 4.10 with $x(1) = -5$

4.59. Problem 4.24 with $y(1) = -5$

4.60. Problem 4.26 with $y(0) = -1$

4.61. Problem 4.31 with $y(0) = -2$

4.62. Problem 4.31 with $y(2) = -2$

4.63. Problem 4.32 with $y(2) = -2$

4.64. Problem 4.36 with $x(1) = 5$

4.65. Problem 4.38 with $x(1) = -2$

Chapter 5

Linear First-Order Differential Equations

METHOD OF SOLUTION

A first-order *linear* differential equation has the form (see Chapter 2)

$$y' + p(x)y = q(x) \tag{5.1}$$

An integrating factor for Eq. (*5.1*) is

$$I(x) = e^{\int p(x)\,dx} \tag{5.2}$$

which depends only on x and is independent of y. When both sides of (*5.1*) are multiplied by $I(x)$, the resulting equation

$$I(x)y' + p(x)I(x)y = I(x)q(x) \tag{5.3}$$

is exact. This equation can be solved by the method described in Chapter 4. A simpler procedure is to rewrite (*5.3*) as

$$\frac{d(yI)}{dx} = Iq(x)$$

integrate both sides of this last equation with respect to x, and then solve the resulting equation for y.

REDUCTION OF BERNOULLI EQUATIONS

A Bernoulli differential equation has the form

$$y' + p(x)y = q(x)y^n \tag{5.4}$$

where n is a real number. The substitution

$$z = y^{1-n} \tag{5.5}$$

transforms (*5.4*) into a linear differential equation in the unknown function $z(x)$.

Solved Problems

5.1. Find an integrating factor for $y' - 3y = 6$.

The differential equation has the form of Eq. (*5.1*), with $p(x) = -3$ and $q(x) = 6$, and is linear. Here

$$\int p(x)\,dx = \int -3\,dx = -3x$$

so (*5.2*) becomes

$$I(x) = e^{\int p(x)\,dx} = e^{-3x} \tag{1}$$

5.2. Solve the differential equation in the previous problem.

Multiplying the differential equation by the integrating factor defined by (*1*) of Problem 5.1, we obtain

$$e^{-3x}y' - 3e^{-3x}y = 6e^{-3x} \qquad \text{or} \qquad \frac{d}{dx}(ye^{-3x}) = 6e^{-3x}$$

Integrating both sides of this last equation with respect to x, we have

$$\int \frac{d}{dx}(ye^{-3x})\,dx = \int 6e^{-3x}\,dx$$

$$ye^{-3x} = -2e^{-3x} + c$$

$$y = ce^{3x} - 2$$

5.3. Find an integrating factor for $y' - 2xy = x$.

The differential equation has the form of Eq. (*5.1*), with $p(x) = -2x$ and $q(x) = x$, and is linear. Here

$$\int p(x)\,dx = \int (-2x)\,dx = -x^2$$

so (*5.2*) becomes

$$I(x) = e^{\int p(x)\,dx} = e^{-x^2} \tag{1}$$

5.4. Solve the differential equation in the previous problem.

Multiplying the differential equation by the integrating factor defined by (*1*) of Problem 5.3, we obtain

$$e^{-x^2}y' - 2xe^{-x^2}y = xe^{-x^2} \qquad \text{or} \qquad \frac{d}{dx}[ye^{-x^2}] = xe^{-x^2}$$

Integrating both sides of this last equation with respect to x, we find that

$$\int \frac{d}{dx}(ye^{-x^2})\,dx = \int xe^{-x^2}\,dx$$

$$ye^{-x^2} = -\frac{1}{2}e^{-x^2} + c$$

$$y = ce^{x^2} - \frac{1}{2}$$

5.5. Find an integrating factor for $y' + (4/x)y = x^4$.

The differential equation has the form of Eq. (*5.1*), with $p(x) = 4/x$ and $q(x) = x^4$, and is linear. Here

$$\int p(x)\,dx = \int \frac{4}{x}\,dx = 4 \ln|x| = \ln x^4$$

so (*5.2*) becomes

$$I(x) = e^{\int p(x)\,dx} = e^{\ln x^4} = x^4 \tag{1}$$

5.6. Solve the differential equation in the previous problem.

Multiplying the differential equation by the integrating factor defined by (*1*) of Problem 5.5, we obtain

$$x^4y' + 4x^3y = x^8 \qquad \text{or} \qquad \frac{d}{dx}(yx^4) = x^8$$

Integrating both sides of this last equation with respect to x, we obtain

$$yx^4 = \frac{1}{9}x^9 + c \qquad \text{or} \qquad y = \frac{c}{x^4} + \frac{1}{9}x^5$$

5.7. Solve $y' + y = \sin x$.

Here $p(x) = 1$; hence $I(x) = e^{\int 1\,dx} = e^x$. Multiplying the differential equation by $I(x)$, we obtain

$$e^x y' + e^x y = e^x \sin x \quad \text{or} \quad \frac{d}{dx}(ye^x) = e^x \sin x$$

Integrating both sides of the last equation with respect to x (to integrate the right side, we use integration by parts twice), we find

$$ye^x = \frac{1}{2}e^x(\sin x - \cos x) + c \quad \text{or} \quad y = ce^{-x} + \frac{1}{2}\sin x - \frac{1}{2}\cos x$$

5.8. Solve the initial-value problem $y' + y = \sin x$; $y(\pi) = 1$.

From Problem 5.7, the solution to the differential equation is

$$y = ce^{-x} + \frac{1}{2}\sin x - \frac{1}{2}\cos x$$

Applying the initial condition directly, we obtain

$$1 = y(\pi) = ce^{-\pi} + \frac{1}{2} \quad \text{or} \quad c = \frac{1}{2}e^{\pi}$$

Thus
$$y = \frac{1}{2}e^{\pi}e^{-x} + \frac{1}{2}\sin x - \frac{1}{2}\cos x = \frac{1}{2}(e^{\pi - x} + \sin x - \cos x)$$

5.9. Solve $y' - 5y = 0$.

Here $p(x) = -5$ and $I(x) = e^{\int(-5)\,dx} = e^{-5x}$. Multiplying the differential equation by $I(x)$, we obtain

$$e^{-5x}y' - 5e^{-5x}y = 0 \quad \text{or} \quad \frac{d}{dx}(ye^{-5x}) = 0$$

Integrating, we obtain $ye^{-5x} = c$, or $y = ce^{5x}$.

Note that the differential equation is also separable. (See Problem 3.4.)

5.10. Solve $\dfrac{dz}{dx} - xz = -x$.

This is a linear differential equation for the unknown function $z(x)$. It has the form of Eq. *(5.1)* with y replaced by z and $p(x) = q(x) = -x$. The integrating factor is

$$I(x) = e^{\int(-x)\,dx} = e^{-x^2/2}$$

Multiplying the differential equation by $I(x)$, we obtain

$$e^{-x^2/2}\frac{dz}{dx} - xe^{-x^2/2}z = -xe^{-x^2/2}$$

or
$$\frac{d}{dx}(ze^{-x^2/2}) = -xe^{-x^2/2}$$

Upon integrating both sides of this last equation, we have

$$ze^{-x^2/2} = e^{-x^2/2} + c$$

whereupon
$$z(x) = ce^{x^2/2} + 1$$

5.11. Solve the initial-value problem $z' - xz = -x$; $z(0) = -4$.

The solution to this differential equation is given in Problem 5.10 as

$$z(x) = 1 + ce^{x^2/2}$$

Applying the initial condition directly, we have

$$-4 = z(0) = 1 + ce^0 = 1 + c$$

or $c = -5$. Thus,

$$z(x) = 1 - 5e^{x^2/2}$$

5.12. Solve $z' - \dfrac{2}{x}z = \dfrac{2}{3}x^4$.

This is a linear differential equation for the unknown function $z(x)$. It has the form of Eq. (*5.1*) with y replaced by z. The integrating factor is

$$I(x) = e^{\int(-2/x)\,dx} = e^{-2\ln|x|} = e^{\ln x^{-2}} = x^{-2}$$

Multiplying the differential equation by $I(x)$, we obtain

$$x^{-2}z' - 2x^{-3}z = \frac{2}{3}x^2$$

or
$$\frac{d}{dx}(x^{-2}z) = \frac{2}{3}x^2$$

Upon integrating both sides of this last equation, we have

$$x^{-2}z = \frac{2}{9}x^3 + c$$

whereupon
$$z(x) = cx^2 + \frac{2}{9}x^5$$

5.13. Solve $\dfrac{dQ}{dt} + \dfrac{2}{10+2t}Q = 4$.

This is a linear differential equation for the unknown function $Q(t)$. It has the form of Eq. (*5.1*) with y replaced by Q, x replaced by t, $p(t) = 2/(10+2t)$, and $q(t) = 4$. The integrating factor is

$$I(t) = e^{\int[2/(10+2t)]\,dt} = e^{\ln|10+2t|} = 10 + 2t \qquad (t > -5)$$

Multiplying the differential equation by $I(t)$, we obtain

$$(10 + 2t)\frac{dQ}{dt} + 2Q = 40 + 8t$$

or
$$\frac{d}{dt}[(10 + 2t)Q] = 40 + 8t$$

Upon integrating both sides of this last equation, we have

$$(10 + 2t)Q = 40t + 4t^2 + c$$

whereupon
$$Q(t) = \frac{40t + 4t^2 + c}{10 + 2t} \qquad (t > -5)$$

5.14. Solve the initial-value problem $\frac{dQ}{dt}+\frac{2}{10+2t}Q=4;\ Q(2)=100.$

The solution to this differential equation is given in Problem 5.13 as

$$Q(t)=\frac{40t+4t^2+c}{10+2t}\qquad (t>-5)$$

Applying the initial condition directly, we have

$$100=Q(2)=\frac{40(2)+4(4)+c}{10+2(2)}$$

or $c=1304$. Thus,

$$Q(t)=\frac{4t^2+40t+1304}{2t+10}\qquad (t>-5)$$

5.15. Solve $\frac{dT}{dt}+kT=100k$, where k denotes a constant.

This is a linear differential equation for the unknown function $T(t)$. It has the form of Eq. (*5.1*) with y replaced by T, x replaced by t, $p(t)=k$, and $q(t)=100k$. The integrating factor is

$$I(t)=e^{\int k\,dt}=e^{kt}$$

Multiplying the differential equation by $I(t)$, we obtain

$$e^{kt}\frac{dT}{dt}+ke^{kt}T=100ke^{kt}$$

or

$$\frac{d}{dt}(Te^{kt})=100ke^{kt}$$

Upon integrating both sides of this last equation, we have

$$Te^{kt}=100e^{kt}+c$$

whereupon

$$T(t)=ce^{-kt}+100$$

5.16. Solve $y'+xy=xy^2$.

This equation is not linear. It is, however, a Bernoulli differential equation having the form of Eq. (*5.4*) with $p(x)=q(x)=x$, and $n=2$. We make the substitution suggested by (*5.5*), namely, $z=y^{1-2}=y^{-1}$, from which follow

$$y=\frac{1}{z}\qquad\text{and}\qquad y'=-\frac{z'}{z^2}$$

Substituting these equations into the differential equation, we obtain

$$-\frac{z'}{z^2}+\frac{x}{z}=\frac{x}{z^2}\qquad\text{or}\qquad z'-xz=-x$$

This last equation is linear. Its solution is found in Problem 5.10 to be $z=ce^{x^2/2}+1$. The solution of the original differential equation is then

$$y=\frac{1}{z}=\frac{1}{ce^{x^2/2}+1}$$

5.17. Solve $y'-\frac{3}{x}y=x^4y^{1/3}$.

This is a Bernoulli differential equation with $p(x)=-3/x$, $q(x)=x^4$, and $n=\frac{1}{3}$. Using Eq. (*5.5*), we make the substitution $z=y^{1-(1/3)}=y^{2/3}$. Thus, $y=z^{3/2}$ and $y'=\frac{3}{2}z^{1/2}z'$. Substituting these values into the differential equation, we obtain

$$\frac{3}{2}z^{1/2}z' - \frac{3}{x}z^{3/2} = x^4 z^{1/2} \qquad \text{or} \qquad z' - \frac{2}{x}z = \frac{2}{3}x^4$$

This last equation is linear. Its solution is found in Problem 5.12 to be $z = cx^2 + \frac{2}{9}x^5$. Since $z = y^{2/3}$, the solution of the original problem is given implicitly by $y^{2/3} = cx^2 + \frac{2}{9}x^5$, or explicitly by $y = \pm(cx^2 + \frac{2}{9}x^5)^{3/2}$.

5.18. Show that the integrating factor found in Problem 5.1 is also an integrating factor as defined in Chapter 4, Eq. (*4.7*).

The differential equation of Problem 5.1 can be rewritten as

$$\frac{dy}{dx} = 3y + 6$$

which has the differential form

$$dy = (3y + 6)\,dx$$

or

$$(3y + 6)\,dx + (-1)\,dy = 0 \tag{1}$$

Multiplying (*1*) by the integrating factor $I(x) = e^{-3x}$, we obtain

$$(3ye^{-3x} + 6e^{-3x})\,dx + (-e^{-3x})\,dy = 0 \tag{2}$$

Setting $\qquad M(x, y) = 3ye^{-3x} + 6e^{-3x} \qquad \text{and} \qquad N(x, y) = -e^{-3x}$

we have

$$\frac{\partial M}{\partial y} = 3e^{-3x} = \frac{\partial N}{\partial x}$$

from which we conclude that (*2*) is an exact differential equation.

5.19. Find the general form of the solution of Eq. (*5.1*).

Multiplying (*5.1*) by (*5.2*), we have

$$e^{\int p(x)\,dx}y' + e^{\int p(x)\,dx}p(x)y = e^{\int p(x)\,dx}q(x) \tag{1}$$

Since

$$\frac{d}{dx}[e^{\int p(x)\,dx}] = e^{\int p(x)\,dx}p(x)$$

it follows from the product rule of differentiation that the left side of (*1*) equals $\frac{d}{dx}[e^{\int p(x)\,dx}y]$. Thus, (*1*) can be rewritten as

$$\frac{d}{dx}[e^{\int p(x)\,dx}y] = e^{\int p(x)\,dx}q(x) \tag{2}$$

Integrating both sides of (*2*) with respect to x, we have

$$\int \frac{d}{dx}[e^{\int p(x)\,dx}y]\,dx = \int e^{\int p(x)\,dx}q(x)\,dx$$

or,

$$e^{\int p(x)\,dx}y + c_1 = \int e^{\int p(x)\,dx}q(x)\,dx \tag{3}$$

Finally, setting $c_1 = -c$ and solving (*3*) for y, we obtain

$$y = ce^{-\int p(x)\,dx} + e^{-\int p(x)\,dx}\int e^{\int p(x)\,dx}q(x)\,dx \tag{4}$$

Solved Problems

In Problems 5.20 through 5.49, solve the given differential equations.

5.20. $\dfrac{dy}{dx} + 5y = 0$

5.21. $\dfrac{dy}{dx} - 5y = 0$

5.22. $\dfrac{dy}{dx} - 0.01y = 0$

5.23. $\dfrac{dy}{dx} + 2xy = 0$

5.24. $y' + 3x^2y = 0$

5.25. $y' - x^2y = 0$

5.26. $y' - 3x^4y = 0$

5.27. $y' + \dfrac{1}{x}y = 0$

5.28. $y' + \dfrac{2}{x}y = 0$

5.29. $y' - \dfrac{2}{x}y = 0$

5.30. $y' - \dfrac{2}{x^2}y = 0$

5.31. $y' - 7y = e^x$

5.32. $y' - 7y = 14x$

5.33. $y' - 7y = \sin 2x$

5.34. $y' + x^2y = x^2$

5.35. $y' - \dfrac{3}{x^2}y = \dfrac{1}{x^2}$

5.36. $y' = \cos x$

5.37. $y' + y = y^2$

5.38. $xy' + y = xy^3$

5.39. $y' + xy = 6x\sqrt{y}$

5.40. $y' + y = y^2$

5.41. $y' + y = y^{-2}$

5.42. $y' + y = y^2e^x$

5.43. $\dfrac{dy}{dt} + 50y = 0$

5.44. $\dfrac{dz}{dt} - \dfrac{1}{2t}z = 0$

5.45. $\dfrac{dN}{dt} = kN$, (k = a constant)

5.46. $\dfrac{dp}{dt} - \dfrac{1}{t}p = t^2 + 3t - 2$

5.47. $\dfrac{dQ}{dt} + \dfrac{2}{20 - t}Q = 4$

5.48. $25\dfrac{dT}{dt} + T = 80e^{-0.04t}$

5.49. $\dfrac{dp}{dz} + \dfrac{2}{z}p = 4$

Solve the following initial-value problems.

5.50. $y' + \dfrac{2}{x}y = x;\ y(1) = 0$

5.51. $y' + 6xy = 0;\ y(\pi) = 5$

5.52. $y' + 2xy = 2x^3;\ y(0) = 1$

5.53. $y' + \frac{2}{x}y = -x^9y^5;\ y(-1) = 2$

5.54. $\frac{dv}{dt} + 2v = 32;\ v(0) = 0$

5.55. $\frac{dq}{dt} + q = 4\cos 2t;\ q(0) = 1$

5.56. $\frac{dN}{dt} + \frac{1}{t}N = t;\ N(2) = 8$

5.57. $\frac{dT}{dt} + 0.069T = 2.07;\ T(0) = -30$

Chapter 6

Applications of First-Order Differential Equations

GROWTH AND DECAY PROBLEMS

Let $N(t)$ denote the amount of substance (or population) that is either growing or decaying. If we assume that dN/dt, the time rate of change of this amount of substance, is proportional to the amount of substance present, then $dN/dt = kN$, or

$$\frac{dN}{dt} - kN = 0 \tag{6.1}$$

where k is the constant of proportionality. (See Problems 6.1–6.7.)

We are assuming that $N(t)$ is a differentiable, hence continuous, function of time. For population problems, where $N(t)$ is actually discrete and integer-valued, this assumption is incorrect. Nonetheless, (*6.1*) still provides a good approximation to the physical laws governing such a system. (See Problem 6.5.)

TEMPERATURE PROBLEMS

Newton's law of cooling, which is equally applicable to heating, states that *the time rate of change of the temperature of a body is proportional to the temperature difference between the body and its surrounding medium.* Let T denote the temperature of the body and let T_m denote the temperature of the surrounding medium. Then the time rate of change of the temperature of the body is dT/dt, and Newton's law of cooling can be formulated as $dT/dt = -k(T - T_m)$, or as

$$\frac{dT}{dt} + kT = kT_m \tag{6.2}$$

where k is a *positive* constant of proportionality. Once k is chosen positive, the minus sign is required in Newton's law to make dT/dt negative in a cooling process, when T is greater than T_m, and positive in a heating process, when T is less than T_m. (See Problems 6.8–6.10.)

FALLING BODY PROBLEMS

Consider a vertically falling body of mass m that is being influenced only by gravity g and an air resistance that is proportional to the velocity of the body. Assume that both gravity and mass remain constant and, for convenience, choose the downward direction as the positive direction.

Newton's second law of motion: *The net force acting on a body is equal to the time rate of change of the momentum of the body; or, for constant mass,*

$$F = m\frac{dv}{dt} \tag{6.3}$$

where F is the net force on the body and v is the velocity of the body, both at time t.

For the problem at hand, there are two forces acting on the body: (1) the force due to gravity given by the weight w of the body, which equals mg, and (2) the force due to air resistance given by

$-kv$, where $k \geq 0$ is a constant of proportionality. The minus sign is required because this force opposes the velocity; that is, it acts in the upward, or negative, direction (see Fig. 6-1). The net force F on the body is, therefore, $F = mg - kv$. Substituting this result into (*6.3*), we obtain

$$mg - kv = m\frac{dv}{dt}$$

or

$$\frac{dv}{dt} + \frac{k}{m}v = g \tag{6.4}$$

as the equation of motion for the body.

If air resistance is negligible or nonexistent, then $k = 0$ and (*6.4*) simplifies to

$$\frac{dv}{dt} = g \tag{6.5}$$

(See Problem 6.11.) When $k > 0$, the limiting velocity v_l is defined by

$$v_l = \frac{mg}{k} \tag{6.6}$$

Caution: Equations (*6.4*), (*6.5*), and (*6.6*) are valid only if the given conditions are satisfied. These equations are not valid if, for example, air resistance is not proportional to velocity but to the velocity squared, or if the upward direction is taken to be the positive direction. (See Problems 6.14 and 6.15.)

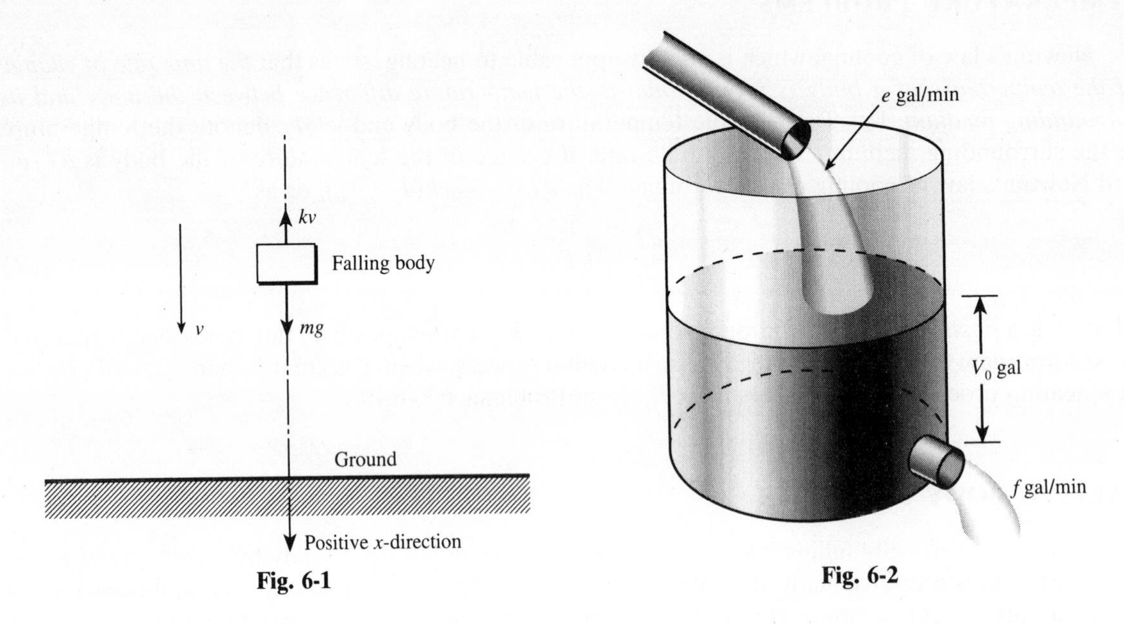

Fig. 6-1 **Fig. 6-2**

DILUTION PROBLEMS

Consider a tank which initially holds V_0 gal of brine that contains a lb of salt. Another brine solution, containing b lb of salt per gallon, is poured into the tank at the rate of e gal/min while, simultaneously, the well-stirred solution leaves the tank at the rate of f gal/min (Fig. 6-2). The problem is to find the amount of salt in the tank at any time t.

Let Q denote the amount (in pounds) of salt in the tank at any time. The time rate of change of Q, dQ/dt, equals the rate at which salt enters the tank minus the rate at which salt leaves the tank. Salt enters the tank at the rate of be lb/min. To determine the rate at which salt leaves the tank, we first calculate the volume of brine in the tank at any time t, which is the initial volume V_0 plus the volume of brine added et minus the volume of brine removed ft. Thus, the volume of brine at any time is

$$V_0 + et - ft \tag{6.7}$$

The concentration of salt in the tank at any time is $Q/(V_0 + et - ft)$, from which it follows that salt leaves the tank at the rate of

$$f\left(\frac{Q}{V_0 + et - ft}\right) \text{ lb/min}$$

Thus,

$$\frac{dQ}{dt} = be - f\left(\frac{Q}{V_0 + et - ft}\right)$$

or

$$\frac{dQ}{dt} + \frac{f}{V_0 + (e - f)t} Q = be \tag{6.8}$$

(See Problems 6.16–6.18.)

ELECTRICAL CIRCUITS

The basic equation governing the amount of current I (in amperes) in a simple RL circuit (Fig. 6-3) consisting of a resistance R (in ohms), an inductor L (in henries), and an electromotive force (abbreviated emf) E (in volts) is

$$\frac{dI}{dt} + \frac{R}{L} I = \frac{E}{L} \tag{6.9}$$

For an RC circuit consisting of a resistance, a capacitance C (in farads), an emf, and no inductance (Fig. 6-4), the equation governing the amount of electrical charge q (in coulombs) on the capacitor is

$$\frac{dq}{dt} + \frac{1}{RC} q = \frac{E}{R} \tag{6.10}$$

The relationship between q and I is

$$I = \frac{dq}{dt} \tag{6.11}$$

(See Problems 6.19–6.22.) For more complex circuits see Chapter 13.

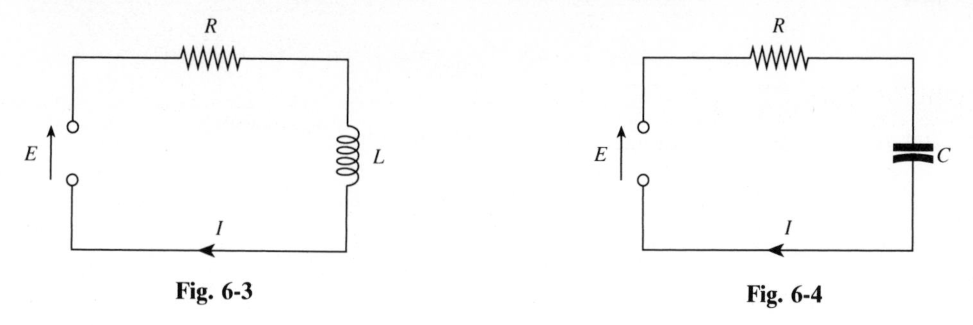

Fig. 6-3　Fig. 6-4

ORTHOGONAL TRAJECTORIES

Consider a one-parameter family of curves in the xy-plane defined by

$$F(x, y, c) = 0 \tag{6.12}$$

where c denotes the parameter. The problem is to find another one-parameter family of curves, called the *orthogonal trajectories* of the family (*6.12*) and given analytically by

$$G(x, y, k) = 0 \tag{6.13}$$

such that every curve in this new family (*6.13*) intersects at right angles every curve in the original family (*6.12*).

We first implicitly differentiate (*6.12*) with respect to x, then eliminate c between this derived equation and (*6.12*). This gives an equation connecting x, y, and y', which we solve for y' to obtain a differential equation of the form

$$\frac{dy}{dx} = f(x, y) \tag{6.14}$$

The orthogonal trajectories of (*6.12*) are the solutions of

$$\frac{dy}{dx} = -\frac{1}{f(x, y)} \tag{6.15}$$

(See Problems 6.23–6.25.)

For many families of curves, one cannot explicitly solve for dy/dx and obtain a differential equation of the form (*6.14*). We do not consider such curves in this book.

Solved Problems

6.1. A person places \$20,000 in a savings account which pays 5 percent interest per annum, compounded continuously. Find (*a*) the amount in the account after three years, and (*b*) the time required for the account to double in value, presuming no withdrawals and no additional deposits.

Let $N(t)$ denote the balance in the account at any time t. Initially, $N(0) = 20{,}000$. The balance in the account grows by the accumulated interest payments, which are proportional to the amount of money in the account. The constant of proportionality is the interest rate. In this case, $k = 0.05$ and Eq. (*6.1*) becomes

$$\frac{dN}{dt} - 0.05N = 0$$

This differential equation is both linear and separable. Its solution is

$$N(t) = ce^{0.05t} \tag{1}$$

At $t = 0$, $N(0) = 20{,}000$, which when substituted into (*1*) yields

$$20{,}000 = ce^{0.05(0)} = c$$

With this value of c, (*1*) becomes

$$N(t) = 20{,}000e^{0.05t} \tag{2}$$

Equation (*2*) gives the dollar balance in the account at any time t.

(*a*) Substituting $t = 3$ into (*2*), we find the balance after three years to be

$$N(3) = 20{,}000e^{0.05(3)} = 20{,}000(1.161834) = \$23{,}236.68$$

(b) We seek the time t at which $N(t) = \$40,000$. Substituting these values into (2) and solving for t, we obtain

$$40{,}000 = 20{,}000e^{0.05t}$$
$$2 = e^{0.05t}$$
$$\ln|2| = 0.05t$$
$$t = \frac{1}{0.05}\ln|2| = 13.86 \text{ years}$$

6.2. A person places \$5000 in an account that accrues interest compounded continuously. Assuming no additional deposits or withdrawals, how much will be in the account after seven years if the interest rate is a constant 8.5 percent for the first four years and a constant 9.25 percent for the last three years?

Let $N(t)$ denote the balance in the account at any time t. Initially, $N(0) = 5000$. For the first four years, $k = 0.085$ and Eq. (6.1) becomes

$$\frac{dN}{dt} - 0.085N = 0$$

Its solution is

$$N(t) = ce^{0.085t} \qquad (0 \le t \le 4) \tag{1}$$

At $t = 0$, $N(0) = 5000$, which when substituted into (1) yields

$$5000 = ce^{0.085(0)} = c$$

and (1) becomes

$$N(t) = 5000e^{0.085t} \qquad (0 \le t \le 4) \tag{2}$$

Substituting $t = 4$ into (2), we find the balance after four years to be

$$N(4) = 5000e^{0.085(4)} = 5000(1.404948) = \$7024.74$$

This amount also represents the beginning balance for the last three-year period.

Over the last three years, the interest rate is 9.25 percent and (6.1) becomes

$$\frac{dN}{dt} - 0.0925N = 0 \qquad (4 \le t \le 7)$$

Its solution is

$$N(t) = ce^{0.0925t} \qquad (4 \le t \le 7) \tag{3}$$

At $t = 4$, $N(4) = \$7024.74$, which when substituted into (3) yields

$$7024.74 = ce^{0.0925(4)} = c(1.447735) \qquad \text{or} \qquad c = 4852.23$$

and (3) becomes

$$N(t) = 4852.23e^{0.095t} \qquad (4 \le t \le 7) \tag{4}$$

Substituting $t = 7$ into (4), we find the balance after seven years to be

$$N(7) = 4852.23e^{0.0925(7)} = 4852.23(1.910758) = \$9271.44$$

6.3. What constant interest rate is required if an initial deposit placed into an account that accrues interest compounded continuously is to double its value in six years?

The balance $N(t)$ in the account at any time t is governed by (6.1)

$$\frac{dN}{dt} - kN = 0$$

which has as its solution

$$N(t) = ce^{kt} \tag{1}$$

We are not given an amount for the initial deposit, so we denote it as N_0. At $t = 0$, $N(0) = N_0$, which when substituted into (*1*) yields

$$N_0 = ce^{k(0)} = c$$

and (*1*) becomes

$$N(t) = N_0 e^{kt} \tag{2}$$

We seek the value of k for which $N = 2N_0$ when $t = 6$. Substituting these values into (*2*) and solving for k, we find

$$2N_0 = N_0 e^{k(6)}$$

$$e^{6k} = 2$$

$$6k = \ln|2|$$

$$k = \frac{1}{6}\ln|2| = 0.1155$$

An interest rate of 11.55 percent is required.

6.4. A bacteria culture is known to grow at a rate proportional to the amount present. After one hour, 1000 strands of the bacteria are observed in the culture; and after four hours, 3000 strands. Find (*a*) an expression for the approximate number of strands of the bacteria present in the culture at any time t and (*b*) the approximate number of strands of the bacteria originally in the culture.

(*a*) Let $N(t)$ denote the number of bacteria strands in the culture at time t. From (*6.1*), $dN/dt - kN = 0$, which is both linear and separable. Its solution is

$$N(t) = ce^{kt} \tag{1}$$

At $t = 1$, $N = 1000$; hence,

$$1000 = ce^{k} \tag{2}$$

At $t = 4$, $N = 3000$; hence,

$$3000 = ce^{4k} \tag{3}$$

Solving (*2*) and (*3*) for k and c, we find

$$k = \frac{1}{3}\ln 3 = 0.366 \quad \text{and} \quad c = 1000e^{-0.366} = 694$$

Substituting these values of k and c into (*1*), we obtain

$$N(t) = 694e^{0.366t} \tag{4}$$

as an expression for the amount of the bacteria present at any time t.

(*b*) We require N at $t = 0$. Substituting $t = 0$ into (*4*), we obtain $N(0) = 694e^{(0.366)(0)} = 694$.

6.5. The population of a certain country is known to increase at a rate proportional to the number of people presently living in the country. If after two years the population has doubled, and after three years the population is 20,000, estimate the number of people initially living in the country.

Let N denote the number of people living in the country at any time t, and let N_0 denote the number of people initially living in the country. Then, from (*6.1*),

$$\frac{dN}{dt} - kN = 0$$

which has the solution

$$N = ce^{kt} \tag{1}$$

At $t = 0$, $N = N_0$; hence, it follows from (*1*) that $N_0 = ce^{k(0)}$, or that $c = N_0$. Thus,

$$N = N_0 e^{kt} \tag{2}$$

At $t = 2$, $N = 2N_0$. Substituting these values into (*2*), we have

$$2N_0 = N_0 e^{2k} \qquad \text{from which} \qquad k = \frac{1}{2}\ln 2 = 0.347$$

Substituting this value into (*2*) gives

$$N = N_0 e^{0.347t} \tag{3}$$

At $t = 3$, $N = 20{,}000$. Substituting these values into (*3*), we obtain

$$20{,}000 = N_0 e^{(0.347)(3)} = N_0(2.832) \qquad \text{or} \qquad N_0 = 7062$$

6.6. A certain radioactive material is known to decay at a rate proportional to the amount present. If initially there is 50 milligrams of the material present and after two hours it is observed that the material has lost 10 percent of its original mass, find (*a*) an expression for the mass of the material remaining at any time *t*, (*b*) the mass of the material after four hours, and (*c*) the time at which the material has decayed to one half of its initial mass.

(*a*) Let N denote the amount of material present at time t. Then, from (*6.1*),

$$\frac{dN}{dt} - kN = 0$$

This differential equation is separable and linear; its solution is

$$N = ce^{kt} \tag{1}$$

At $t = 0$, we are given that $N = 50$. Therefore, from (*1*), $50 = ce^{k(0)}$, or $c = 50$. Thus,

$$N = 50e^{kt} \tag{2}$$

At $t = 2$, 10 percent of the original mass of 50 mg, or 5 mg, has decayed. Hence, at $t = 2$, $N = 50 - 5 = 45$. Substituting these values into (*2*) and solving for k, we have

$$45 = 50e^{2k} \qquad \text{or} \qquad k = \frac{1}{2}\ln\frac{45}{50} = -0.053$$

Substituting this value into (*2*), we obtain the amount of mass present at any time t as

$$N = 50e^{-0.053t} \tag{3}$$

where t is measured in hours.

(*b*) We require N at $t = 4$. Substituting $t = 4$ into (*3*) and then solving for N, we find that

$$N = 50e^{(-0.053)(4)} = 50(0.809) = 40.5 \text{ mg}$$

(*c*) We require t when $N = 50/2 = 25$. Substituting $N = 25$ into (*3*) and solving for t, we find

$$25 = 50e^{-0.053t} \qquad \text{or} \qquad -0.053t = \ln\frac{1}{2} \qquad \text{or} \qquad t = 13 \text{ hours}$$

The time required to reduce a decaying material to one half its original mass is called the *half-life* of the material. For this problem, the half-life is 13 hours.

6.7. Five mice in a stable population of 500 are intentionally infected with a contagious disease to test a theory of epidemic spread that postulates the rate of change in the infected population is proportional to the product of the number of mice who have the disease with the number that are disease free. Assuming the theory is correct, how long will it take half the population to contract the disease?

Let $N(t)$ denote the number of mice with the disease at time t. We are given that $N(0) = 5$, and it follows that $500 - N(t)$ is the number of mice without the disease at time t. The theory predicts that

$$\frac{dN}{dt} = kN(500 - N) \tag{1}$$

where k is a constant of proportionality. This equation is different from (6.1) because the rate of change is no longer proportional to just the number of mice who have the disease. Equation (1) has the differential form

$$\frac{dN}{N(500 - N)} - k\,dt = 0 \tag{2}$$

which is separable. Using partial fraction decomposition, we have

$$\frac{1}{N(500 - N)} = \frac{1/500}{N} + \frac{1/500}{500 - N}$$

hence (2) may be rewritten as

$$\frac{1}{500}\left(\frac{1}{N} + \frac{1}{500 - N}\right) dN - k\,dt = 0$$

Its solution is

$$\frac{1}{500}\int \left(\frac{1}{N} + \frac{1}{500 - N}\right) dN - \int k\,dt = c$$

or

$$\frac{1}{500}(\ln |N| - \ln |500 - N|) - kt = c$$

which may be rewritten as

$$\ln \left|\frac{N}{500 - N}\right| = 500(c + kt)$$

$$\frac{N}{500 - N} = e^{500(c+kt)} \tag{3}$$

But $e^{500(c+kt)} = e^{500c}e^{kt}$. Setting $c_1 = e^{500c}$, we can write (3) as

$$\frac{N}{500 - N} = c_1 e^{500kt} \tag{4}$$

At $t = 0$, $N = 5$. Substituting these values into (4), we find

$$\frac{5}{495} = c_1 e^{500k(0)} = c_1$$

so $c_1 = 1/99$ and (4) becomes

$$\frac{N}{500 - N} = \frac{1}{99} e^{500kt} \tag{5}$$

We could solve (5) for N, but this is not necessary. We seek a value of t when $N = 250$, one-half the population. Substituting $N = 250$ into (5) and solving for t, we obtain

$$1 = \frac{1}{99} e^{500kt}$$

$$99 = e^{500kt}$$

$$\ln 99 = 500kt$$

or $t = 0.00919/k$ time units. Without additional information, we cannot obtain a numerical value for the constant of proportionality k or be more definitive about t.

6.8. A metal bar at a temperature of 100° F is placed in a room at a constant temperature of 0° F. If after 20 minutes the temperature of the bar is 50° F, find (*a*) the time it will take the bar to reach a temperature of 25° F and (*b*) the temperature of the bar after 10 minutes.

Use Eq. (*6.2*) with $T_m = 0$; the medium here is the room which is being held at a constant temperature of 0° F. Thus we have

$$\frac{dT}{dt} + kT = 0$$

whose solution is

$$T = ce^{-kt} \tag{1}$$

Since $T = 100$ at $t = 0$ (the temperature of the bar is initially 100° F), it follows from (*1*) that $100 = ce^{-k(0)}$ or $100 = c$. Substituting this value into (*1*), we obtain

$$T = 100e^{-kt} \tag{2}$$

At $t = 20$, we are given that $T = 50$; hence, from (*2*),

$$50 = 100e^{-20k} \qquad \text{from which} \qquad k = \frac{-1}{20}\ln\frac{50}{100} = \frac{-1}{20}(-0.693) = 0.035$$

Substituting this value into (*2*), we obtain the temperature of the bar at any time t as

$$T = 100e^{-0.035t} \tag{3}$$

(*a*) We require t when $T = 25$. Substituting $T = 25$ into (*3*), we have

$$25 = 100e^{-0.035t} \qquad \text{or} \qquad -0.035t = \ln\frac{1}{4}$$

Solving, we find that $t = 39.6$ min.

(*b*) We require T when $t = 10$. Substituting $t = 10$ into (*3*) and then solving for T, we find that

$$T = 100e^{(-0.035)(10)} = 100(0.705) = 70.5^\circ \text{ F}$$

It should be noted that since Newton's law is valid only for small temperature differences, the above calculations represent only a first approximation to the physical situation.

6.9. A body at a temperature of 50° F is placed outdoors where the temperature is 100° F. If after 5 minutes the temperature of the body is 60° F, find (*a*) how long it will take the body to reach a temperature of 75° F and (*b*) the temperature of the body after 20 minutes.

Using (*6.2*) with $T_m = 100$ (the surrounding medium is the outside air), we have

$$\frac{dT}{dt} + kT = 100k$$

This differential equation is linear. Its solution is given in Problem 5.15 as

$$T = ce^{-kt} + 100 \tag{1}$$

Since $T = 50$ when $t = 0$, it follows from (*1*) that $50 = ce^{-k(0)} + 100$, or $c = -50$. Substituting this value into (*1*), we obtain

$$T = -50e^{-kt} + 100 \tag{2}$$

At $t = 5$, we are given that $T = 60$; hence, from (*2*), $60 = -50e^{-5k} + 100$. Solving for k, we obtain

$$-40 = -50e^{-5k} \quad \text{or} \quad k = \frac{-1}{5}\ln\frac{40}{50} = \frac{-1}{5}(-0.223) = 0.045$$

Substituting this value into (*2*), we obtain the temperature of the body at any time t as

$$T = -50e^{-0.045t} + 100 \tag{3}$$

(*a*) We require t when $T = 75$. Substituting $T = 75$ into (*3*), we have

$$75 = -50e^{-0.045t} + 100 \quad \text{or} \quad e^{-0.045t} = \frac{1}{2}$$

Solving for t, we find

$$-0.045t = \ln\frac{1}{2} \quad \text{or} \quad t = 15.4 \text{ min}$$

(*b*) We require T when $t = 20$. Substituting $t = 20$ into (*3*) and then solving for T, we find

$$T = -50e^{(-0.045)(20)} + 100 = -50(0.41) + 100 = 79.5^\circ \text{ F}$$

6.10. A body at an unknown temperature is placed in a room which is held at a constant temperature of 30° F. If after 10 minutes the temperature of the body is 0° F and after 20 minutes the temperature of the body is 15° F, find the unknown initial temperature.

From (*6.2*),

$$\frac{dT}{dt} + kT = 30k$$

Solving, we obtain

$$T = ce^{-kt} + 30 \tag{1}$$

At $t = 10$, we are given that $T = 0$. Hence, from (*1*),

$$0 = ce^{-10k} + 30 \quad \text{or} \quad ce^{-10k} = -30 \tag{2}$$

At $t = 20$, we are given that $T = 15$. Hence, from (*1*) again,

$$15 = ce^{-20k} + 30 \quad \text{or} \quad ce^{-20k} = -15 \tag{3}$$

Solving (*2*) and (*3*) for k and c, we find

$$k = \frac{1}{10}\ln 2 = 0.069 \quad \text{and} \quad c = -30e^{10k} = -30(2) = -60$$

Substituting these values into (*1*), we have for the temperature of the body at any time t

$$T = -60e^{-0.069t} + 30 \tag{4}$$

Since we require T at the initial time $t = 0$, it follows from (4) that

$$T_0 = -60e^{(-0.069)(0)} + 30 = -60 + 30 = -30° \text{ F}$$

6.11. A body of mass 5 slugs is dropped from a height of 100 ft with zero velocity. Assuming no air resistance, find (a) an expression for the velocity of the body at any time t, (b) an expression for the position of the body at any time t, and (c) the time required to reach the ground.

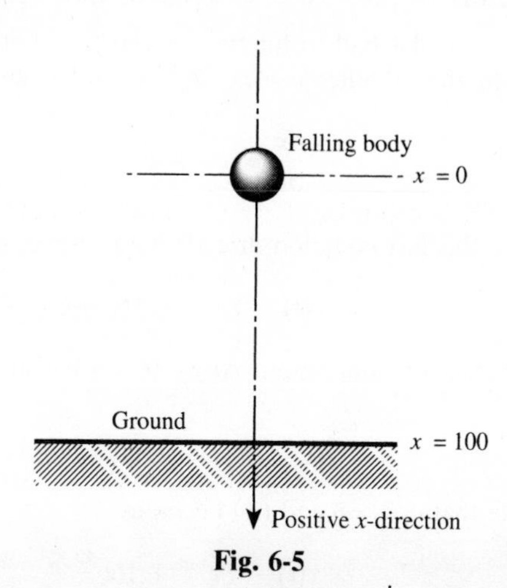

Fig. 6-5

(a) Choose the coordinate system as in Fig. 6-5. Then, since there is no air resistance, (6.5) applies: $dv/dt = g$. This differential equation is linear or, in differential form, separable; its solution is $v = gt + c$. When $t = 0$, $v = 0$ (initially the body has zero velocity); hence $0 = g(0) + c$, or $c = 0$. Thus, $v = gt$ or, assuming $g = 32 \text{ ft/sec}^2$,

$$v = 32t \tag{1}$$

(b) Recall that velocity is the time rate of change of displacement, designated here by x. Hence, $v = dx/dt$, and (1) becomes $dx/dt = 32t$. This differential equation is also both linear and separable; its solution is

$$x = 16t^2 + c_1 \tag{2}$$

But at $t = 0$, $x = 0$ (see Fig. 6-5). Thus, $0 = (16)(0)^2 + c_1$, or $c_1 = 0$. Substituting this value into (2), we have

$$x = 16t^2 \tag{3}$$

(c) We require t when $x = 100$. From (2), $t = \sqrt{(100)/(16)} = 2.5$ sec.

6.12. A steel ball weighing 2 lb is dropped from a height of 3000 ft with no velocity. As it falls, the ball encounters air resistance numerically equal to $v/8$ (in pounds), where v denotes the velocity of the ball (in feet per second). Find (a) the limiting velocity for the ball and (b) the time required for the ball to hit the ground.

Locate the coordinate system as in Fig. 6-5 with the ground now situated at $x = 3000$. Here $w = 2$ lb and $k = 1/8$. Assuming gravity g is 32 ft/sec^2, we have from the formula $w = mg$ that $2 = m(32)$ or that the mass of the ball is $m = 1/16$ slug. Equation (6.4) becomes

$$\frac{dv}{dt} + 2v = 32$$

which has as its solution

$$v(t) = ce^{-2t} + 16 \tag{1}$$

At $t = 0$, we are given that $v = 0$. Substituting these values into (1), we obtain

$$0 = ce^{-2(0)} + 16 = c + 16$$

from which we conclude that $c = -16$ and (1) becomes

$$v(t) = -16e^{-2t} + 16 \tag{2}$$

(a) From (1) or (2), we see that as $t \to \infty$, $v \to 16$ so the limiting velocity is 16 ft/sec^2.

(b) To find the time it takes for the ball to hit the ground ($x = 3000$), we need an expression for the position of the ball at any time t. Since $v = dx/dt$, (2) can be rewritten as

$$\frac{dx}{dt} = -16e^{-2t} + 16$$

Integrating both sides of this last equation directly with respect to t, we have

$$x(t) = 8e^{-2t} + 16t + c_1 \tag{3}$$

where c_1 denotes a constant of integration. At $t = 0$, $x = 0$. Substituting these values into (3), we obtain

$$0 = 8e^{-2(0)} + 16(0) + c_1 = 8 + c_1$$

from which we conclude that $c_1 = -8$ and (3) becomes

$$x(t) = 8e^{-2t} + 16t - 8 \tag{4}$$

The ball hits the ground when $x(t) = 3000$. Substituting this value into (4), we have

$$3000 = 8e^{-2t} + 16t - 8$$

or

$$376 = e^{-2t} + 2t \tag{5}$$

Although (5) cannot be solved explicitly for t, we can approximate the solution by trial and error, substituting different values of t into (5) until we locate a solution to the degree of accuracy we need. Alternatively, we note that for any large value of t, the negative exponential term will be negligible. A good approximation is obtained by setting $2t = 376$ or $t = 188$ sec. For this value of t, the exponential is essentially zero.

6.13. A body weighing 64 lb is dropped from a height of 100 ft with an initial velocity of 10 ft/sec. Assume that the air resistance is proportional to the velocity of the body. If the limiting velocity is known to be 128 ft/sec, find (a) an expression for the velocity of the body at any time t and (b) an expression for the position of the body at any time t.

(a) Locate the coordinate system as in Fig. 6-5. Here $w = 64$ lb. Since $w = mg$, it follows that $mg = 64$, or $m = 2$ slugs. Given that $v_l = 128$ ft/sec, it follows from (6.6) that $128 = 64/k$, or $k = \frac{1}{2}$. Substituting these values into (6.4), we obtain the linear differential equation

$$\frac{dv}{dt} + \frac{1}{4}v = 32$$

which has the solution

$$v = ce^{-t/4} + 128 \tag{1}$$

At $t = 0$, we are given that $v = 10$. Substituting these values into (1), we have $10 = ce^0 + 128$, or $c = -118$. The velocity at any time t is given by

$$v = -118e^{-t/4} + 128 \tag{2}$$

(*b*) Since $v = dx/dt$, where x is displacement, (*2*) can be rewritten as

$$\frac{dx}{dt} = -118e^{-t/4} + 128$$

This last equation, in differential form, is separable; its solution is

$$x = 472e^{-t/4} + 128t + c_1 \tag{3}$$

At $t = 0$, we have $x = 0$ (see Fig. 6-5). Thus, (*3*) gives

$$0 = 472e^0 + (128)(0) + c_1 \qquad \text{or} \qquad c_1 = -472$$

The displacement at any time t is then given by

$$x = 472e^{-t/4} + 128t - 472$$

6.14. A body of mass m is thrown vertically into the air with an initial velocity v_0. If the body encounters an air resistance proportional to its velocity, find (*a*) the equation of motion in the coordinate system of Fig. 6-6, (*b*) an expression for the velocity of the body at any time t, and (*c*) the time at which the body reaches its maximum height.

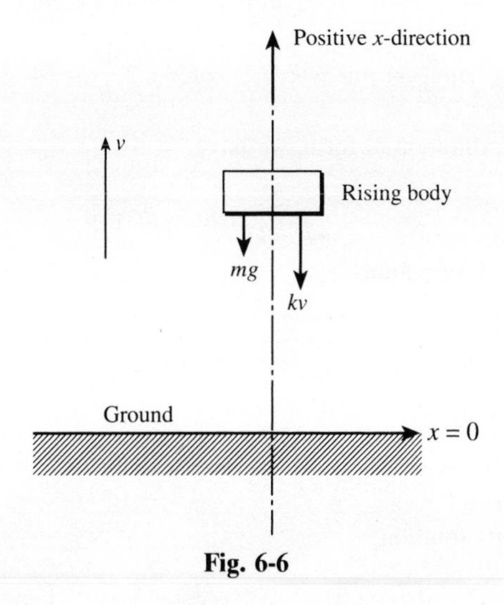

Fig. 6-6

(*a*) In this coordinate system, Eq. (*6.4*) may not be the equation of motion. To derive the appropriate equation, we note that there are two forces on the body: (1) the force due to the gravity given by mg and (2) the force due to air resistance given by kv, which will impede the velocity of the body. Since both of these forces act in the downward or negative direction, the net force on the body is $-mg - kv$. Using (*6.3*) and rearranging, we obtain

$$\frac{dv}{dt} + \frac{k}{m}v = -g \tag{1}$$

as the equation of motion.

(*b*) Equation (*1*) is a linear differential equation, and its solution is $v = ce^{-(k/m)t} - mg/k$. At $t = 0$, $v = v_0$; hence $v_0 = ce^{-(k/m)0} - (mg/k)$, or $c = v_0 + (mg/k)$. The velocity of the body at any time t is

$$v = \left(v_0 + \frac{mg}{k}\right)e^{-(k/m)t} - \frac{mg}{k} \tag{2}$$

(*c*) The body reaches its maximum height when $v = 0$. Thus, we require t when $v = 0$. Substituting

$v = 0$ into (2) and solving for t, we find

$$0 = \left(v_0 + \frac{mg}{k}\right)e^{-(k/m)t} - \frac{mg}{k}$$

$$e^{-(k/m)t} = \frac{1}{1 + \dfrac{v_0 k}{mg}}$$

$$-(k/m)t = \ln\left(\frac{1}{1 + \dfrac{v_0 k}{mg}}\right)$$

$$t = \frac{m}{k}\ln\left(1 + \frac{v_0 k}{mg}\right)$$

6.15. A body of mass 2 slugs is dropped with no initial velocity and encounters an air resistance that is proportional to the square of its velocity. Find an expression for the velocity of the body at any time t.

The force due to air resistance is $-kv^2$, so that Newton's second law of motion becomes

$$m\frac{dv}{dt} = mg - kv^2 \qquad \text{or} \qquad 2\frac{dv}{dt} = 64 - kv^2$$

Rewriting this equation in differential form, we have

$$\frac{2}{64 - kv^2}dv - dt = 0 \tag{1}$$

which is separable. By partial fractions,

$$\frac{2}{64 - kv^2} = \frac{2}{(8 - \sqrt{k}\,v)(8 + \sqrt{k}\,v)} = \frac{\frac{1}{8}}{8 - \sqrt{k}\,v} + \frac{\frac{1}{8}}{8 + \sqrt{k}\,v}$$

Hence (1) can be rewritten as

$$\frac{1}{8}\left(\frac{1}{8 - \sqrt{k}\,v} + \frac{1}{8 + \sqrt{k}\,v}\right)dv - dt = 0$$

This last equation has as its solution

$$\frac{1}{8}\int\left(\frac{1}{8 - \sqrt{k}\,v} + \frac{1}{8 + \sqrt{k}\,v}\right)dv - \int dt = c$$

or

$$\frac{1}{8}\left[-\frac{1}{\sqrt{k}}\ln|8 - \sqrt{k}\,v| + \frac{1}{\sqrt{k}}\ln|8 + \sqrt{k}\,v|\right] - t = c$$

which can be rewritten as

$$\ln\left|\frac{8 + \sqrt{k}\,v}{8 - \sqrt{k}\,v}\right| = 8\sqrt{k}\,t + 8\sqrt{k}\,c$$

or

$$\frac{8 + \sqrt{k}\,v}{8 - \sqrt{k}\,v} = c_1 e^{8\sqrt{k}t} \qquad (c_1 = \pm e^{8\sqrt{k}c})$$

At $t = 0$, we are given that $v = 0$. This implies $c_1 = 1$, and the velocity is given by

$$\frac{8 + \sqrt{k}\,v}{8 - \sqrt{k}\,v} = e^{8\sqrt{k}t} \qquad \text{or} \qquad v = \frac{8}{\sqrt{k}}\tanh 4\sqrt{k}\,t$$

Note that without additional information, we cannot obtain a numerical value for the constant k.

6.16. A tank initially holds 100 gal of a brine solution containing 20 lb of salt. At $t = 0$, fresh water is poured into the tank at the rate of 5 gal/min, while the well-stirred mixture leaves the tank at the same rate. Find the amount of salt in the tank at any time t.

Here, $V_0 = 100$, $a = 20$, $b = 0$, and $e = f = 5$. Equation (*6.8*) becomes

$$\frac{dQ}{dt} + \frac{1}{20}Q = 0$$

The solution of this linear equation is

$$Q = ce^{-t/20} \tag{1}$$

At $t = 0$, we are given that $Q = a = 20$. Substituting these values into (*1*), we find that $c = 20$, so that (*1*) can be rewritten as $Q = 20e^{-t/20}$. Note that as $t \to \infty$, $Q \to 0$ as it should, since only fresh water is being added.

6.17. A tank initially holds 100 gal of a brine solution containing 1 lb of salt. At $t = 0$ another brine solution containing 1 lb of salt per gallon is poured into the tank at the rate of 3 gal/min, while the well-stirred mixture leaves the tank at the same rate. Find (*a*) the amount of salt in the tank at any time t and (*b*) the time at which the mixture in the tank contains 2 lb of salt.

(*a*) Here $V_0 = 100$, $a = 1$, $b = 1$, and $e = f = 3$; hence, (*6.8*) becomes

$$\frac{dQ}{dt} + 0.03Q = 3$$

The solution to this linear differential equation is

$$Q = ce^{-0.03t} + 100 \tag{1}$$

At $t = 0$, $Q = a = 1$. Substituting these values into (*1*), we find $1 = ce^0 + 100$, or $c = -99$. Then (*1*) can be rewritten as

$$Q = -99e^{-0.03t} + 100 \tag{2}$$

(*b*) We require t when $Q = 2$. Substituting $Q = 2$ into (*2*), we obtain

$$2 = -99e^{-0.03t} + 100 \qquad \text{or} \qquad e^{-0.03t} = \frac{98}{99}$$

from which

$$t = -\frac{1}{0.03}\ln\frac{98}{99} = 0.338 \text{ min}$$

6.18. A 50-gal tank initially contains 10 gal of fresh water. At $t = 0$, a brine solution containing 1 lb of salt per gallon is poured into the tank at the rate of 4 gal/min, while the well-stirred mixture leaves the tank at the rate of 2 gal/min. Find (*a*) the amount of time required for overflow to occur and (*b*) the amount of salt in the tank at the moment of overflow.

(*a*) Here $a = 0$, $b = 1$, $e = 4$, $f = 2$, and $V_0 = 10$. The volume of brine in the tank at any time t is given by (*6.7*) as $V_0 + et - ft = 10 + 2t$. We require t when $10 + 2t = 50$; hence, $t = 20$ min.

(*b*) For this problem, (*6.8*) becomes

$$\frac{dQ}{dt} + \frac{2}{10 + 2t}Q = 4$$

This is a linear equation; its solution is given in Problem 5.13 as

$$Q = \frac{40t + 4t^2 + c}{10 + 2t} \tag{1}$$

At $t = 0$, $Q = a = 0$. Substituting these values into (*1*), we find that $c = 0$. We require Q at the moment of overflow, which from part (*a*) is $t = 20$. Thus,

$$Q = \frac{40(20) + 4(20)^2}{10 + 2(20)} = 48 \text{ lb}$$

6.19. An RL circuit has an emf of 5 volts, a resistance of 50 ohms, an inductance of 1 henry, and no initial current. Find the current in the circuit at any time t.

Here $E = 5$, $R = 50$, and $L = 1$; hence (*6.9*) becomes

$$\frac{dI}{dt} + 50I = 5$$

This equation is linear; its solution is

$$I = ce^{-50t} + \frac{1}{10}$$

At $t = 0$, $I = 0$; thus, $0 = ce^{-50(0)} + \frac{1}{10}$, or $c = -\frac{1}{10}$. The current at any time t is then

$$I = -\frac{1}{10}e^{-50t} + \frac{1}{10} \tag{1}$$

The quantity $-\frac{1}{10}e^{-50t}$ in (*1*) is called the *transient current*, since this quantity goes to zero ("dies out") as $t \to \infty$. The quantity $\frac{1}{10}$ in (*1*) is called the *steady-state current*. As $t \to \infty$, the current I approaches the value of the steady-state current.

6.20. An RL circuit has an emf given (in volts) by $3 \sin 2t$, a resistance of 10 ohms, an inductance of 0.5 henry, and an initial current of 6 amperes. Find the current in the circuit at any time t.

Here, $E = 3 \sin 2t$, $R = 10$, and $L = 0.5$; hence (*6.9*) becomes

$$\frac{dI}{dt} + 20I = 6 \sin 2t$$

This equation is linear, with solution (see Chapter 5)

$$\int d(Ie^{20t}) = \int 6e^{20t} \sin 2t \, dt$$

Carrying out the integrations (the second integral requires two integrations by parts), we obtain

$$I = ce^{-20t} + \frac{30}{101} \sin 2t - \frac{3}{101} \cos 2t$$

At $t = 0$, $I = 6$; hence,

$$6 = ce^{-20(0)} + \frac{30}{101} \sin 2(0) - \frac{3}{101} \cos 2(0) \quad \text{or} \quad 6 = c - \frac{3}{101}$$

whence $c = 609/101$. The current at any time t is

$$I = \frac{609}{101} e^{-20t} + \frac{30}{101} \sin 2t - \frac{3}{101} \cos 2t$$

As in Problem 6.18, the current is the sum of a transient current, here $(609/101)e^{-20t}$, and a steady-state current,

$$\frac{30}{101}\sin 2t - \frac{3}{101}\cos 2t$$

6.21. Rewrite the steady-state current of Problem 6.20 in the form $A \sin(2t - \phi)$. The angle ϕ is called the *phase angle.*

Since $A \sin(2t - \phi) = A(\sin 2t \cos \phi - \cos 2t \sin \phi)$, we require

$$I_s = \frac{30}{101}\sin 2t - \frac{3}{101}\cos 2t = A\cos\phi \sin 2t - A\sin\phi\cos 2t$$

Thus, $A\cos\phi = \dfrac{30}{101}$ and $A\sin\phi = \dfrac{3}{101}$. It now follows that

$$\left(\frac{30}{101}\right)^2 + \left(\frac{3}{101}\right)^2 = A^2\cos^2\phi + A^2\sin^2\phi = A^2(\cos^2\phi + \sin^2\phi) = A^2$$

and
$$\tan\phi = \frac{A\sin\phi}{A\cos\phi} = \left(\frac{3}{101}\right)\Big/\left(\frac{30}{101}\right) = \frac{1}{10}$$

Consequently, I_s has the required form if

$$A = \sqrt{\frac{909}{(101)^2}} = \frac{3}{\sqrt{101}} \quad \text{and} \quad \phi = \arctan\frac{1}{10} = 0.0997 \text{ radians}$$

6.22. An RC circuit has an emf given (in volts) by $400\cos 2t$, a resistance of 100 ohms, and a capacitance of 10^{-2} farad. Initially there is no charge on the capacitor. Find the current in the circuit at any time t.

We first find the charge q and then use (*6.11*) to obtain the current. Here, $E = 400\cos 2t$, $R = 100$, and $C = 10^{-2}$; hence (*6.10*) becomes

$$\frac{dq}{dt} + q = 4\cos 2t$$

This equation is linear, and its solution is (two integrations by parts are required)

$$q = ce^{-t} + \frac{8}{5}\sin 2t + \frac{4}{5}\cos 2t$$

At $t = 0$, $q = 0$; hence,

$$0 = ce^{-(0)} + \frac{8}{5}\sin 2(0) + \frac{4}{5}\cos 2(0) \quad \text{or} \quad c = -\frac{4}{5}$$

Thus
$$q = -\frac{4}{5}e^{-t} + \frac{8}{5}\sin 2t + \frac{4}{5}\cos 2t$$

and using (*6.11*), we obtain

$$I = \frac{dq}{dt} = \frac{4}{5}e^{-t} + \frac{16}{5}\cos 2t - \frac{8}{5}\sin 2t$$

6.23. Find the orthogonal trajectories of the family of curves $x^2 + y^2 = c^2$.

The family, which is given by (*6.12*) with $F(x, y, c) = x^2 + y^2 - c^2$, consists of circles with centers at the origin and radii c. Implicitly differentiating the given equation with respect to x, we obtain

$$2x + 2yy' = 0 \qquad \text{or} \qquad \frac{dy}{dx} = -\frac{x}{y}$$

Here $f(x, y) = -x/y$, so that (*6.15*) becomes

$$\frac{dy}{dx} = \frac{y}{x}$$

This equation is linear (and, in differential form, separable); its solution is

$$y = kx \tag{1}$$

which represents the orthogonal trajectories.

In Fig. 6-7 some members of the family of circles are shown in solid lines and some members of the family (*1*), which are straight lines through the origin, are shown in dashed lines. Observe that each straight line intersects each circle at right angles.

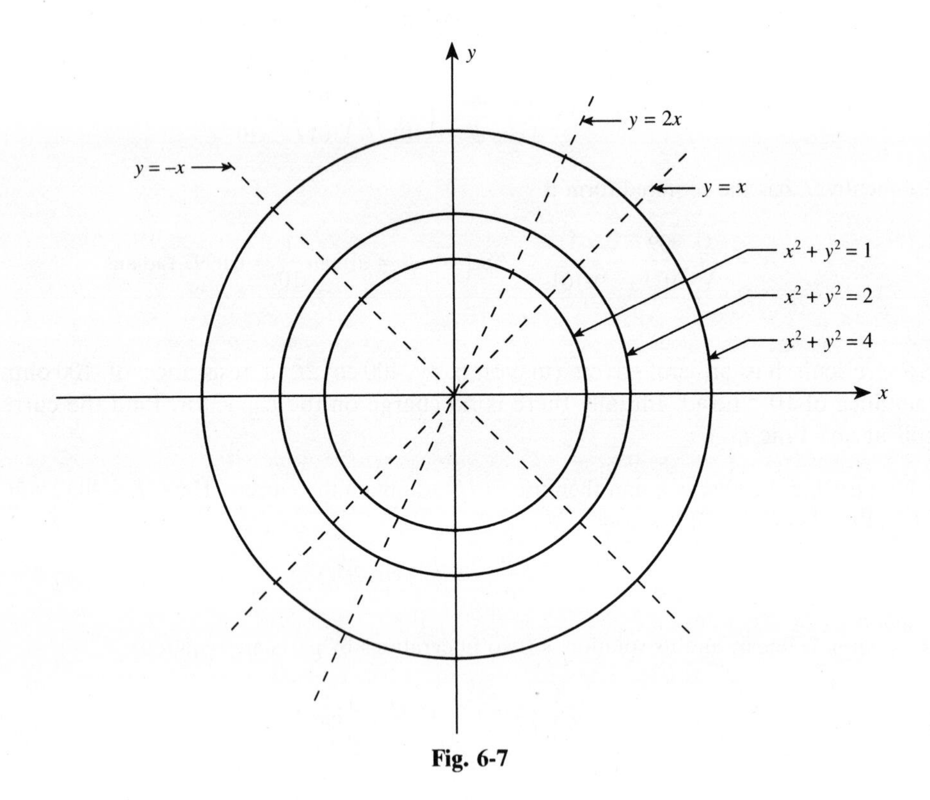

Fig. 6-7

6.24. Find the orthogonal trajectories of the family of curves $y = cx^2$.

The family, which is given by (*6.12*) with $F(x, y, c) = y - cx^2$, consists of parabolas symmetric about the y-axis with vertices at the origin. Differentiating the given equation with respect to x, we obtain $dy/dx = 2cx$. To eliminate c, we observe, from the given equation, that $c = y/x^2$; hence, $dy/dx = 2y/x$. Here $f(x, y) = 2y/x$, so (*6.15*) becomes

$$\frac{dy}{dx} = \frac{-x}{2y} \qquad \text{or} \qquad x\,dx + 2y\,dy = 0$$

The solution of this separable equation is $\frac{1}{2}x^2 + y^2 = k$. These orthogonal trajectories are ellipses. Some

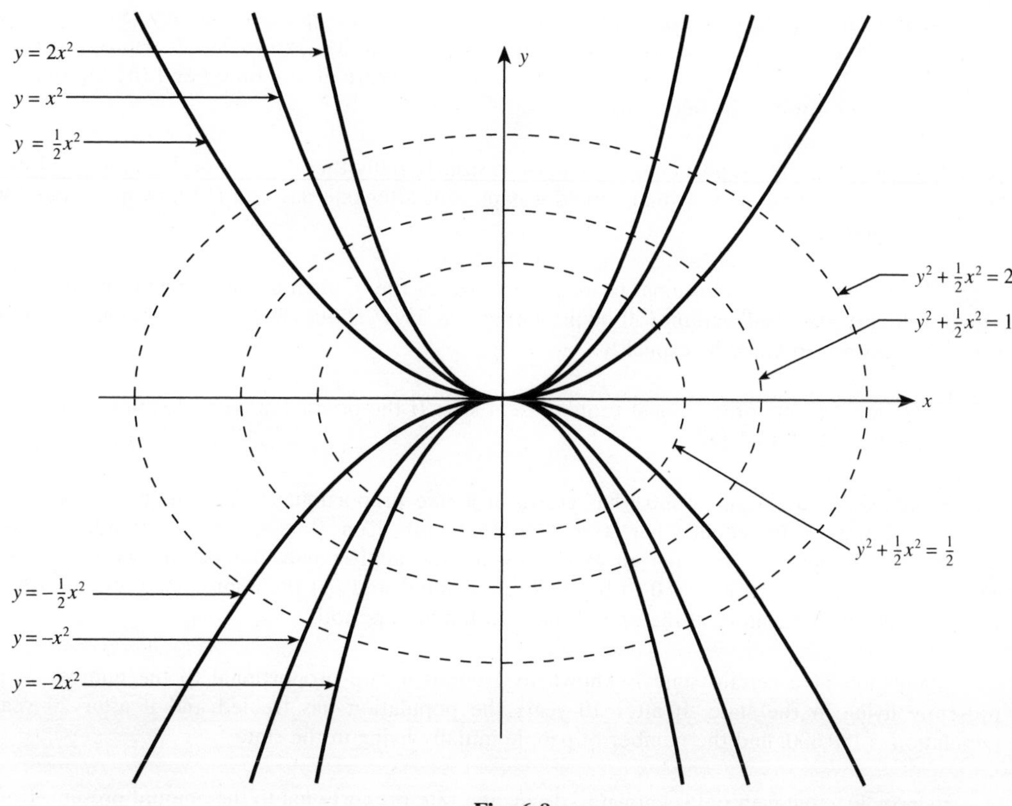

Fig. 6-8

members of this family, along with some members of the original family of parabolas, are shown in Fig. 6-8. Note that each ellipse intersects each parabola at right angles.

6.25. Find the orthogonal trajectories of the family of curves $x^2 + y^2 = cx$.

Here, $F(x, y, c) = x^2 + y^2 - cx$. Implicitly differentiating the given equation with respect to x, we obtain

$$2x + 2y\frac{dy}{dx} = c$$

Eliminating c between this equation and $x^2 + y^2 - cx = 0$, we find

$$2x + 2y\frac{dy}{dx} = \frac{x^2 + y^2}{x} \quad \text{or} \quad \frac{dy}{dx} = \frac{y^2 - x^2}{2xy}$$

Here $f(x, y) = (y^2 - x^2)/2xy$, so (*6.15*) becomes

$$\frac{dy}{dx} = \frac{2xy}{x^2 - y^2}$$

This equation is homogeneous, and its solution (see Problem 3.14) gives the orthogonal trajectories as $x^2 + y^2 = ky$.

Supplementary Problems

6.26. Bacteria grow in a nutrient solution at a rate proportional to the amount present. Initially, there are 250 strands of the bacteria in the solution which grows to 800 strands after seven hours. Find (*a*) an expression for the approximate number of strands in the culture at any time t and (*b*) the time needed for the bacteria to grow to 1600 strands.

6.27. Bacteria grow in a culture at a rate proportional to the amount present. Initially, 300 strands of the bacteria are in the culture and after two hours that number has grown by 20 percent. Find (*a*) an expression for the approximate number of strands in the culture at any time t and (*b*) the time needed for the bacteria to double its initial size.

6.28. A mold grows at a rate proportional to its present size. Initially there is 2 oz of this mold, and two days later there is 3 oz. Find (*a*) how much mold was present after one day and (*b*) how much mold will be present in ten days.

6.29. A mold grows at a rate proportional to its present size. If the original amount doubles in one day, what proportion of the original amount will be present in five days? *Hint:* Designate the initial amount by N_0. It is not necessary to know N_0 explicitly.

6.30. A yeast grows at a rate proportional to its present size. If the original amount doubles in two hours, in how many hours will it triple?

6.31. The population of a certain country has grown at a rate proportional to the number of people in the country. At present, the country has 80 million inhabitants. Ten years ago it had 70 million. Assuming that this trend continues, find (*a*) an expression for the approximate number of people living in the country at any time t (taking $t = 0$ to be the present time) and (*b*) the approximate number of people who will inhabit the country at the end of the next ten-year period.

6.32. The population of a certain state is known to grow at a rate proportional to the number of people presently living in the state. If after 10 years the population has trebled and if after 20 years the population is 150,000, find the number of people initially living in the state.

6.33. A certain radioactive material is known to decay at a rate proportional to the amount present. If initially there are 100 milligrams of the material present and if after two years it is observed that 5 percent of the original mass has decayed, find (*a*) an expression for the mass at any time t and (*b*) the time necessary for 10 percent of the original mass to have decayed.

6.34. A certain radioactive material is known to decay at a rate proportional to the amount present. If after one hour it is observed that 10 percent of the material has decayed, find the half-life of the material. *Hint:* Designate the initial mass of the material by N_0. It is not necessary to know N_0 explicitly.

6.35. Find $N(t)$ for the situation described in Problem 6.7.

6.36. A depositor places \$10,000 in a certificate of deposit which pays 6 percent interest per annum, compounded continuously. How much will be in the account at the end of seven years, assuming no additional deposits or withdrawals?

6.37. How much will be in the account described in the previous problem if the interest rate is $7\frac{1}{2}$ percent instead?

6.38. A depositor places \$5000 in an account established for a child at birth. Assuming no additional deposits or withdrawals, how much will the child have upon reaching the age of 21 if the bank pays 5 percent interest per annum compounded continuously for the entire time period?

6.39. Determine the interest rate required to double an investment in eight years under continuous compounding.

6.40. Determine the interest rate required to triple an investment in ten years under continuous compounding.

6.41. How long will it take a bank deposit to triple in value if interest is compounded continuously at a constant rate of $5\frac{1}{4}$ percent per annum?

6.42. How long will it take a bank deposit to double in value if interest is compounded continuously at a constant rate of $8\frac{3}{4}$ percent per annum?

6.43. A depositor currently has $6000 and plans to invest it in an account that accrues interest continuously. What interest rate must the bank pay if the depositor needs to have $10,000 in four years?

6.44. A depositor currently has $8000 and plans to invest it in an account that accrues interest continuously at the rate of $6\frac{1}{4}$ percent. How long will it take for the account to grow to $13,500?

6.45. A body at a temperature of 0° F is placed in a room whose temperature is kept at 100° F. If after 10 minutes the temperature of the body is 25° F, find (*a*) the time required for the body to reach a temperature of 50° F, and (*b*) the temperature of the body after 20 minutes.

6.46. A body of unknown temperature is placed in a refrigerator at a constant temperature of 0° F. If after 20 minutes the temperature of the body is 40° F and after 40 minutes the temperature of the body is 20° F, find the initial temperature of the body.

6.47. A body at a temperature of 50° F is placed in an oven whose temperature is kept at 150° F. If after 10 minutes the temperature of the body is 75° F, find the time required for the body to reach a temperature of 100° F.

6.48. A hot pie that was cooked at a constant temperature of 325° F is taken directly from an oven and placed outdoors in the shade to cool on a day when the air temperature in the shade is 85° F. After 5 minutes in the shade, the temperature of the pie had been reduced to 250° F. Determine (*a*) the temperature of the pie after 20 minutes and (*b*) the time required for the pie to reach 275 °F.

6.49. A cup of tea is prepared in a preheated cup with hot water so that the temperature of both the cup and the brewing tea is initially 190° F. The cup is then left to cool in a room kept at a constant 72° F. Two minutes later, the temperature of the tea is 150° F. Determine (*a*) the temperature of the tea after 5 minutes and (*b*) the time required for the tea to reach 100° F.

6.50. A bar of iron, previously heated to 1200° C, is cooled in a large bath of water maintained at a constant temperature of 50° C. The bar cools by 200° in the first minute. How much longer will it take to cool a second 200°?

6.51. A body of mass 3 slugs is dropped from a height of 500 ft in *a* with zero velocity. Assuming no air resistance, find (*a*) an expression for the velocity of the body at any time t and (*b*) an expression for the position of the body at any time t with respect to the coordinate system described in Fig. 6-5.

6.52. (*a*) Determine the time required for the body described in the previous problem to hit the ground. (*b*) How long would it take if instead the mass of the body was 10 slugs?

6.53. A body is dropped from a height of 300 ft with an initial velocity of 30 ft/sec. Assuming no air resistance, find (*a*) an expression for the velocity of the body at any time t and (*b*) the time required for the body to hit the ground.

6.54. A body of mass 2 slugs is dropped from a height of 450 ft with an initial velocity of 10 ft/sec. Assuming no air resistance, find (*a*) an expression for the velocity of the body at any time t and (*b*) the time required for the body to hit the ground.

6.55. A body is propelled straight up with an initial velocity of 500 ft/sec in a vacuum with no air resistance. How long will it take the body to return to the ground?

6.56. A ball is propelled straight up with an initial velocity of 250 ft/sec in a vacuum with no air resistance. How high will it go?

6.57. A body of mass m is thrown vertically into the air with an initial velocity v_0. The body encounters no air resistance. Find (*a*) the equation of motion in the coordinate system of Fig. 6-6, (*b*) an expression for the velocity of the body at any time t, (*c*) the time t_m at which the body reaches its maximum height, (*d*) an expression for the position of the body at any time t, and (*e*) the maximum height attained by the body.

6.58. Redo Problem 6.51 assuming there is air resistance which creates a force on the body equal to $-2v$ lb.

6.59. Redo Problem 6.54 assuming there is air resistance which creates a force on the body equal to $-\frac{1}{2}v$ lb.

6.60. A ball of mass 5 slugs is dropped from a height of 1000 ft. Find the limiting velocity of the ball if it encounters a force due to air resistance equal to $-\frac{1}{2}v$.

6.61. A body of mass 2 kg is dropped from a height of 200 m. Find the limiting velocity of the body if it encounters a resistance force equal to $-50v$.

6.62. A body of mass 10 slugs is dropped from a height of 1000 ft with no initial velocity. The body encounters an air resistance proportional to its velocity. If the limiting velocity is known to be 320 ft/sec, find (*a*) an expression for the velocity of the body at any time t, (*b*) an expression for the position of the body at any time t, and (*c*) the time required for the body to attain a velocity of 160 ft/sec.

6.63. A body weighing 8 lb is dropped from a great height with no initial velocity. As it falls, the body encounters a force due to air resistance proportional to its velocity. If the limiting velocity of this body is 4 ft/sec, find (*a*) an expression for the velocity of the body at any time t and (*b*) an expression for the position of the body at any time t.

6.64. A body weighing 160 lb is dropped 2000 ft above ground with no initial velocity. As it falls, the body encounters a force due to air resistance proportional to its velocity. If the limiting velocity of this body is 320 ft/sec, find (*a*) an expression for the velocity of the body at any time t and (*b*) an expression for the position of the body at any time t.

6.65. A tank initially holds 10 gal of fresh water. At $t = 0$, a brine solution containing $\frac{1}{2}$ lb of salt per gallon is poured into the tank at a rate of 2 gal/min, while the well-stirred mixture leaves the tank at the same rate. Find (*a*) the amount and (*b*) the concentration of salt in the tank at any time t.

6.66. A tank initially holds 80 gal of a brine solution containing $\frac{1}{8}$ lb of salt per gallon. At $t = 0$, another brine solution containing 1 lb of salt per gallon is poured into the tank at the rate of 4 gal/min, while the well-stirred mixture leaves the tank at the rate of 8 gal/min. Find the amount of salt in the tank when the tank contains exactly 40 gal of solution.

6.67. A tank contains 100 gal of brine made by dissolving 80 lb of salt in water. Pure water runs into the tank at the rate of 4 gal/min, and the well-stirred mixture runs out at the same rate. Find (*a*) the amount of salt in the tank at any time t and (*b*) the time required for half the salt to leave the tank.

6.68. A tank contains 100 gal of brine made by dissolving 60 lb of salt in water. Salt water containing 1 lb of salt per gallon runs in at the rate of 2 gal/min and the well-stirred mixture runs out at the same rate of 3 gal/min. Find the amount of salt in the tank after 30 minutes.

6.69. A tank contains 40 l of solution containing 2 g of substance per liter. Salt water containing 3 g of this substance per liter runs in at the rate of 4 l/min and the well-stirred mixture runs out at the same rate. Find the amount of substance in the tank after 15 minutes.

6.70. A tank contains 40 l of a chemical solution prepared by dissolving 80 g of a soluble substance in fresh water. Fluid containing 2 g of this substance per liter runs in at the rate of 3 l/min and the well-stirred mixture runs out at the same rate. Find the amount of substance in the tank after 20 minutes.

6.71. An RC circuit has an emf of 5 volts, a resistance of 10 ohms, a capacitance of 10^{-2} farad, and initially a charge of 5 coulombs on the capacitor. Find (a) the transient current and (b) the steady-state current.

6.72. An RC circuit has an emf of 100 volts, a resistance of 5 ohms, a capacitance of 0.02 farad, and an initial charge on the capacitor of 5 coulombs. Find (a) an expression for the charge on the capacitor at any time t and (b) the current in the circuit at any time t.

6.73. An RC circuit has no applied emf, a resistance of 10 ohms, a capacitance of 0.04 farad, and an initial charge on the capacitor of 10 coulombs. Find (a) an expression for the charge on the capacitor at any time t and (b) the current in the circuit at any time t.

6.74. A RC circuit has an emf of $10 \sin t$ volts, a resistance of 100 ohms, a capacitance of 0.005 farad, and no initial charge on the capacitor. Find (a) the charge on the capacitor at any time t and (b) the steady-state current.

6.75. A RC circuit has an emf of $300 \cos 2t$ volts, a resistance of 150 ohms, a capacitance of $1/6 \times 10^{-2}$ farad, and an initial charge on the capacitor of 5 coulombs. Find (a) the charge on the capacitor at any time t and (b) the steady-state current.

6.76. A RL circuit has an emf of 5 volts, a resistance of 50 ohms, an inductance of 1 henry, and no initial current. Find (a) the current in the circuit at any time t and (b) its steady-state component.

6.77. A RL circuit has no applied emf, a resistance of 50 ohms, an inductance of 2 henries, and an initial current of 10 amperes. Find (a) the current in the circuit at any time t and (b) its transient component.

6.78. A RL circuit has a resistance of 10 ohms, an inductance of 1.5 henries, an applied emf of 9 volts, and an initial current of 6 amperes. Find (a) the current in the circuit at any time t and (b) its transient component.

6.79. An RL circuit has an emf given (in volts) by $4 \sin t$, a resistance of 100 ohms, an inductance of 4 henries, and no initial current. Find the current at any time t.

6.80. The steady-state current in a circuit is known to be $\frac{5}{17} \sin t - \frac{3}{17} \cos t$. Rewrite this current in the form $A \sin (t - \phi)$.

6.81. Rewrite the steady-state current of Problem 6.21 in the form $A \cos (2t + \phi)$. *Hint:* Use the identity $\cos (x + y) \equiv \cos x \cos y - \sin x \sin y$.

6.82. Find the orthogonal trajectories of the family of curves $x^2 - y^2 = c^2$.

6.83. Find the orthogonal trajectories of the family of curves $y = ce^x$.

6.84. Find the orthogonal trajectories of the family of curves $x^2 - y^2 = cx$.

6.85. Find the orthogonal trajectories of the family of curves $x^2 + y^2 = cy$.

6.86. Find the orthogonal trajectories of the family of curves $y^2 = 4cx$.

6.87. One hundred strands of bacteria are placed in a nutrient solution in which a plentiful supply of food is constantly provided but space is limited. The competition for space will force the bacteria population to stabilize at 1000 strands. Under these conditions, the growth rate of bacteria is proportional to the product of the amount of bacteria present in the culture with the difference between the maximum population the solution can sustain and the current population. Estimate the amount of bacteria in the solution at any time t if it is known that there were 200 strands of bacteria in the solution after seven hours.

6.88. A new product is to be test marketed by giving it free to 1000 people in a city of one million inhabitants, which is assumed to remain constant for the period of the test. It is further assumed that the rate of product adoption will be proportional to the number of people who have it with the number who do not. Estimate as a function of time the number of people who will adopt the product if it is known that 3000 people have adopted the product after four weeks.

6.89. A body of mass 1 slug is dropped with an initial velocity of 1 ft/sec and encounters a force due to air resistance given exactly by $-8v^2$. Find the velocity at any time t.

Chapter 7

Linear Differential Equations: Theory of Solutions

LINEAR DIFFERENTIAL EQUATIONS

An nth-order linear differential equation has the form

$$b_n(x)y^{(n)} + b_{n-1}(x)y^{(n-1)} + \cdots + b_2(x)y'' + b_1(x)y' + b_0(x)y = g(x) \tag{7.1}$$

where $g(x)$ and the coefficients $b_j(x)$ $(j = 0, 1, 2, \ldots, n)$ depend solely on the variable x. In other words, they do *not* depend on y or on any derivative of y.

If $g(x) \equiv 0$, then Eq. (*7.1*) is *homogeneous*; if not, (*7.1*) is *nonhomogeneous*. A linear differential equation has *constant coefficients* if all the coefficients $b_j(x)$ in (*7.1*) are constants; if one or more of these coefficients is not constant, (*7.1*) has *variable coefficients*.

Theorem 7.1. Consider the initial-value problem given by the linear differential equation (*7.1*) and the n initial conditions

$$y(x_0) = c_0, \qquad y'(x_0) = c_1, \qquad y''(x_0) = c_2, \ldots, y^{(n-1)}(x_0) = c_{n-1} \tag{7.2}$$

If $g(x)$ and $b_j(x)$ $(j = 0, 1, 2, \ldots, n)$ are continuous in some interval $\mathscr{I}$ containing x_0 and if $b_n(x) \neq 0$ in $\mathscr{I}$, then the initial-value problem given by (*7.1*) and (*7.2*) has a unique (only one) solution defined throughout $\mathscr{I}$.

When the conditions on $b_n(x)$ in Theorem 7.1 hold, we can divide Eq. (*7.1*) by $b_n(x)$ to get

$$y^{(n)} + a_{n-1}y^{(n-1)} + \cdots + a_2(x)y'' + a_1(x)y' + a_0(x)y = \phi(x) \tag{7.3}$$

where $a_j(x) = b_j(x)/b_n(x)$ $(j = 0, 1, \ldots, n-1)$ and $\phi(x) = g(x)/b_n(x)$.

Let us define the differential operator $\mathbf{L}(y)$ by

$$\mathbf{L}(y) \equiv y^{(n)} + a_{n-1}(x)y^{(n-1)} + \cdots + a_2(x)y'' + a_1(x)y' + a_0(x)y \tag{7.4}$$

where $a_i(x)$ $(i = 0, 1, 2, \ldots, n-1)$ is continuous on some interval of interest. Then (*7.3*) can be rewritten as

$$\mathbf{L}(y) = \phi(x) \tag{7.5}$$

and, in particular, a linear *homogeneous* differential equation can be expressed as

$$\mathbf{L}(y) = 0 \tag{7.6}$$

LINEARLY INDEPENDENT SOLUTIONS

A set of functions $\{y_1(x), y_2(x), \ldots, y_n(x)\}$ is *linearly dependent* on $a \leq x \leq b$ if there exist constants $c_1, c_2, \ldots, c_n$, *not all zero*, such that

$$c_1 y_1(x) + c_2 y_2(x) + \cdots + c_n y_n(x) \equiv 0 \tag{7.7}$$

on $a \leq x \leq b$.

Example 7.1. The set $\{x, 5x, 1, \sin x\}$ is linearly dependent on $[-1, 1]$ since there exist constants $c_1 = -5$, $c_2 = 1$, $c_3 = 0$, and $c_4 = 0$, *not all zero*, such that (*7.7*) is satisfied. In particular,

$$-5 \cdot x + 1 \cdot 5x + 0 \cdot 1 + 0 \cdot \sin x \equiv 0$$

Note that $c_1 = c_2 = \cdots = c_n = 0$ is a set of constants that always satisfies (*7.7*). A set of functions is linearly dependent if there exists *another* set of constants, *not all zero,* that also satisfies (7.7). If the *only* solution to (7.7) is $c_1 = c_2 = \cdots = c_n = 0$, then the set of functions $\{y_1(x), y_2(x), \ldots, y_n(x)\}$ is *linearly independent* on $a \le x \le b$.

Theorem 7.2. The nth-order linear *homogeneous* differential equation $\mathbf{L}(y) = 0$ always has n linearly independent solutions. If $y_1(x), y_2(x), \ldots, y_n(x)$ represent these solutions, then the general solution of $\mathbf{L}(y) = 0$ is

$$y(x) = c_1 y_1(x) + c_2 y_2(x) + \cdots + c_n y_n(x) \tag{7.8}$$

where $c_1, c_2, \ldots, c_n$ denote arbitrary constants.

THE WRONSKIAN

The *Wronskian* of a set of functions $\{z_1(x), z_2(x), \ldots, z_n(x)\}$ on the interval $a \le x \le b$, having the property that each function possesses $n - 1$ derivatives on this interval, is the determinant

$$W(z_1, z_2, \ldots, z_n) = \begin{vmatrix} z_1 & z_2 & \cdots & z_n \\ z_1' & z_2' & \cdots & z_n' \\ z_1'' & z_2'' & \cdots & z_n'' \\ \vdots & \vdots & & \vdots \\ z_1^{(n-1)} & z_2^{(n-1)} & \cdots & z_n^{(n-1)} \end{vmatrix}$$

Theorem 7.3. If the Wronskian of a set of n functions defined on the interval $a \le x \le b$ is nonzero for at least one point in this interval, then the set of functions is linearly independent there. If the Wronskian is identically zero on this interval and if each of the functions is a solution to the same linear differential equation, then the set of functions is linearly dependent.

Caution: Theorem 7.3 is silent when the Wronskian is identically zero *and* the functions are not known to be solutions of the same linear differential equation. In this case, one must test directly whether Eq. (7.7) is satisfied.

NONHOMOGENEOUS EQUATIONS

Let y_p denote any *particular* solution of Eq. (7.5) (see Chapter 2) and let y_h (henceforth called the *homogeneous* or *complementary solution*) represent the *general* solution of the associated homogeneous equation $\mathbf{L}(y) = 0$.

Theorem 7.4. The general solution to $\mathbf{L}(y) = \phi(x)$ is

$$y = y_h + y_p \tag{7.9}$$

Solved Problems

7.1. State the order of each of the following differential equations and determine whether any are linear:

(*a*) $2xy'' + x^2y' - (\sin x)y = 2$ (*b*) $yy''' + xy' + y = x^2$

(*c*) $y'' - y = 0$ (*d*) $3y' + xy = e^{-x^2}$

(*e*) $2e^x y''' + e^x y'' = 1$ (*f*) $\dfrac{d^4y}{dx^4} + y^4 = 0$

(*g*) $y'' + \sqrt{y'} + y = x^2$ (*h*) $y' + 2y + 3 = 0$

(*a*) Second-order. Here $b_2(x) = 2x$, $b_1(x) = x^2$, $b_0(x) = -\sin x$, and $g(x) = 2$. Since none of these terms depends on y or any derivative of y, the differential equation is linear.

(*b*) Third-order. Since $b_3 = y$, which does depend on y, the differential equation is nonlinear.

(*c*) Second-order. Here $b_2(x) = 1$, $b_1(x) = 0$, $b_0(x) = 1$, and $g(x) = 0$. None of these terms depends on y or any derivative of y; hence the differential equation is linear.

(*d*) First-order. Here $b_1(x) = 3$, $b_0(x) = x$, and $g(x) = e^{-x^2}$; hence the differential equation is linear. (See also Chapter 5.)

(*e*) Third-order. Here $b_3(x) = 2e^x$, $b_2(x) = e^x$, $b_1(x) = b_0(x) = 0$, and $g(x) = 1$. None of these terms depends on y or any of its derivatives, so the equation is linear.

(*f*) Fourth-order. The equation is nonlinear because y is raised to a power higher than unity.

(*g*) Second-order. The equation is nonlinear because the first derivative of y is raised to a power other than unity, here the one-half power.

(*h*) First-order. Here $b_1(x) = 1$, $b_0(x) = 2$, and $g(x) = -3$. None of these terms depends on y or any of its derivatives, so the equation is linear.

7.2. Which of the linear differential equations given in Problem 7.1 are homogeneous?

Using the results of Problem 7.1, we see that the only linear differential equation having $g(x) \equiv 0$ is (*c*), so this is the only one that is homogeneous. Equations (*a*), (*d*), (*e*), and (*h*) are nonhomogeneous linear differential equations.

7.3. Which of the linear differential equations given in Problem 7.1 have constant coefficients?

In their present forms, only (*c*) and (*h*) have constant coefficients, for only in these equations are *all* the coefficients constants. Equation (*e*) can be transformed into one having constant coefficients by multiplying it by e^{-x}. The equation then becomes

$$2y''' + y'' = e^{-x}$$

7.4. Find the general form of a linear differential equation of (*a*) order two and (*b*) order one.

(*a*) For a second-order differential equation, (*7.1*) becomes

$$b_2(x)y'' + b_1(x)y' + b_0(x)y = g(x)$$

If $b_2(x) \neq 0$, we can divide through by it, in which case (*7.3*) takes the form

$$y'' + a_1(x)y' + a_0(x)y = \phi(x)$$

(*b*) For a first-order differential equation, (*7.1*) becomes

If $b_1(x) \neq 0$, we can divide through by it, in which case (*7.3*) takes the form

$$y' + a_0(x)y = \phi(x)$$

This last equation is identical to (*5.1*) with $p(x) = a_0(x)$ and $q(x) = \phi(x)$.

7.5. Find the Wronskian of the set $\{e^x, e^{-x}\}$.

$$W(e^x, e^{-x}) = \begin{vmatrix} e^x & e^{-x} \\ \dfrac{de^x}{dx} & \dfrac{de^{-x}}{dx} \end{vmatrix} = \begin{vmatrix} e^x & e^{-x} \\ e^x & -e^{-x} \end{vmatrix}$$

$$= e^x(-e^{-x}) - e^{-x}(e^x) = -2$$

7.6. Find the Wronskian of the set $\{\sin 3x, \cos 3x\}$.

$$W(\sin 3x, \cos 3x) = \begin{vmatrix} \sin 3x & \cos 3x \\ \dfrac{d(\sin 3x)}{dx} & \dfrac{d(\cos 3x)}{dx} \end{vmatrix} = \begin{vmatrix} \sin 3x & \cos 3x \\ 3\cos 3x & -3\sin 3x \end{vmatrix}$$

$$= -3(\sin^2 3x + \cos^2 3x) = -3$$

7.7. Find the Wronskian of the set $\{x, x^2, x^3\}$.

$$W(x, x^2, x^3) = \begin{vmatrix} x & x^2 & x^3 \\ \dfrac{d(x)}{dx} & \dfrac{d(x^2)}{dx} & \dfrac{d(x^3)}{dx} \\ \dfrac{d^2(x)}{dx^2} & \dfrac{d^2(x^2)}{dx^2} & \dfrac{d^2(x^3)}{dx^2} \end{vmatrix}$$

$$= \begin{vmatrix} x & x^2 & x^3 \\ 1 & 2x & 3x^2 \\ 0 & 2 & 6x \end{vmatrix} = 2x^3$$

This example shows that the Wronskian is in general a nonconstant function.

7.8. Find the Wronskian of the set $\{1-x, 1+x, 1-3x\}$.

$$W(1-x, 1+x, 1-3x) = \begin{vmatrix} 1-x & 1+x & 1-3x \\ \dfrac{d(1-x)}{dx} & \dfrac{d(1+x)}{dx} & \dfrac{d(1-3x)}{dx} \\ \dfrac{d^2(1-x)}{dx^2} & \dfrac{d^2(1+x)}{dx^2} & \dfrac{d^2(1-3x)}{dx^2} \end{vmatrix}$$

$$= \begin{vmatrix} 1-x & 1+x & 1-3x \\ -1 & 1 & -3 \\ 0 & 0 & 0 \end{vmatrix} = 0$$

7.9. Determine whether the set $\{e^x, e^{-x}\}$ is linearly dependent on $(-\infty, \infty)$.

The Wronskian of this set was found in Problem 7.5 to be -2. Since it is nonzero for at least one point in the interval of interest (in fact, it is nonzero at every point in the interval), it follows from Theorem 7.3 that the set is linearly independent.

7.10. Redo Problem 7.9 by testing directly how Eq. (7.7) is satisfied.

Consider the equation

$$c_1 e^x + c_2 e^{-x} \equiv 0 \tag{1}$$

We must determine whether there exist values of c_1 and c_2, *not both zero,* that will satisfy (*1*). Rewriting (*1*), we have $c_2 e^{-x} \equiv -c_1 e^x$ or

$$c_2 \equiv -c_1 e^{2x} \tag{2}$$

For any nonzero value of c_1, the left side of (*2*) is a constant whereas the right side is not; hence the equality in (*2*) is not valid. It follows that the *only* solution to (*2*), and therefore to (*1*), is $c_1 = c_2 = 0$. Thus, the set is not linearly dependent; rather it is linearly independent.

7.11. Is the set $\{x^2, x, 1\}$ linearly dependent on $(-\infty, \infty)$?

The Wronskian of this set was found in Problem 7.7 to be $2x^3$. Since it is nonzero for at least one point in the interval of interest (in particular, at $x = 3$, $W = 54 \neq 0$), it follows from Theorem 7.3 that the set is linearly independent.

7.12. Redo Problem 7.11 by testing directly how Eq. (*7.7*) is satisfied.

Consider the equation

$$c_1 x^2 + c_2 x + c_3 \equiv 0 \tag{1}$$

Since this equation is valid for all x only if $c_1 = c_2 = c_3 = 0$, the given set is linearly independent. Note that if any of the c's were not zero, then the quadratic equation (*1*) could hold for at most two values of x, the roots of the equation, *and not for all x.*

7.13. Determine whether the set $\{1 - x, 1 + x, 1 - 3x\}$ is linearly dependent on $(-\infty, \infty)$.

The Wronskian of this set was found in Problem 7.8 to be identically zero. In this case, Theorem 7.3 provides no information, and we must test directly how Eq. (*7.7*) is satisfied.

Consider the equation

$$c_1(1 - x) + c_2(1 + x) + c_3(1 - 3x) \equiv 0 \tag{1}$$

which can be rewritten as

$$(-c_1 + c_2 - 3c_3)x + (c_1 + c_2 + c_3) \equiv 0$$

This linear equation can be satisfied for all x only if both coefficients are zero. Thus,

$$-c_1 + c_2 - 3c_3 = 0 \qquad \text{and} \qquad c_1 + c_2 + c_3 = 0$$

Solving these equations simultaneously, we find that $c_1 = -2c_3$, $c_2 = c_3$, with c_3 arbitrary. Choosing $c_3 = 1$ (any other nonzero number would do), we obtain $c_1 = -2$, $c_2 = 1$, and $c_3 = 1$ as a set of constants, not all zero, that satisfy (*1*). Thus, the given set of functions is linearly dependent.

7.14. Redo Problem 7.13 knowing that all three functions of the given set are solutions to the differential equation $y'' = 0$.

The Wronskian is identically zero *and* all functions in the set are solutions to the same linear differential equation, so it now follows from Theorem 7.3 that the set is linearly dependent.

7.15. Find the general solution of $y'' + 9y = 0$ if it is known that two solutions are

$$y_1(x) = \sin 3x \qquad \text{and} \qquad y_2(x) = \cos 2x$$

The Wronskian of the two solutions was found in Problem 7.6 to be -3, which is nonzero everywhere. It follows, first from Theorem 7.3, that the two solutions are linearly independent and, then from Theorem 7.2 that the general solution is

$$y(x) = c_1 \sin 3x + c_2 \cos 2x$$

7.16. Find the general solution of $y'' - y = 0$ if it is known that two solutions are

$$y_1(x) = e^x \quad \text{and} \quad y_2(x) = e^{-x}$$

It was shown in both Problems 7.9 and 7.10 that these two functions are linearly independent. It follows from Theorem 7.2 that the general solution is

$$y(x) = c_1 e^x + c_2 e^{-x}$$

7.17. Two solutions of $y'' - 2y' + y = 0$ are e^{-x} and $5e^{-x}$. Is the general solution $y = c_1 e^{-x} + c_2 5e^{-x}$?

We calculate

$$W(e^{-x}, 5e^{-x}) = \begin{vmatrix} e^{-x} & 5e^{-x} \\ -e^{-x} & -5e^{-x} \end{vmatrix} \equiv 0$$

Therefore the functions e^{-x} and $5e^{-x}$ are linearly dependent (see Theorem 7.3), and we conclude from Theorem 7.2 that $y = c_1 e^{-x} + c_2 5e^{-x}$ is *not* the general solution.

7.18. Two solutions of $y'' - 2y' + y = 0$ are e^x and xe^x. Is the general solution $y = c_1 e^x + c_2 xe^x$?

We have

$$W(e^x, xe^x) = \begin{vmatrix} e^x & xe^x \\ e^x & e^x + xe^x \end{vmatrix} = e^{2x} \not\equiv 0$$

It follows, first from Theorem 7.3, that the two particular solutions are linearly independent and then from Theorem 7.2, that the general solution is

$$y = c_1 e^x + c_2 xe^x$$

7.19. Three solutions of $y''' = 0$ are x^2, x, and 1. Is the general solution $y = c_1 x^2 + c_2 x + c_3$?

Yes. It was shown in Problems 7.11 and 7.12 that three solutions are linearly independent, so the result is immediate from Theorem 7.3.

7.20. Two solutions of $y''' - 6y'' + 11y' - 6y = 0$ are e^x and e^{2x}. Is the general solution $y = c_1 e^x + c_2 e^{2x}$?

No. Theorem 7.2 states that the general solution of a *third*-order linear homogeneous differential equation is a combination of *three* linearly independent solutions, not two.

7.21. Use the results of Problem 7.16 to find the general solution of

$$y'' - y = 2 \sin x$$

if it is known that $-\sin x$ is a particular solution.

We are given that $y_p = -\sin x$, and we know from Problem 7.16 that the general solution to the associated homogeneous differential equation is $y_h = c_1 e^x + c_2 e^{-x}$. It follows from Theorem 7.4 that the general solution to the given nonhomogeneous differential equation is

$$y = y_h + y_p = c_1 e^x + c_2 e^{-x} - \sin x$$

7.22. Use the results of Problem 7.18 to find the general solution of

$$y'' - 2y' + y = x^2$$

if it is known that $x^2 + 4x + 6$ is a particular solution.

We have from Problem 7.18 that the general solution to the associated homogeneous differential equation is

$$y_h = c_1e^x + c_2xe^x$$

Since we are given that $y_p = x^2 + 4x + 6$, it follows from Theorem 7.4 that

$$y = y_h + y_p = c_1e^x + c_2xe^x + x^2 + 4x + 6$$

7.23. Use the results of Problem 7.18 to find the general solution of

$$y'' - 2y' + y = e^{3x}$$

if it is known that $\frac{1}{4}e^{3x}$ is a particular solution.

We have from Problem 7.18 that the general solution to the associated homogeneous differential equation is

$$y_h = c_1e^x + c_2xe^x$$

In addition, we are given that $y_p = \frac{1}{4}e^{3x}$. It follows directly from Theorem 7.4 that

$$y = y_h + y_p = c_1e^x + c_2xe^x + \frac{1}{4}e^{3x}$$

7.24. Determine whether the set $\{x^3, |x^3|\}$ is linearly dependent on $[-1, 1]$.

Consider the equation

$$c_1x^3 + c_2|x^3| \equiv 0 \tag{1}$$

Recall that $|x^3| = x^3$ if $x \geq 0$ and $|x^3| = -x^3$ if $x < 0$. Thus, when $x \geq 0$, (*1*) becomes

$$c_1x^3 + c_2x^3 \equiv 0 \tag{2}$$

whereas when $x < 0$, (*1*) becomes

$$c_1x^3 - c_2x^3 \equiv 0 \tag{3}$$

Solving (*2*) and (*3*) simultaneously for c_1 and c_2, we find that the *only* solution is $c_1 = c_2 = 0$. The given set is, therefore, linearly independent.

7.25. Find $W(x^3, |x^3|)$ on $[-1, 1]$.

We have

$$|x^3| = \begin{cases} x^3 & \text{if } x \geq 0 \\ -x^3 & \text{if } x < 0 \end{cases} \qquad \frac{d(|x^3|)}{dx} = \begin{cases} 3x^2 & \text{if } x > 0 \\ 0 & \text{if } x = 0 \\ -3x^2 & \text{if } x < 0 \end{cases}$$

Then, for $x > 0$,

$$W(x^3, |x^3|) = \begin{vmatrix} x^3 & x^3 \\ 3x^2 & 3x^2 \end{vmatrix} \equiv 0$$

For $x < 0$,

$$W(x^3, |x^3|) = \begin{vmatrix} x^3 & -x^3 \\ 3x^2 & -3x^2 \end{vmatrix} \equiv 0$$

For $x = 0$,

$$W(x^3, |x^3|) = \begin{vmatrix} 0 & 0 \\ 0 & 0 \end{vmatrix} = 0$$

Thus, $W(x^3, |x^3|) \equiv 0$ on $[-1, 1]$.

7.26. Do the results of Problems 7.24 and 7.25 contradict Theorem 7.3?

No. Since the Wronskian of two linearly independent functions is identically zero, it follows from Theorem 7.3 that these two functions, x^3 and $|x^3|$, are *not* both solutions of the *same* linear homogeneous differential equation of the form $\mathbf{L}(y) = 0$.

7.27. Two solutions of $y'' - (2/x)y' = 0$ on $[-1, 1]$ are $y = x^3$ and $y = |x^3|$. Does this result contradict the solution to Problem 7.26?

No. Although $W(x^3, |x^3|) \equiv 0$ and both $y = x^3$ and $y = |x^3|$ are linearly independent solutions of the same linear homogeneous differential equation $y'' - (2/x)y' = 0$, this differential equation is not of the form $\mathbf{L}(y) = 0$. The coefficient $-2/x$ is discontinuous at $x = 0$.

7.28. The initial-value problem $y' = 2\sqrt{|y|}$; $y(0) = 0$ has the two solutions $y = x\,|x|$ and $y \equiv 0$. Does this result violate Theorem 7.1?

No. Here $\phi = 2\sqrt{|y|}$, which depends on y; therefore, the differential equation is not linear and Theorem 7.1 does not apply.

7.29. Determine all solutions of the initial-value problem $y'' + e^x y' + (x + 1)y = 0$; $y(1) = 0$, $y'(1) = 0$.

Here, $b_2(x) = 1$, $b_1(x) = e^x$, $b_0(x) = x + 1$, and $g(x) \equiv 0$ satisfy the hypotheses of Theorem 7.1; thus, the solution to the initial-value problem is unique. By inspection, $y \equiv 0$ is a solution. It follows that $y \equiv 0$ is the only solution.

7.30. Show that the second-order operator $\mathbf{L}(y)$ is linear; that is

$$\mathbf{L}(c_1 y_1 + c_2 y_2) = c_1 \mathbf{L}(y_1) + c_2 \mathbf{L}(y_2)$$

where c_1 and c_2 are arbitrary constants and y_1 and y_2 are arbitrary n-times differentiable functions.

In general,

$$\mathbf{L}(y) = y'' + a_1(x)y' + a_0(x)y$$

Thus

$$\begin{aligned}\mathbf{L}(c_1 y_1 + c_2 y_2) &= (c_1 y_1 + c_2 y_2)'' + a_1(x)(c_1 y_1 + c_2 y_2)' + a_0(x)(c_1 y_1 + c_2 y_2)\\ &= c_1 y_1'' + c_2 y_2'' + a_1(x)c_1 y_1' + a_1(x)c_2 y_2' + a_0(x)c_1 y_1 + a_0(x)c_2 y_2\\ &= c_1[y_1'' + a_1(x)y_1' + a_0(x)y_1] + c_2[y_2'' + a_1(x)y_2' + a_0(x)y_2]\\ &= c_1\mathbf{L}(y_1) + c_2\mathbf{L}(y_2)\end{aligned}$$

7.31. Prove the *principle of superposition* for homogeneous linear differential equations; that is, if y_1 and y_2 are two solutions of $\mathbf{L}(y) = 0$, then $c_1 y_1 + c_2 y_2$ is also a solution of $\mathbf{L}(y) = 0$ for any two constants c_1 and c_2.

Let y_1 and y_2 be two solutions of $\mathbf{L}(y) = 0$; that is, $\mathbf{L}(y_1) = 0$ and $\mathbf{L}(y_2) = 0$. Using the results of Problem 7.30, it follows that

$$\mathbf{L}(c_1 y_1 + c_2 y_2) = c_1\mathbf{L}(y_1) + c_2\mathbf{L}(y_2) = c_1(0) + c_2(0) = 0$$

Thus, $c_1 y_1 + c_2 y_2$ is also a solution of $\mathbf{L}(y) = 0$.

7.32. Prove Theorem 7.4.

Since $\mathbf{L}(y_h) = 0$ and $\mathbf{L}(y_p) = \phi(x)$, it follows from the linearity of $\mathbf{L}$ that

$$\mathbf{L}(y) = \mathbf{L}(y_h + y_p) = \mathbf{L}(y_h) + \mathbf{L}(y_p) = 0 + \phi(x) = \phi(x)$$

Thus, y is a solution.

To prove that it is the general solution, we must show that every solution of $\mathbf{L}(y) = \phi(x)$ is of the form (7.9). Let y be any solution of $\mathbf{L}(y) = \phi(x)$ and set $z = y - y_p$. Then

$$\mathbf{L}(z) = \mathbf{L}(y - y_p) = \mathbf{L}(y) - \mathbf{L}(y_p) = \phi(x) - \phi(x) = 0$$

so that z is a solution to the homogeneous equation $\mathbf{L}(y) = 0$. Since $z = y - y_p$, it follows that $y = z + y_p$, where z is a solution of $\mathbf{L}(y) = 0$.

Supplementary Problems

7.33. Determine which of the following differential equations are linear:

(a) $y'' + xy' + 2y = 0$

(b) $y''' - y = x$

(c) $y' + 5y = 0$

(d) $y^{(4)} + x^2y''' + xy'' - e^xy' + 2y = x^2 + x + 1$

(e) $y'' + 2xy' + y = 4xy^2$

(f) $y' - 2y = xy$

(g) $y'' + yy' = x^2$

(h) $y''' + (x^2 - 1)y'' - 2y' + y = 5\sin x$

(i) $y' + y(\sin x) = x$

(j) $y' + x(\sin y) = x$

(k) $y'' + e^y = 0$

(l) $y'' + e^x = 0$

7.34. Determine which of the linear differential equations in Problem 7.33 are homogeneous.

7.35. Determine which of the linear differential equations in Problem 7.33 have constant coefficients.

In Problems 7.36 through 7.49, find the Wronskians of the given sets of functions and, where appropriate, use that information to determine whether the given sets are linearly independent.

7.36. $\{3x, 4x\}$

7.37. $\{x^2, x\}$

7.38. $\{x^3, x^2\}$

7.39. $\{x^3, x\}$

7.40. $\{x^2, 5\}$

7.41. $\{x^2, -x^2\}$

7.42. $\{e^{2x}, e^{-2x}\}$

7.43. $\{e^{2x}, e^{3x}\}$

7.44. $\{3e^{2x}, 5e^{2x}\}$

7.45. $\{x, 1, 2x - 7\}$

7.46. $\{x + 1, x^2 + x, 2x^2 - x - 3\}$

7.47. $\{x^2, x^3, x^4\}$

7.48. $\{e^{-x}, e^x, e^{2x}\}$

7.49. $\{\sin x, 2\cos x, 3\sin x + \cos x\}$

7.50. Prove directly that the set given in Problem 7.36 is linearly dependent.

7.51. Prove directly that the set given in Problem 7.41 is linearly dependent.

7.52. Prove directly that the set given in Problem 7.44 is linearly dependent.

7.53. Prove directly that the set given in Problem 7.45 is linearly dependent.

7.54. Prove directly that the set given in Problem 7.46 is linearly dependent.

7.55. Prove directly that the set given in Problem 7.49 is linearly dependent.

7.56. Using the results of Problem 7.42, construct the general solution of $y'' - 4y = 0$.

7.57. Using the results of Problem 7.43, construct the general solution of $y'' - 5y' + 6y = 0$.

7.58. What can one say about the general solution of $y'' + 16y = 0$ if two particular solutions are known to be $y_1 = \sin 4x$ and $y_2 = \cos 4x$?

7.59. What can one say about the general solution of $y'' - 8y' = 0$ if two particular solutions are known to be $y_1 = e^{8x}$ and $y_2 = 1$?

7.60. What can one say about the general solution of $y'' + y' = 0$ if two particular solutions are known to be $y_1 = 8$ and $y_2 = 1$?

7.61. What can one say about the general solution of $y''' - y'' = 0$ if two particular solutions are known to be $y_1 = x$ and $y_2 = e^x$?

7.62. What can one say about the general solution of $y''' + y'' + y' + y = 0$ if three particular solutions are known to be the functions given in Problem 7.49?

7.63. What can one say about the general solution of $y''' - 2y'' - y' + 2y = 0$ if three particular solutions are known to be the functions given in Problem 7.48?

7.64. What can one say about the general solution of $d^5y/dx^5 = 0$ if three particular solutions are known to be the functions given in Problem 7.47?

7.65. Find the general solution of $y'' + y = x^2$, if one solution is $y = x^2 - 2$, and if two solutions of $y'' + y = 0$ are $\sin x$ and $\cos x$.

7.66. Find the general solution of $y'' - y = x^2$, if one solution is $y = -x^2 - 2$, and if two solutions of $y'' - y = 0$ are e^x and $3e^x$.

7.67. Find the general solution of $y''' - y'' - y' + y = 5$, if one solution is $y = 5$, and if three solutions of $y''' - y'' - y' + y = 0$ are e^x, e^{-x}, and xe^x.

7.68. The initial-value problem $y' - (2/x)y = 0$; $y(0) = 0$ has two solutions $y \equiv 0$ and $y = x^2$. Why doesn't this result violate Theorem 7.1?

7.69. Does Theorem 7.1 apply to the initial-value problem $y' - (2/x)y = 0$; $y(1) = 3$?

7.70. The initial-value problem $xy' - 2y = 0$; $y(0) = 0$ has two solutions $y \equiv 0$ and $y = x^2$. Why doesn't this result violate Theorem 7.1?

Chapter 8

Second-Order Linear Homogeneous Differential Equations with Constant Coefficients

THE CHARACTERISTIC EQUATION

Corresponding to the differential equation

$$y'' + a_1 y' + a_0 y = 0 \tag{8.1}$$

in which a_1 and a_0 are constants, is the algebraic equation

$$\lambda^2 + a_1\lambda + a_0 = 0 \tag{8.2}$$

which is obtained from Eq. (*8.1*) by replacing y'', y', and y by λ^2, λ^1, and $\lambda^0 = 1$, respectively. Equation (*8.2*) is called the *characteristic equation* of (*8.1*).

Example 8.1. The characteristic equation of $y'' + 3y' - 4y = 0$ is $\lambda^2 + 3\lambda - 4 = 0$; the characteristic equation of $y'' - 2y' + y = 0$ is $\lambda^2 - 2\lambda + 1 = 0$.

Characteristic equations for differential equations having dependent variables other than y are obtained analogously, by replacing the jth derivative of the dependent variable by λ^j $(j = 0, 1, 2)$.

The characteristic equation can be factored into

$$(\lambda - \lambda_1)(\lambda - \lambda_2) = 0 \tag{8.3}$$

THE GENERAL SOLUTION

The general solution of (*8.1*) is obtained directly from the roots of (*8.3*). There are three cases to consider.

Case 1. λ_1 and λ_2 both real and distinct. Two linearly independent solutions are $e^{\lambda_1 x}$ and $e^{\lambda_2 x}$, and the general solution is (Theorem 7.2)

$$y = c_1 e^{\lambda_1 x} + c_2 e^{\lambda_2 x} \tag{8.4}$$

In the special case $\lambda_2 = -\lambda_1$, the solution (*8.4*) can be rewritten as $y = k_1 \cosh \lambda_1 x + k_2 \sinh \lambda_1 x$.

Case 2. $\lambda_1 = a + ib$, a complex number. Since a_1 and a_0 in (*8.1*) and (*8.2*) are assumed real, the roots of (*8.2*) must appear in conjugate pairs; thus, the other root is $\lambda_2 = a - ib$. Two linearly independent solutions are $e^{(a+ib)x}$ and $e^{(a-ib)x}$, and the general complex solution is

$$y = d_1 e^{(a+ib)x} + d_2 e^{(a-ib)x} \tag{8.5}$$

which is algebraically equivalent to (see Problem 8.16)

$$y = c_1 e^{ax} \cos bx + c_2 e^{ax} \sin bx \tag{8.6}$$

Case 3. $\lambda_1 = \lambda_2$. Two linearly independent solutions are $e^{\lambda_1 x}$ and $xe^{\lambda_1 x}$, and the general solution is

$$y = c_1 e^{\lambda_1 x} + c_2 x e^{\lambda_1 x} \tag{8.7}$$

Warning: The above solutions *are not valid* if the differential equation is not linear or does not have constant coefficients. Consider, for example, the equation $y'' - x^2 y = 0$. The roots of the characteristic equation are $\lambda_1 = x$ and $\lambda_2 = -x$, but the solution is *not*

$$y = c_1 e^{(x)x} + c_2 e^{(-x)x} = c_1 e^{x^2} + c_2 e^{-x^2}$$

Linear equations with variable coefficients are considered in Chapters 23 and 24.

Solved Problems

8.1. Solve $y'' - y' - 2y = 0$.

The characteristic equation is $\lambda^2 - \lambda - 2 = 0$, which can be factored into $(\lambda + 1)(\lambda - 2) = 0$. Since the roots $\lambda_1 = -1$ and $\lambda_2 = 2$ are real and distinct, the solution is given by *(8.4)* as

$$y = c_1 e^{-x} + c_2 e^{2x}$$

8.2. Solve $y'' - 7y' = 0$.

The characteristic equation is $\lambda^2 - 7\lambda = 0$, which can be factored into $(\lambda - 0)(\lambda - 7) = 0$. Since the roots $\lambda_1 = 0$ and $\lambda_2 = 7$ are real and distinct, the solution is given by *(8.4)* as

$$y = c_1 e^{0x} + c_2 e^{7x} = c_1 + c_2 e^{7x}$$

8.3. Solve $y'' - 5y = 0$.

The characteristic equation is $\lambda^2 - 5 = 0$, which can be factored into $(\lambda - \sqrt{5})(\lambda + \sqrt{5}) = 0$. Since the roots $\lambda_1 = \sqrt{5}$ and $\lambda_2 = -\sqrt{5}$ are real and distinct, the solution is given by *(8.4)* as

$$y = c_1 e^{\sqrt{5}x} + c_2 e^{-\sqrt{5}x}$$

8.4. Rewrite the solution of Problem 8.3 in terms of hyperbolic functions.

Using the results of Problem 8.3 with the identities

$$e^{\lambda x} = \cosh \lambda x + \sinh \lambda x \quad \text{and} \quad e^{-\lambda x} = \cosh \lambda x - \sinh \lambda x$$

we obtain,

$$\begin{aligned} y &= c_1 e^{\sqrt{5}x} + c_2 e^{-\sqrt{5}x} \\ &= c_1(\cosh \sqrt{5}\,x + \sinh \sqrt{5}\,x) + c_2(\cosh \sqrt{5}\,x - \sinh \sqrt{5}\,x) \\ &= (c_1 + c_2)\cosh \sqrt{5}\,x + (c_1 - c_2)\sinh \sqrt{5}\,x \\ &= k_1 \cosh \sqrt{5}\,x + k_2 \sinh \sqrt{5}\,x \end{aligned}$$

where $k_1 = c_1 + c_2$ and $k_2 = c_1 - c_2$.

8.5. Solve $\ddot{y} + 10\dot{y} + 21y = 0$.

Here the independent variable is t. The characteristic equation is

$$\lambda^2 + 10\lambda + 21 = 0$$

which can be factored into

$$(\lambda + 3)(\lambda + 7) = 0$$

The roots $\lambda_1 = -3$ and $\lambda_2 = -7$ are real and distinct, so the general solution is

$$y = c_1 e^{-3t} + c_2 e^{-7t}$$

8.6. Solve $\ddot{x} - 0.01x = 0$.

The characteristic equation is

$$\lambda^2 - 0.01 = 0$$

which can be factored into

$$(\lambda - 0.1)(\lambda + 0.1) = 0$$

The roots $\lambda_1 = 0.1$ and $\lambda_2 = -0.1$ are real and distinct, so the general solution is

$$y = c_1 e^{0.1t} + c_2 e^{-0.1t}$$

or, equivalently,

$$y = k_1 \cosh 0.1t + k_2 \sinh 0.1t$$

8.7. Solve $y'' + 4y' + 5y = 0$.

The characteristic equation is

$$\lambda^2 + 4\lambda + 5 = 0$$

Using the quadratic formula, we find its roots to be

$$\lambda = \frac{-(4) \pm \sqrt{(4)^2 - 4(5)}}{2} = -2 \pm i$$

These roots are a complex conjugate pair, so the general solution is given by (*8.6*) (with $a = -2$ and $b = 1$) as

$$y = c_1 e^{-2x} \cos x + c_2 e^{-2x} \sin x$$

8.8. Solve $y'' + 4y = 0$.

The characteristic equation is

$$\lambda^2 + 4\lambda = 0$$

which can be factored into

$$(\lambda - 2i)(\lambda + 2i) = 0$$

These roots are a complex conjugate pair, so the general solution is given by (*8.6*) (with $a = 0$ and $b = 2$) as

$$y = c_1 \cos 2x + c_2 \sin 2x$$

8.9. Solve $y'' - 3y' + 4y = 0$.

The characteristic equation is

$$\lambda^2 - 3\lambda + 4 = 0$$

Using the quadratic formula, we find its roots to be

$$\lambda = \frac{-(-3) \pm \sqrt{(-3)^2 - 4(4)}}{2} = \frac{3}{2} \pm i\frac{\sqrt{7}}{2}$$

These roots are a complex conjugate pair, so the general solution is given by (*8.6*) as

$$y = c_1 e^{(3/2)x} \cos \frac{\sqrt{7}}{2}x + c_2 e^{(3/2)x} \sin \frac{\sqrt{7}}{2}x$$

8.10. Solve $\ddot{y} - 6\dot{y} + 25y = 0$.

The characteristic equation is

$$\lambda^2 - 6\lambda + 25 = 0$$

Using the quadratic formula, we find its roots to be

$$\lambda = \frac{-(-6) \pm \sqrt{(-6)^2 - 4(25)}}{2} = 3 \pm i4$$

These roots are a complex conjugate pair, so the general solution is

$$y = c_1 e^{3t} \cos 4t + c_2 e^{3t} \sin 4t$$

8.11. Solve $\dfrac{d^2I}{dt^2} + 20\dfrac{dI}{dt} + 200I = 0$.

The characteristic equation is

$$\lambda^2 + 20\lambda + 200 = 0$$

Using the quadratic formula, we find its roots to be

$$\lambda = \frac{-(20) \pm \sqrt{(20)^2 - 4(200)}}{2} = -10 \pm i10$$

These roots are a complex conjugate pair, so the general solution is

$$I = c_1 e^{-10t} \cos 10t + c_2 e^{-10t} \sin 10t$$

8.12. Solve $y'' - 8y' + 16y = 0$.

The characteristic equation is

$$\lambda^2 - 8\lambda + 16 = 0$$

which can be factored into

$$(\lambda - 4)^2 = 0$$

The roots $\lambda_1 = \lambda_2 = 4$ are real and equal, so the general solution is given by (*8.7*) as

$$y = c_1 e^{4x} + c_2 x e^{4x}$$

8.13. Solve $y'' = 0$.

The characteristic equation is $\lambda^2 = 0$, which has roots $\lambda_1 = \lambda_2 = 0$. The solution is given by (*8.7*) as

$$y = c_1 e^{0x} + c_2 x e^{0x} = c_1 + c_2 x$$

8.14. Solve $\ddot{x} + 4\dot{x} + 4x = 0$.

The characteristic equation is

$$\lambda^2 + 4\lambda + 4 = 0$$

which can be factored into

$$(\lambda + 2)^2 = 0$$

The roots $\lambda_1 = \lambda_2 = -2$ are real and equal, so the general solution is

$$x = c_1 e^{-2t} + c_2 t e^{-2t}$$

8.15. Solve $100\dfrac{d^2N}{dt^2} - 20\dfrac{dN}{dt} + N = 0$.

Dividing both sides of the differential equation by 100, to force the coefficient of the highest derivative to be unity, we obtain

$$\frac{d^2N}{dt^2} - 0.2\frac{dN}{dt} + 0.01N = 0$$

Its characteristic equation is

$$\lambda^2 - 0.2\lambda + 0.01 = 0$$

which can be factored into

$$(\lambda - 0.1)^2 = 0$$

The roots $\lambda_1 = \lambda_2 = 0.1$ are real and equal, so the general solution is

$$N = c_1 e^{-0.1t} + c_2 t e^{-0.1t}$$

8.16. Prove that *(8.6)* is algebraically equivalent to *(8.5)*.

Using Euler's relations

$$e^{ibx} = \cos bx + i \sin bx \qquad e^{-ibx} = \cos bx - i \sin bx$$

we can rewrite *(8.5)* as

$$\begin{aligned} y &= d_1 e^{ax} e^{ibx} + d_2 e^{ax} e^{-ibx} = e^{ax}(d_1 e^{ibx} + d_2 e^{-ibx}) \\ &= e^{ax}[d_1(\cos bx + i \sin bx) + d_2(\cos bx - i \sin bx)] \\ &= e^{ax}[(d_1 + d_2)\cos bx + i(d_1 - d_2)\sin bx] \\ &= c_1 e^{ax} \cos bx + c_2 e^{ax} \sin bx \qquad (1) \end{aligned}$$

where $c_1 = d_1 + d_2$ and $c_2 = i(d_1 - d_2)$.

Equation *(1)* is real if and only if c_1 and c_2 are both real, which occurs if and only if d_1 and d_2 are complex conjugates. Since we are interested in the general *real* solution to *(8.1)*, we restrict d_1 and d_2 to be a conjugate pair.

Supplementary Problems

Solve the following differential equations.

8.17. $y'' - y = 0$

8.18. $y'' - y' - 30y = 0$

8.19. $y'' - 2y' + y = 0$

8.20. $y'' + y = 0$

8.21. $y'' + 2y' + 2y = 0$

8.22. $y'' - 7y = 0$

8.23. $y'' + 6y' + 9y = 0$

8.24. $y'' + 2y' + 3y = 0$

8.25. $y'' - 3y' - 5y = 0$

8.26. $y'' + y' + \frac{1}{4}y = 0$

8.27. $\ddot{x} - 20\dot{x} + 64x = 0$

8.28. $\ddot{x} + 60\dot{x} + 500x = 0$

8.29. $\ddot{x} - 3\dot{x} + x = 0$

8.30. $\ddot{x} - 10\dot{x} + 25x = 0$

8.31. $\ddot{x} + 25x = 0$

8.32. $\ddot{x} + 25\dot{x} = 0$

8.33. $\ddot{x} + \dot{x} + 2x = 0$

8.34. $\ddot{u} - 2\dot{u} + 4u = 0$

8.35. $\ddot{u} - 4\dot{u} + 2u = 0$

8.36. $\ddot{u} - 36\dot{u} = 0$

8.37. $\ddot{u} - 36u = 0$

8.38. $\dfrac{d^2Q}{dt^2} - 5\dfrac{dQ}{dt} + 7Q = 0$

8.39. $\frac{d^2Q}{dt^2} - 7\frac{dQ}{dt} + 5Q = 0$

8.40. $\frac{d^2P}{dt^2} - 18\frac{dP}{dt} + 81P = 0$

8.41. $\frac{d^2P}{dx^2} + 2\frac{dP}{dx} + 9P = 0$

8.42. $\frac{d^2N}{dx^2} + 5\frac{dN}{dx} - 24N = 0$

8.43. $\frac{d^2N}{dx^2} + 5\frac{dN}{dx} + 24N = 0$

8.44. $\frac{d^2T}{d\theta^2} + 30\frac{dT}{d\theta} + 225T = 0$

8.45. $\frac{d^2R}{d\theta^2} + 5\frac{dR}{d\theta} = 0$

Chapter 9

*n*th-Order Linear Homogeneous Differential Equations with Constant Coefficients

THE CHARACTERISTIC EQUATION

The characteristic equation of the differential equation

$$y^{(n)} + a_{n-1}y^{(n-1)} + \cdots + a_1 y' + a_0 y = 0 \tag{9.1}$$

with constant coefficients a_j $(j = 0, 1, \ldots, n-1)$ is

$$\lambda^n + a_{n-1}\lambda^{n-1} + \cdots + a_1\lambda + a_0 = 0 \tag{9.2}$$

The characteristic equation (*9.2*) is obtained from (*9.1*) by replacing $y^{(j)}$ by λ^j $(j = 0, 1, \ldots, n-1)$. Characteristic equations for differential equations having dependent variables other than y are obtained analogously, by replacing the jth derivative of the dependent variable by λ^j $(j = 0, 1, \ldots, n-1)$.

Example 9.1. The characteristic equation of $y^{(4)} - 3y''' + 2y'' - y = 0$ is $\lambda^4 - 3\lambda^3 + 2\lambda^2 - 1 = 0$. The characteristic equation of

$$\frac{d^5x}{dt^5} - 3\frac{d^3x}{dt^3} + 5\frac{dx}{dt} - 7x = 0$$

is

$$\lambda^5 - 3\lambda^3 + 5\lambda - 7 = 0$$

Caution: Characteristic equations are only defined for linear homogeneous differential equations with constant coefficients.

THE GENERAL SOLUTION

The roots of the characteristic equation determine the solution of (*9.1*). If the roots $\lambda_1, \lambda_2, \ldots, \lambda_n$ are all real and distinct, the solution is

$$y = c_1 e^{\lambda_1 x} + c_2 e^{\lambda_2 x} + \cdots + c_n e^{\lambda_n x} \tag{9.3}$$

If the roots are distinct, but some are complex, then the solution is again given by (*9.3*). As in Chapter 8, those terms involving complex exponentials can be combined to yield terms involving sines and cosines. If λ_k is a root of multiplicity p [that is, if $(\lambda - \lambda_k)^p$ is a factor of the characteristic equation, but $(\lambda - \lambda_k)^{p+1}$ is not] then there will be p linearly independent solutions associated with λ_k given by $e^{\lambda_k x}, xe^{\lambda_k x}, x^2 e^{\lambda_k x}, \ldots, x^{p-1}e^{\lambda_k x}$. These solutions are combined in the usual way with the solutions associated with the other roots to obtain the complete solution.

In theory it is always possible to factor the characteristic equation, but in practice this can be extremely difficult, especially for differential equations of high order. In such cases, one must often use numerical techniques to approximate the solutions. See Chapters 27 and 28.

Solved Problems

9.1. Solve $y''' - 6y'' + 11y' - 6y = 0$.

The characteristic equation is $\lambda^3 - 6\lambda^2 + 11\lambda - 6 = 0$, which can be factored into

$$(\lambda - 1)(\lambda - 2)(\lambda - 3) = 0$$

The roots are $\lambda_1 = 1$, $\lambda_2 = 2$, and $\lambda_3 = 3$; hence the solution is

$$y = c_1e^x + c_2e^{2x} + c_3e^{3x}$$

9.2. Solve $y^{(4)} - 9y'' + 20y = 0$.

The characteristic equation is $\lambda^4 - 9\lambda^2 + 20 = 0$, which can be factored into

$$(\lambda - 2)(\lambda + 2)(\lambda - \sqrt{5})(\lambda + \sqrt{5}) = 0$$

The roots are $\lambda_1 = 2$, $\lambda_2 = -2$, $\lambda_3 = \sqrt{5}$, and $\lambda_4 = -\sqrt{5}$; hence the solution is

$$\begin{aligned} y &= c_1e^{2x} + c_2e^{-2x} + c_3e^{\sqrt{5}x} + c_4e^{-\sqrt{5}x} \\ &= k_1 \cosh 2x + k_2 \sinh 2x + k_3 \cosh \sqrt{5}\,x + k_4 \sinh \sqrt{5}\,x \end{aligned}$$

9.3. Solve $y' - 5y = 0$.

The characteristic equation is $\lambda - 5 = 0$, which has the single root $\lambda_1 = 5$. The solution is then $y = c_1e^{5x}$. (Compare this result with Problem 5.9.)

9.4. Solve $y''' - 6y'' + 2y' + 36y = 0$.

The characteristic equation, $\lambda^3 - 6\lambda^2 + 2\lambda + 36 = 0$, has roots $\lambda_1 = -2$, $\lambda_2 = 4 + i\sqrt{2}$, and $\lambda_3 = 4 - i\sqrt{2}$. The solution is

$$y = c_1e^{-2x} + d_2e^{(4+i\sqrt{2})x} + d_3e^{(4-i\sqrt{2})x}$$

which can be rewritten, using Euler's relations (see Problem 8.16) as

$$y = c_1e^{-2x} + c_2e^{4x}\cos\sqrt{2}\,x + c_3e^{4x}\sin\sqrt{2}\,x$$

9.5. Solve $\dfrac{d^4x}{dt^4} - 4\dfrac{d^3x}{dt^3} + 7\dfrac{d^2x}{dt^2} - 4\dfrac{dx}{dt} + 6x = 0$.

The characteristic equation, $\lambda^4 - 4\lambda^3 + 7\lambda^2 - 4\lambda + 6 = 0$, has roots $\lambda_1 = 2 + i\sqrt{2}$, $\lambda_2 = 2 - i\sqrt{2}$, $\lambda_3 = i$, and $\lambda_4 = -i$. The solution is

$$x = d_1e^{(2+i\sqrt{2})t} + d_2e^{(2-i\sqrt{2})t} + d_3e^{it} + d_4e^{-it}$$

If, using Euler's relations, we combine the first two terms and then similarly combine the last two terms, we can rewrite the solution as

$$x = c_1e^{2t}\cos\sqrt{2}\,t + c_2e^{2x}\sin\sqrt{2}\,t + c_3\cos t + c_4\sin t$$

9.6. Solve $y^{(4)} + 8y''' + 24y'' + 32y' + 16y = 0$.

The characteristic equation, $\lambda^4 + 8\lambda^3 + 24\lambda^2 + 32\lambda + 16 = 0$, can be factored into $(\lambda + 2)^4 = 0$. Here

$\lambda_1 = -2$ is a root of multiplicity four; hence the solution is

$$y = c_1e^{-2x} + c_2xe^{-2x} + c_3x^2e^{-2x} + c_4x^3e^{-2x}$$

9.7. Solve $\dfrac{d^5P}{dt^5} - \dfrac{d^4P}{dt^4} - 2\dfrac{d^3P}{dt^3} + 2\dfrac{d^2P}{dt^2} + \dfrac{dP}{dt} - P = 0.$

The characteristic equation can be factored into $(\lambda-1)^3(\lambda+1)^2 = 0$; hence, $\lambda_1 = 1$ is a root of multiplicity three and $\lambda_2 = -1$ is a root of multiplicity two. The solution is

$$P = c_1e^t + c_2te^t + c_3t^2e^t + c_4e^{-t} + c_5te^{-t}$$

9.8. Solve $\dfrac{d^4Q}{dx^4} - 8\dfrac{d^3Q}{dx^3} + 32\dfrac{d^2Q}{dx^2} - 64\dfrac{dQ}{dx} + 64Q = 0.$

The characteristic equation has roots $2 \pm i2$ and $2 \pm i2$; hence $\lambda_1 = 2 + i2$ and $\lambda_2 = 2 - i2$ are both roots of multiplicity two. The solution is

$$\begin{aligned}
Q &= d_1e^{(2+i2)x} + d_2xe^{(2+i2)x} + d_3e^{(2-i2)x} + d_4xe^{(2-i2)x} \\
&= e^{2x}(d_1e^{i2x} + d_3e^{-i2x}) + xe^{2x}(d_2e^{i2x} + d_4e^{-i2x}) \\
&= e^{2x}(c_1\cos 2x + c_3\sin 2x) + xe^{2x}(c_2\cos 2x + c_4\sin 2x) \\
&= (c_1 + c_2x)e^{2x}\cos 2x + (c_3 + c_4x)e^{2x}\sin 2x
\end{aligned}$$

9.9. Find the general solution to a fourth-order linear homogeneous differential equation for $y(x)$ with real numbers as coefficients if one solution is known to be x^3e^{4x}.

If x^3e^{4x} is a solution, then so too are x^2e^{4x}, xe^{4x}, and e^{4x}. We now have four linearly independent solutions to a fourth-order linear, homogeneous differential equation, so we can write the general solution as

$$y(x) = c_4x^3e^{4x} + c_3x^2e^{4x} + c_2xe^{4x} + c_1e^{4x}$$

9.10. Determine the differential equation described in Problem 9.9.

The characteristic equation of a fourth-order differential equation is a fourth-degree polynomial having exactly four roots. Because x^3e^{4x} is a solution, we know that $\lambda = 4$ is a root of multiplicity four of the corresponding characteristic equation, so the characteristic equation must be $(\lambda-4)^4 = 0$, or

$$\lambda^4 - 16\lambda^3 + 96\lambda^2 - 256\lambda + 256 = 0$$

The associated differential equation is

$$y^{(4)} - 16y''' + 96y'' - 256y' + 256y = 0$$

9.11. Find the general solution to a third-order linear homogeneous differential equation for $y(x)$ with real numbers as coefficients if two solutions are known to be e^{-2x} and $\sin 3x$.

If $\sin 3x$ is a solution, then so too is $\cos 3x$. Together with e^{-2x}, we have three linearly independent solutions to a third-order linear, homogeneous differential equation, and we can write the general solution as

$$y(x) = c_1e^{-2x} + c_2\cos 3x + c_3\sin 3x$$

9.12. Determine the differential equation described in Problem 9.11.

The characteristic equation of a third-order differential equation must have three roots. Because

e^{-2x} and $\sin 3x$ are solutions, we know that $\lambda = -2$ and $\lambda = \pm i3$ are roots of the corresponding characteristic equation, so this equation must be

$$(\lambda + 2)(\lambda - i3)(\lambda + i3) = 0$$

or

$$\lambda^3 + 2\lambda^2 + 9\lambda + 18 = 0$$

The associated differential equation is

$$y''' + 2y'' + 9y' + 18y = 0$$

9.13. Find the general solution to a sixth-order linear homogeneous differential equation for $y(x)$ with real numbers as coefficients if one solution is known to be $x^2e^{7x}\cos 5x$.

If $x^2e^{7x}\cos 5x$ is a solution, then so too are $xe^{7x}\cos 5x$ and $e^{7x}\cos 5x$. Furthermore, because complex roots of a characteristic equation come in conjugate pairs, every solution containing a cosine term is matched with another solution containing a sine term. Consequently, $x^2e^{7x}\sin 5x$, $xe^{7x}\sin 5x$, and $e^{7x}\sin 5x$ are also solutions. We now have six linearly independent solutions to a sixth-order linear, homogeneous differential equation, so we can write the general solution as

$$y(x) = c_1x^2e^{7x}\cos 5x + c_2x^2e^{7x}\sin 5x + c_3xe^{7x}\cos 5x + c_4xe^{7x}\sin 5x + c_5e^{7x}\cos 5x + c_6e^{7x}\sin 5x$$

9.14. Redo Problem 9.13 if the differential equation has order 8.

An eighth-order linear differential equation possesses eight linearly independent solutions, and since we can only identify six of them, as we did in Problem 9.13, we do not have enough information to solve the problem. We can say that the solution to Problem 9.13 will be *part* of the solution to this problem.

9.15. Solve $\dfrac{d^4y}{dx^4} - 4\dfrac{d^3y}{dx^3} - 5\dfrac{d^2y}{dx^2} + 36\dfrac{dy}{dx} - 36y = 0$ if one solution is xe^{2x}.

If xe^{2x} is a solution, then so too is e^{2x} which implies that $(\lambda - 2)^2$ is a factor of the characteristic equation $\lambda^4 - 4\lambda^3 - 5\lambda^2 + 36\lambda - 36 = 0$. Now,

$$\frac{\lambda^4 - 4\lambda^3 - 5\lambda^2 + 36\lambda - 36}{(\lambda - 2)^2} = \lambda^2 - 9$$

so two other roots of the characteristic equation are $\lambda = \pm 3$, with corresponding solutions e^{3x} and e^{-3x}. Having identified four linearly independent solutions to the given fourth-order linear differential equation, we can write the general solution as

$$y(x) = c_1e^{2x} + c_2xe^{2x} + c_3e^{3x} + c_4e^{-3x}$$

Supplementary Problems

In Problems 9.16 through 9.34, solve the given differential equations.

9.16. $y''' - 2y'' - y' + 2y = 0$

9.17. $y''' - y'' - y' + y = 0$

9.18. $y''' - 3y'' + 3y' - y = 0$

9.19. $y''' - y'' + y' - y = 0$

9.20. $y^{(4)} + 2y'' + y = 0$

9.21. $y^{(4)} - y = 0$

9.22. $y^{(4)} + 2y''' - 2y' - y = 0$

9.23. $y^{(4)} - 4y'' + 16y' + 32y = 0$

9.24. $y^{(4)} + 5y''' = 0$

9.25. $y^{(4)} + 2y''' + 3y'' + 2y' + y = 0$

9.26. $y^{(6)} - 5y^{(4)} + 16y''' + 36y'' - 16y' - 32y = 0$

9.27. $\dfrac{d^4x}{dt^4} + 4\dfrac{d^3x}{dt^3} + 6\dfrac{d^2x}{dt^2} + 4\dfrac{dx}{dt} + x = 0$

9.28. $\dfrac{d^3x}{dt^3} = 0$

9.29. $\dfrac{d^4x}{dt^4} + 10\dfrac{d^2x}{dt^2} + 9x = 0$

9.30. $\dfrac{d^3x}{dt^3} - 5\dfrac{d^2x}{dt^2} + 25\dfrac{dx}{dt} - 125x = 0$

9.31. $q^{(4)} + q'' - 2q = 0$

9.32. $q^{(4)} - 3q'' + 2q = 0$

9.33. $N''' - 12N'' - 28N + 480N = 0$

9.34. $\dfrac{d^5r}{d\theta^5} + 5\dfrac{d^4r}{d\theta^4} + 10\dfrac{d^3r}{d\theta^3} + 10\dfrac{d^2r}{d\theta^2} + 5\dfrac{dr}{d\theta} + r = 0$

In Problems 9.35 through 9.41, a complete set of roots is given for the characteristic equation of an nth-order linear homogeneous differential equation in $y(x)$ with real numbers as coefficients. Determine the general solution of the differential equation.

9.35. $2, 8, -14$

9.36. $0, \pm i19$

9.37. $0, 0, 2 \pm i9$

9.38. $2 \pm i9, 2 \pm i9$

9.39. $5, 5, 5, -5, -5$

9.40. $\pm i6, \pm i6, \pm i6$

9.41. $-3 \pm i, -3 \pm i, 3 \pm i, 3 \pm i$

9.42. Determine the differential equation associated with the roots given in Problem 9.35.

9.43. Determine the differential equation associated with the roots given in Problem 9.36.

9.44. Determine the differential equation associated with the roots given in Problem 9.37.

9.45. Determine the differential equation associated with the roots given in Problem 9.38.

9.46. Determine the differential equation associated with the roots given in Problem 9.39.

9.47. Find the general solution to a fourth-order linear homogeneous differential equation for $y(x)$ with real numbers as coefficients if one solution is known to be x^3e^{-x}.

9.48. Find the general solution to a fourth-order linear homogeneous differential equation for $y(x)$ with real numbers as coefficients if two solutions are $\cos 4x$ and $\sin 3x$.

9.49. Find the general solution to a fourth-order linear homogeneous differential equation for $y(x)$ with real numbers as coefficients if one solution is $x \cos 4x$.

9.50. Find the general solution to a fourth-order linear homogeneous differential equation for $y(x)$ with real numbers as coefficients if two solutions are xe^{2x} and xe^{5x}.

Chapter 10

The Method of Undetermined Coefficients

The general solution to the linear differential equation $\mathbf{L}(y) = \phi(x)$ is given by Theorem 7.4 as $y = y_h + y_p$ where y_p denotes one solution to the differential equation and y_h is the general solution to the associated homogeneous equation, $\mathbf{L}(y) = 0$. Methods for obtaining y_h when the differential equation has constant coefficients are given in Chapters 9 and 10. In this chapter and the next, we give methods for obtaining a particular solution y_p *once* y_h *is known.*

SIMPLE FORM OF THE METHOD

The *method of undetermined coefficients* is applicable only if $\phi(x)$ and *all* of its derivatives can be written in terms of the same *finite* set of linearly independent functions, which we denote by $\{y_1(x), y_2(x), \ldots, y_n(x)\}$. The method is initiated by assuming a particular solution of the form

$$y_p(x) = A_1 y_1(x) + A_2 y_2(x) + \cdots + A_n y_n(x)$$

where $A_1, A_2, \ldots, A_n$ denote arbitrary multiplicative constants. These arbitrary constants are then evaluated by substituting the proposed solution into the given differential equation and equating the coefficients of like terms.

Case 1. $\phi(x) = p_n(x)$, an nth-degree polynomial in x. Assume a solution of the form

$$y_p = A_n x^n + A_{n-1}x^{n-1} + \cdots + A_1 x + A_0 \tag{10.1}$$

where A_j $(j = 0, 1, 2, \ldots, n)$ is a constant to be determined.

Case 2. $\phi(x) = ke^{\alpha x}$ where k and α are known constants. Assume a solution of the form

$$y_p = Ae^{\alpha x} \tag{10.2}$$

where A is a constant to be determined.

Case 3. $\phi(x) = k_1 \sin \beta x + k_2 \cos \beta x$ where k_1, k_2, and β are known constants. Assume a solution of the form

$$y_p = A \sin \beta x + B \cos \beta x \tag{10.3}$$

where A and B are constants to be determined.

Note: (*10.3*) in its entirety is assumed even when k_1 or k_2 is zero, because the derivatives of sines or cosines involve both sines and cosines.

GENERALIZATIONS

If $\phi(x)$ is the product of terms considered in Cases 1 through 3, take y_p to be the product of the corresponding assumed solutions and algebraically combine arbitrary constants where possible. In particular, if $\phi(x) = e^{\alpha x}p_n(x)$ is the product of a polynomial with an exponential, assume

$$y_p = e^{\alpha x}(A_nx^n + A_{n-1}x^{n-1} + \cdots + A_1x + A_0) \tag{10.4}$$

where A_j is as in Case 1. If, instead, $\phi(x) = e^{\alpha x}p_n(x)\sin\beta x$ is the product of a polynomial, exponential, and sine term, or if $\phi(x) = e^{\alpha x}p_n(x)\cos\beta x$ is the product of a polynomial, exponential, and cosine term, then assume

$$y_p = e^{\alpha x}\sin\beta x(A_nx^n + \cdots + A_1x + A_0) + e^{\alpha x}\cos\beta x(B_nx^n + \cdots + B_1x + B_0) \tag{10.5}$$

where A_j and B_j $(j = 0, 1, \ldots, n)$ are constants which still must be determined.

If $\phi(x)$ is the sum (or difference) of terms already considered, then we take y_p to be the sum (or difference) of the corresponding assumed solutions and algebraically combine arbitrary constants where possible.

MODIFICATIONS

If any term of the assumed solution, disregarding multiplicative constants, is also a term of y_h (the homogeneous solution), then the assumed solution must be modified by multiplying it by x^m, where m is the smallest positive integer such that the product of x^m with the assumed solution has no terms in common with y_h.

LIMITATIONS OF THE METHOD

In general, if $\phi(x)$ is not one of the types of functions considered above, or if the differential equation *does not have constant coefficients,* then the method given in Chapter 11 is preferred.

Solved Problems

10.1. Solve $y'' - y' - 2y = 4x^2$.

From Problem 8.1, $y_h = c_1e^{-x} + c_2e^{2x}$. Here $\phi(x) = 4x^2$, a second-degree polynomial. Using *(10.1)*, we assume that

$$y_p = A_2x^2 + A_1x + A_0 \tag{1}$$

Thus, $y_p' = 2A_2x + A_1$ and $y_p'' = 2A_2$. Substituting these results into the differential equation, we have

$$2A_2 - (2A_2x + A_1) - 2(A_2x^2 + A_1x + A_0) = 4x^2$$

or, equivalently,

$$(-2A_2)x^2 + (-2A_2 - 2A_1)x + (2A_2 - A_1 - 2A_0) = 4x^2 + (0)x + 0$$

Equating the coefficients of like powers of x, we obtain

$$-2A_2 = 4 \qquad -2A_2 - 2A_1 = 0 \qquad 2A_2 - A_1 - 2A_0 = 0$$

Solving this system, we find that $A_2 = -2$, $A_1 = 2$, and $A_0 = -3$. Hence *(1)* becomes

$$y_p = -2x^2 + 2x - 3$$

and the general solution is

$$y = y_h + y_p = c_1 e^{-x} + c_2 e^{2x} - 2x^2 + 2x - 3$$

10.2. Solve $y'' - y' - 2y = e^{3x}$.

From Problem 8.1, $y_h = c_1 e^{-x} + c_2 e^{2x}$. Here $\phi(x)$ has the form displayed in Case 2 with $k = 1$ and $\alpha = 3$. Using (*10.2*), we assume that

$$y_p = Ae^{3x} \tag{1}$$

Thus, $y_p' = 3Ae^{3x}$ and $y_p'' = 9Ae^{3x}$. Substituting these results into the differential equation, we have

$$9Ae^{3x} - 3Ae^{3x} - 2Ae^{3x} = e^{3x} \quad \text{or} \quad 4Ae^{3x} = e^{3x}$$

It follows that $4A = 1$, or $A = \frac{1}{4}$, so that (*1*) becomes $y_p = \frac{1}{4}e^{3x}$. The general solution then is

$$y = c_1 e^{-x} + c_2 e^{2x} + \frac{1}{4} e^{3x}$$

10.3. Solve $y'' - y' - 2y = \sin 2x$.

Again by Problem 8.1, $y_h = c_1 e^{-x} + c_2 e^{2x}$. Here $\phi(x)$ has the form displayed in Case 3 with $k_1 = 1$, $k_2 = 0$, and $\beta = 2$. Using (*10.3*), we assume that

$$y_p = A \sin 2x + B \cos 2x \tag{1}$$

Thus, $y_p' = 2A \cos 2x - 2B \sin 2x$ and $y_p'' = -4A \sin 2x - 4B \cos 2x$. Substituting these results into the differential equation, we have

$$(-4A \sin 2x - 4B \cos 2x) - (2A \cos 2x - 2B \sin 2x) - 2(A \sin 2x + B \cos 2x) = \sin 2x$$

or, equivalently,

$$(-6A + 2B) \sin 2x + (-6B - 2A) \cos 2x = (1) \sin 2x + (0) \cos 2x$$

Equating coefficients of like terms, we obtain

$$-6A + 2B = 1 \qquad -2A - 6B = 0$$

Solving this system, we find that $A = -3/20$ and $B = 1/20$. Then from (*1*),

$$y_p = -\frac{3}{20} \sin 2x + \frac{1}{20} \cos 2x$$

and the general solution is

$$y = c_1 e^{-x} + c_2 e^{2x} - \frac{3}{20} \sin 2x + \frac{1}{20} \cos 2x$$

10.4. Solve $\ddot{y} - 6\dot{y} + 25y = 2 \sin \frac{t}{2} - \cos \frac{t}{2}$.

From Problem 8.10,

$$y_h = c_1 e^{3t} \cos 4t + c_2 e^{3t} \sin 4t$$

Here $\phi(t)$ has the form displayed in Case 3 with the independent variable t replacing x, $k_1 = 2$, $k_2 = -1$, and $\beta = \frac{1}{2}$. Using (*10.3*), with t replacing x, we assume that

$$y_p = A \sin\frac{t}{2} + B \cos\frac{t}{2} \tag{1}$$

Consequently,

$$\dot{y}_p = \frac{A}{2}\cos\frac{t}{2} - \frac{B}{2}\sin\frac{t}{2}$$

and

$$\ddot{y}_p = -\frac{A}{4}\sin\frac{t}{2} - \frac{B}{4}\cos\frac{t}{2}$$

Substituting these results into the differential equation, we obtain

$$\left(-\frac{A}{4}\sin\frac{t}{2} - \frac{B}{4}\cos\frac{t}{4}\right) - 6\left(\frac{A}{2}\cos\frac{t}{2} - \frac{B}{2}\sin\frac{t}{2}\right) + 25\left(A\sin\frac{t}{2} + B\cos\frac{t}{2}\right) = 2\sin\frac{t}{2} - \cos\frac{t}{2}$$

or, equivalently,

$$\left(\frac{99}{4}A + 3B\right)\sin\frac{t}{2} + \left(-3A + \frac{99}{4}B\right)\cos\frac{t}{2} = 2\sin\frac{t}{2} - \cos\frac{t}{2}$$

Equating coefficients of like terms, we have

$$\frac{99}{4}A + 3B = 2; \qquad -3A + \frac{99}{4}B = -1$$

It follows that $A = 56/663$ and $B = -20/663$, so that (*1*) becomes

$$y_p = \frac{56}{663}\sin\frac{t}{2} - \frac{20}{663}\cos\frac{t}{2}$$

The general solution is

$$y = y_h + y_p = c_1e^{3t}\cos 4t + c_2e^{3t}\sin 4t + \frac{56}{663}\sin\frac{t}{2} - \frac{20}{663}\cos\frac{t}{2}$$

10.5. Solve $\ddot{y} - 6\dot{y} + 25y = 64e^{-t}$.

From Problem 8.10,

$$y_h = c_1e^{3t}\cos 4t + c_2e^{3t}\sin 4t$$

Here $\phi(t)$ has the form displayed in Case 2 with the independent variable t replacing x, $k = 64$ and $\alpha = -1$. Using (*10.2*), with t replacing x, we assume that

$$y_p = Ae^{-t} \tag{1}$$

Consequently, $\dot{y}_p = -Ae^{-t}$ and $\ddot{y}_p = Ae^{-t}$. Substituting these results into the differential equation, we obtain

$$Ae^{-t} - 6(-Ae^{-t}) + 25(Ae^{-t}) = 64e^{-t}$$

or, equivalently, $32Ae^{-t} = 64e^{-t}$. It follows that $32A = 64$ or $A = 2$, so that (*1*) becomes $y_p = 2e^{-t}$. The general solution is

$$y = y_h + y_p = c_1e^{3t}\cos 4t + c_2e^{3t}\sin 4t + 2e^{-t}$$

10.6. Solve $\ddot{y} - 6\dot{y} + 25y = 50t^3 - 36t^2 - 63t + 18$.

Again by Problem 8.10,

$$y_h = c_1e^{3t}\cos 4t + c_2e^{3t}\sin 4t$$

Here $\phi(t)$ is a third-degree polynomial in t. Using (*10.1*) with t replacing x, we assume that

$$y_p = A_3t^3 + A_2t^2 + A_1t + A_0 \tag{1}$$

Consequently,

$$\dot{y}_p = 3A_3t^2 + 2A_2t + A_1$$

and

$$\ddot{y}_p = 6A_3t + 2A_2$$

Substituting these results into the differential equation, we obtain

$$(6A_3t + 2A_2) - 6(3A_3t^2 + 2A_2t + A_1) + 25(A_3t^3 + A_2t^2 + A_1t + A_0) = 50t^3 - 36t^2 - 63t + 18$$

or, equivalently,

$$(25A_3)t^3 + (-18A_3 + 25A_2)t^2 + (6A_3 - 12A_2 + 25A_1) + (2A_2 - 6A_1 + 25A_0) = 50t^3 - 36t^2 - 63t + 18$$

Equating coefficients of like powers of t, we have

$$25A_3 = 50; \qquad -18A_3 + 25A_2 = -36; \qquad 6A_3 - 12A_2 + 25A_1 = -63; \qquad 2A_2 - 6A_1 + 25A_0 = 18$$

Solving these four algebraic equations simultaneously, we obtain $A_3 = 2$, $A_2 = 0$, $A_1 = -3$, and $A_0 = 0$, so that *(1)* becomes

$$y_p = 2t^3 - 3t$$

The general solution is

$$y = y_h + y_p = c_1e^{3t}\cos 4t + c_2e^{3t}\sin 4t + 2t^3 - 3t$$

10.7. Solve $y''' - 6y'' + 11y' - 6y = 2xe^{-x}$.

From Problem 9.1, $y_h = c_1e^x + c_2e^{2x} + c_3e^{3x}$. Here $\phi(x) = e^{\alpha x}p_n(x)$, where $\alpha = -1$ and $p_n(x) = 2x$, a first-degree polynomial. Using Eq. *(10.4)*, we assume that $y_p = e^{-x}(A_1x + A_0)$, or

$$y_p = A_1xe^{-x} + A_0e^{-x} \tag{1}$$

Thus,

$$y_p' = -A_1xe^{-x} + A_1e^{-x} - A_0e^{-x}$$

$$y_p'' = A_1xe^{-x} - 2A_1e^{-x} + A_0e^{-x}$$

$$y_p''' = -A_1xe^{-x} + 3A_1e^{-x} - A_0e^{-x}$$

Substituting these results into the differential equation and simplifying, we obtain

$$-24A_1xe^{-x} + (26A_1 - 24A_0)e^{-x} = 2xe^{-x} + (0)e^{-x}$$

Equating coefficients of like terms, we have

$$-24A_1 = 2 \qquad 26A_1 - 24A_0 = 0$$

from which $A_1 = -1/12$ and $A_0 = -13/144$.

Equation *(1)* becomes

$$y_p = -\frac{1}{12}xe^{-x} - \frac{13}{144}e^{-x}$$

and the general solution is

$$y = c_1e^x + c_2e^{2x} + c_3e^{3x} - \frac{1}{12}xe^{-x} - \frac{13}{144}e^{-x}$$

10.8. Determine the form of a particular solution for $y'' = 9x^2 + 2x - 1$.

Here $\phi(x) = 9x^2 + 2x - 1$, and the solution of the associated homogeneous differential equation $y'' = 0$ is $y_h = c_1x + c_0$. Since $\phi(x)$ is a second-degree polynomial, we first try $y_p = A_2x^2 + A_1x + A_0$. Note, however, that this assumed solution has terms, disregarding multiplicative constants, in common with y_h: in particular, the first-power term and the constant term. Hence, we must determine the smallest positive integer m such that $x^m(A_2x^2 + A_1x + A_0)$ has no terms in common with y_h.

For $m = 1$, we obtain

$$x(A_2x^2 + A_1x + A_0) = A_2x^3 + A_1x^2 + A_0x$$

which still has a first-power term in common with y_h. For $m = 2$, we obtain

$$x^2(A_2x^2 + A_1x + A_0) = A_2x^4 + A_1x^3 + A_0x^2$$

which has no terms in common with y_h; therefore, we assume an expression of this form for y_p.

10.9. Solve $y'' = 9x^2 + 2x - 1$.

Using the results of Problem 10.8, we have $y_h = c_1 x + c_0$ and we assume

$$y_p = A_2 x^4 + A_1 x^3 + A_0 x^2 \tag{1}$$

Substituting (*1*) into the differential equation, we obtain

$$12A_2 x^2 + 6A_1 x + 2A_0 = 9x^2 + 2x - 1$$

from which $A_2 = 3/4$, $A_1 = 1/3$, and $A_0 = -1/2$. Then (*1*) becomes

$$y_p = \frac{3}{4}x^4 + \frac{1}{3}x^3 - \frac{1}{2}x^2$$

and the general solution is

$$y = c_1 x + c_0 + \frac{3}{4}x^4 + \frac{1}{3}x^3 - \frac{1}{2}x^2$$

The solution also can be obtained simply by twice integrating both sides of the differential equation with respect to x.

10.10. Solve $y' - 5y = 2e^{5x}$.

From Problem 9.3, $y_h = c_1 e^{5x}$. Since $\phi(x) = 2e^{5x}$, it would follow from Eq. (*10.2*) that the guess for y_p should be $y_p = A_0 e^{5x}$. Note, however, that this y_p has exactly the same form as y_h; therefore, we must modify y_p. Multiplying y_p by x ($m = 1$), we obtain

$$y_p = A_0 x e^{5x} \tag{1}$$

As this expression has no terms in common with y_h, it is a candidate for the particular solution. Substituting (*1*) and $y_p' = A_0 e^{5x} + 5A_0 x e^{5x}$ into the differential equation and simplifying, we obtain $A_0 e^{5x} = 2e^{5x}$, from which $A_0 = 2$. Equation (*1*) becomes $y_p = 2xe^{5x}$, and the general solution is $y = (c_1 + 2x)e^{5x}$.

10.11. Determine the form of a particular solution of

$$y' - 5y = (x - 1)\sin x + (x + 1)\cos x$$

Here $\phi(x) = (x - 1)\sin x + (x + 1)\cos x$, and from Problem 9.3, we know that the solution to the associated homogeneous problem $y' - 5y = 0$ is $y_h = c_1 e^{5x}$. An assumed solution for $(x - 1)\sin x$ is given by Eq. (*10.5*) (with $\alpha = 0$) as

$$(A_1 x + A_0)\sin x + (B_1 x + B_0)\cos x$$

and an assumed solution for $(x + 1)\cos x$ is given also by Eq. (*10.5*) as

$$(C_1 x + C_0)\sin x + (D_1 x + D_0)\cos x$$

(Note that we have used C and D in the last expression, since the constants A and B already have been used.) We therefore take

$$y_p = (A_1 x + A_0)\sin x + (B_1 x + B_0)\cos x + (C_1 x + C_0)\sin x + (D_1 x + D_0)\cos x$$

Combining like terms, we arrive at

$$y_p = (E_1 x + E_0)\sin x + (F_1 x + F_0)\cos x$$

as the assumed solution, where $E_j = A_j + C_j$ and $F_j = B_j + D_j$ ($j = 0, 1$).

10.12. Solve $y' - 5y = (x - 1)\sin x + (x + 1)\cos x$.

From Problem 9.3, $y_h = c_1 e^{5x}$. Using the results of Problem 10.11, we assume that

$$y_p = (E_1 x + E_0)\sin x + (F_1 x + F_0)\cos x \tag{1}$$

Thus, $$y_p' = (E_1 - F_1 x - F_0)\sin x + (E_1 x + E_0 + F_1)\cos x$$

Substituting these values into the differential equation and simplifying, we obtain

$$(-5E_1 - F_1)x\sin x + (-5E_0 + E_1 - F_0)\sin x + (-5F_1 + E_1)x\cos x + (-5F_0 + E_0 + F_1)\cos x$$
$$= (1)x\sin x + (-1)\sin x + (1)x\cos x + (1)\cos x$$

Equating coefficients of like terms, we have

$$\begin{aligned} -5E_1 \qquad\qquad - F_1 &= 1 \\ -5E_0 + E_1 - F_0 \qquad &= -1 \\ E_1 \qquad - 5F_1 &= 1 \\ E_0 \qquad - 5F_0 + F_1 &= 1 \end{aligned}$$

Solving, we obtain $E_1 = -2/13$, $E_0 = 71/338$, $F_1 = -3/13$, and $F_0 = -69/338$. Then, from (1),

$$y_p = \left(-\frac{2}{13}x + \frac{71}{338}\right)\sin x + \left(-\frac{3}{13}x - \frac{69}{338}\right)\cos x$$

and the general solution is

$$y = c_1 e^{5x} + \left(\frac{-2}{13}x + \frac{71}{338}\right)\sin x - \left(\frac{3}{13}x + \frac{69}{338}\right)\cos x$$

10.13. Solve $y' - 5y = 3e^x - 2x + 1$.

From Problem 9.3, $y_h = c_1 e^{5x}$. Here, we can write $\phi(x)$ as the sum of two manageable functions: $\phi(x) = (3e^x) + (-2x + 1)$. For the term $3e^x$ we would assume a solution of the form Ae^x; for the term $-2x + 1$ we would assume a solution of the form $B_1 x + B_0$. Thus, we try

$$y_p = Ae^x + B_1 x + B_0 \tag{1}$$

Substituting (1) into the differential equation and simplifying, we obtain

$$(-4A)e^x + (-5B_1)x + (B_1 - 5B_0) = (3)e^x + (-2)x + (1)$$

Equating coefficients of like terms, we find that $A = -3/4$, $B_1 = 2/5$, and $B_0 = -3/25$. Hence, (1) becomes

$$y_p = -\frac{3}{4}e^x + \frac{2}{5}x - \frac{3}{25}$$

and the general solution is

$$y = c_1 e^{5x} - \frac{3}{4}e^x + \frac{2}{5}x - \frac{3}{25}$$

10.14. Solve $y' - 5y = x^2 e^x - xe^{5x}$.

From Problem 9.3, $y_h = c_1 e^{5x}$. Here $\phi(x) = x^2 e^x - xe^{5x}$, which is the difference of two terms, each in manageable form. For $x^2 e^x$ we would assume a solution of the form

$$e^x(A_2 x^2 + A_1 x + A_0) \tag{1}$$

For xe^{5x} we would try initially a solution of the form

$$e^{5x}(B_1 x + B_0) = B_1 xe^{5x} + B_0 e^{5x}$$

But this supposed solution would have, disregarding multiplicative constants, the term e^{5x} in common with y_h. We are led, therefore, to the modified expression

$$xe^{5x}(B_1 x + B_0) = e^{5x}(B_1 x^2 + B_0 x) \tag{2}$$

We now take y_p to be the sum of (*1*) and (*2*):

$$y_p = e^x(A_2x^2 + A_1x + A_0) + e^{5x}(B_1x^2 + B_0x) \tag{3}$$

Substituting (*3*) into the differential equation and simplifying, we obtain

$$e^x[(-4A_2)x^2 + (2A_2 - 4A_1)x + (A_1 - 4A_0)] + e^{5x}[(2B_1)x + B_0]$$
$$= e^x[(1)x^2 + (0)x + (0)] + e^{5x}[(-1)x + (0)]$$

Equating coefficients of like terms, we have

$$-4A_2 = 1 \qquad 2A_2 - 4A_1 = 0 \qquad A_1 - 4A_0 = 0$$
$$2B_1 = -1 \qquad B_0 = 0$$

from which

$$A_2 = -\frac{1}{4} \qquad A_1 = -\frac{1}{8} \qquad A_0 = -\frac{1}{32}$$
$$B_1 = -\frac{1}{2} \qquad B_0 = 0$$

Equation (*3*) then gives

$$y_p = e^x\left(-\frac{1}{4}x^2 - \frac{1}{8}x - \frac{1}{32}\right) - \frac{1}{2}x^2e^{5x}$$

and the general solution is

$$y = c_1e^{5x} + e^x\left(-\frac{1}{4}x^2 - \frac{1}{8}x - \frac{1}{32}\right) - \frac{1}{2}x^2e^{5x}$$

Supplementary Problems

In Problems 10.15 through 10.26, determine the form of a particular solution to $\mathbf{L}(y) = \phi(x)$ for $\phi(x)$ as given if the solution to the associated homogeneous equation $\mathbf{L}(y) = 0$ is $y_h = c_1e^{2x} + c_2e^{3x}$.

10.15. $\phi(x) = 2x - 7$

10.16. $\phi(x) = -3x^2$

10.17. $\phi(x) = 132x^2 - 388x + 1077$

10.18. $\phi(x) = 0.5e^{-2x}$

10.19. $\phi(x) = 13e^{5x}$

10.20. $\phi(x) = 4e^{2x}$

10.21. $\phi(x) = 2\cos 3x$

10.22. $\phi(x) = \frac{1}{2}\cos 3x - 3\sin 3x$

10.23. $\phi(x) = x\cos 3x$

10.24. $\phi(x) = 2x + 3e^{8x}$

10.25. $\phi(x) = 2xe^{5x}$

10.26. $\phi(x) = 2xe^{3x}$

In Problems 10.27 through 10.36, determine the form of a particular solution to $\mathbf{L}(y) = \phi(x)$ for $\phi(x)$ as given if the solution to the associated homogeneous equation $\mathbf{L}(y) = 0$ is $y_h = c_1e^{5x}\cos 3x + c_2e^{5x}\sin 3x$.

10.27. $\phi(x) = 2e^{3x}$

10.28. $\phi(x) = xe^{3x}$

10.29. $\phi(x) = -23e^{5x}$

10.30. $\phi(x) = (x^2 - 7)e^{5x}$

10.31. $\phi(x) = 5 \cos \sqrt{2}\, x$

10.32. $\phi(x) = x^2 \sin \sqrt{2}\, x$

10.33. $\phi(x) = -\cos 3x$

10.34. $\phi(x) = 2 \sin 4x - \cos 7x$

10.35. $\phi(x) = 31e^{-x} \cos 3x$

10.36. $\phi(x) = -\frac{1}{6} e^{5x} \cos 3x$

In Problems 10.37 through 10.43, determine the form of a particular solution to $\mathbf{L}(x) = \phi(t)$ for $\phi(t)$ as given if the solution to the associated homogeneous equation $\mathbf{L}(x) = 0$ is $x_h = c_1 + c_2 e^t + c_3 t e^t$.

10.37. $\phi(t) = t$

10.38. $\phi(t) = 2t^2 - 3t + 82$

10.39. $\phi(t) = te^{-2t} + 3$

10.40. $\phi(t) = -6e^t$

10.41. $\phi(t) = te^t$

10.42. $\phi(t) = 3 + t \cos t$

10.43. $\phi(t) = te^{2t} \cos 3t$

In Problems 10.44 through 10.52, find the general solutions to the given differential equations.

10.44. $y'' - 2y' + y = x^2 - 1$

10.45. $y'' - 2y' + y = 3e^{2x}$

10.46. $y'' - 2y' + y = 4 \cos x$

10.47. $y'' - 2y' + y = 3e^x$

10.48. $y'' - 2y' + y = xe^x$

10.49. $y' - y = e^x$

10.50. $y' - y = xe^{2x} + 1$

10.51. $y' - y = \sin x + \cos 2x$

10.52. $y''' - 3y'' + 3y' - y = e^x + 1$

Chapter 11

Variation of Parameters

Variation of parameters is another method (see Chapter 10) for finding a particular solution of the nth-order linear differential equation

$$\mathbf{L}(y) = \phi(x) \tag{11.1}$$

once the solution of the associated homogeneous equation $\mathbf{L}(y) = 0$ is known. Recall from Theorem 7.2 that if $y_1(x), y_2(x), \ldots, y_n(x)$ are n linearly independent solutions of $\mathbf{L}(y) = 0$, then the general solution of $\mathbf{L}(y) = 0$ is

$$y_h = c_1 y_1(x) + c_2 y_2(x) + \cdots + c_n y_n(x) \tag{11.2}$$

THE METHOD

A particular solution of $\mathbf{L}(y) = \phi(x)$ has the form

$$y_p = v_1 y_1 + v_2 y_2 + \cdots + v_n y_n \tag{11.3}$$

where $y_i = y_i(x)$ $(i = 1, 2, \ldots, n)$ is given in Eq. (*11.2*) and v_i $(i = 1, 2, \ldots, n)$ is an unknown function of x which still must be determined.

To find v_i, first solve the following linear equations simultaneously for v_i':

$$\begin{aligned}
v_1' y_1 + v_2' y_2 + \cdots + v_n' y_n &= 0 \\
v_1' y_1' + v_2' y_2' + \cdots + v_n' y_n' &= 0 \\
&\vdots \\
v_1' y_1^{(n-2)} + v_2' y_2^{(n-2)} + \cdots + v_n' y_n^{(n-2)} &= 0 \\
v_1' y_1^{(n-1)} + v_2' y_2^{(n-1)} + \cdots + v_n' y_n^{(n-1)} &= \phi(x)
\end{aligned} \tag{11.4}$$

Then integrate each v_i' to obtain v_i, disregarding all constants of integration. This is permissible because we are seeking only *one* particular solution.

Example 11.1. For the special case $n = 3$, Eqs. (*11.4*) reduce to

$$\begin{aligned}
v_1' y_1 + v_2' y_2 + v_3' y_3 &= 0 \\
v_1' y_1' + v_2' y_2' + v_3' y_3' &= 0 \\
v_1' y_1'' + v_2' y_2'' + v_3' y_3'' &= \phi(x)
\end{aligned} \tag{11.5}$$

For the case $n = 2$, Eqs. (*11.4*) become

$$\begin{aligned}
v_1' y_1 + v_2' y_2 &= 0 \\
v_1' y_1' + v_2' y_2' &= \phi(x)
\end{aligned} \tag{11.6}$$

and for the case $n = 1$, we obtain the single equation

$$v_1' y_1 = \phi(x) \tag{11.7}$$

Since $y_1(x), y_2(x), \ldots, y_n(x)$ are n linearly independent solutions of the same equation $\mathbf{L}(y) = 0$, their Wronskian is not zero (Theorem 7.3). This means that the system (*11.4*) has a nonzero determinant and can be solved uniquely for $v_1'(x), v_2'(x), \ldots, v_n'(x)$.

SCOPE OF THE METHOD

The method of variation of parameters can be applied to *all* linear differential equations. It is therefore more powerful than the method of undetermined coefficients, which is restricted to linear differential equations with constant coefficients and particular forms of $\phi(x)$. Nonetheless, in those cases where both methods are applicable, the method of undetermined coefficients is usually the more efficient and, hence, preferable.

As a practical matter, the integration of $v_i'(x)$ may be impossible to perform. In such an event, other methods (in particular, numerical techniques) must be employed.

Solved Problems

11.1. Solve $y''' + y' = \sec x$.

This is a third-order equation with

$$y_h = c_1 + c_2 \cos x + c_3 \sin x$$

(see Chapter 9); it follows from Eq. (*11.3*) that

$$y_p = v_1 + v_2 \cos x + v_3 \sin x \tag{1}$$

Here $y_1 = 1$, $y_2 = \cos x$, $y_3 = \sin x$, and $\phi(x) = \sec x$, so (*11.5*) becomes

$$v_1'(1) + v_2'(\cos x) + v_3'(\sin x) = 0$$

$$v_1'(0) + v_2'(-\sin x) + v_3'(\cos x) = 0$$

$$v_1'(0) + v_2'(-\cos x) + v_3'(-\sin x) = \sec x$$

Solving this set of equations simultaneously, we obtain $v_1' = \sec x$, $v_2' = -1$, and $v_3' = -\tan x$. Thus,

$$v_1 = \int v_1'\,dx = \int \sec x\,dx = \ln|\sec x + \tan x|$$

$$v_2 = \int v_2'\,dx = \int -1\,dx = -x$$

$$v_3 = \int v_3'\,dx = \int -\tan x\,dx = -\int \frac{\sin x}{\cos x}\,dx = \ln|\cos x|$$

Substituting these values into (*1*), we obtain

$$y_p = \ln|\sec x + \tan x| - x\cos x + (\sin x)\ln|\cos x|$$

The general solution is therefore

$$y = y_h + y_p = c_1 + c_2\cos x + c_3 \sin x + \ln|\sec x + \tan x| - x\cos x + (\sin x)\ln|\cos x|$$

11.2. Solve $y''' - 3y'' + 2y' = \dfrac{e^x}{1+e^{-x}}$.

This is a third-order equation with

$$y_h = c_1 + c_2 e^x + c_3 e^{2x}$$

(see Chapter 9); it follows from Eq. (*11.3*) that

$$y_p = v_1 + v_2 e^x + v_3 e^{2x} \tag{1}$$

Here $y_1 = 1$, $y_2 = e^x$, $y_3 = e^{2x}$, and $\phi(x) = e^x/(1+e^{-x})$, so Eq. (*11.5*) becomes

$$v_1'(1) + v_2'(e^x) + v_3'(e^{2x}) = 0$$
$$v_1'(0) + v_2'(e^x) + v_3'(2e^{2x}) = 0$$
$$v_1'(0) + v_2'(e^x) + v_3'(4e^{2x}) = \frac{e^x}{1+e^{-x}}$$

Solving this set of equations simultaneously, we obtain

$$v_1' = \frac{1}{2}\left(\frac{e^x}{1+e^{-x}}\right)$$
$$v_2' = \frac{-1}{1+e^{-x}}$$
$$v_3' = \frac{1}{2}\left(\frac{e^{-x}}{1+e^{-x}}\right)$$

Thus, using the substitutions $u = e^x + 1$ and $w = 1 + e^{-x}$, we find that

$$\begin{aligned} v_1 &= \frac{1}{2}\int \frac{e^x}{1+e^{-x}}\,dx = \frac{1}{2}\int \frac{e^x}{e^x+1}e^x\,dx \\ &= \frac{1}{2}\int \frac{u-1}{u}\,du = \frac{1}{2}u - \frac{1}{2}\ln|u| \\ &= \frac{1}{2}(e^x+1) - \frac{1}{2}\ln(e^x+1) \end{aligned}$$

$$\begin{aligned} v_2 &= \int \frac{-1}{1+e^{-x}}\,dx = -\int \frac{e^x}{e^x+1}\,dx \\ &= -\int \frac{du}{u} = -\ln|u| = -\ln(e^x+1) \end{aligned}$$

$$v_3 = \frac{1}{2}\int \frac{e^{-x}}{1+e^{-x}}\,dx = -\frac{1}{2}\int \frac{dw}{w} = -\frac{1}{2}\ln|w| = -\frac{1}{2}\ln(1+e^{-x})$$

Substituting these values into (*1*), we obtain

$$y_p = \left[\frac{1}{2}(e^x+1) - \frac{1}{2}\ln(e^x+1)\right] + [-\ln(e^x+1)]e^x + \left[-\frac{1}{2}\ln(1+e^{-x})\right]e^{2x}$$

The general solution is

$$y = y_h + y_p = c_1 + c_2e^x + c_3e^{2x} + \frac{1}{2}(e^x+1) - \frac{1}{2}\ln(e^x+1) - e^x\ln(e^x+1) - \frac{1}{2}e^{2x}\ln(1+e^{-x})$$

This solution can be simplified. We first note that

$$\ln(1+e^{-x}) = \ln[e^{-x}(e^x+1)] = \ln e^{-x} + \ln(e^x+1) = -1 + \ln(e^x+1)$$

so
$$-\frac{1}{2}e^{2x}\ln(1+e^{-x}) = -\frac{1}{2}e^{2x}[-1+\ln(e^x+1)] = \frac{1}{2}e^{2x} - \frac{1}{2}e^{2x}\ln(e^x+1)$$

Then, combining like terms, we have

$$\begin{aligned} y &= \left(c_1 + \frac{1}{2}\right) + \left(c_2 + \frac{1}{2}\right)e^x + \left(c_3 + \frac{1}{2}\right)e^{2x} + \left[-\frac{1}{2} - e^x - \frac{1}{2}e^{2x}\right]\ln(e^x+1) \\ &= c_4 + c_5e^x + c_6e^{2x} - \frac{1}{2}[1 + 2e^x + (e^x)^2]\ln(e^x+1) \\ &= c_4 + c_5e^x + c_6e^{2x} - \frac{1}{2}(e^x+1)^2\ln(e^x+1) \quad \left(\text{with } c_4 = c_1 + \frac{1}{2},\quad c_5 = c_2 + \frac{1}{2},\quad c_6 = c_3 + \frac{1}{2}\right) \end{aligned}$$

11.3. Solve $y'' - 2y' + y = \dfrac{e^x}{x}$.

Here $n = 2$ and $y_h = c_1e^x + c_2xe^x$; hence,

$$y_p = v_1e^x + v_2xe^x \tag{1}$$

Since $y_1 = e^x$, $y_2 = xe^x$, and $\phi(x) = e^x/x$, it follows from Eq. (*11.6*) that

$$v_1'(e^x) + v_2'(xe^x) = 0$$

$$v_1'(e^x) + v_2'(e^x + xe^x) = \frac{e^x}{x}$$

Solving this set of equations simultaneously, we obtain $v_1' = -1$ and $v_2' = 1/x$. Thus,

$$v_1 = \int v_1'\,dx = \int -1\,dx = -x$$

$$v_2 = \int v_2'\,dx = \int \frac{1}{x}\,dx = \ln|x|$$

Substituting these values into (*1*), we obtain

$$y_p = -xe^x + xe^x \ln|x|$$

The general solution is therefore,

$$y = y_h + y_p = c_1e^x + c_2xe^x - xe^x + xe^x \ln|x|$$
$$= c_1e^x + c_3xe^x + xe^x \ln|x| \qquad (c_3 = c_2 - 1)$$

11.4. Solve $y'' - y' - 2y = e^{3x}$.

Here $n = 2$ and $y_h = c_1e^{-x} + c_2e^{2x}$; hence,

$$y_p = v_1e^{-x} + v_2e^{2x} \tag{1}$$

Since $y_1 = e^{-x}$, $y_2 = e^{2x}$, and $\phi(x) = e^{3x}$, it follows from Eq. (*11.6*) that

$$v_1'(e^{-x}) + v_2'(e^{2x}) = 0$$

$$v_1'(-e^{-x}) + v_2'(2e^{2x}) = e^{3x}$$

Solving this set of equations simultaneously, we obtain $v_1' = -e^{4x}/3$ and $v_2' = e^x/3$, from which $v_1 = -e^{4x}/12$ and $v_2 = e^x/3$. Substituting these results into (*1*), we obtain

$$y_p = -\frac{1}{12}e^{4x}e^{-x} + \frac{1}{3}e^xe^{2x} = -\frac{1}{12}e^{3x} + \frac{1}{3}e^{3x} = \frac{1}{4}e^{3x}$$

The general solution is, therefore,

$$y = c_1e^{-x} + c_2e^{2x} + \frac{1}{4}e^{3x}$$

(Compare with Problem 10.2.)

11.5. Solve $\ddot{x} + 4x = \sin^2 2t$.

This is a second-order equation for $x(t)$ with

$$x_h = c_1\cos 2t + c_2\sin 2t$$

It follows from Eq. (*11.3*) that

$$x_p = v_1\cos 2t + v_2\sin 2t \tag{1}$$

where v_1 and v_2 are now functions of t. Here $x_1 = \cos 2t$, $x_2 = \sin 2t$ are two linearly independent solutions of the associated homogeneous differential equation and $\phi(t) = \sin^2 2t$, so Eq. (11.6), with x replacing y, becomes

$$v_1' \cos 2t + v_2' \sin 2t = 0$$

$$v_1'(-2 \sin 2t) + v_2'(2 \cos 2t) = \sin^2 2t$$

The solution of this set of equations is

$$v_1' = -\frac{1}{2} \sin^3 2t$$

$$v_2' = \frac{1}{2} \sin^2 2t \cos 2t$$

Thus,
$$v_1 = -\frac{1}{2} \int \sin^3 2t \, dt = \frac{1}{4} \cos 2t - \frac{1}{12} \cos^3 2t$$

$$v_2 = \frac{1}{2} \int \sin^2 2t \cos 2t \, dt = \frac{1}{12} \sin^3 2t$$

Substituting these values into (1), we obtain

$$x_p = \left[\frac{1}{4} \cos 2t - \frac{1}{12} \cos^3 2t\right] \cos 2t + \left[\frac{1}{12} \sin^3 2t\right] \sin 2t$$

$$= \frac{1}{4} \cos^2 2t - \frac{1}{12} (\cos^4 2t - \sin^4 2t)$$

$$= \frac{1}{4} \cos^2 2t - \frac{1}{12} (\cos^2 2t - \sin^2 2t)(\cos^2 2t + \sin^2 2t)$$

$$= \frac{1}{6} \cos^2 2t + \frac{1}{12} \sin^2 2t$$

because $\cos^2 2t + \sin^2 2t = 1$. The general solution is

$$x = x_h + x_p = c_1 \cos 2t + c_2 \sin 2t + \frac{1}{6} \cos^2 2t + \frac{1}{12} \sin^2 2t$$

11.6. Solve $t^2 \dfrac{d^2N}{dt^2} - 2t \dfrac{dN}{dt} + 2N = t \ln t$ if it is known that two linearly independent solutions of the associated homogeneous differential equation are t and t^2.

We first write the differential equation in standard form, with unity as the coefficient of the highest derivative. Dividing the equation by t^2, we obtain

$$\frac{d^2N}{dt^2} - \frac{2}{t}\frac{dN}{dt} + \frac{2}{t^2} N = \frac{1}{t} \ln t$$

with $\phi(t) = (1/t) \ln t$. We are given $N_1 = t$ and $N_2 = t^2$ as two linearly independent solutions of the associated second-order homogeneous equation. It follows from Theorem 7.2 that

$$N_h = c_1 t + c_2 t^2$$

We assume, therefore, that

$$N_p = v_1 t + v_2 t^2 \tag{1}$$

Equations (*11.6*), with N replacing y, become

$$v_1'(t) + v_2'(t^2) = 0$$

$$v_1'(1) + v_2'(2t) = \frac{1}{t}\ln t$$

The solution of this set of equations is

$$v_1' = -\frac{1}{t}\ln t \qquad \text{and} \qquad v_2' = \frac{1}{t^2}\ln t$$

Thus,

$$v_1 = -\int \frac{1}{t}\ln t\,dt = -\frac{1}{2}\ln^2 t$$

$$v_2 = \int \frac{1}{t^2}\ln t\,dt = -\frac{1}{t}\ln t - \frac{1}{t}$$

and (*1*) becomes

$$N_p = \left[-\frac{1}{2}\ln^2 t\right]t + \left[-\frac{1}{t}\ln t - \frac{1}{t}\right]t^2 = -\frac{t}{2}\ln^2 t - t\ln t - t$$

The general solution is

$$N = N_h + N_p = c_1 t + c_2 t^2 - \frac{t}{2}\ln^2 t - t\ln t - t$$

$$= c_3 t + c_2 t^2 - \frac{t}{2}\ln^2 t - t\ln t \qquad (\text{with } c_3 = c_1 - 1)$$

11.7. Solve $y' + \dfrac{4}{x}y = x^4$.

Here $n = 1$ and (from Chapter 5) $y_h = c_1 x^{-4}$; hence,

$$y_p = v_1 x^{-4} \tag{1}$$

Since $y_1 = x^{-4}$ and $\phi(x) = x^4$, Eq. (*11.7*) becomes $v_1' x^{-4} = x^4$, from which we obtain $v_1' = x^8$ and $v_1 = x^9/9$. Equation (*1*) now becomes $y_p = x^5/9$, and the general solution is therefore

$$y = c_1 x^{-4} + \frac{1}{9}x^5$$

(Compare with Problem 5.6.)

11.8. Solve $y^{(4)} = 5x$ by variation of parameters.

Here $n = 4$ and $y_h = c_1 + c_2 x + c_3 x^2 + c_4 x^3$; hence,

$$y_p = v_1 + v_2 x + v_3 x^2 + v_4 x^3 \tag{1}$$

Since $y_1 = 1$, $y_2 = x$, $y_3 = x^2$, $y_4 = x^3$, and $\phi(x) = 5x$, it follows from Eq. (*11.4*), with $n = 4$, that

$$v_1'(1) + v_2'(x) + v_3'(x^2) + v_4'(x^3) = 0$$

$$v_1'(0) + v_2'(1) + v_3'(2x) + v_4'(3x^2) = 0$$

$$v_1'(0) + v_2'(0) + v_3'(2) + v_4'(6x) = 0$$

$$v_1'(0) + v_2'(0) + v_3'(0) + v_4'(6) = 5x$$

Solving this set of equations simultaneously, we obtain

$$v_1' = -\frac{5}{6}x^4 \qquad v_2' = \frac{5}{2}x^3 \qquad v_3' = -\frac{5}{2}x^2 \qquad v_4' = \frac{5}{6}x$$

whence

$$v_1 = -\frac{1}{6}x^5 \qquad v_2 = \frac{5}{8}x^4 \qquad v_3 = -\frac{5}{6}x^3 \qquad v_4 = \frac{5}{12}x^2$$

Then, from (*1*),

$$y_p = -\frac{1}{6}x^5 + \frac{5}{8}x^4(x) - \frac{5}{6}x^3(x^2) + \frac{5}{12}x^2(x^3) = \frac{1}{24}x^5$$

and the general solution is

$$y_h = c_1 + c_2 x + c_3 x^2 + c_4 x^3 + \frac{1}{24}x^5$$

The solution also can be obtained simply by integrating both sides of the differential equation four times with respect to x.

Supplementary Problems

Use variation of parameters to find the general solutions of the following differential equations:

11.9. $y'' - 2y' + y = \dfrac{e^x}{x^5}$

11.10. $y'' + y = \sec x$

11.11. $y'' - y' - 2y = e^{3x}$

11.12. $y'' - 60y' + 900y = 5e^{10x}$

11.13. $y'' - 7y' = -3$

11.14. $y'' + \dfrac{1}{x}y' - \dfrac{1}{x^2}y = \ln x$ if two solutions to the associated homogeneous problem are known to be x and $1/x$.

11.15. $x^2y'' - xy' = x^3e^x$ if two solutions to the associated homogeneous problem are known to be 1 and x^2.

11.16. $y' - \dfrac{1}{x}y = x^2$

11.17. $y' + 2xy = x$

11.18. $y''' = 12$

11.19. $\ddot{x} - 2\dot{x} + x = \dfrac{e^t}{t^3}$

11.20. $\ddot{x} - 6\dot{x} + 9x = \dfrac{e^{3t}}{t^2}$

11.21. $\ddot{x} + 4x = 4\sec^2 2t$

11.22. $\ddot{x} - 4\dot{x} + 3x = \dfrac{e^t}{1 + e^t}$

11.23. $(t^2 - 1)\ddot{x} - 2t\dot{x} + 2x = (t^2 - 1)^2$ if two solutions to the associated homogeneous equation are known to be t and $t^2 + 1$.

11.24. $(t^2 + t)\ddot{x} + (2 - t^2)\dot{x} - (2 + t)x = t(t + 1)^2$ if two solutions to the associated homogeneous equation are known to be e^t and $1/t$.

11.25 $\dddot{r} - 3\ddot{r} + 3\dot{r} - r = \dfrac{e^t}{t}$

11.26. $\dddot{r} + 6\ddot{r} + 12\dot{r} + 8r = 12e^{-2t}$

11.27. $\dddot{z} - 5\ddot{z} + 25\dot{z} - 125z = 1000$

11.28. $\dfrac{d^3z}{d\theta^3} - 3\dfrac{d^2z}{d\theta^2} + 2\dfrac{dz}{d\theta} = \dfrac{e^{3\theta}}{1 + e^\theta}$

11.29. $t^3\dddot{y} + 3t^2\ddot{y} = 1$ if three linearly independent solutions to the associated homogeneous equation are known to be $1/t$, 1, and t.

11.30. $y^{(5)} - 4y^{(3)} = 32e^{2x}$

Chapter 12

Initial-Value Problems

Initial-value problems are solved by applying the initial conditions to the general solution of the differential equation. It must be emphasized that the initial conditions are applied *only* to the general solution and *not* to the homogeneous solution y_h, even though it is y_h that possesses all the arbitrary constants that must be evaluated. The one exception is when the general solution is the homogeneous solution; that is, when the differential equation under consideration is itself homogeneous.

Solved Problems

12.1. Solve $y'' - y' - 2y = 4x^2$; $y(0) = 1$, $y'(0) = 4$.

The general solution of the differential equation is given in Problem 10.1 as

$$y = c_1e^{-x} + c_2e^{2x} - 2x^2 + 2x - 3 \tag{1}$$

Therefore,

$$y' = -c_1e^{-x} + 2c_2e^{2x} - 4x + 2 \tag{2}$$

Applying the first initial condition to (*1*), we obtain

$$y(0) = c_1e^{-(0)} + c_2e^{2(0)} - 2(0)^2 + 2(0) - 3 = 1 \qquad \text{or} \qquad c_1 + c_2 = 4 \tag{3}$$

Applying the second initial condition to (*2*), we obtain

$$y'(0) = -c_1e^{-(0)} + 2c_2e^{2(0)} - 4(0) + 2 = 4 \qquad \text{or} \qquad -c_1 + 2c_2 = 2 \tag{4}$$

Solving (*3*) and (*4*) simultaneously, we find that $c_1 = 2$ and $c_2 = 2$. Substituting these values into (*1*), we obtain the solution of the initial-value problem as

$$y = 2e^{-x} + 2e^{2x} - 2x^2 + 2x - 3$$

12.2. Solve $y'' - 2y' + y = \dfrac{e^x}{x}$; $y(1) = 0$, $y'(1) = 1$.

The general solution of the differential equation is given in Problem 11.3 as

$$y = c_1e^x + c_3xe^x + xe^x \ln|x| \tag{1}$$

Therefore,

$$y' = c_1e^x + c_3e^x + c_3xe^x + e^x \ln|x| + xe^x \ln|x| + e^x \tag{2}$$

Applying the first initial condition to (*1*), we obtain

$$y(1) = c_1e^1 + c_3(1)e^1 + (1)e^1 \ln 1 = 0$$

or (noting that $\ln 1 = 0$),

$$c_1e + c_3e = 0 \tag{3}$$

Applying the second initial condition to (2), we obtain

$$y'(1) = c_1e^1 + c_3e^1 + c_3(1)e^1 + e^1 \ln 1 + (1)e^1 \ln 1 + e^1 = 1$$

or

$$c_1e + 2c_3e = 1 - e \tag{4}$$

Solving (3) and (4) simultaneously, we find that $c_1 = -c_3 = (e-1)/e$. Substituting these values into (1), we obtain the solution of the initial-value problem as

$$y = e^{x-1}(e-1)(1-x) + xe^x \ln|x|$$

12.3. Solve $y'' + 4y' + 8y = \sin x$; $y(0) = 1$, $y'(0) = 0$.

Here $y_h = e^{-2x}(c_1 \cos 2x + c_2 \sin 2x)$, and, by the method of undetermined coefficients,

$$y_p = \frac{7}{65}\sin x - \frac{4}{65}\cos x$$

Thus, the general solution to the differential equation is

$$y = e^{-2x}(c_1 \cos 2x + c_2 \sin 2x) + \frac{7}{65}\sin x - \frac{4}{65}\cos x \tag{1}$$

Therefore,

$$y' = -2e^{-2x}(c_1 \cos 2x + c_2 \sin 2x) + e^{-2x}(-2c_1 \sin 2x + 2c_2 \cos 2x) + \frac{7}{65}\cos x + \frac{4}{65}\sin x \tag{2}$$

Applying the first initial condition to (1), we obtain

$$c_1 = \frac{69}{65} \tag{3}$$

Applying the second initial condition to (2), we obtain

$$-2c_1 + 2c_2 = -\frac{7}{65} \tag{4}$$

Solving (3) and (4) simultaneously, we find that $c_1 = 69/65$ and $c_2 = 131/130$. Substituting these values into (1), we obtain the solution of the initial-value problem as

$$y = e^{-2x}\left(\frac{69}{65}\cos 2x + \frac{131}{130}\sin 2x\right) + \frac{7}{65}\sin x - \frac{4}{65}\cos x$$

12.4. Solve $y''' - 6y'' + 11y' - 6y = 0$; $y(\pi) = 0$, $y'(\pi) = 0$, $y''(\pi) = 1$.

From Problem 9.1, we have

$$y_h = c_1e^x + c_2e^{2x} + c_3e^{3x} \tag{1}$$

$$y_h' = c_1e^x + 2c_2e^{2x} + 3c_3e^{3x}$$

$$y_h'' = c_1e^x + 4c_2e^{2x} + 9c_3e^{3x}$$

Since the given differential equation is homogeneous, y_h is also the general solution. Applying each initial condition separately, we obtain

$$y(\pi) = c_1 e^{\pi} + c_2 e^{2\pi} + c_3 e^{3\pi} = 0$$

$$y'(\pi) = c_1 e^{\pi} + 2c_2 e^{2\pi} + 3c_3 e^{3\pi} = 0$$

$$y''(\pi) = c_1 e^{\pi} + 4c_2 e^{2\pi} + 9c_3 e^{3\pi} = 1$$

Solving these equations simultaneously, we find

$$c_1 = \frac{1}{2}e^{-\pi} \qquad c_2 = -e^{-2\pi} \qquad c_3 = \frac{1}{2}e^{-3\pi}$$

Substituting these values into the first equation (*1*), we obtain

$$y = \frac{1}{2}e^{(x-\pi)} - e^{2(x-\pi)} + \frac{1}{2}e^{3(x-\pi)}$$

12.5. Solve $\ddot{x} + 4x = \sin^2 2t$; $x(0) = 0$, $\dot{x}(0) = 0$.

The general solution of the differential equation is given in Problem 11.5 as

$$x = c_1 \cos 2t + c_2 \sin 2t + \frac{1}{6}\cos^2 2t + \frac{1}{12}\sin^2 2t \tag{1}$$

Therefore,

$$\dot{x} = -2c_1 \sin 2t + 2c_2 \cos 2t - \frac{1}{3}\cos 2t \sin 2t \tag{2}$$

Applying the first initial condition to (*1*), we obtain

$$x(0) = c_1 + \frac{1}{6} = 0$$

Hence $c_1 = -1/6$. Applying the second initial condition to (*2*), we obtain

$$\dot{x}(0) = 2c_2 = 0$$

Hence $c_2 = 0$. The solution to the initial-value problem is

$$x = -\frac{1}{6}\cos 2t + \frac{1}{6}\cos^2 2t + \frac{1}{12}\sin^2 2t$$

12.6. Solve $\ddot{x} + 4x = \sin^2 2t$; $x(\pi/8) = 0$, $\dot{x}(\pi/8) = 0$.

The general solution of the differential equation and the derivative of the solution are as given in (*1*) and (*2*) of Problem 12.5. Applying the first initial condition, we obtain

$$0 = x\left(\frac{\pi}{8}\right) = c_1 \cos\frac{\pi}{4} + c_2 \sin\frac{\pi}{4} + \frac{1}{6}\cos^2\frac{\pi}{4} + \frac{1}{12}\sin^2\frac{\pi}{4}$$

$$= c_1\frac{\sqrt{2}}{2} + c_2\frac{\sqrt{2}}{2} + \frac{1}{6}\left(\frac{1}{2}\right) + \frac{1}{12}\left(\frac{1}{2}\right)$$

or

$$c_1 + c_2 = -\frac{\sqrt{2}}{8} \tag{1}$$

Applying the second initial condition, we obtain

$$0 = \dot{x}\left(\frac{\pi}{8}\right) = -2c_1 \sin\frac{\pi}{4} + 2c_2 \cos\frac{\pi}{4} - \frac{1}{3}\cos\frac{\pi}{4}\sin\frac{\pi}{4}$$

$$= -2c_1\frac{\sqrt{2}}{2} + 2c_2\frac{\sqrt{2}}{2} - \frac{1}{3}\left(\frac{\sqrt{2}}{2}\right)\left(\frac{\sqrt{2}}{2}\right)$$

or

$$-c_1 + c_2 = \frac{\sqrt{2}}{12} \tag{2}$$

Solving (*1*) and (*2*) simultaneously, we find that

$$c_1 = -\frac{5}{48}\sqrt{2} \quad \text{and} \quad c_2 = -\frac{1}{48}\sqrt{2}$$

whereupon, the solution to the initial-value problem becomes

$$x = -\frac{5}{48}\sqrt{2}\cos 2t - \frac{1}{48}\sqrt{2}\sin 2t + \frac{1}{6}\cos^2 2t + \frac{1}{12}\sin^2 2t$$

Supplementary Problems

Solve the following initial-value problems.

12.7. $y'' - y' - 2y = e^{3x}$; $y(0) = 1$, $y'(0) = 2$

12.8. $y'' - y' - 2y = e^{3x}$; $y(0) = 2$, $y'(0) = 1$

12.9. $y'' - y' - 2y = 0$; $y(0) = 2$, $y'(0) = 1$

12.10. $y'' - y' - 2y = e^{3x}$; $y(1) = 2$, $y'(1) = 1$

12.11. $y'' + y = x$; $y(1) = 0$, $y'(1) = 1$

12.12. $y'' + 4y = \sin^2 2x$; $y(\pi) = 0$, $y'(\pi) = 0$

12.13. $y'' + y = 0$; $y(2) = 0$, $y'(2) = 0$

12.14. $y''' = 12$; $y(1) = 0$, $y'(1) = 0$, $y''(1) = 0$

12.15. $\ddot{y} + 2\dot{y} + 2y = \sin 2t + \cos 2t$; $y(0) = 0$, $\dot{y}(0) = 1$

Chapter 13

Applications of Second-Order Linear Differential Equations

SPRING PROBLEMS

The simple spring system shown in Fig. 13-1 consists of a mass m attached to the lower end of a spring that is itself suspended vertically from a mounting. The system is in its *equilibrium position* when it is at rest. The mass is set in motion by one or more of the following means: displacing the mass from its equilibrium position, providing it with an initial velocity, or subjecting it to an external force $F(t)$.

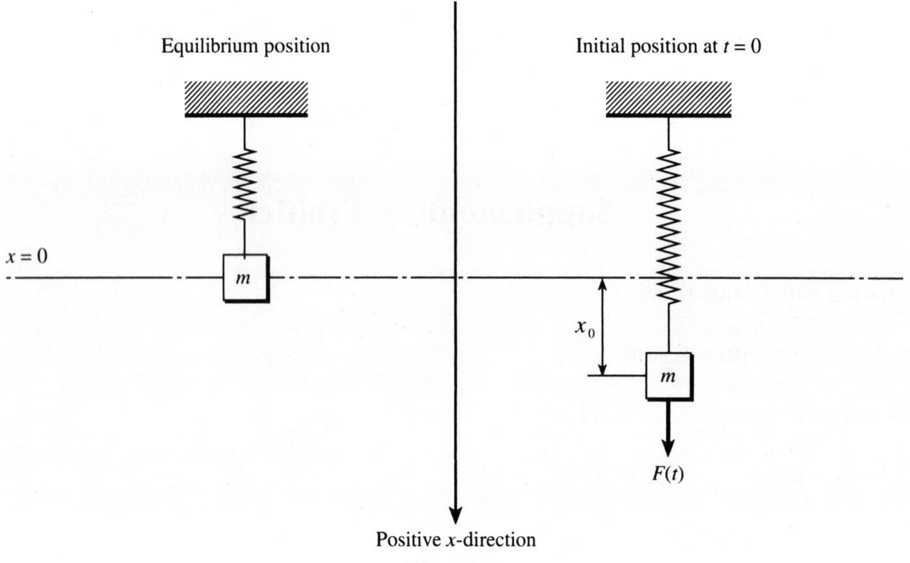

Fig. 13-1

Hooke's law: *The restoring force F of a spring is equal and opposite to the forces applied to the spring and is proportional to the extension (contraction) l of the spring as a result of the applied force; that is, $F = -kl$, where k denotes the constant of proportionality, generally called the spring constant.*

Example 13.1. A steel ball weighing 128 lb is suspended from a spring, whereupon the spring is stretched 2 ft from its natural length. The applied force responsible for the 2-ft displacement is the weight of the ball, 128 lb. Thus, $F = -128$ lb. Hooke's law then gives $-128 = -k(2)$, or $k = 64$ lb/ft.

For convenience, we choose the downward direction as the positive direction and take the origin to be the center of gravity of the mass in the equilibrium position. We assume that the mass of the spring is negligible and can be neglected and that air resistance, when present, is proportional to the velocity of the mass. Thus, at any time t, there are three forces acting on the system: (1) $F(t)$, measured in the positive direction; (2) a restoring force given by Hooke's law as $F_s = -kx$, $k > 0$; and (3) a force due to air resistance given by $F_a = -a\dot{x}$, $a > 0$, where a is the constant of proportionality. Note that the restoring force F_s always acts in a direction that will tend to return the system to the equilibrium position: if the mass is below the equilibrium position, then x is positive and $-kx$ is negative; whereas if the mass is above the equilibrium position, then x is negative and $-kx$ is positive. Also note that because $a > 0$ the force F_a due to air resistance acts in the opposite direction of the velocity and thus tends to retard, or damp, the motion of the mass.

It now follows from Newton's second law (see Chapter 6) that $m\ddot{x} = -kx - a\dot{x} + F(t)$, or

$$\ddot{x} + \frac{a}{m}\dot{x} + \frac{k}{m}x = \frac{F(t)}{m} \tag{13.1}$$

If the system starts at $t = 0$ with an initial velocity v_0 and from an initial position x_0, we also have the initial conditions

$$x(0) = x_0 \qquad \dot{x}(0) = v_0 \tag{13.2}$$

(See Problems 13.1–13.10.)

The force of gravity does not explicitly appear in *(13.1)*, but it is present nonetheless. We automatically compensated for this force by measuring distance from the equilibrium position of the spring. If one wishes to exhibit gravity explicitly, then distance must be measured from the bottom end of the *natural length* of the spring. That is, the motion of a vibrating spring can be given by

$$\ddot{x} + \frac{a}{m}\dot{x} + \frac{k}{m}x = g + \frac{F(t)}{m}$$

if the origin, $x = 0$, is the terminal point of the unstretched spring before the mass m is attached.

ELECTRICAL CIRCUIT PROBLEMS

The simple electrical circuit shown in Fig. 13-2 consists of a resistor R in ohms; a capacitor C in farads; an inductor L in henries; and an electromotive force (emf) $E(t)$ in volts, usually a battery or a generator, all connected in series. The current I flowing through the circuit is measured in amperes and the charge q on the capacitor is measured in coulombs.

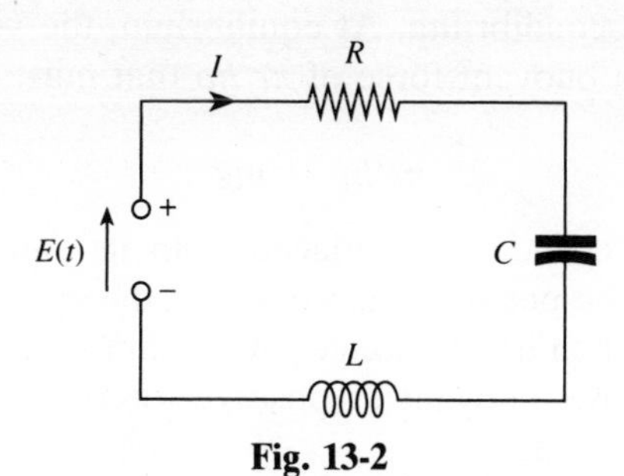

Fig. 13-2

Kirchhoff's loop law: *The algebraic sum of the voltage drops in a simple closed electric circuit is zero.*

It is known that the voltage drops across a resistor, a capacitor, and an inductor are respectively RI, $(1/C)q$, and $L(dI/dt)$ where q is the charge on the capacitor. The voltage drop across an emf is $-E(t)$. Thus, from Kirchhoff's loop law, we have

$$RI + L\frac{dI}{dt} + \frac{1}{C}q - E(t) = 0 \tag{13.3}$$

The relationship between q and I is

$$I = \frac{dq}{dt} \qquad \frac{dI}{dt} = \frac{d^2q}{dt^2} \tag{13.4}$$

Substituting these values into *(13.3)*, we obtain

$$\frac{d^2q}{dt^2} + \frac{R}{L}\frac{dq}{dt} + \frac{1}{LC}q = \frac{1}{L}E(t) \tag{13.5}$$

The initial conditions for q are

$$q(0) = q_0 \qquad \left.\frac{dq}{dt}\right|_{t=0} = I(0) = I_0 \tag{13.6}$$

To obtain a differential equation for the current, we differentiate Eq. *(13.3)* with respect to t

and then substitute Eq. (*13.4*) directly into the resulting equation to obtain

$$\frac{d^2I}{dt^2} + \frac{R}{L}\frac{dI}{dt} + \frac{1}{LC}I = \frac{1}{L}\frac{dE(t)}{dt} \tag{13.7}$$

The first initial condition is $I(0) = I_0$. The second initial condition is obtained from Eq. (*13.3*) by solving for dI/dt and then setting $t = 0$. Thus,

$$\left.\frac{dI}{dt}\right|_{t=0} = \frac{1}{L}E(0) - \frac{R}{L}I_0 - \frac{1}{LC}q_0 \tag{13.8}$$

An expression for the current can be gotten either by solving Eq. (*13.7*) directly or by solving Eq. (*13.5*) for the charge and then differentiating that expression. (See Problems 13.12–13.16.)

BUOYANCY PROBLEMS

Consider a body of mass m submerged either partially or totally in a liquid of weight density ρ. Such a body experiences two forces, a downward force due to gravity and a counter force governed by:

Archimedes' principle: *A body in liquid experiences a buoyant upward force equal to the weight of the liquid displaced by that body.*

Equilibrium occurs when the buoyant force of the displaced liquid equals the force of gravity on the body. Figure 13-3 depicts the situation for a cylinder of radius r and height H where h units of cylinder height are submerged at equilibrium. At equilibrium, the volume of water displaced by the cylinder is $\pi r^2 h$, which provides a buoyant force of $\pi r^2 h\rho$ that must equal the weight of the cylinder mg. Thus,

$$\pi r^2 h\rho = mg \tag{13.9}$$

Motion will occur when the cylinder is displaced from its equilibrium position. We arbitrarily take the upward direction to be the positive x-direction. If the cylinder is raised out of the water by $x(t)$ units, as shown in Fig. 13-3, then it is no longer in equilibrium. The downward or negative force on such a body remains mg but the buoyant or positive force is reduced to $\pi r^2[h - x(t)]\rho$. It now follows from Newton's second law that

$$m\ddot{x} = \pi r^2[h - x(t)]\rho - mg$$

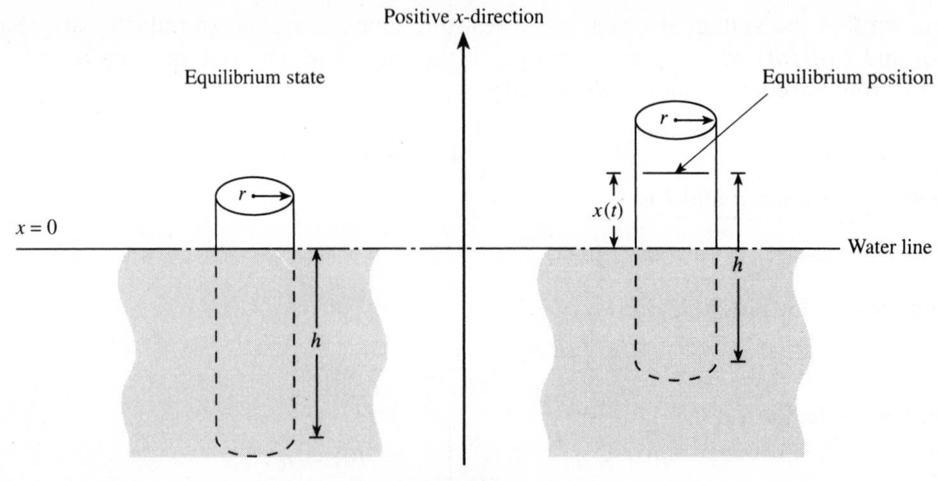

Fig. 13-3

Substituting (*13.9*) into this last equation, we can simplify it to

$$m\ddot{x} = -\pi r^2 x(t)\rho$$

or

$$\ddot{x} + \frac{\pi r^2 \rho}{m}x = 0 \tag{13.10}$$

(See Problems 13.19–13.24.)

CLASSIFYING SOLUTIONS

Vibrating springs, simple electrical circuits, and floating bodies are all governed by second-order linear differential equations with constant coefficients of the form

$$\ddot{x} + a_1\dot{x} + a_0 x = f(t) \tag{13.11}$$

For vibrating spring problems defined by Eq. (*13.1*), $a_1 = a/m$, $a_0 = k/m$, and $f(t) = F(t)/m$. For buoyancy problems defined by Eq. (*13.10*), $a_1 = 0$, $a_0 = \pi r^2 \rho/m$, and $f(t) \equiv 0$. For electrical circuit problems, the independent variable x is replaced either by q in Eq. (*13.5*) or I in Eq. (*13.7*).

The motion or current in all of these systems is classified as *free* and *undamped* when $f(t) \equiv 0$ and $a_1 = 0$. It is classified as *free* and *damped* when $f(t)$ is identically zero but a_1 is not zero. For damped motion, there are three separate cases to consider, according as the roots of the associated characteristic equation (see Chapter 8) are (1) real and distinct, (2) equal, or (3) complex conjugate. These cases are respectively classified as (1) *overdamped,* (2) *critically damped,* and (3) *oscillatory damped* (or, in electrical problems, *underdamped*). If $f(t)$ is not identically zero, the motion or current is classified as *forced.*

A motion or current is *transient* if it "dies out" (that is, goes to zero) as $t \to \infty$. A *steady-state* motion or current is one that is not transient and does not become unbounded. Free damped systems always yield transient motions, while forced damped systems (assuming the external force to be sinusoidal) yield both transient and steady-state motions.

Free undamped motion defined by Eq. (*13.11*) with $a_1 = 0$ and $f(t) \equiv 0$ always has solutions of the form

$$x(t) = c_1 \cos \omega t + c_2 \sin \omega t \tag{13.12}$$

which defines *simple harmonic motion.* Here c_1, c_2, and ω are constants with ω often referred to as *circular frequency.* The *natural frequency* f is

$$f = \frac{\omega}{2\pi}$$

and it represents the number of complete oscillations per time unit undertaken by the solution. The *period* of the system of the time required to complete one oscillation is

$$T = \frac{1}{f}$$

Equation (*13.12*) has the alternate form

$$x(t) = (-1)^k A \cos(\omega t - \phi) \tag{13.13}$$

where the *amplitude* $A = \sqrt{c_1^2 + c_2^2}$, the *phase angle* $\phi = \arctan(c_2/c_1)$, and k is zero when c_1 is positive and unity when c_1 is negative.

Solved Problems

13.1. A steel ball weighing 128 lb is suspended from a spring, whereupon the spring is stretched 2 ft from its natural length. The ball is started in motion with no initial velocity by displacing it

6 in above the equilibrium position. Assuming no air resistance, find (*a*) an expression for the position of the ball at any time *t*, and (*b*) the position of the ball at $t = \pi/12$ sec.

(*a*) The equation of motion is governed by Eq. (*13.1*). There is no externally applied force, so $F(t) = 0$, and no resistance from the surrounding medium, so $a = 0$. The motion is free and undamped. Here $g = 32$ ft/sec^2, $m = 128/32 = 4$ slugs, and it follows from Example 13.1 that $k = 64$ lb/ft. Equation (*13.1*) becomes $\ddot{x} + 16x = 0$. The roots of its characteristic equation are $\lambda = \pm 4i$, so its solution is

$$x(t) = c_1 \cos 4t + c_2 \sin 4t \tag{1}$$

At $t = 0$, the position of the ball is $x_0 = -\frac{1}{2}$ ft (the minus sign is required because the ball is initially displaced *above* the equilibrium position, which is in the *negative* direction). Applying this initial condition to (*1*), we find that

$$-\frac{1}{2} = x(0) = c_1 \cos 0 + c_2 \sin 0 = c_1$$

so (*1*) becomes

$$x(t) = -\frac{1}{2}\cos 4t + c_2 \sin 4t \tag{2}$$

The initial velocity is given as $v_0 = 0$ ft/sec. Differentiating (*2*), we obtain

$$v(t) = \dot{x}(t) = 2\sin 4t + 4c_2 \cos 4t$$

whereupon

$$0 = v(0) = 2\sin 0 + 4c_2 \cos 0 = 4c_2$$

Thus, $c_2 = 0$, and (*2*) simplifies to

$$x(t) = -\frac{1}{2}\cos 4t \tag{3}$$

as the equation of motion of the steel ball at any time *t*.

(*b*) At $t = \pi/12$,

$$x\left(\frac{\pi}{12}\right) = -\frac{1}{2}\cos\frac{4\pi}{12} = -\frac{1}{4}\text{ ft}$$

13.2. A mass of 2 kg is suspended from a spring with a known spring constant of 10 N/m and allowed to come to rest. It is then set in motion by giving it an initial velocity of 150 cm/sec. Find an expression for the motion of the mass, assuming no air resistance.

The equation of motion is governed by Eq. (*13.1*) and represents free undamped motion because there is no externally applied force on the mass, $F(t) = 0$, and no resistance from the surrounding medium, $a = 0$. The mass and the spring constant are given as $m = 2$ kg and $k = 10$ N/m, respectively, so Eq. (*13.1*) becomes $\ddot{x} + 5x = 0$. The roots of its characteristic equation are purely imaginary, so its solution is

$$x(t) = c_1 \cos\sqrt{5}\,t + c_2 \sin\sqrt{5}\,t \tag{1}$$

At $t = 0$, the position of the ball is at the equilibrium position $x_0 = 0$ m. Applying this initial condition to (*1*), we find that

$$0 = x(0) = c_1 \cos 0 + c_2 \sin 0 = c_1$$

whereupon (*1*) becomes

$$x(t) = c_2 \sin\sqrt{5}\,t \tag{2}$$

The initial velocity is given as $v_0 = 150$ cm/sec $= 1.5$ m/sec. Differentiating (*2*), we obtain

$$v(t) = \dot{x}(t) = \sqrt{5}\,c_2 \cos\sqrt{5}\,t$$

whereupon,

$$1.5 = v(0) = \sqrt{5}\,c_2 \cos 0 = \sqrt{5}\,c_2 \qquad c_2 = \frac{1.5}{\sqrt{5}} = 0.6708$$

and (*2*) simplifies to

$$x(t) = 0.6708 \sin\sqrt{5}\,t \tag{3}$$

as the position of the mass at any time *t*.

13.3. Determine the circular frequency, natural frequency, and period for the simple harmonic motion described in Problem 13.2.

Circular frequency: $\omega = \sqrt{5} = 2.236$ cycles/sec $= 2.236$ Hz

Natural frequency: $f = \omega/2\pi = \dfrac{\sqrt{5}}{2\pi} = 0.3559$ Hz

Period: $T = 1/f = \dfrac{2\pi}{\sqrt{5}} = 2.81$ sec

13.4. Determine the circular frequency, natural frequency, and period for the simple harmonic motion described in Problem 13.1.

Circular frequency: $\omega = 4$ cycles/sec $= 4$ Hz

Natural frequency: $f = 4/2\pi = 0.6366$ Hz

Period: $T = 1/f = \pi/2 = 1.57$ sec

13.5. A 10-kg mass is attached to a spring, stretching it 0.7 m from its natural length. The mass is started in motion from the equilibrium position with an initial velocity of 1 m/sec in the upward direction. Find the subsequent motion, if the force due to air resistance is $-90\dot{x}$ N.

Taking $g = 9.8$ m/sec^2, we have $w = mg = 98$ N and $k = w/l = 140$ N/m. Furthermore, $a = 90$ and $F(t) \equiv 0$ (there is no external force). Equation (*13.1*) becomes

$$\ddot{x} + 9\dot{x} + 14x = 0 \tag{1}$$

The roots of the associated characteristic equation are $\lambda_1 = -2$ and $\lambda_2 = -7$, which are real and distinct; hence this problem is an example of overdamped motion. The solution of (*1*) is

$$x = c_1 e^{-2t} + c_2 e^{-7t}$$

The initial conditions are $x(0) = 0$ (the mass starts at the equilibrium position) and $\dot{x}(0) = -1$ (the initial velocity is in the negative direction). Applying these conditions, we find that $c_1 = -c_2 = -\frac{1}{5}$, so that $x = \frac{1}{5}(e^{-7t} - e^{-2t})$. Note that $x \to 0$ as $t \to \infty$; thus, the motion is transient.

13.6. A mass of 1/4 slug is attached to a spring, whereupon the spring is stretched 1.28 ft from its natural length. The mass is started in motion from the equilibrium position with an initial velocity of 4 ft/sec in the downward direction. Find the subsequent motion of the mass if the force due to air resistance is $-2\dot{x}$ lb.

Here $m = 1/4$, $a = 2$, $F(t) \equiv 0$ (there is no external force), and, from Hooke's law, $k = mg/l = (1/4)(32)/1.28 = 6.25$. Equation (*13.1*) becomes

$$\ddot{x} + 8\dot{x} + 25x = 0 \tag{1}$$

The roots of the associated characteristic equation are $\lambda_1 = -4 + i3$ and $\lambda_2 = -4 - i3$, which are complex conjugates; hence this problem is an example of oscillatory damped motion. The solution of (*1*) is

$$x = e^{-4t}(c_1 \cos 3t + c_2 \sin 3t)$$

The initial conditions are $x(0) = 0$ and $\dot{x}(0) = 4$. Applying these conditions, we find that $c_1 = 0$ and $c_2 = \frac{4}{3}$; thus, $x = \frac{4}{3}e^{-4t} \sin 3t$. Since $x \to 0$ as $t \to \infty$, the motion is transient.

13.7. A mass of 1/4 slug is attached to a spring having a spring constant of 1 lb/ft. The mass is started in motion by initially displacing it 2 ft in the downward direction and giving it an initial velocity of 2 ft/sec in the upward direction. Find the subsequent motion of the mass, if the force due to air resistance is $-1\dot{x}$ lb.

Here $m = 1/4$, $a = 1$, $k = 1$, and $F(t) \equiv 0$. Equation (*13.1*) becomes

$$\ddot{x} + 4\dot{x} + 4x = 0 \tag{1}$$

The roots of the associated characteristic equation are $\lambda_1 = \lambda_2 = -2$, which are equal; hence this problem is an example of critically damped motion. The solution of (*1*) is

$$x = c_1 e^{-2t} + c_2 t e^{-2t}$$

The initial conditions are $x(0) = 2$ and $\dot{x}(0) = -2$ (the initial velocity is in the negative direction). Applying these conditions, we find that $c_1 = c_2 = 2$. Thus,

$$x = 2e^{-2t} + 2te^{-2t}$$

Since $x \to 0$ as $t \to \infty$, the motion is transient.

13.8. Show that the types of motions that result from free damped problems are completely determined by the quantity $a^2 - 4km$.

For free damped motions $F(t) \equiv 0$ and Eq. (*13.1*) becomes

$$\ddot{x} + \frac{a}{m}\dot{x} + \frac{k}{m}x = 0$$

The roots of the associated characteristic equation are

$$\lambda_1 = \frac{-a + \sqrt{a^2 - 4km}}{2m} \qquad \lambda_2 = \frac{-a - \sqrt{a^2 - 4km}}{2m}$$

If $a^2 - 4km > 0$, the roots are real and distinct; if $a^2 - 4km = 0$, the roots are equal; if $a^2 - 4km < 0$, the roots are complex conjugates. The corresponding motions are, respectively, overdamped, critically damped, and oscillatory damped. Since the real parts of both roots are always negative, the resulting motion in all three cases is transient. (For overdamped motion, we need only note that $\sqrt{a^2 - 4km} < a$, whereas for the other two cases the real parts are both $-a/2m$.)

13.9. A 10-kg mass is attached to a spring having a spring constant of 140 N/m. The mass is started in motion from the equilibrium position with an initial velocity of 1 m/sec in the upward direction and with an applied external force $F(t) = 5 \sin t$. Find the subsequent motion of the mass if the force due to air resistance is $-90\dot{x}$ N.

Here $m = 10$, $k = 140$, $a = 90$, and $F(t) = 5 \sin t$. The equation of motion, (*13.1*), becomes

$$\ddot{x} + 9\dot{x} + 14x = \frac{1}{2}\sin t \tag{1}$$

The general solution to the associated homogeneous equation $\ddot{x} + 9\dot{x} + 14x = 0$ is (see Problem 13.5)

$$x_h = c_1 e^{-2t} + c_2 e^{-7t}$$

Using the method of undetermined coefficients (see Chapter 10), we find

$$x_p = \frac{13}{500}\sin t - \frac{9}{500}\cos t \tag{2}$$

The general solution of (*1*) is therefore

$$x = x_h + x_p = c_1 e^{-2t} + c_2 e^{-7t} + \frac{13}{500}\sin t - \frac{9}{500}\cos t$$

Applying the initial conditions, $x(0) = 0$ and $\dot{x}(0) = -1$, we obtain

$$x = \frac{1}{500}(-90e^{-2t} + 99e^{-7t} + 13\sin t - 9\cos t)$$

Note that the exponential terms, which come from x_h and hence represent an associated free overdamped motion, quickly die out. These terms are the transient part of the solution. The terms coming from x_p, however, do not die out as $t \to \infty$; they are the steady-state part of the solution.

13.10. A 128-lb weight is attached to a spring having a spring constant of 64 lb/ft. The weight is started in motion with no initial velocity by displacing it 6 in above the equilibrium position and by simultaneously applying to the weight an external force $F(t) = 8 \sin 4t$. Assuming no air resistance, find the subsequent motion of the weight.

Here $m = 4$, $k = 64$, $a = 0$, and $F(t) = 8 \sin 4t$; hence, Eq. (*13.1*) becomes

$$\ddot{x} + 16x = 2 \sin 4t \tag{1}$$

This problem is, therefore, an example of forced undamped motion. The solution to the associated homogeneous equation is

$$x_h = c_1 \cos 4t + c_2 \sin 4t$$

A particular solution is found by the method of undetermined coefficients (the modification described in Chapter 10 is necessary here): $x_p = -\frac{1}{4}t \cos 4t$. The solution to (*1*) is then

$$x = c_1 \cos 4t + c_2 \sin 4t - \frac{1}{4} t \cos 4t$$

Applying the initial conditions, $x(0) = -\frac{1}{2}$ and $\dot{x}(0) = 0$, we obtain

$$x = -\frac{1}{2} \cos 4t + \frac{1}{16} \sin 4t - \frac{1}{4} t \cos 4t$$

Note that $|x| \to \infty$ as $t \to \infty$. This phenomenon is called *pure resonance.* It is due to the forcing function $F(t)$ having the same circular frequency as that of the associated free undamped system.

13.11. Write the steady-state motion found in Problem 13.9 in the form specified by Eq. (*13.13*).

The steady-state displacement is given by (*2*) of Problem 13.9 as

$$x(t) = -\frac{9}{500} \cos t + \frac{13}{500} \sin t$$

Its circular frequency is $\omega = 1$. Here

$$A = \sqrt{\left(\frac{13}{500}\right)^2 + \left(-\frac{9}{500}\right)^2} = 0.0316$$

and

$$\phi = \arctan \frac{13/500}{-9/500} = -0.965 \text{ radians}$$

The coefficient of the cosine term in the steady-state displacement is negative, so $k = 1$, and Eq. (*13.13*) becomes

$$x(t) = -0.0316 \cos (t + 0.965)$$

13.12. An RCL circuit connected in series has $R = 180$ ohms, $C = 1/280$ farad, $L = 20$ henries, and an applied voltage $E(t) = 10 \sin t$. Assuming no initial charge on the capacitor, but an initial current of 1 ampere at $t = 0$ when the voltage is first applied, find the subsequent charge on the capacitor.

Substituting the given quantities into Eq. (*13.5*), we obtain

$$\ddot{q} + 9\dot{q} + 14q = \frac{1}{2} \sin t$$

This equation is identical in form to (*1*) of Problem 13.9; hence, the solution must be identical in form to the solution of that equation. Thus,

$$q = c_1 e^{-2t} + c_2 e^{-7t} + \frac{13}{500}\sin t - \frac{9}{500}\cos t$$

Applying the initial conditions $q(0) = 0$ and $\dot{q}(0) = 1$, we obtain $c_1 = 110/500$ and $c_2 = -101/500$. Hence,

$$q = \frac{1}{500}(110e^{-2t} - 101e^{-7t} + 13\sin t - 9\cos t)$$

As in Problem 13.9, the solution is the sum of transient and steady-state terms.

13.13. An RCL circuit connected in series has $R = 10$ ohms, $C = 10^{-2}$ farad, $L = \frac{1}{2}$ henry, and an applied voltage $E = 12$ volts. Assuming no initial current and no initial charge at $t = 0$ when the voltage is first applied, find the subsequent current in the system.

Substituting the given values into Eq. (*13.7*), we obtain the homogeneous equation [since $E(t) = 12$, $dE/dt = 0$]

$$\frac{d^2I}{dt^2} + 20\frac{dI}{dt} + 200I = 0$$

The roots of the associated characteristic equation are $\lambda_1 = -10 + 10i$ and $\lambda_2 = -10 - 10i$; hence, this is an example of a free underdamped system for the current. The solution is

$$I = e^{-10t}(c_1\cos 10t + c_2\sin 10t) \tag{1}$$

The initial conditions are $I(0) = 0$ and, from Eq. (*13.8*),

$$\left.\frac{dI}{dt}\right|_{t=0} = \frac{12}{1/2} - \left(\frac{10}{1/2}\right)(0) - \frac{1}{(1/2)(10^{-2})}(0) = 24$$

Applying these conditions to (*1*), we obtain $c_1 = 0$ and $c_2 = \frac{12}{5}$; thus, $I = \frac{12}{5}e^{-10t}\sin 10t$, which is completely transient.

13.14. Solve Problem 13.13 by first finding the charge on the capacitor.

We first solve for the charge q and then use $I = dq/dt$ to obtain the current. Substituting the values given in Problem 13.13 into Eq. (*13.5*), we have $\ddot{q} + 20\dot{q} + 200q = 24$, which represents a forced system for the charge, in contrast to the free damped system obtained in Problem 13.13 for the current. Using the method of undetermined coefficients to find a particular solution, we obtain the general solution

$$q = e^{-10t}(c_1\cos 10t + c_2\sin 10t) + \frac{3}{25}$$

Initial conditions for the charge are $q(0) = 0$ and $\dot{q}(0) = 0$; applying them, we obtain $c_1 = c_2 = -3/25$. Therefore,

$$q = -e^{-10t}\left(\frac{3}{25}\cos 10t + \frac{3}{25}\sin 10t\right) + \frac{3}{25}$$

and

$$I = \frac{dq}{dt} = \frac{12}{5}e^{-10t}\sin 10t$$

as before.

Note that although the current is completely transient, the charge on the capacitor is the sum of both transient and steady-state terms.

13.15. An RCL circuit connected in series has a resistance of 5 ohms, an inductance of 0.05 henry, a capacitor of 4×10^{-4} farad, and an applied alternating emf of $200\cos 100t$ volts. Find an expression for the current flowing through this circuit if the initial current and the initial charge on the capacitor are both zero.

Here $R/L = 5/0.05 = 100$, $1/(LC) = 1/[0.05(4 \times 10^{-4})] = 50{,}000$, and

$$\frac{1}{L}\frac{dE(t)}{dt} = \frac{1}{0.05}200(-100 \sin 100t) = -400{,}000 \sin 100t$$

so Eq. (*13.7*) becomes

$$\frac{d^2I}{dt^2} + 100\frac{dI}{dt} + 50{,}000I = -400{,}000 \sin 100t$$

The roots of its characteristic equation are $-50 \pm 50\sqrt{19}\,i$, hence the solution to the associated homogeneous problem is

$$I_h = c_1e^{-50t} \cos 50\sqrt{19}\,t + c_2e^{-50t} \sin 50\sqrt{19}\,t$$

Using the method of undetermined coefficients, we find a particular solution to be

$$I_p = \frac{40}{17}\cos 100t - \frac{160}{17}\sin 100t$$

so the general solution is

$$I = I_h + I_p = c_1e^{-50t} \cos 50\sqrt{19}\,t + c_2e^{-50t} \sin 50\sqrt{19}\,t + \frac{40}{17}\cos 100t - \frac{160}{17}\sin 100t \tag{1}$$

The initial conditions are $I(0) = 0$ and, from Eq. (*13.8*),

$$\left.\frac{dI}{dt}\right|_{t=0} = \frac{200}{0.05} - \frac{5}{0.05}(0) - \frac{1}{0.05(4 \times 10^{-4})}(0) = 4000$$

Applying the first of these conditions to (*1*) directly, we obtain

$$0 = I(0) = c_1(1) + c_2(0) + \frac{40}{17}$$

or $c_1 = -40/17 = -2.35$. Substituting this value into (*1*) and then differentiating, we find that

$$\frac{dI}{dt} = -2.35(-50e^{-50t} \cos 50\sqrt{19}\,t - 50\sqrt{19}\,e^{-50t} \sin 50\sqrt{19}\,t)$$

$$+ c_2(-50e^{-50t} \sin 50\sqrt{19}\,t + 50\sqrt{19}\,e^{-50t} \cos 50\sqrt{19}\,t) - \frac{4000}{17}\sin 100t - \frac{16{,}000}{17}\cos 100t$$

whereupon
$$4000 = \left.\frac{dI}{dt}\right|_{t=0} = -2.35(-50) + c_2(50\sqrt{19}) - \frac{16{,}000}{17}$$

and $c_2 = 22.13$. Equation (*1*) becomes

$$I = -2.35e^{-50t} \cos 50\sqrt{19}\,t + 22.13e^{-50t} \sin 50\sqrt{19}\,t + \frac{40}{17}\cos 100t - \frac{160}{17}\sin 100t$$

13.16. Solve Problem 13.15 by first finding the charge on the capacitor.

Substituting the values given in Problem 13.15 into Eq. (*13.5*), we obtain

$$\frac{d^2q}{dt^2} + 100\frac{dq}{dt} + 50{,}000q = 4000 \cos 100t$$

The associated homogeneous equation is identical in form to the one in Problem 13.15, so it has the same solution (with I_h replaced by q_h). Using the method of undetermined coefficients, we find a particular solution to be

$$q_p = \frac{16}{170}\cos 100t + \frac{4}{170}\sin 100t$$

so the general solution is

$$q = q_h + q_p = c_1 e^{-50t}\cos 50\sqrt{19}\,t + c_2 e^{-50t}\sin 50\sqrt{19}\,t + \frac{16}{170}\cos 100t + \frac{4}{170}\sin 100t \qquad (1)$$

The initial conditions on the charge are $q(0) = 0$ and

$$\left.\frac{dq}{dt}\right|_{t=0} = I(0) = 0$$

Applying the first of these conditions to (1) directly, we obtain

$$0 = q(0) = c_1(1) + c_2(0) + \frac{16}{170}$$

or $c_1 = -16/170 = -0.0941$. Substituting this value into (1) and then differentiating, we find that

$$\frac{dq}{dt} = -0.0941(-50e^{-50t}\cos 50\sqrt{19}\,t - 50\sqrt{19}\,e^{-50t}\sin 50\sqrt{19}\,t)$$

$$+ c_2(-50e^{-50t}\sin 50\sqrt{19}\,t + 50\sqrt{19}\,e^{-50t}\cos 50\sqrt{19}\,t) - \frac{160}{17}\sin 100t + \frac{40}{17}\cos 100t \qquad (2)$$

whereupon
$$0 = \left.\frac{dq}{dt}\right|_{t=0} = -0.0941(-50) + c_2(50\sqrt{19}) + \frac{40}{17}$$

and $c_2 = -0.0324$. Substituting this value into (2) and simplifying, we obtain as before

$$I(t) = \frac{dq}{dt} = -2.35e^{-50t}\cos 50\sqrt{19}\,t + 22.13e^{-50t}\sin 50\sqrt{19}\,t + \frac{40}{17}\cos 100t - \frac{160}{17}\sin 100t \qquad (3)$$

13.17. Determine the circular frequency, the natural frequency, and the period of the steady-state current found in Problem 13.16.

The current is given by (3) of Problem 13.16. As $t \to \infty$, the exponential terms tend to zero, so the steady-state current is

$$I(t) = \frac{40}{17}\cos 100t - \frac{160}{17}\sin 100t$$

Circular frequency: $\omega = 100$ Hz

Natural frequency: $f = \omega/2\pi = 100/2\pi = 15.92$ Hz

Period: $T = 1/f = 2\pi/100 = 0.063$ sec

13.18. Write the steady-state current found in Problem 13.17 in the form specified by Eq. (13.13).

The amplitude is

$$A = \sqrt{\left(\frac{40}{17}\right)^2 + \left(-\frac{160}{17}\right)^2} = 9.701$$

and the phase angle is

$$\phi = \arctan\frac{-160/17}{40/17} = -1.326 \text{ radians}$$

The circular frequency is $\omega = 100$. The coefficient of the cosine term is positive, so $k = 0$ and Eq. (13.13) becomes

$$I_s(t) = 9.701\cos(100t + 1.326)$$

13.19. Determine whether a cylinder of radius 4 in, height 10 in, and weight 15 lb can float in a deep pool of water of weight density 62.5 lb/ft^3.

Let h denote the length (in feet) of the submerged portion of the cylinder at equilibrium. With $r = \frac{1}{3}$ ft, it follows from Eq. (*13.9*) that

$$h = \frac{mg}{\pi r^2 \rho} = \frac{15}{\pi\left(\frac{1}{3}\right)^2 62.5} = 0.688 \text{ ft} = 8.25 \text{ in}$$

Thus, the cylinder will float with $10 - 8.25 = 1.75$ in of length above the water line at equilibrium.

13.20. Determine an expression for the motion of the cylinder described in Problem 13.19 if it is released with 20 percent of its length above the water line with a velocity of 5 ft/sec in the downward direction.

Here $r = \frac{1}{3}$ ft, $\rho = 62.5$ lb/ft^3, $m = 15/32$ slugs and Eq. (*13.10*) becomes

$$\ddot{x} + 46.5421x = 0$$

The roots of the associated characteristic equation are $\pm\sqrt{46.5421}\, i = \pm 6.82i$; the general solution of the differential equation is

$$x(t) = c_1 \cos 6.82t + c_2 \sin 6.82t \tag{1}$$

At $t = 0$, 20 percent of the 10-in length of the cylinder, or 2 in, is out of the water. Using the results of Problem 13.19, we know that the equilibrium position has 1.75 in above the water, so at $t = 0$, the cylinder is raised 1/4 in or 1/48 ft above its equilibrium position. In the context of Fig. 13-3, $x(0) = 1/48$ ft. The initial velocity is 5 ft/sec in the downward or *negative* direction in the coordinate system of Fig. 13-3, so $\dot{x}(0) = -5$. Applying these initial conditions to (*1*), we find that

$$c_1 = \frac{1}{48} = 0.021 \quad \text{and} \quad c_2 = \frac{-5}{6.82} = -0.73$$

Equation (*1*) becomes

$$x(t) = 0.021 \cos 6.82t - 0.73 \sin 6.82t$$

13.21. Determine whether a cylinder of diameter 10 cm, height 15 cm, and weight 19.6 N can float in a deep pool of water of weight density 980 dynes/cm^3.

Let h denote the length (in centimeters) of the submerged portion of the cylinder at equilibrium. With $r = 5$ cm and $mg = 19.6 \text{ N} = 1.96 \times 10^6$ dynes, it follows from Eq. (*13.9*) that

$$h = \frac{mg}{\pi r^2 \rho} = \frac{1.96 \times 10^6}{\pi (5)^2 (980)} = 25.5 \text{ cm}$$

Since this is more height than the cylinder possesses, the cylinder cannot displace sufficient water to float and will sink to the bottom of the pool.

13.22. Determine whether a cylinder of diameter 10 cm, height 15 cm, and weight 19.6 N can float in a deep pool of liquid having weight density 2450 dynes/cm^3.

Let h denote the length of the submerged portion of the cylinder at equilibrium. With $r = 5$ cm and $mg = 19.6 \text{ N} = 1.96 \times 10^6$ dynes, it follows from Eq. (*13.9*) that

$$h = \frac{mg}{\pi r^2 \rho} = \frac{1.96 \times 10^6}{\pi (5)^2 (2450)} = 10.2 \text{ cm}$$

Thus, the cylinder will float with $15 - 10.2 = 4.8$ cm of length above the liquid at equilibrium.

13.23. Determine an expression for the motion of the cylinder described in Problem 13.22 if it is released at rest with 12 cm of its length fully submerged.

Here $r = 5$ cm, $\rho = 2450$ dynes/cm^3, $m = 19.6/9.8 = 2$ kg $= 2000$ g, and Eq. (*13.10*) becomes

$$\ddot{x} + 96.21x = 0$$

The roots of the associated characteristic equation are $\pm\sqrt{96.21}\,i = \pm 9.81i$; the general solution of the differential equation is

$$x(t) = c_1 \cos 9.81t + c_2 \sin 9.81t \tag{1}$$

At $t = 0$, 12 cm of the length of the cylinder is submerged. Using the results of Problem 13.22, we know that the equilibrium position has 10.2 cm submerged, so at $t = 0$, the cylinder is submerged $12 - 10.2 = 1.8$ cm *below* its equilibrium position. In the context of Fig. 13-3, $x(0) = -1.8$ cm with a negative sign indicating that the equilibrium line is submerged. The cylinder begins at rest, so its initial velocity is $\dot{x}(0) = 0$. Applying these initial conditions to (*1*), we find that $c_1 = -1.8$ and $c_2 = 0$. Equation (*1*) becomes

$$x(t) = -1.8 \cos 9.81t$$

13.24. A solid cylinder partially submerged in water having weight density 62.5 lb/ft^3, with its axis vertical, oscillates up and down within a period of 0.6 sec. Determine the diameter of the cylinder if it weighs 2 lb.

With $\rho = 62.5$ lb/ft^3 and $m = 2/32$ slugs, Eq. (*13.10*) becomes

$$\ddot{x} + 1000\pi r^2 x = 0$$

which has as its general solution

$$x(t) = c_1 \cos \sqrt{1000\pi}\, rt + c_2 \sin \sqrt{1000\pi}\, rt \tag{1}$$

Its circular frequency is $\omega = r\sqrt{1000\pi}$; its natural frequency is $f = \omega/2\pi = r\sqrt{250/\pi} = 8.92r$; its period is $T = 1/f = 1/8.92r$. We are given $0.6 = T = 1/8.92r$, thus $r = 0.187$ ft $= 2.24$ in with a diameter of 4.48 in.

13.25. A prism whose cross section is an equilateral triangle with sides of length l floats in a pool of liquid of weight density ρ with its height parallel to the vertical axis. The prism is set in motion by displacing it from its equilibrium position (see Fig. 13-4) and giving it an initial velocity. Determine the differential equation governing the subsequent motion of this prism.

Equilibrium occurs when the buoyant force of the displaced liquid equals the force of gravity on the body. The area of an equilateral triangle with sides of length l is $A = \sqrt{3}\,l^2/4$. For the prism depicted

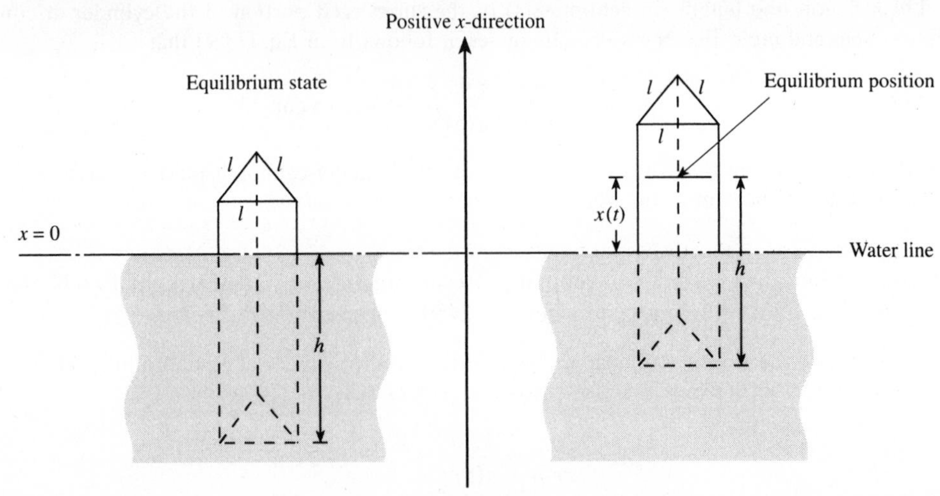

Fig. 13-4

in Fig. 13-4, with h units of height submerged at equilibrium, the volume of water displaced at equilibrium is $\sqrt{3}\,l^2h/4$, providing a buoyant force of $\sqrt{3}\,l^2h\rho/4$. By Archimedes' principle, this buoyant force at equilibrium must equal the weight of the prism mg; hence,

$$\sqrt{3}\,l^2h\rho/4 = mg \tag{1}$$

We arbitrarily take the upward direction to be the positive x-direction. If the prism is raised out of the water by $x(t)$ units, as shown in Fig. 13-4, then it is no longer in equilibrium. The downward or negative force on such a body remains mg but the buoyant or positive force is reduced to $\sqrt{3}\,l^2[h - x(t)]\rho/4$. It now follows from Newton's second law that

$$m\ddot{x} = \frac{\sqrt{3}\,l^2[h - x(t)]\rho}{4} - mg$$

Substituting (*1*) into this last equation, we simplify it to

$$\ddot{x} + \frac{\sqrt{3}\,l^2\rho}{4m}x = 0$$

Supplementary Problems

13.26. A 10-lb weight is suspended from a spring and stretches it 2 in from its natural length. Find the spring constant.

13.27. A mass of 0.4 slug is hung onto a spring and stretches it 9 in from its natural length. Find the spring constant.

13.28. A mass of 0.4 g is hung onto a spring and stretches it 3 cm from its natural length. Find the spring constant.

13.29. A mass of 0.3 kg is hung onto a spring and stretches it 15 cm from its natural length. Find the spring constant.

13.30. A 20-lb weight is suspended from the end of a vertical spring having a spring constant of 40 lb/ft and is allowed to reach equilibrium. It is then set into motion by stretching the spring 2 in from its equilibrium position and releasing the mass from rest. Find the position of the weight at any time t if there is no external force and no air resistance.

13.31. Solve Problem 13.30 if the weight is set in motion by compressing the spring by 2 in from its equilibrium position and giving it an initial velocity of 2 ft/sec in the downward direction.

13.32. A 20-g mass is suspended from the end of a vertical spring having a spring constant of 2880 dynes/cm and is allowed to reach equilibrium. It is then set into motion by stretching the spring 3 cm from its equilibrium position and releasing the mass with an initial velocity of 10 cm/sec in the downward direction. Find the position of the mass at any time t if there is no external force and no air resistance.

13.33. A 32-lb weight is attached to a spring, stretching it 8 ft from its natural length. The weight is started in motion by displacing it 1 ft in the upward direction and by giving it an initial velocity of 2 ft/sec in the downward direction. Find the subsequent motion of the weight, if the medium offers negligible resistance.

13.34. Determine (*a*) the circular frequency, (*b*) the natural frequency, and (*c*) the period for the vibrations described in Problem 13.31.

13.35. Determine (*a*) the circular frequency, (*b*) the natural frequency, and (*c*) the period for the vibrations described in Problem 13.32.

13.36. Determine (*a*) the circular frequency, (*b*) the natural frequency, and (*c*) the period for the vibrations described in Problem 13.33.

13.37. Find the solution to Eq. (*13.1*) with initial conditions given by Eq. (*13.2*) when the vibrations are free and undamped.

13.38. A $\frac{1}{4}$-slug mass is hung onto a spring, whereupon the spring is stretched 6 in from its natural length. The mass is then started in motion from the equilibrium position with an initial velocity of 4 ft/sec in the upward direction. Find the subsequent motion of the mass, if the force due to air resistance is $-2\dot{x}$ lb.

13.39. A $\frac{1}{2}$-slug mass is attached to a spring so that the spring is stretched 2 ft from its natural length. The mass is started in motion with no initial velocity by displacing it $\frac{1}{2}$ ft in the upward direction. Find the subsequent motion of the mass, if the medium offers a resistance of $-4\dot{x}$ lb.

13.40. A $\frac{1}{2}$-slug mass is attached to a spring having a spring constant of 6 lb/ft. The mass is set into motion by displacing it 6 in below its equilibrium position with no initial velocity. Find the subsequent motion of the mass, if the force due to the medium is $-4\dot{x}$ lb.

13.41. A $\frac{1}{2}$-kg mass is attached to a spring having a spring constant of 8 N/m. The mass is set into motion by displacing it 10 cm above its equilibrium position with an initial velocity of 2 m/sec in the upward direction. Find the subsequent motion of the mass if the surrounding medium offers a resistance of $-4\dot{x}$ N.

13.42. Solve Problem 13.41 if instead the spring constant is 8.01 N/m.

13.43. Solve Problem 13.41 if instead the spring constant is 7.99 N/m.

13.44. A 1-slug mass is attached to a spring having a spring constant of 8 lb/ft. The mass is initially set into motion from the equilibrium position with no initial velocity by applying an external force $F(t) = 16\cos 4t$. Find the subsequent motion of the mass, if the force due to air resistance is $-4\dot{x}$ lb.

13.45. A 64-lb weight is attached to a spring whereupon the spring is stretched 1.28 ft and allowed to come to rest. The weight is set into motion by applying an external force $F(t) = 4\sin 2t$. Find the subsequent motion of the weight if the surrounding medium offers a negligible resistance.

13.46. A 128-lb weight is attached to a spring whereupon the spring is stretched 2 ft and allowed to come to rest. The weight is set into motion from rest by displacing the spring 6 in above its equilibrium position and also by applying an external force $F(t) = 8\sin 4t$. Find the subsequent motion of the weight if the surrounding medium offers a negligible resistance.

13.47. Solve Problem 13.38 if, in addition, the mass is subjected to an externally applied force $F(t) = 16\sin 8t$.

13.48. A 16-lb weight is attached to a spring whereupon the spring is stretched 1.6 ft and allowed to come to rest. The weight is set into motion from rest by displacing the spring 9 in above its equilibrium position and also by applying an external force $F(t) = 5\cos 2t$. Find the subsequent motion of the weight if the surrounding medium offers a resistance of $-2\dot{x}$ lb.

13.49. Write the steady-state portion of the motion found in Problem 13.48 in the form specified by Eq. (*13.13*).

13.50. A $\frac{1}{2}$-kg mass is attached to a spring having a spring constant of 6 N/m and allowed to come to rest. The mass is set into motion by applying an external force $F(t) = 24\cos 3t - 33\sin 3t$. Find the subsequent motion of the mass if the surrounding medium offers a resistance of $-3\dot{x}$ N.

13.51. Write the steady-state portion of the motion found in Problem 13.50 in the form of Eq. (*13.13*).

13.52. An RCL circuit connected in series with $R = 6$ ohms, $C = 0.02$ farad, and $L = 0.1$ henry has an applied voltage $E(t) = 6$ volts. Assuming no initial current and no initial charge at $t = 0$ when the voltage is first applied, find the subsequent charge on the capacitor and the current in the circuit.

13.53. An RCL circuit connected in series with a resistance of 5 ohms, a condenser of capacitance 4×10^{-4} farad, and an inductance of 0.05 henry has an applied emf $E(t) = 110$ volts. Assuming no initial current and no initial charge on the capacitor, find expressions for the current flowing through the circuit and the charge on the capacitor at any time t.

13.54. An RCL circuit connected in series with $R = 6$ ohms, $C = 0.02$ farad, and $L = 0.1$ henry has no applied voltage. Find the subsequent current in the circuit if the initial charge on the capacitor is $\frac{1}{10}$ coulomb and the initial current is zero.

13.55. An RCL circuit connected in series with a resistance of 1000 ohm, a condenser of capacitance 4×10^{-6} farad, and an inductance of 1 henry has an applied emf $E(t) = 24$ volts. Assuming no initial current and no initial charge on the capacitor, find an expression for the current flowing through the circuit at any time t.

13.56. An RCL circuit connected in series with a resistance of 4 ohms, a capacitor of 1/26 farad, and an inductance of 1/2 henry has an applied voltage $E(t) = 16 \cos 2t$. Assuming no initial current and no initial charge on the capacitor, find an expression for the current flowing through the circuit at any time t.

13.57. Determine the steady-state current in the circuit described in Problem 13.56 and write it in the form of Eq. (*13.13*).

13.58. An RCL circuit connected in series with a resistance of 16 ohms, a capacitor of 0.02 farad, and an inductance of 2 henries has an applied voltage $E(t) = 100 \sin 3t$. Assuming no initial current and no initial charge on the capacitor, find an expression for the current flowing through the circuit at any time t.

13.59. Determine the steady-state current in the circuit described in Problem 13.56 and write it in the form of Eq. (*13.13*).

13.60. An RCL circuit connected in series with a resistance of 20 ohms, a capacitor of 10^{-4} farad, and an inductance of 0.05 henry has an applied voltage $E(t) = 100 \cos 200t$. Assuming no initial current and no initial charge on the capacitor, find an expression for the current flowing through the circuit at any time t.

13.61. Determine the steady-state current in the circuit described in Problem 13.60 and write it in the form of Eq. (*13.13*).

13.62. An RCL circuit connected in series with a resistance of 2 ohms, a capacitor of 1/260 farad, and an inductance of 0.1 henry has an applied voltage $E(t) = 100 \sin 60t$. Assuming no initial current and no initial charge on the capacitor, find an expression for the charge on the capacitor at any time t.

13.63. Determine the steady-state charge on the capacitor in the circuit described in Problem 13.62 and write it in the form of Eq. (*13.13*).

13.64. An RCL circuit connected in series has $R = 5$ ohms, $C = 10^{-2}$ farad, $L = \frac{1}{8}$ henry, and no applied voltage. Find the subsequent steady-state current in the circuit. *Hint:* Initial conditions are not needed.

13.65. An RCL circuit connected in series with $R = 5$ ohms, $C = 10^{-2}$ farad, and $L = \frac{1}{8}$ henry has applied voltage $E(t) = \sin t$. Find the steady-state current in the circuit. *Hint:* Initial conditions are not needed.

13.66. Determine the equilibrium position of a cylinder of radius 3 in, height 20 in, and weight 5π lb that is floating with its axis vertical in a deep pool of water of weight density 62.5 lb/ft^3.

13.67. Find an expression for the motion of the cylinder described in Problem 13.66 if it is disturbed from its equilibrium position by submerging an additional 2 in of height below the water line and with a velocity of 1 ft/sec in the downward direction.

13.68. Write the harmonic motion of the cylinder described in Problem 13.67 in the form of Eq. (*13.13*).

13.69. Determine the equilibrium position of a cylinder of radius 2 ft, height 4 ft, and weight 600 lb that is floating with its axis vertical in a deep pool of water of weight density 62.5 lb/ft^3.

13.70. Find an expression for the motion of the cylinder described in Problem 13.69 if it is released from rest with 1 ft of its height submerged in water.

13.71. Determine (a) the circular frequency, (b) the natural frequency, and (c) the period for the vibrations described in Problem 13.70.

13.72. Determine (a) the circular frequency, (b) the natural frequency, and (c) the period for the vibrations described in Problem 13.67.

13.73. Determine the equilibrium position of a cylinder of radius 3 cm, height 10 cm, and mass 700 g that is floating with its axis vertical in a deep pool of water of mass density 1 g/cm^3.

13.74. Solve Problem 13.73 if the liquid is not water but another substance with mass density 2 g/cm^3.

13.75. Determine the equilibrium position of a cylinder of radius 30 cm, height 500 cm, and weight 2.5×10^7 dynes that is floating with its axis vertical in a deep pool of water of weight density 980 dynes/cm^3.

13.76. Find an expression for the motion of the cylinder described in Problem 13.75 if it is set in motion from its equilibrium position by striking it to produce an initial velocity of 50 cm/sec in the downward direction.

13.77. Find the general solution to Eq. (*13.10*) and determine its period.

13.78. Determine the radius of a cylinder weighing 5 lb with its axis vertical that oscillates in a pool of deep water ($\rho = 62.5$ lb/ft^3) with a period of 0.75 sec. *Hint:* Use the results of Problem 13.77.

13.79. Determine the weight of a cylinder having a diameter of 1 ft with its axis vertical that oscillates in a pool of deep water ($\rho = 62.5$ lb/ft^3) with a period of 2 sec. Hint: Use the results of Problem 13.77.

13.80. A rectangular box of width w, length l, and height h floats in a pool of liquid of weight density ρ with its height parallel to the vertical axis. The box is set into motion by displacing it x_0 units from its equilibrium position and giving it an initial velocity of v_0. Determine the differential equation governing the subsequent motion of the box.

13.81. Determine (a) the period of oscillations for the motion described in Problem 13.80 and (b) the change in that period if the length of the box is doubled.

Chapter 14

The Laplace Transform

DEFINITION

Let $f(x)$ be defined for $0 \le x < \infty$ and let s denote an arbitrary real variable. The *Laplace transform of* $f(x)$, designated by either $\mathscr{L}\{f(x)\}$ or $F(s)$, is

$$\mathscr{L}\{f(x)\} = F(s) = \int_0^\infty e^{-sx} f(x)\,dx \tag{14.1}$$

for all values of s for which the improper integral converges. Convergence occurs when the limit

$$\lim_{R\to\infty} \int_0^R e^{-sx} f(x)\,dx \tag{14.2}$$

exists. If this limit does not exist, the improper integral diverges and $f(x)$ has no Laplace transform. When evaluating the integral in Eq. (*14.1*), the variable s is treated as a constant because the integration is with respect to x.

The Laplace transforms for a number of elementary functions are calculated in Problems 14.4 through 14.8; additional transforms are given in Appendix A.

PROPERTIES OF LAPLACE TRANSFORMS

Property 14.1 (Linearity). If $\mathscr{L}\{f(x)\} = F(s)$ and $\mathscr{L}\{g(x)\} = G(s)$, then for any two constants c_1 and c_2

$$\mathscr{L}\{c_1 f(x) + c_2 g(x)\} = c_1 \mathscr{L}\{f(x)\} + c_2 \mathscr{L}\{g(x)\} = c_1 F(s) + c_2 G(s) \tag{14.3}$$

Property 14.2. If $\mathscr{L}\{f(x)\} = F(s)$, then for any constant a

$$\mathscr{L}\{e^{ax} f(x)\} = F(s-a) \tag{14.4}$$

Property 14.3. If $\mathscr{L}\{f(x)\} = F(s)$, then for any positive integer n

$$\mathscr{L}\{x^n f(x)\} = (-1)^n \frac{d^n}{ds^n}[F(s)] \tag{14.5}$$

Property 14.4. If $\mathscr{L}\{f(x)\} = F(s)$ and if $\lim\limits_{\substack{x\to 0\\ x>0}} \dfrac{f(x)}{x}$ exists, then

$$\mathscr{L}\left\{\frac{1}{x} f(x)\right\} = \int_s^\infty F(t)\,dt \tag{14.6}$$

Property 14.5. If $\mathscr{L}\{f(x)\} = F(s)$, then

$$\mathscr{L}\left\{\int_0^x f(t)\,dt\right\} = \frac{1}{s} F(s) \tag{14.7}$$

Property 14.6. If $f(x)$ is periodic with period ω, that is, $f(x+\omega) = f(x)$, then

$$\mathscr{L}\{f(x)\} = \frac{\int_0^\omega e^{-sx} f(x)\,dx}{1 - e^{-\omega s}} \tag{14.8}$$

FUNCTIONS OF OTHER INDEPENDENT VARIABLES

For consistency only, the definition of the Laplace transform and its properties, Eqs. (*14.1*) through (*14.8*), are presented for functions of x. They are equally applicable for functions of any independent variable and are generated by replacing the variable x in the above equations by any variable of interest. In particular, the counterpart of Eq. (*14.1*) for the Laplace transform of a function of t is

$$\mathscr{L}\{f(t)\} = F(s) = \int_0^\infty e^{-st} f(t)\,dt$$

Solved Problems

14.1. Determine whether the improper integral $\int_2^\infty \frac{1}{x^2}\,dx$ converges.

Since

$$\lim_{R\to\infty} \int_2^R \frac{1}{x^2}\,dx = \lim_{R\to\infty} \left(-\frac{1}{x}\right)\Bigg|_2^R = \lim_{R\to\infty} \left(-\frac{1}{R} + \frac{1}{2}\right) = \frac{1}{2}$$

the improper integral converges to the value $\frac{1}{2}$.

14.2. Determine whether the improper integral $\int_9^\infty \frac{1}{x}\,dx$ converges.

Since

$$\lim_{R\to\infty} \int_9^R \frac{1}{x}\,dx = \lim_{R\to\infty} \ln|x|\,\Bigg|_9^R = \lim_{R\to\infty} (\ln R - \ln 9) = \infty$$

the improper integral diverges.

14.3. Determine those values of s for which the improper integral $\int_0^\infty e^{-sx}\,dx$ converges.

For $s = 0$,

$$\int_0^\infty e^{-sx}\,dx = \int_0^\infty e^{-(0)(x)}\,dx = \lim_{R\to\infty} \int_0^R (1)\,dx = \lim_{R\to\infty} x\,\Bigg|_0^R = \lim_{R\to\infty} R = \infty$$

hence the integral diverges. For $s \neq 0$,

$$\int_0^\infty e^{-sx}\,dx = \lim_{R\to\infty} \int_0^R e^{-sx}\,dx = \lim_{R\to\infty} \left[-\frac{1}{s} e^{-sx}\right]_{x=0}^{x=R}$$

$$= \lim_{R\to\infty} \left(\frac{-1}{s} e^{-sR} + \frac{1}{s}\right)$$

When $s < 0$, $-sR > 0$; hence the limit is ∞ and the integral diverges. When $s > 0$, $-sR < 0$; hence, the limit is $1/s$ and the integral converges.

14.4. Find the Laplace transform of $f(x) \equiv 1$.

Using Eq. (*14.1*) and the results of Problem 14.3, we have

$$F(s) = \mathscr{L}\{1\} = \int_0^\infty e^{-sx}(1)\,dx = \frac{1}{s} \qquad (\text{for } s > 0)$$

(See also entry 1 in Appendix A.)

14.5. Find the Laplace transform of $f(x) = x^2$.

Using Eq. (*14.1*) and integration by parts twice, we find that

$$F(s) = \mathscr{L}\{x^2\} = \int_0^\infty e^{-sx}x^2\,dx = \lim_{R\to\infty}\int_0^R x^2 e^{-sx}\,dx$$

$$= \lim_{R\to\infty}\left[-\frac{x^2}{s}e^{-sx} - \frac{2x}{s^2}e^{-sx} - \frac{2}{s^3}e^{-sx}\right]_{x=0}^{x=R}$$

$$= \lim_{R\to\infty}\left(-\frac{R^2}{s}e^{-sR} - \frac{2R}{s^2}e^{-sR} - \frac{2}{s^3}e^{-sR} + \frac{2}{s^3}\right)$$

For $s<0$, $\lim_{R\to\infty}[-(R^2/s)e^{-sR}] = \infty$, and the improper integral diverges. For $s>0$, it follows from repeated use of L'Hôpital's rule that

$$\lim_{R\to\infty}\left(-\frac{R^2}{s}e^{-sR}\right) = \lim_{R\to\infty}\left(\frac{-R^2}{se^{sR}}\right) = \lim_{R\to\infty}\left(\frac{-2R}{s^2e^{sR}}\right)$$

$$= \lim_{R\to\infty}\left(\frac{-2}{s^3e^{sR}}\right) = 0$$

$$\lim_{R\to\infty}\left(-\frac{2R}{s}e^{-sR}\right) = \lim_{R\to\infty}\left(\frac{-2R}{se^{sR}}\right) = \lim_{R\to\infty}\left(\frac{-2}{s^2e^{sR}}\right) = 0$$

Also, $\lim_{R\to\infty}[-(2/s^3)e^{-sR}] = 0$ directly; hence the integral converges, and $F(s) = 2/s^3$. For the special case $s = 0$, we have

$$\int_0^\infty e^{-sx}x^2\,dx = \int_0^\infty e^{-s(0)}x^2\,dx = \lim_{R\to\infty}\int_0^R x^2\,dx = \lim_{R\to\infty}\frac{R^3}{3} = \infty$$

Finally, combining all cases, we obtain $\mathscr{L}\{x^2\} = 2/s^3$, $s>0$. (See also entry 3 in Appendix A.)

14.6. Find $\mathscr{L}\{e^{ax}\}$.

Using Eq. (*14.1*), we obtain

$$F(s) = \mathscr{L}\{e^{ax}\} = \int_0^\infty e^{-sx}e^{ax}\,dx = \lim_{R\to\infty}\int_0^R e^{(a-s)x}\,dx$$

$$= \lim_{R\to\infty}\left[\frac{e^{(a-s)x}}{a-s}\right]_{x=0}^{x=R} = \lim_{R\to\infty}\left[\frac{e^{(a-s)R}-1}{a-s}\right]$$

$$= \frac{1}{s-a} \qquad (\text{for } s > a)$$

Note that when $s \le a$, the improper integral diverges. (See also entry 7 in Appendix A.)

14.7. Find $\mathscr{L}\{\sin ax\}$.

Using Eq. (*14.1*) and integration by parts twice, we obtain

$$\mathscr{L}\{\sin ax\} = \int_0^\infty e^{-sx}\sin ax\,dx = \lim_{R\to\infty}\int_0^R e^{-sx}\sin ax\,dx$$

$$= \lim_{R\to\infty}\left[\frac{-se^{-sx}\sin ax}{s^2+a^2} - \frac{ae^{-sx}\cos ax}{s^2+a^2}\right]_{x=0}^{x=R}$$

$$= \lim_{R\to\infty}\left[\frac{-se^{-sR}\sin aR}{s^2+a^2} - \frac{ae^{-sR}\cos aR}{s^2+a^2} + \frac{a}{s^2+a^2}\right]$$

$$= \frac{a}{s^2+a^2} \qquad (\text{for } s > 0)$$

(See also entry 8 in Appendix A.)

14.8. Find the Laplace transform of $f(x) = \begin{cases} e^x & x \le 2 \\ 3 & x > 2 \end{cases}$.

$$\mathscr{L}\{f(x)\} = \int_0^\infty e^{-sx} f(x)\,dx = \int_0^2 e^{-sx} e^x\,dx + \int_2^\infty e^{-sx}(3)\,dx$$

$$= \int_0^2 e^{(1-s)x}\,dx + 3\lim_{R\to\infty}\int_2^R e^{-sx}\,dx = \frac{e^{(1-s)x}}{1-s}\Bigg|_{x=0}^{x=2} - \frac{3}{s}\lim_{R\to\infty} e^{-sx}\Bigg|_{x=2}^{x=R}$$

$$= \frac{e^{2(1-s)}}{1-s} - \frac{1}{1-s} - \frac{3}{s}\lim_{R\to\infty}[e^{-Rs} - e^{-2s}] = \frac{1-e^{-2(s-1)}}{s-1} + \frac{3}{s}e^{-2s} \qquad (\text{for } s > 0)$$

14.9. Find the Laplace transform of the function graphed in Fig. 14-1.

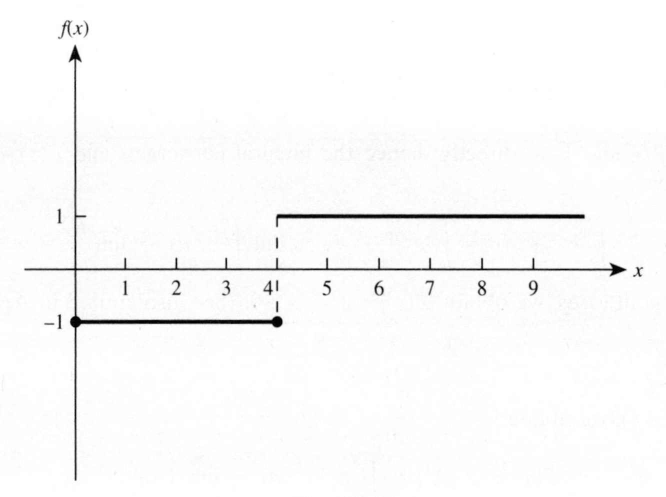

Fig. 14-1

$$f(x) = \begin{cases} -1 & x \le 4 \\ 1 & x > 4 \end{cases}$$

$$\mathscr{L}\{f(x)\} = \int_0^\infty e^{-sx} f(x)\,dx = \int_0^4 e^{-sx}(-1)\,dx + \int_4^\infty e^{-sx}(1)\,dx$$

$$= \frac{e^{-sx}}{s}\Bigg|_{x=0}^{x=4} + \lim_{R\to\infty}\int_4^R e^{-sx}\,dx$$

$$= \frac{e^{-4s}}{s} - \frac{1}{s} + \lim_{R\to\infty}\left(\frac{-1}{s}e^{-Rs} + \frac{1}{s}e^{-4s}\right)$$

$$= \frac{2e^{-4s}}{s} - \frac{1}{s} \qquad (\text{for } s > 0)$$

14.10. Find the Laplace transform of $f(x) = 3 + 2x^2$.

Using Property 14.1 with the results of Problems 14.4 and 14.5, or alternatively, entries 1 and 3

($n = 3$) of Appendix A, we have

$$F(s) = \mathscr{L}\{3 + 2x^2\} = 3\mathscr{L}\{1\} + 2\mathscr{L}\{x^2\}$$
$$= 3\left(\frac{1}{s}\right) + 2\left(\frac{2}{s^3}\right) = \frac{3}{s} + \frac{4}{s^3}$$

14.11. Find the Laplace transform of $f(x) = 5\sin 3x - 17e^{-2x}$.

Using Property 14.1 with the results of Problems 14.6 ($a = -2$) and 14.7 ($a = 3$), or alternatively, entries 7 and 8 of Appendix A, we have

$$F(s) = \mathscr{L}\{5\sin 3x - 17e^{-2x}\} = 5\mathscr{L}\{\sin 3x\} - 17\mathscr{L}\{e^{-2x}\}$$
$$= 5\left(\frac{3}{s^2 + (3)^2}\right) - 17\left(\frac{1}{s - (-2)}\right) = \frac{15}{s^2 + 9} - \frac{17}{s + 2}$$

14.12. Find the Laplace transform of $f(x) = 2\sin x + 3\cos 2x$.

Using Property 14.1 with entries 8 ($a = 1$) and 9 ($a = 2$) of Appendix A, we have

$$F(s) = \mathscr{L}\{2\sin x + 3\cos 2x\} = 2\mathscr{L}\{\sin x\} + 3\mathscr{L}\{\cos 2x\}$$
$$= 2\frac{1}{s^2 + 1} + 3\frac{s}{s^2 + 4} = \frac{2}{s^2 + 1} + \frac{3s}{s^2 + 4}$$

14.13. Find the Laplace transform of $f(x) = 2x^2 - 3x + 4$.

Using Property 14.1 repeatedly with entries 1, 2, and 3 ($n = 3$) of Appendix A, we have

$$F(s) = \mathscr{L}\{2x^2 - 3x + 4\} = 2\mathscr{L}\{x^2\} - 3\mathscr{L}\{x\} + 4\mathscr{L}\{1\}$$
$$= 2\left(\frac{2}{s^3}\right) - 3\left(\frac{1}{s^2}\right) + 4\left(\frac{1}{s}\right) = \frac{4}{s^3} - \frac{3}{s^2} + \frac{4}{s}$$

14.14. Find $\mathscr{L}\{xe^{4x}\}$.

This problem can be done three ways.

(*a*) Using entry 14 of Appendix A with $n = 2$ and $a = 4$, we have directly that

$$\mathscr{L}\{xe^{4x}\} = \frac{1}{(s - 4)^2}$$

(*b*) Set $f(x) = x$. Using Property 14.2 with $a = 4$ and entry 2 of Appendix A, we have

$$F(s) = \mathscr{L}\{f(x)\} = \mathscr{L}\{x\} = \frac{1}{s^2}$$

and
$$\mathscr{L}\{e^{4x}x\} = F(s - 4) = \frac{1}{(s - 4)^2}$$

(*c*) Set $f(x) = e^{4x}$. Using Property 14.3 with $n = 1$ and the results of Problem 14.6, or alternatively, entry 7 of Appendix A with $a = 4$, we find that

$$F(s) = \mathscr{L}\{f(x)\} = \mathscr{L}\{e^{4x}\} = \frac{1}{s-4}$$

and

$$\mathscr{L}\{xe^{4x}\} = -F'(s) = -\frac{d}{ds}\left(\frac{1}{s-4}\right) = \frac{1}{(s-4)^2}$$

14.15. Find $\mathscr{L}\{e^{-2x} \sin 5x\}$.

This problem can be done two ways.

(a) Using entry 15 of Appendix A with $b = -2$ and $a = 5$, we have directly that

$$\mathscr{L}\{e^{-2x} \sin 5x\} = \frac{5}{[s-(-2)]^2 + (5)^2} = \frac{5}{(s+2)^2 + 25}$$

(b) Set $f(x) = \sin 5x$. Using Property 14.2 with $a = -2$ and the results of Problem 14.7, or alternatively, entry 8 of Appendix A with $a = 5$, we have

$$F(s) = \mathscr{L}\{f(x)\} = \mathscr{L}\{\sin 5x\} = \frac{5}{s^2 + 25}$$

and

$$\mathscr{L}\{e^{-2x} \sin 5x\} = F(s-(-2)) = F(s+2) = \frac{5}{(s+2)^2 + 25}$$

14.16. Find $\mathscr{L}\{x \cos \sqrt{7}\, x\}$.

This problem can be done two ways.

(a) Using entry 13 of Appendix A with $a = \sqrt{7}$, we have directly that

$$\mathscr{L}\{x \cos \sqrt{7}\, x\} = \frac{s^2 - (\sqrt{7})^2}{[s^2 + (\sqrt{7})^2]^2} = \frac{s^2 - 7}{(s^2 + 7)^2}$$

(b) Set $f(x) = \cos \sqrt{7}\, x$. Using Property 14.3 with $n = 1$ and entry 9 of Appendix A with $a = \sqrt{7}$, we have

$$F(s) = \mathscr{L}\{\cos \sqrt{7}\, x\} = \frac{s}{s^2 + (\sqrt{7})^2} = \frac{s}{s^2 + 7}$$

and

$$\mathscr{L}\{x \cos \sqrt{7}\, x\} = -\frac{d}{ds}\left(\frac{s}{s^2+7}\right) = \frac{s^2-7}{(s^2+7)^2}$$

14.17. Find $\mathscr{L}\{e^{-x} x \cos 2x\}$.

Let $f(x) = x \cos 2x$. From entry 13 of Appendix A with $a = 2$, we obtain

$$F(s) = \frac{s^2 - 4}{(s^2 + 4)^2}$$

Then, from Property 14.2 with $a = -1$,

$$\mathscr{L}\{e^{-x} x \cos 2x\} = F(s+1) = \frac{(s+1)^2 - 4}{[(s+1)^2 + 4]^2}$$

14.18. Find $\mathscr{L}\{x^{7/2}\}$.

Define $f(x) = \sqrt{x}$. Then $x^{7/2} = x^3\sqrt{x} = x^3 f(x)$ and, from entry 4 of Appendix A, we obtain

$$F(s) = \mathscr{L}\{f(x)\} = \mathscr{L}\{\sqrt{x}\} = \frac{1}{2}\sqrt{\pi}\, s^{-3/2}$$

It now follows from Property 14.3 with $n = 3$ that

$$\mathscr{L}\{x^3\sqrt{x}\} = (-1)^3 \frac{d^3}{ds^3}\left(\frac{1}{2}\sqrt{\pi}\, s^{-3/2}\right) = \frac{105}{16}\sqrt{\pi}\, s^{-9/2}$$

which agrees with entry 6 of Appendix A for $n = 4$.

14.19. Find $\mathscr{L}\left\{\frac{\sin 3x}{x}\right\}$.

Taking $f(x) = \sin 3x$, we find from entry 8 of Appendix A with $a = 3$ that

$$F(s) = \frac{3}{s^2 + 9} \qquad \text{or} \qquad F(t) = \frac{3}{t^2 + 9}$$

Then, using Property 14.4, we obtain

$$\begin{aligned}
\mathscr{L}\left\{\frac{\sin 3x}{x}\right\} &= \int_s^\infty \frac{3}{t^2 + 9}\, dt = \lim_{R\to\infty} \int_s^R \frac{3}{t^2 + 9}\, dt \\
&= \lim_{R\to\infty} \arctan \frac{t}{3}\Big|_s^R \\
&= \lim_{R\to\infty} \left(\arctan \frac{R}{3} - \arctan \frac{s}{3}\right) \\
&= \frac{\pi}{2} - \arctan \frac{s}{3}
\end{aligned}$$

14.20. Find $\mathscr{L}\left\{\int_0^x \sinh 2t\, dt\right\}$.

Taking $f(t) = \sinh 2t$, we have $f(x) = \sinh 2x$. It now follows from entry 10 of Appendix A with $a = 2$ that $F(s) = 2/(s^2 - 4)$, and then, from Property 14.5 that

$$\mathscr{L}\left\{\int_0^x \sinh 2t\, dt\right\} = \frac{1}{s}\left(\frac{2}{s^2 - 4}\right) = \frac{2}{s(s^2 - 4)}$$

14.21. Prove that if $f(x + \omega) = -f(x)$, then

$$\mathscr{L}\{f(x)\} = \frac{\int_0^\omega e^{-sx} f(x)\, dx}{1 + e^{-\omega s}} \tag{1}$$

Since

$$f(x + 2\omega) = f[(x + \omega) + \omega] = -f(x + \omega) = -[-f(x)] = f(x)$$

$f(x)$ is periodic with period 2ω. Then, using Property 14.6 with ω replaced by 2ω, we have

$$\mathscr{L}\{f(x)\} = \frac{\int_0^{2\omega} e^{-sx}f(x)\,dx}{1-e^{-2\omega s}} = \frac{\int_0^{\omega} e^{-sx}f(x)\,dx + \int_{\omega}^{2\omega} e^{-sx}f(x)\,dx}{1-e^{-2\omega s}}$$

Substituting $y = x - \omega$ into the second integral, we find that

$$\int_{\omega}^{2\omega} e^{-sx}f(x)\,dx = \int_0^{\omega} e^{-s(y+\omega)}f(y+\omega)\,dy = e^{-\omega s}\int_0^{\omega} e^{-sy}[-f(y)]\,dy$$

$$= -e^{-\omega s}\int_0^{\omega} e^{-sy}f(y)\,dy$$

The last integral, upon changing the dummy variable of integration back to x, equals

$$-e^{-\omega s}\int_0^{\omega} e^{-sx}f(x)\,dx$$

Thus,
$$\mathscr{L}\{f(x)\} = \frac{(1-e^{-\omega s})\int_0^{\omega} e^{-sx}f(x)\,dx}{1-e^{-2\omega s}}$$

$$= \frac{(1-e^{-\omega s})\int_0^{\omega} e^{-sx}f(x)\,dx}{(1-e^{-\omega s})(1+e^{-\omega s})} = \frac{\int_0^{\omega} e^{-sx}f(x)\,dx}{1+e^{-\omega s}}$$

14.22. Find $\mathscr{L}\{f(x)\}$ for the square wave shown in Fig. 14-2.

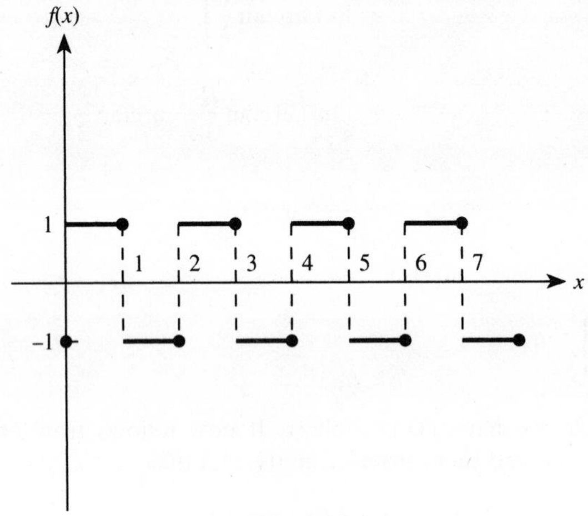

Fig. 14-2

This problem can be done two ways.

(*a*) Note that $f(x)$ is periodic with period $\omega = 2$, and in the interval $0 < x \le 2$ it can be defined analytically by

$$f(x) = \begin{cases} 1 & 0 < x \le 1 \\ -1 & 1 < x \le 2 \end{cases}$$

From Eq. (*14.8*), we have

$$\mathscr{L}\{f(x)\} = \frac{\int_0^{2} e^{-sx}f(x)\,dx}{1-e^{-2s}}$$

Since
$$\int_0^2 e^{-sx}f(x)\,dx = \int_0^1 e^{-sx}(1)\,dx + \int_1^2 e^{-sx}(-1)\,dx$$
$$= \frac{1}{s}(e^{-2s} - 2e^{-s} + 1) = \frac{1}{s}(e^{-s} - 1)^2$$

it follows that
$$F(s) = \frac{(e^{-s} - 1)^2}{s(1 - e^{-2s})} = \frac{(1 - e^{-s})^2}{s(1 - e^{-s})(1 + e^{-s})} = \frac{1 - e^{-s}}{s(1 + e^{-s})}$$
$$= \left[\frac{e^{s/2}}{e^{s/2}}\right]\left[\frac{1 - e^{-s}}{s(1 + e^{-s})}\right] = \frac{e^{s/2} - e^{-s/2}}{s(e^{s/2} + e^{-s/2})} = \frac{1}{s}\tanh\frac{s}{2}$$

(*b*) The square wave $f(x)$ also satisfies the equation $f(x+1) = -f(x)$. Thus, using (*1*) of Problem 14.21 with $\omega = 1$, we obtain
$$\mathscr{L}\{f(x)\} = \frac{\int_0^1 e^{-sx}f(x)\,dx}{1 + e^{-s}} = \frac{\int_0^1 e^{-sx}(1)\,dx}{1 + e^{-s}}$$
$$= \frac{(1/s)(1 - e^{-s})}{1 + e^{-s}} = \frac{1}{s}\tanh\frac{s}{2}$$

14.23. Find the Laplace transform of the function graphed in Fig. 14-3.

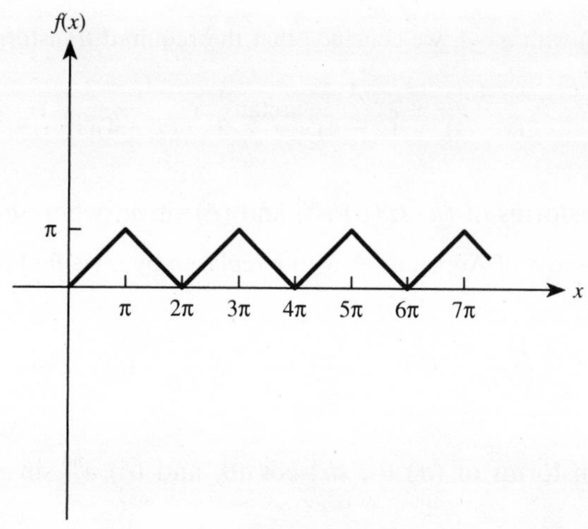

Fig. 14-3

Note that $f(x)$ is periodic with period $\omega = 2\pi$, and in the interval $0 \le x < 2\pi$ it can be defined analytically by
$$f(x) = \begin{cases} x & 0 \le x \le \pi \\ 2\pi - x & \pi \le x < 2\pi \end{cases}$$

From Eq. (*14.8*), we have
$$\mathscr{L}\{f(x)\} = \frac{\int_0^{2\pi} e^{-sx}f(x)\,dx}{1 - e^{-2\pi s}}$$

Since
$$\int_0^{2\pi} e^{-sx}f(x)\,dx = \int_0^{\pi} e^{-sx}x\,dx + \int_{\pi}^{2\pi} e^{-sx}(2\pi - x)\,dx$$
$$= \frac{1}{s^2}(e^{-2\pi s} - 2e^{-\pi s} + 1) = \frac{1}{s^2}(e^{-\pi s} - 1)^2$$

it follows that

$$\mathscr{L}\{f(x)\} = \frac{(1/s^2)(e^{-\pi s} - 1)^2}{1 - e^{-2\pi s}} = \frac{(1/s^2)(e^{-\pi s} - 1)^2}{(1 - e^{-\pi s})(1 + e^{-\pi s})}$$

$$= \frac{1}{s^2}\left(\frac{1 - e^{-\pi s}}{1 + e^{-\pi s}}\right) = \frac{1}{s^2}\tanh\frac{\pi s}{2}$$

14.24. Find $\mathscr{L}\left\{e^{4x}x\int_0^x \frac{1}{t}e^{-4t}\sin 3t\,dt\right\}$.

Using Eq. (*14.4*) with $a = -4$ on the results of Problem 14.19, we obtain

$$\mathscr{L}\left\{\frac{1}{x}e^{-4x}\sin 3x\right\} = \frac{\pi}{2} - \arctan\frac{s+4}{3}$$

It now follows from Eq. (*14.7*) that

$$\mathscr{L}\left\{\int_0^x \frac{1}{t}e^{-4t}\sin 3t\,dt\right\} = \frac{\pi}{2s} - \frac{1}{s}\arctan\frac{s+4}{3}$$

and then from Property 14.3 with $n = 1$,

$$\mathscr{L}\left\{x\int_0^x \frac{1}{t}e^{-4t}\sin 3t\,dt\right\} = \frac{\pi}{2s^2} - \frac{1}{s^2}\arctan\frac{s+4}{3} + \frac{3}{s[9 + (s+4)^2]}$$

Finally, using Eq. (*14.4*) with $a = 4$, we conclude that the required transform is

$$\frac{\pi}{2(s-4)^2} - \frac{1}{(s-4)^2}\arctan\frac{s}{3} + \frac{3}{(s-4)(s^2+9)}$$

14.25. Find the Laplace transforms of (*a*) t, (*b*) e^{at}, and (*c*) $\sin at$, where a denotes a constant.

Using entries 2, 7, and 8 of Appendix A with x replaced by t, we find the Laplace transforms to be, respectively,

$$(a)\quad \mathscr{L}\{t\} = \frac{1}{s^2} \qquad (b)\quad \mathscr{L}\{e^{at}\} = \frac{1}{s-a} \qquad (c)\quad \mathscr{L}\{\sin at\} = \frac{a}{s^2 + a^2}$$

14.26. Find the Laplace transforms of (*a*) θ^2, (*b*) $\cos a\theta$, and (*c*) $e^{b\theta}\sin a\theta$, where a and b denote constants.

Using entries 3 (with $n = 3$), 9, and 15 of Appendix A with x replaced by θ, we find the Laplace transforms to be, respectively,

$$(a)\quad \mathscr{L}\{\theta^2\} = \frac{2}{s^3} \qquad (b)\quad \mathscr{L}\{\cos a\theta\} = \frac{s}{s^2 + a^2} \qquad (c)\quad \mathscr{L}\{e^{b\theta}\sin a\theta\} = \frac{a}{(s-b)^2 + a^2}$$

Supplementary Problems

In Problems 14.27 through 14.42, find the Laplace transforms of the given functions using Eq. (*14.1*).

14.27. $f(x) = 3$

14.28. $f(x) = \sqrt{5}$

14.29. $f(x) = e^{2x}$

14.30. $f(x) = e^{-6x}$

14.31. $f(x) = x$

14.32. $f(x) = -8x$

14.33. $f(x) = \cos 3x$

14.34. $f(x) = \cos 4x$

14.35. $f(x) = \cos bx$, where b denotes a constant

14.36. $f(x) = xe^{-8x}$

14.37. $f(x) = xe^{bx}$, where b denotes a constant

14.38. $f(x) = x^3$

14.39. $f(x) = \begin{cases} x & 0 \le x \le 2 \\ 2 & x > 2 \end{cases}$

14.40. $f(x) = \begin{cases} 1 & 0 \le x \le 1 \\ e^x & 1 < x \le 4 \\ 0 & x > 4 \end{cases}$

14.41. $f(x)$ in Fig. 14-4

14.42. $f(x)$ in Fig. 14-5

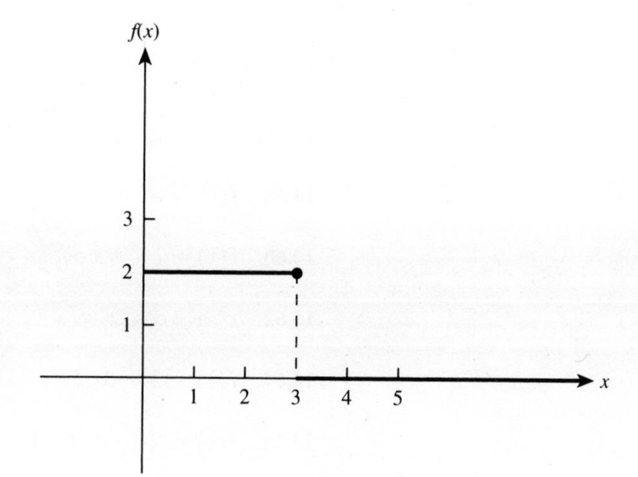

Fig. 14-4

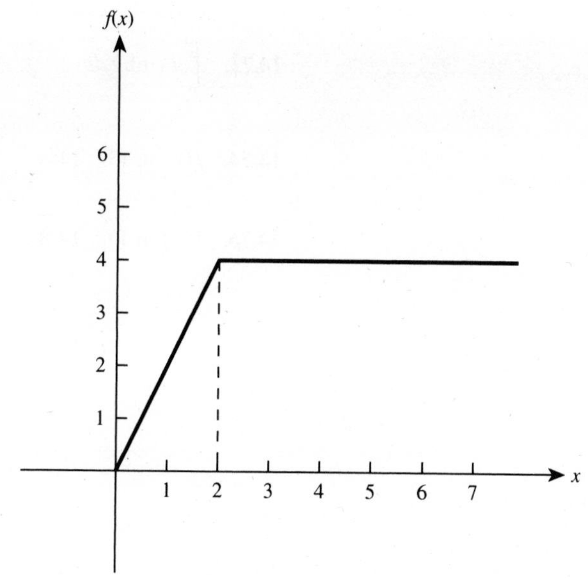

Fig. 14-5

In Problems 14.43 through 14.76, use Appendix A and the Properties 14.1 through 14.6, where appropriate, to find the Laplace transforms of the given functions.

14.43. $f(x) = x^7$

14.44. $f(x) = x \cos 3x$

14.45. $f(x) = x^5 e^{-x}$

14.46. $f(x) = \dfrac{1}{\sqrt{x}}$

14.47. $f(x) = \dfrac{1}{3} e^{-x/3}$

14.48. $f(x) = 5e^{-x/3}$

14.49. $f(x) = 2 \sin^2 \sqrt{3}\, x$

14.50. $f(x) = 8e^{-5x}$

14.51. $f(x) = 3 \sin \dfrac{x}{2}$

14.52. $f(x) = -\cos \sqrt{19}\, x$

14.53. $f(x) = -1.8\sqrt{x}$

14.54. $f(x) = e^{-x} \sin 2x$

14.55. $f(x) = e^{x} \sin 2x$

14.56. $f(x) = e^{x} \cos 2x$

14.57. $f(x) = e^{3x} \cos 2x$

14.58. $f(x) = e^{3x} \cos 5x$

14.59. $f(x) = e^{5x}\sqrt{x}$

14.60. $f(x) = e^{-5x}\sqrt{x}$

14.61. $f(x) = e^{-2x} \sin^2 x$

14.62. $x^3 + 3 \cos 2x$

14.63. $5e^{2x} + 7e^{-x}$

14.64. $f(x) = 2 + 3x$

14.65. $f(x) = 3 - 4x^2$

14.66. $f(x) = 2x + 5 \sin 3x$

14.67. $f(x) = 2 \cos 3x - \sin 3x$

14.68. $2x^2 \cosh x$

14.69. $2x^2 e^{-x} \cosh x$

14.70. $x^2 \sin 4x$

14.71. $\sqrt{x}\, e^{2x}$

14.72. $\displaystyle\int_0^x t \sinh t \, dt$

14.73. $\displaystyle\int_0^x e^{3t} \cos t \, dt$

14.74. $f(x)$ in Fig. 14-6

14.75. $f(x)$ in Fig. 14-7

14.76. $f(x)$ in Fig. 14-8

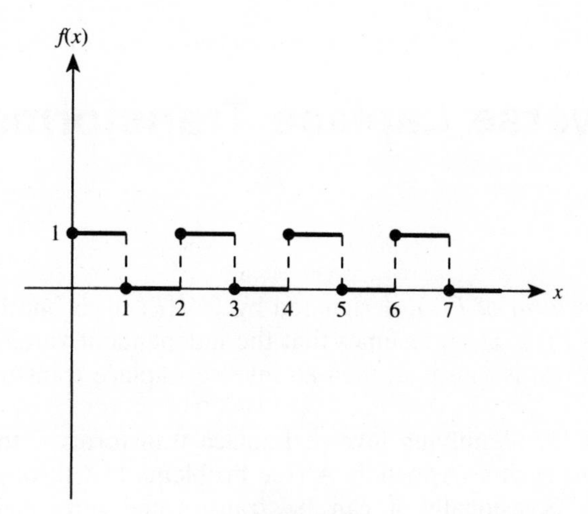

Fig. 14-6

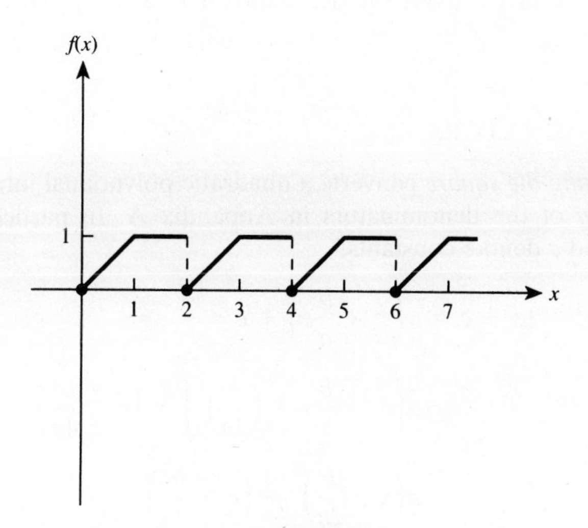

Fig. 14-7

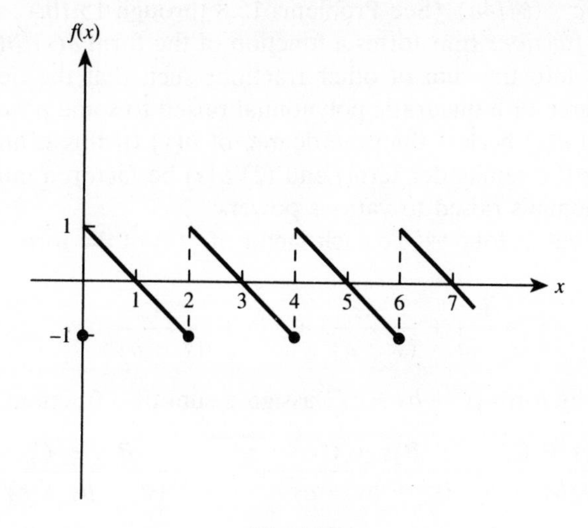

Fig. 14-8

Chapter 15

Inverse Laplace Transforms

DEFINITION

An *inverse Laplace transform* of $F(s)$, designated by $\mathscr{L}^{-1}\{F(s)\}$, is another function $f(x)$ having the property that $\mathscr{L}\{f(x)\} = F(s)$. This presumes that the independent variable of interest is x. If the independent variable of interest is t instead, then an inverse Laplace transform of $F(s)$ is $f(t)$ where $\mathscr{L}\{f(t)\} = F(s)$.

The simplest technique for identifying inverse Laplace transforms is to recognize them, either from memory or from a table such as Appendix A (see Problems 15.1 through 15.3). If $F(s)$ is not in a recognizable form, then occasionally it can be transformed into such a form by algebraic manipulation. Observe from Appendix A that almost all Laplace transforms are quotients. The recommended procedure is to first convert the denominator to a form that appears in Appendix A and then the numerator.

MANIPULATING DENOMINATORS

The method of *completing the square* converts a quadratic polynomial into the sum of squares, a form that appears in many of the denominators in Appendix A. In particular, for the quadratic $as^2 + bs + c$, where a, b, and c denote constants,

$$\begin{aligned} as^2 + bs + c &= a\left(s^2 + \frac{b}{a}s\right) + c \\ &= a\left[s^2 + \frac{b}{a}s + \left(\frac{b}{2a}\right)^2\right] + \left[c - \frac{b^2}{4a}\right] \\ &= a\left(s + \frac{b}{2a}\right)^2 + \left(c - \frac{b^2}{4a}\right) \\ &= a(s+k)^2 + h^2 \end{aligned}$$

where $k = b/2a$ and $h = \sqrt{c - (b^2/4a)}$. (See Problems 15.8 through 15.10.)

The method of *partial fractions* transforms a function of the form $a(s)/b(s)$, where both $a(s)$ and $b(s)$ are polynomials in s, into the sum of other fractions such that the denominator of each new fraction is either a first-degree or a quadratic polynomial raised to some power. The method requires only that (1) the degree of $a(s)$ be less than the degree of $b(s)$ (if this is not the case, first perform long division, and consider the remainder term) and (2) $b(s)$ be factored into the product of distinct linear and quadratic polynomials raised to various powers.

The method is carried out as follows. To each factor of $b(s)$ of the form $(s-a)^m$, assign a sum of m fractions, of the form

$$\frac{A_1}{s-a} + \frac{A_2}{(s-a)^2} + \cdots + \frac{A_m}{(s-a)^m}$$

To each factor of $b(s)$ of the form $(s^2 + bs + c)^p$, assign a sum of p fractions, of the form

$$\frac{B_1 s + C_1}{s^2 + bs + c} + \frac{B_2 s + C_2}{(s^2 + bs + c)^2} + \cdots + \frac{B_p s + C_p}{(s^2 + bs + c)^p}$$

Here A_i, B_j, and C_k $(i = 1, 2, \ldots, m;\ j, k = 1, 2, \ldots, p)$ are constants which still must be determined.

Set the original fraction $a(s)/b(s)$ equal to the sum of the new fractions just constructed. Clear the resulting equation of fractions and then equate coefficients of like powers of s, thereby obtaining a set of simultaneous linear equations in the unknown constants A_i, B_j, and C_k. Finally, solve these equations for A_i, B_j, and C_k. (See Problems 15.11 through 15.14.)

MANIPULATING NUMERATORS

A factor $s - a$ in the numerator may be written in terms of the factor $s - b$, where both a and b are constants, through the identity $s - a = (s - b) + (b - a)$. The multiplicative constant a in the numerator may be written explicitly in terms of the multiplicative constant b through the identity

$$a = \frac{a}{b}(b)$$

Both identities generate recognizable inverse Laplace transforms when they are combined with:

Property 15.1 (Linearity). If the inverse Laplace transforms of two functions $F(s)$ and $G(s)$ exist, then for any constants c_1 and c_2,

$$\mathscr{L}^{-1}\{c_1F(s) + c_2G(s)\} = c_1\mathscr{L}^{-1}\{F(s)\} + c_2\mathscr{L}^{-1}\{G(s)\}$$

(See Problems 15.4 through 15.7.)

Solved Problems

15.1. Find $\mathscr{L}^{-1}\left\{\frac{1}{s}\right\}$.

Here $F(s) = 1/s$. From either Problem 14.4 or entry 1 of Appendix A, we have $\mathscr{L}\{1\} = 1/s$. Therefore, $\mathscr{L}^{-1}\{1/s\} = 1$.

15.2. Find $\mathscr{L}^{-1}\left\{\frac{1}{s-8}\right\}$.

From either Problem 14.6 or entry 7 of Appendix A with $a = 8$, we have

$$\mathscr{L}\{e^{8x}\} = \frac{1}{s-8}$$

Therefore,

$$\mathscr{L}^{-1}\left\{\frac{1}{s-8}\right\} = e^{8x}$$

15.3. Find $\mathscr{L}^{-1}\left\{\frac{s}{s^2+6}\right\}$.

From entry 9 of Appendix A with $a = \sqrt{6}$, we have

$$\mathscr{L}\{\cos\sqrt{6}\,x\} = \frac{s}{s^2 + (\sqrt{6})^2} = \frac{s}{s^2+6}$$

Therefore,

$$\mathscr{L}^{-1}\left\{\frac{s}{s^2+6}\right\} = \cos\sqrt{6}\,x$$

15.4. Find $\mathscr{L}^{-1}\left\{\frac{5s}{(s^2+1)^2}\right\}$.

The given function is similar in form to entry 12 of Appendix A. The denominators become identical if we take $a = 1$. Manipulating the numerator of the given function and using Property 15.1, we obtain

$$\mathscr{L}^{-1}\left\{\frac{5s}{(s^2+1)^2}\right\} = \mathscr{L}^{-1}\left\{\frac{\frac{5}{2}(2s)}{(s^2+1)^2}\right\} = \frac{5}{2}\mathscr{L}^{-1}\left\{\frac{2s}{(s^2+1)^2}\right\} = \frac{5}{2}x \sin x$$

15.5. Find $\mathscr{L}^{-1}\left\{\frac{1}{\sqrt{s}}\right\}$.

The given function is similar in form to entry 5 of Appendix A. Their denominators are identical; manipulating the numerator of the given function and using Property 15.1, we obtain

$$\mathscr{L}^{-1}\left\{\frac{1}{\sqrt{s}}\right\} = \mathscr{L}^{-1}\left\{\frac{1}{\sqrt{\pi}}\frac{\sqrt{\pi}}{\sqrt{s}}\right\} = \frac{1}{\sqrt{\pi}}\mathscr{L}^{-1}\left\{\frac{\sqrt{\pi}}{\sqrt{s}}\right\} = \frac{1}{\sqrt{\pi}}\frac{1}{\sqrt{x}}$$

15.6. Find $\mathscr{L}^{-1}\left\{\frac{s+1}{s^2-9}\right\}$.

The denominator of this function is identical to the denominators of entries 10 and 11 of Appendix A with $a = 3$. Using Property 15.1 followed by a simple algebraic manipulation, we obtain

$$\mathscr{L}^{-1}\left\{\frac{s+1}{s^2-9}\right\} = \mathscr{L}^{-1}\left\{\frac{s}{s^2-9}\right\} + \mathscr{L}^{-1}\left\{\frac{1}{s^2-9}\right\} = \cosh 3x + \mathscr{L}^{-1}\left\{\frac{1}{3}\left(\frac{3}{s^2-(3)^2}\right)\right\}$$

$$= \cosh 3x + \frac{1}{3}\mathscr{L}^{-1}\left\{\frac{3}{s^2-(3)^2}\right\} = \cosh 3x + \frac{1}{3}\sinh 3x$$

15.7. Find $\mathscr{L}^{-1}\left\{\frac{s}{(s-2)^2+9}\right\}$.

The denominator of this function is identical to the denominators of entries 15 and 16 of Appendix A with $a = 3$ and $b = 2$. Both the given function and entry 16 have the *variable* s in their numerators, so they are the most closely matched. Manipulating the numerator of the given function and using Property 15.1, we obtain

$$\mathscr{L}^{-1}\left\{\frac{s}{(s-2)^2+9}\right\} = \mathscr{L}^{-1}\left\{\frac{(s-2)+2}{(s-2)^2+9}\right\} = \mathscr{L}^{-1}\left\{\frac{s-2}{(s-2)^2+9}\right\} + \mathscr{L}^{-1}\left\{\frac{2}{(s-2)^2+9}\right\}$$

$$= e^{2x}\cos 3x + \mathscr{L}^{-1}\left\{\frac{2}{(s-2)^2+9}\right\} = e^{2x}\cos 3x + \mathscr{L}^{-1}\left\{\frac{2}{3}\left(\frac{3}{(s-2)^2+9}\right)\right\}$$

$$= e^{2x}\cos 3x + \frac{2}{3}\mathscr{L}^{-1}\left\{\frac{3}{(s-2)^2+9}\right\} = e^{2x}\cos 3x + \frac{2}{3}e^{2x}\sin 3x$$

15.8. Find $\mathscr{L}^{-1}\left\{\frac{1}{s^2-2s+9}\right\}$.

No function of this form appears in Appendix A. But, by completing the square, we obtain

$$s^2 - 2s + 9 = (s^2 - 2s + 1) + (9 - 1) = (s-1)^2 + (\sqrt{8})^2$$

Hence,

$$\frac{1}{s^2-2s+9} = \frac{1}{(s-1)^2+(\sqrt{8})^2} = \left(\frac{1}{\sqrt{8}}\right)\frac{\sqrt{8}}{(s-1)^2+(\sqrt{8})^2}$$

Then, using Property 15.1 and entry 15 of Appendix A with $a = \sqrt{8}$ and $b = 1$, we find that

$$\mathcal{L}^{-1}\left\{\frac{1}{s^2 - 2s + 9}\right\} = \frac{1}{\sqrt{8}}\mathcal{L}^{-1}\left\{\frac{\sqrt{8}}{(s-1)^2 + (\sqrt{8})^2}\right\} = \frac{1}{\sqrt{8}}e^x \sin\sqrt{8}\,x$$

15.9. Find $\mathcal{L}^{-1}\left\{\dfrac{s+4}{s^2+4s+8}\right\}$.

No function of this form appears in Appendix A. Completing the square in the denominator, we have

$$s^2 + 4s + 8 = (s^2 + 4s + 4) + (8 - 4) = (s+2)^2 + (2)^2$$

Hence,

$$\frac{s+4}{s^2+4s+8} = \frac{s+4}{(s+2)^2 + (2)^2}$$

This expression also is not found in Appendix A. However, if we rewrite the numerator as $s + 4 = (s+2) + 2$ and then decompose the fraction, we have

$$\frac{s+4}{s^2+4s+8} = \frac{s+2}{(s+2)^2+(2)^2} + \frac{2}{(s+2)^2+(2)^2}$$

Then, from entries 15 and 16 of Appendix A,

$$\mathcal{L}^{-1}\left\{\frac{s+4}{s^2+4s+8}\right\} = \mathcal{L}^{-1}\left\{\frac{s+2}{(s+2)^2+(2)^2}\right\} + \mathcal{L}^{-1}\left\{\frac{2}{(s+2)^2+(2)^2}\right\}$$

$$= e^{-2x}\cos 2x + e^{-2x}\sin 2x$$

15.10. Find $\mathcal{L}^{-1}\left\{\dfrac{s+2}{s^2-3s+4}\right\}$.

No function of this form appears in Appendix A. Completing the square in the denominator, we obtain

$$s^2 - 3s + 4 = \left(s^2 - 3s + \frac{9}{4}\right) + \left(4 - \frac{9}{4}\right) = \left(s - \frac{3}{2}\right)^2 + \left(\frac{\sqrt{7}}{2}\right)^2$$

so that

$$\frac{s+2}{s^2-3s+4} = \frac{s+2}{\left(s-\frac{3}{2}\right)^2 + \left(\frac{\sqrt{7}}{2}\right)^2}$$

We now rewrite the numerator as

$$s + 2 = s - \frac{3}{2} + \frac{7}{2} = \left(s - \frac{3}{2}\right) + \sqrt{7}\left(\frac{\sqrt{7}}{2}\right)$$

so that

$$\frac{s+2}{s^2-3s+4} = \frac{s-\frac{3}{2}}{\left(s-\frac{3}{2}\right)^2+\left(\frac{\sqrt{7}}{2}\right)^2} + \sqrt{7}\frac{\frac{\sqrt{7}}{2}}{\left(s-\frac{3}{2}\right)^2+\left(\frac{\sqrt{7}}{2}\right)^2}$$

Then,

$$\mathcal{L}^{-1}\left\{\frac{s+2}{s^2-3s+4}\right\} = \mathcal{L}^{-1}\left\{\frac{s-\frac{3}{2}}{\left(s-\frac{3}{2}\right)^2+\left(\frac{\sqrt{7}}{2}\right)^2}\right\} + \sqrt{7}\,\mathcal{L}^{-1}\left\{\frac{\frac{\sqrt{7}}{2}}{\left(s-\frac{3}{2}\right)^2+\left(\frac{\sqrt{7}}{2}\right)^2}\right\}$$

$$= e^{(3/2)x}\cos\frac{\sqrt{7}}{2}x + \sqrt{7}\,e^{(3/2)x}\sin\frac{\sqrt{7}}{2}x$$

15.11. Use partial fractions to decompose $\dfrac{1}{(s+1)(s^2+1)}$.

To the linear factor $s+1$, we associate the fraction $A/(s+1)$; whereas to the quadratic factor s^2+1, we associate the fraction $(Bs+C)/(s^2+1)$. We then set

$$\frac{1}{(s+1)(s^2+1)} \equiv \frac{A}{s+1} + \frac{Bs+C}{s^2+1} \tag{1}$$

Clearing fractions, we obtain

$$1 \equiv A(s^2+1) + (Bs+C)(s+1) \tag{2}$$

or $$s^2(0) + s(0) + 1 \equiv s^2(A+B) + s(B+C) + (A+C)$$

Equating coefficients of like powers of s, we conclude that $A+B=0$, $B+C=0$, and $A+C=1$. The solution of this set of equations is $A=\frac{1}{2}$, $B=-\frac{1}{2}$, and $C=\frac{1}{2}$. Substituting these values into (*1*), we obtain the partial-fractions decomposition

$$\frac{1}{(s+1)(s^2+1)} \equiv \frac{\frac{1}{2}}{s+1} + \frac{-\frac{1}{2}s+\frac{1}{2}}{s^2+1}$$

The following is an alternative procedure for finding the constants A, B, and C in (*1*). Since (*2*) must hold for all s, it must in particular hold for $s=-1$. Substituting this value into (*2*), we immediately find $A=\frac{1}{2}$. Equation (*2*) must also hold for $s=0$. Substituting this value along with $A=\frac{1}{2}$ into (*2*), we obtain $C=\frac{1}{2}$. Finally, substituting any other value of s into (*2*), we find that $B=-\frac{1}{2}$.

15.12. Use partial fractions to decompose $\dfrac{1}{(s^2+1)(s^2+4s+8)}$.

To the quadratic factors s^2+1 and s^2+4s+8, we associate the fractions $(As+B)/(s^2+1)$ and $(Cs+D)/(s^2+4s+8)$. We set

$$\frac{1}{(s^2+1)(s^2+4s+8)} \equiv \frac{As+B}{s^2+1} + \frac{Cs+D}{s^2+4s+8} \tag{1}$$

and clear fractions to obtain

$$1 \equiv (As+B)(s^2+4s+8) + (Cs+D)(s^2+1)$$

or $$s^3(0) + s^2(0) + s(0) + 1 \equiv s^3(A+C) + s^2(4A+B+D) + s(8A+4B+C) + (8B+D)$$

Equating coefficients of like powers of s, we obtain $A+C=0$, $4A+B+D=0$, $8A+4B+C=0$, and $8B+D=1$. The solution of this set of equations is

$$A=-\frac{4}{65} \qquad B=\frac{7}{65} \qquad C=\frac{4}{65} \qquad D=\frac{9}{65}$$

Therefore, $$\frac{1}{(s^2+1)(s^2+4s+8)} \equiv \frac{-\frac{4}{65}s+\frac{7}{65}}{s^2+1} + \frac{\frac{4}{65}s+\frac{9}{65}}{s^2+4s+8}$$

15.13. Use partial fractions to decompose $\dfrac{s+3}{(s-2)(s+1)}$.

To the linear factors $s-2$ and $s+1$, we associate respectively the fractions $A/(s-2)$ and $B/(s+1)$. We set

$$\frac{s+3}{(s-2)(s+1)} \equiv \frac{A}{s-2} + \frac{B}{s+1}$$

and, upon clearing fractions, obtain

$$s + 3 \equiv A(s + 1) + B(s - 2) \tag{1}$$

To find A and B, we use the alternative procedure suggested in Problem 15.11. Substituting $s = -1$ and then $s = 2$ into (*1*), we immediately obtain $A = 5/3$ and $B = -2/3$. Thus,

$$\frac{s+3}{(s-2)(s+1)} \equiv \frac{5/3}{s-2} - \frac{2/3}{s+1}$$

15.14. Use partial fractions to decompose $\dfrac{8}{s^3(s^2 - s - 2)}$.

Note that $s^2 - s - 2$ factors into $(s-2)(s+1)$. To the factor $s^3 = (s-0)^3$, which is a linear polynomial raised to the third power, we associate the sum $A_1/s + A_2/s^2 + A_3/s^3$. To the linear factors $(s-2)$ and $(s+1)$, we associate the fractions $B/(s-2)$ and $C/(s+1)$. Then

$$\frac{8}{s^3(s^2 - s - 2)} \equiv \frac{A_1}{s} + \frac{A_2}{s^2} + \frac{A_3}{s^3} + \frac{B}{s-2} + \frac{C}{s+1}$$

or, clearing fractions,

$$8 \equiv A_1 s^2(s-2)(s+1) + A_2 s(s-2)(s+1) + A_3(s-2)(s+1) + Bs^3(s+1) + Cs^3(s-2)$$

Letting $s = -1$, 2, and 0, consecutively, we obtain, respectively, $C = 8/3$, $B = 1/3$, and $A_3 = -4$. Then choosing $s = 1$ and $s = -2$, and simplifying, we obtain the equations $A_1 + A_2 = -1$ and $2A_1 - A_2 = -8$, which have the solutions $A_1 = -3$ and $A_2 = 2$. Note that any other two values for s (not -1, 2, or 0) will also do; the resulting equations may be different, but the solution will be identical. Finally,

$$\frac{2}{s^3(s^2 - s - 2)} \equiv -\frac{3}{s} + \frac{2}{s^2} - \frac{4}{s^3} + \frac{1/3}{s-2} + \frac{8/3}{s+1}$$

15.15. Find $\mathscr{L}^{-1}\left\{\dfrac{s+3}{(s-2)(s+1)}\right\}$.

No function of this form appears in Appendix A. Using the results of Problem 15.13 and Property 15.1, we obtain

$$\mathscr{L}^{-1}\left\{\frac{s+3}{(s-2)(s+1)}\right\} = \frac{5}{3}\mathscr{L}^{-1}\left\{\frac{1}{s-2}\right\} - \frac{2}{3}\mathscr{L}^{-1}\left\{\frac{1}{s+1}\right\}$$

$$= \frac{5}{3}e^{2x} - \frac{2}{3}e^{-x}$$

15.16. Find $\mathscr{L}^{-1}\left\{\dfrac{8}{s^3(s^2 - s - 2)}\right\}$.

No function of this form appears in Appendix A. Using the results of Problem 15.14 and Property 15.1, we obtain

$$\mathscr{L}^{-1}\left\{\frac{8}{s^3(s^2 - s - 2)}\right\} = -3\mathscr{L}^{-1}\left\{\frac{1}{s}\right\} + 2\mathscr{L}^{-1}\left\{\frac{1}{s^2}\right\}$$

$$- 2\mathscr{L}^{-1}\left\{\frac{2}{s^3}\right\} + \frac{1}{3}\mathscr{L}^{-1}\left\{\frac{1}{s-2}\right\} + \frac{8}{3}\mathscr{L}^{-1}\left\{\frac{1}{s+1}\right\}$$

$$= -3 + 2x - 2x^2 + \frac{1}{3}e^{2x} + \frac{8}{3}e^{-x}$$

15.17. Find $\mathscr{L}^{-1}\left\{\dfrac{1}{(s+1)(s^2+1)}\right\}$.

Using the results of Problem 15.11, and noting that

$$\frac{-\frac{1}{2}s+\frac{1}{2}}{s^2+1} = -\frac{1}{2}\left(\frac{s}{s^2+1}\right) + \frac{1}{2}\left(\frac{1}{s^2+1}\right)$$

we find that

$$\mathscr{L}^{-1}\left\{\frac{1}{(s+1)(s^2+1)}\right\} = \frac{1}{2}\mathscr{L}^{-1}\left\{\frac{1}{s+1}\right\} - \frac{1}{2}\mathscr{L}^{-1}\left\{\frac{s}{s^2+1}\right\} + \frac{1}{2}\mathscr{L}^{-1}\left\{\frac{1}{s^2+1}\right\}$$

$$= \frac{1}{2}e^{-x} - \frac{1}{2}\cos x + \frac{1}{2}\sin x$$

15.18. Find $\mathscr{L}^{-1}\left\{\dfrac{1}{(s^2+1)(s^2+4s+8)}\right\}$.

From Problem 15.12, we have

$$\mathscr{L}^{-1}\left\{\frac{1}{(s^2+1)(s^2+4s+8)}\right\} = \mathscr{L}^{-1}\left\{\frac{-\frac{4}{65}s+\frac{7}{65}}{s^2+1}\right\} + \mathscr{L}^{-1}\left\{\frac{\frac{4}{65}s+\frac{9}{65}}{s^2+4s+8}\right\}$$

The first term can be evaluated easily if we note that

$$\frac{-\frac{4}{65}s+\frac{7}{65}}{s^2+1} = \left(-\frac{4}{65}\right)\frac{s}{s^2+1} + \left(\frac{7}{65}\right)\frac{1}{s^2+1}$$

To evaluate the second inverse transform, we must first complete the square in the denominator, $s^2+4s+8=(s+2)^2+(2)^2$, and then note that

$$\frac{\frac{4}{65}s+\frac{9}{65}}{s^2+4s+8} = \frac{4}{65}\left[\frac{s+2}{(s+2)^2+(2)^2}\right] + \frac{1}{130}\left[\frac{2}{(s+2)^2+(2)^2}\right]$$

Therefore,

$$\mathscr{L}^{-1}\left\{\frac{1}{(s^2+1)(s^2+4s+8)}\right\} = -\frac{4}{65}\mathscr{L}^{-1}\left\{\frac{s}{s^2+1}\right\} + \frac{7}{65}\mathscr{L}^{-1}\left\{\frac{1}{s^2+1}\right\}$$

$$+ \frac{4}{65}\mathscr{L}^{-1}\left\{\frac{s+2}{(s+2)^2+(2)^2}\right\} + \frac{1}{130}\mathscr{L}^{-1}\left\{\frac{2}{(s+2)^2+(2)^2}\right\}$$

$$= -\frac{4}{65}\cos x + \frac{7}{65}\sin x + \frac{4}{65}e^{-2x}\cos 2x + \frac{1}{130}e^{-2x}\sin 2x$$

15.19. Find $\mathscr{L}^{-1}\left\{\dfrac{1}{s(s^2+4)}\right\}$.

By the method of partial fractions, we obtain

$$\frac{1}{s(s^2+4)} \equiv \frac{1/4}{s} + \frac{(-1/4)s}{s^2+4}$$

Thus,

$$\mathscr{L}^{-1}\left\{\frac{1}{s(s^2+4)}\right\} = \frac{1}{4}\mathscr{L}^{-1}\left\{\frac{1}{s}\right\} - \frac{1}{4}\mathscr{L}^{-1}\left\{\frac{s}{s^2+4}\right\} = \frac{1}{4} - \frac{1}{4}\cos 2x$$

Supplementary Problems

Find the inverse Laplace transforms, as a function of x, of the following functions:

15.20. $\dfrac{1}{s^2}$

15.21. $\dfrac{2}{s^2}$

15.22. $\dfrac{2}{s^3}$

15.23. $\dfrac{1}{s^3}$

15.24. $\dfrac{1}{s^4}$

15.25. $\dfrac{1}{s+2}$

15.26. $\dfrac{-2}{s-2}$

15.27. $\dfrac{12}{3s+9}$

15.28. $\dfrac{1}{2s-3}$

15.29. $\dfrac{1}{(s-2)^3}$

15.30. $\dfrac{12}{(s+5)^4}$

15.31. $\dfrac{3s^2}{(s^2+1)^2}$

15.32. $\dfrac{s^2}{(s^2+3)^2}$

15.33. $\dfrac{1}{s^2+4}$

15.34. $\dfrac{2}{(s-2)^2+9}$

15.35. $\dfrac{s}{(s+1)^2+5}$

15.36. $\dfrac{2s+1}{(s-1)^2+7}$

15.37. $\dfrac{1}{2s^2+1}$

15.38. $\dfrac{1}{s^2-2s+2}$

15.39. $\dfrac{s+3}{s^2+2s+5}$

15.40. $\dfrac{s}{s^2-s+17/4}$

15.41. $\dfrac{s+1}{s^2+3s+5}$

15.42. $\dfrac{2s^2}{(s-1)(s^2+1)}$

15.43. $\dfrac{1}{s^2-1}$

15.44. $\dfrac{2}{(s^2+1)(s-1)^2}$

15.45. $\dfrac{s+2}{s^3}$

15.46. $\dfrac{-s+6}{s^3}$

15.47. $\dfrac{s^3+3s}{s^6}$

15.48. $\dfrac{12+15\sqrt{s}}{s^4}$

15.49. $\dfrac{2s-13}{s(s^2-4s+13)}$

15.50. $\dfrac{2(s-1)}{s^2-s+1}$

15.51. $\dfrac{s}{(s^2+9)^2}$

15.52. $\dfrac{1}{2(s-1)(s^2-s-1)} = \dfrac{1/2}{(s-1)(s^2-s-1)}$

15.53. $\dfrac{s}{2s^2+4s+5/2} = \dfrac{(1/2)s}{s^2+2s+5/4}$

Chapter 16

Convolutions and the Unit Step Function

CONVOLUTIONS

The *convolution* of two functions $f(x)$ and $g(x)$ is

$$f(x) * g(x) = \int_0^x f(t)g(x - t)\, dt \tag{16.1}$$

Theorem 16.1. $f(x) * g(x) = g(x) * f(x)$.

Theorem 16.2 (Convolution theorem). If $\mathscr{L}\{f(x)\} = F(s)$ and $\mathscr{L}\{g(x)\} = G(s)$, then

$$\mathscr{L}\{f(x) * g(x)\} = \mathscr{L}\{f(x)\}\mathscr{L}\{g(x)\} = F(s)G(s)$$

It follows directly from these two theorems that

$$\mathscr{L}^{-1}\{F(s)G(s)\} = f(x) * g(x) = g(x) * f(x) \tag{16.2}$$

If one of the two convolutions in Eq. (*16.2*) is simpler to calculate, then that convolution is chosen when determining the inverse Laplace transform of a product.

UNIT STEP FUNCTION

The *unit step function* $u(x)$ is defined as

$$u(x) = \begin{cases} 0 & x < 0 \\ 1 & x \geq 0 \end{cases}$$

As an immediate consequence of the definition, we have for any number c,

$$u(x - c) = \begin{cases} 0 & x < c \\ 1 & x \geq c \end{cases}$$

The graph of $u(x - c)$ is given in Fig. 16-1.

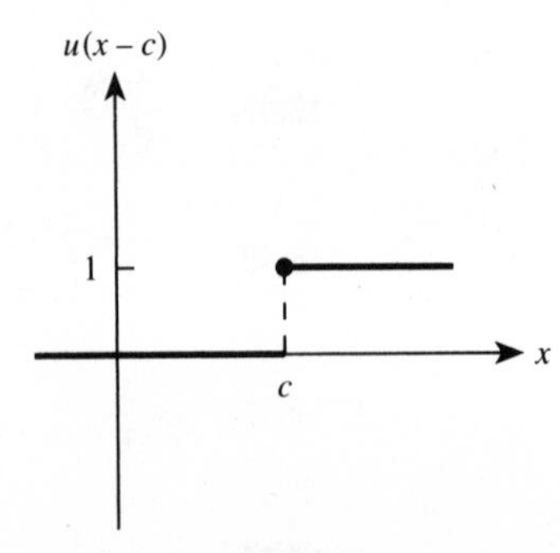

Fig. 16-1

Theorem 16.3. $\mathscr{L}\{u(x-c)\} = \dfrac{1}{s}e^{-cs}$.

TRANSLATIONS

Given a function $f(x)$ defined for $x \geq 0$, the function

$$u(x-c)f(x-c) = \begin{cases} 0 & x < c \\ f(x-c) & x \geq c \end{cases}$$

represents a shift, or translation, of the function $f(x)$ by c units in the positive x-direction. For example, if $f(x)$ is given graphically by Fig. 16-2, then $u(x-c)f(x-c)$ is given graphically by Fig. 16-3.

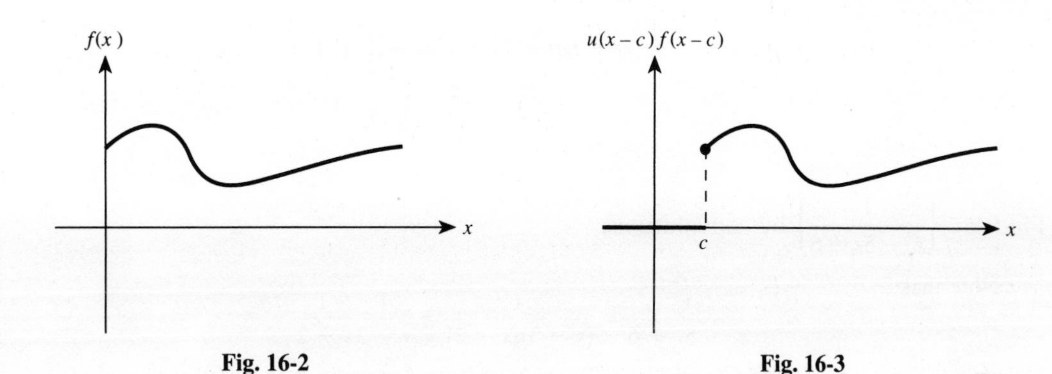

Fig. 16-2 **Fig. 16-3**

Theorem 16.4. If $F(s) = \mathscr{L}\{f(x)\}$, then

$$\mathscr{L}\{u(x-c)f(x-c)\} = e^{-cs}F(s)$$

Conversely,

$$\mathscr{L}^{-1}\{e^{-cs}F(s)\} = u(x-c)f(x-c) = \begin{cases} 0 & x < c \\ f(x-c) & x \geq c \end{cases}$$

Solved Problems

16.1. Find $f(x) * g(x)$ when $f(x) = e^{3x}$ and $g(x) = e^{2x}$.

Here $f(t) = e^{3t}$, $g(x-t) = e^{2(x-t)}$, and

$$f(x) * g(x) = \int_0^x e^{3t}e^{2(x-t)}\,dt = \int_0^x e^{3t}e^{2x}e^{-2t}\,dt$$

$$= e^{2x}\int_0^x e^t\,dt = e^{2x}\left[e^t\right]_{t=0}^{t=x} = e^{2x}(e^x - 1) = e^{3x} - e^{2x}$$

16.2. Find $g(x)*f(x)$ for the two functions in Problem 16.1 and verify Theorem 16.1.

With $f(x-t)=e^{3(x-t)}$ and $g(t)=e^{2t}$,

$$g(x)*f(x) = \int_0^x g(t)f(x-t)\,dt = \int_0^x e^{2t}e^{3(x-t)}\,dt$$

$$= e^{3x}\int_0^x e^{-t}\,dt = e^{3x}\left[-e^{-t}\right]_{t=0}^{t=x}$$

$$= e^{3x}(-e^{-x}+1) = e^{3x} - e^{2x}$$

which, from Problem 16.1, equals $f(x)*g(x)$.

16.3. Find $f(x)*g(x)$ when $f(x)=x$ and $g(x)=x^2$.

Here $f(t)=t$ and $g(x-t)=(x-t)^2=x^2-2xt+t^2$. Thus,

$$f(x)*g(x) = \int_0^x t(x^2-2xt+t^2)\,dt$$

$$= x^2\int_0^x t\,dt - 2x\int_0^x t^2\,dt + \int_0^x t^3\,dt$$

$$= x^2\frac{x^2}{2} - 2x\frac{x^3}{3} + \frac{x^4}{4} = \frac{1}{12}x^4$$

16.4. Find $\mathscr{L}^{-1}\left\{\frac{1}{s^2-5s+6}\right\}$ by convolutions.

Note that

$$\frac{1}{s^2-5s+6} = \frac{1}{(s-3)(s-2)} = \frac{1}{s-3}\frac{1}{s-2}$$

Defining $F(s)=1/(s-3)$ and $G(s)=1/(s-2)$, we have from Appendix A that $f(x)=e^{3x}$ and $g(x)=e^{2x}$. It follows from Eq. (*16.2*) and the results of Problem 16.1 that

$$\mathscr{L}^{-1}\left\{\frac{1}{s^2-5s+6}\right\} = f(x)*g(x) = e^{3x}*e^{2x} = e^{3x}-e^{2x}$$

16.5. Find $\mathscr{L}^{-1}\left\{\frac{6}{s^2-1}\right\}$ by convolutions.

Note that

$$\mathscr{L}^{-1}\left\{\frac{6}{s^2-1}\right\} = \mathscr{L}^{-1}\left\{\frac{6}{(s-1)(s+1)}\right\} = 6\mathscr{L}^{-1}\left\{\frac{1}{(s-1)}\frac{1}{(s+1)}\right\}$$

Defining $F(s)=1/(s-1)$ and $G(s)=1/(s+1)$, we have from Appendix A that $f(x)=e^{x}$ and $g(x)=e^{-x}$. It follows from Eq. (*16.2*) that

$$\mathscr{L}^{-1}\left\{\frac{6}{s^2-1}\right\} = 6\mathscr{L}^{-1}\{F(s)G(s)\} = 6e^{x}*e^{-x}$$

$$= 6\int_0^x e^{t}e^{-(x-t)}\,dt = 6e^{-x}\int_0^x e^{2t}\,dt$$

$$= 6e^{-x}\left[\frac{e^{2x}-1}{2}\right] = 3e^{x} - 3e^{-x}$$

16.6. Find $\mathscr{L}^{-1}\left\{\frac{1}{s(s^2+4)}\right\}$ by convolutions.

Note that

$$\frac{1}{s(s^2+4)} = \frac{1}{s}\frac{1}{s^2+4}$$

Defining $F(s) = 1/s$ and $G(s) = 1/(s^2 + 4)$, we have from Appendix A that $f(x) = 1$ and $g(x) = \frac{1}{2}\sin 2x$. It now follows from Eq. (*16.2*) that

$$\mathscr{L}^{-1}\left\{\frac{1}{s(s^2+4)}\right\} = \mathscr{L}^{-1}\{F(s)G(s)\} = g(x) * f(x)$$

$$= \int_0^x g(t)f(x-t)\,dt = \int_0^x \left(\frac{1}{2}\sin 2t\right)(1)\,dt$$

$$= \frac{1}{4}(1 - \cos 2x)$$

See also Problem 15.19.

16.7. Find $\mathscr{L}^{-1}\left\{\frac{1}{(s-1)^2}\right\}$ by convolutions.

If we define $F(s) = G(s) = 1/(s-1)$, then $f(x) = g(x) = e^x$ and

$$\mathscr{L}^{-1}\left\{\frac{1}{(s-1)^2}\right\} = \mathscr{L}^{-1}\{F(s)G(s)\} = f(x) * g(x)$$

$$= \int_0^x f(t)g(x-t)\,dt = \int_0^x e^t e^{x-t}\,dt$$

$$= e^x \int_0^x (1)\,dt = xe^x$$

16.8. Use the definition of the Laplace transform to find $\mathscr{L}\{u(x-c)\}$ and thereby prove Theorem 16.3.

It follows directly from Eq. (*14.1*) that

$$\mathscr{L}\{u(x-c)\} = \int_0^\infty e^{-sx}u(x-c)\,dx = \int_0^c e^{-sx}(0)\,dx + \int_c^\infty e^{-sx}(1)\,dx$$

$$= \int_c^\infty e^{-sx}\,dx = \lim_{R\to\infty}\int_c^R e^{-sx}\,dx = \lim_{R\to\infty}\frac{e^{-sR} - e^{-sc}}{-s}$$

$$= \frac{1}{s}e^{-sc} \qquad (\text{if } s > 0)$$

16.9. Graph the function $f(x) = u(x-2) - u(x-3)$.

Note that

$$u(x-2) = \begin{cases} 0 & x < 2 \\ 1 & x \geq 2 \end{cases} \qquad \text{and} \qquad u(x-3) = \begin{cases} 0 & x < 3 \\ 1 & x \geq 3 \end{cases}$$

Thus,

$$f(x) = u(x-2) - u(x-3) = \begin{cases} 0 - 0 = 0 & x < 2 \\ 1 - 0 = 1 & 2 \leq x < 3 \\ 1 - 1 = 0 & x \geq 3 \end{cases}$$

the graph of which is given in Fig. 16-4.

16.10. Graph the function $f(x) = 5 - 5u(x-8)$ for $x \geq 0$.

Note that

$$5u(x-8) = \begin{cases} 0 & x < 8 \\ 5 & x \geq 8 \end{cases}$$

Thus

$$f(x) = 5 - 5u(x-8) = \begin{cases} 5 & x < 8 \\ 0 & x \geq 8 \end{cases}$$

The graph of this function when $x \geq 0$ is given in Fig. 16-5.

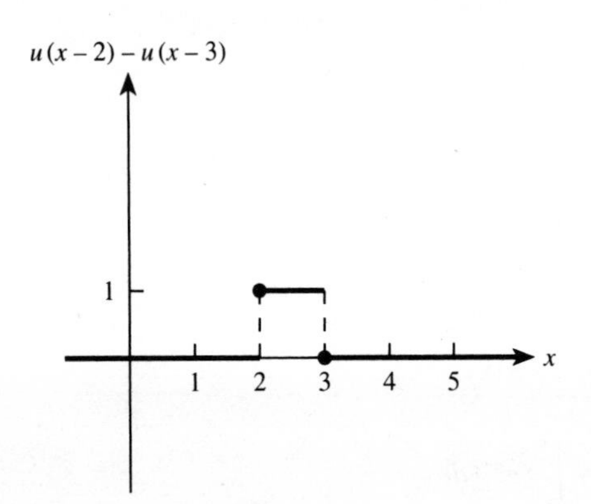

Fig. 16-4

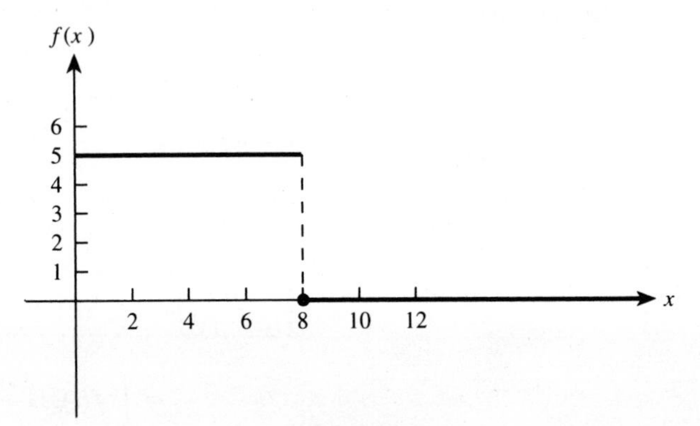

Fig. 16-5

16.11. Use the unit step function to give an analytic representation of the function $f(x)$ graphed in Fig. 16-6.

Note that $f(x)$ is the function $g(x) = x$, $x \geq 0$, translated four units in the positive x-direction. Thus, $f(x) = u(x-4)g(x-4) = (x-4)u(x-4)$.

16.12. Use the unit step function to give an analytic description of the function $g(x)$ graphed on the interval $[0, \infty)$ in Fig. 16-7 if on the subinterval $[0, a)$ the graph is identical to Fig. 16-2.

Let $f(x)$ represent the function graphed in Fig. 16-2. Then $g(x) = f(x)[1 - u(x-a)]$.

16.13. Find $\mathscr{L}\{g(x)\}$ if $g(x) = \begin{cases} 0 & x < 4 \\ (x-4)^2 & x \geq 4 \end{cases}$.

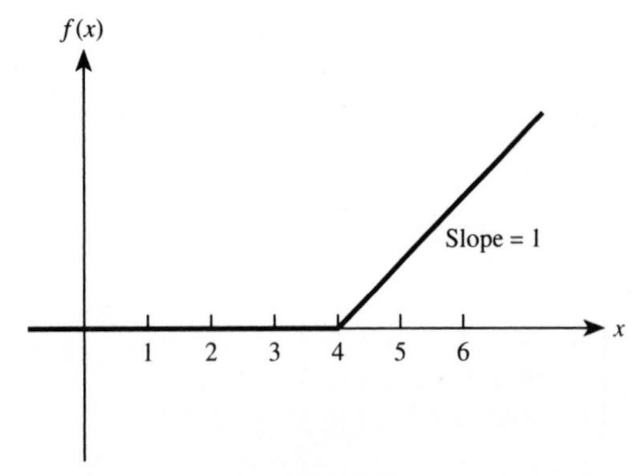

Fig. 16-6

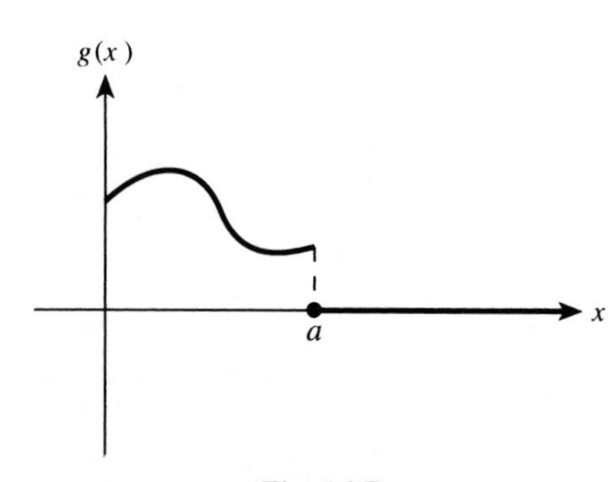

Fig. 16-7

If we define $f(x) = x^2$, then $g(x)$ can be given compactly as $g(x) = u(x-4)f(x-4) = u(x-4)(x-4)^2$. Then, noting that $\mathscr{L}\{f(x)\} = F(s) = 2/s^3$ and using Theorem 16.4, we conclude that

$$\mathscr{L}\{g(x)\} = \mathscr{L}\{u(x-4)(x-4)^2\} = e^{-4s}\frac{2}{s^3}$$

16.14. Find $\mathscr{L}\{g(x)\}$ if $g(x) = \begin{cases} 0 & x < 4 \\ x^2 & x \geq 4 \end{cases}$.

We first determine a function $f(x)$ such that $f(x-4) = x^2$. Once this has been done, $g(x)$ can be written as $g(x) = u(x-4)f(x-4)$ and Theorem 16.4 can be applied. Now, $f(x-4) = x^2$ only if

$$f(x) = f(x+4-4) = (x+4)^2 = x^2 + 8x + 16$$

Since
$$\mathscr{L}\{f(x)\} = \mathscr{L}\{x^2\} + 8\mathscr{L}\{x\} + 16\mathscr{L}\{1\} = \frac{2}{s^3} + \frac{8}{s^2} + \frac{16}{s}$$

it follows that

$$\mathscr{L}\{g(x)\} = \mathscr{L}\{u(x-4)f(x-4)\} = e^{-4s}\left(\frac{2}{s^3} + \frac{8}{s^2} + \frac{16}{s}\right)$$

16.15. Prove Theorem 16.1.

Making the substitution $\tau = x - t$ in the right-hand side of Eq. (*16.1*), we have

$$\begin{aligned} f(x) * g(x) &= \int_0^x f(t)g(x-t)\,dt = \int_x^0 f(x-\tau)g(\tau)(-d\tau) \\ &= -\int_x^0 g(\tau)f(x-\tau)\,d\tau = \int_0^x g(\tau)f(x-\tau)\,d\tau \\ &= g(x) * f(x) \end{aligned}$$

16.16. Prove that $f(x) * [g(x) + h(x)] = f(x) * g(x) + f(x) * h(x)$.

$$\begin{aligned} f(x) * [g(x) + h(x)] &= \int_0^x f(t)[g(x-t) + h(x-t)]\,dt \\ &= \int_0^x [f(t)g(x-t) + f(t)h(x-t)]\,dt \\ &= \int_0^x f(t)g(x-t)\,dt + \int_0^x f(t)h(x-t)\,dt \\ &= f(x) * g(x) + f(x) * h(x) \end{aligned}$$

Supplementary Problems

16.17. Find $x * x$.

16.18. Find $2 * x$.

16.19. Find $4x * e^{2x}$.

16.20. Find $e^{4x} * e^{-2x}$.

16.21. Find $x * e^x$.

16.22. Find $x * xe^{-x}$.

16.23. Find $3 * \sin 2x$.

16.24. Find $x * \cos x$.

In Problems 16.25 through 16.32, use convolutions to find the inverse Laplace transforms of the given functions.

16.25. $\dfrac{1}{(s-1)(s-2)}$

16.26. $\dfrac{1}{(s)(s)}$

16.27. $\dfrac{2}{s(s+1)}$

16.28. $\dfrac{1}{s^2+3s-40}$

16.29. $\dfrac{3}{s^2(s^2+3)}$

16.30. $\dfrac{1}{s(s^2+4)}$ with $F(s) = 1/s^2$ and $G(s) = s/(s^2+4)$. Compare with Problem 16.6.

16.31. $\dfrac{9}{s(s^2+9)}$

16.32. $\dfrac{9}{s^2(s^2+9)}$

16.33 Graph $f(x) = 2u(x-2) - u(x-4)$.

16.34. Graph $f(x) = u(x-2) - 2u(x-3) + u(x-4)$.

16.35. Use the unit step function to give an analytic representation for the function graphed in Fig. 16-8.

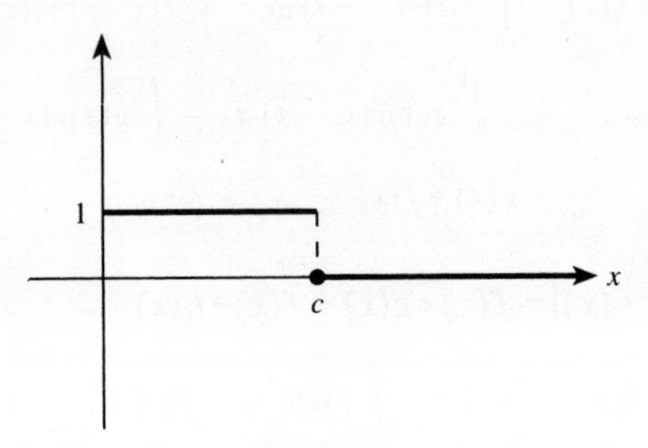

Fig. 16-8

16.36. Graph $f(x) = u(x-\pi)\cos 2(x-\pi)$.

16.37. Graph $f(x) = \dfrac{1}{2}(x-1)^2 u(x-1)$.

In Problems 16.38 through 16.45, find $\mathscr{L}\{g(x)\}$ for the given functions.

16.38. $g(x) = \begin{cases} 0 & x < 1 \\ \sin(x-1) & x \geq 1 \end{cases}$

16.39. $g(x) = \begin{cases} 0 & x < 3 \\ x-3 & x \geq 3 \end{cases}$

16.40. $g(x) = \begin{cases} 0 & x < 3 \\ x & x \geq 3 \end{cases}$

16.41. $g(x) = \begin{cases} 0 & x < 3 \\ x+1 & x \geq 3 \end{cases}$

16.42. $g(x) = \begin{cases} 0 & x < 5 \\ e^{x-5} & x \geq 5 \end{cases}$

16.43. $g(x) = \begin{cases} 0 & x < 5 \\ e^{x} & x \geq 5 \end{cases}$

16.44. $g(x) = \begin{cases} 0 & x < 2 \\ e^{x-5} & x \geq 2 \end{cases}$

16.45. $g(x) = \begin{cases} 0 & x < 2 \\ x^3+1 & x \geq 2 \end{cases}$

In Problems 16.46 through 16.52, determine the inverse Laplace transforms of the given functions.

16.46. $\dfrac{s}{s^2+4}e^{-3s}$

16.47. $\dfrac{1}{s^2+4}e^{-5s}$

16.48. $\dfrac{1}{s^2+4}e^{-\pi s}$

16.49. $\dfrac{2}{s-3}e^{-2s}$

16.50. $\dfrac{8}{s+3}e^{-s}$

16.51. $\dfrac{1}{s^3}e^{-2s}$

16.52. $\dfrac{1}{s^2}e^{-\pi s}$

16.53. Prove that for any constant k, $[kf(x)]*g(x)=k[f(x)*g(x)]$.

Chapter 17

Solutions of Linear Differential Equations with Constant Coefficients by Laplace Transforms

LAPLACE TRANSFORMS OF DERIVATIVES

Denote $\mathscr{L}\{y(x)\}$ by $Y(s)$. Then under very broad conditions, the Laplace transform of the nth-derivative $(n = 1, 2, 3, \ldots)$ of $y(x)$ is

$$\mathscr{L}\left\{\frac{d^n y}{dx^n}\right\} = s^n Y(s) - s^{n-1}y(0) - s^{n-2}y'(0) - \cdots - sy^{(n-2)}(0) - y^{(n-1)}(0) \tag{17.1}$$

If the initial conditions on $y(x)$ at $x = 0$ are given by

$$y(0) = c_0, \qquad y'(0) = c_1, \ldots, y^{(n-1)}(0) = c_{n-1} \tag{17.2}$$

then (*17.1*) can be rewritten as

$$\mathscr{L}\left\{\frac{d^n y}{dx^n}\right\} = s^n Y(s) - c_0 s^{n-1} - c_1 s^{n-2} - \cdots - c_{n-2}s - c_{n-1} \tag{17.3}$$

For the special cases of $n = 1$ and $n = 2$, Eq. (*17.3*) simplifies to

$$\mathscr{L}\{y'(x)\} = sY(s) - c_0 \tag{17.4}$$

$$\mathscr{L}\{y''(x)\} = s^2 Y(s) - c_0 s - c_1 \tag{17.5}$$

SOLUTIONS OF DIFFERENTIAL EQUATIONS

Laplace transforms are used to solve initial-value problems given by the nth-order linear differential equation with constant coefficients

$$b_n \frac{d^n y}{dx^n} + b_{n-1}\frac{d^{n-1}y}{dx^{n-1}} + \cdots + b_1 \frac{dy}{dx} + b_0 y = g(x) \tag{17.6}$$

together with the initial conditions specified in Eq. (*17.2*). First, take the Laplace transform of both sides of Eq. (*17.6*), thereby obtaining an algebraic equation for $Y(s)$. Then solve for $Y(s)$ *algebraically,* and finally take inverse Laplace transforms to obtain $y(x) = \mathscr{L}^{-1}\{Y(s)\}$.

Unlike previous methods, where first the differential equation is solved and then the initial conditions are applied to evaluate the arbitrary constants, the Laplace transform method solves the entire initial-value problem in one step. There are two exceptions: when no initial conditions are specified and when the initial conditions are not at $x = 0$. In these situations, c_0 through c_n in Eqs. (*17.2*) and (*17.3*) remain arbitrary and the solution to differential equation (*17.6*) is found in terms of these constants. They are then evaluated separately when appropriate subsidiary conditions are provided. (See Problems 17.11 through 17.13.)

Solved Problems

17.1. Solve $y' - 5y = 0$; $y(0) = 2$.

Taking the Laplace transform of both sides of this differential equation and using Property 14.1, we obtain $\mathscr{L}\{y'\} - 5\mathscr{L}\{y\} = \mathscr{L}\{0\}$. Then, using Eq. (*17.4*) with $c_0 = 2$, we find

$$[sY(s) - 2] - 5Y(s) = 0 \qquad \text{from which} \qquad Y(s) = \frac{2}{s-5}$$

Finally, taking the inverse Laplace transform of $Y(s)$, we obtain

$$y(x) = \mathscr{L}^{-1}\{Y(s)\} = \mathscr{L}^{-1}\left\{\frac{2}{s-5}\right\} = 2\mathscr{L}^{-1}\left\{\frac{1}{s-5}\right\} = 2e^{5x}$$

17.2. Solve $y' - 5y = e^{5x}$; $y(0) = 0$.

Taking the Laplace transform of both sides of this differential equation and using Property 14.1, we find that $\mathscr{L}\{y'\} - 5\mathscr{L}\{y\} = \mathscr{L}\{e^{5x}\}$. Then, using Appendix A and Eq. (*17.4*) with $c_0 = 0$, we obtain

$$[sY(s) - 0] - 5Y(s) = \frac{1}{s-5} \qquad \text{from which} \qquad Y(s) = \frac{1}{(s-5)^2}$$

Finally, taking the inverse Laplace transform of $Y(s)$, we obtain

$$y(x) = \mathscr{L}^{-1}\{Y(s)\} = \mathscr{L}^{-1}\left\{\frac{1}{(s-5)^2}\right\} = xe^{5x}$$

(see Appendix A, entry 14).

17.3. Solve $y' + y = \sin x$; $y(0) = 1$.

Taking the Laplace transform of both sides of the differential equation, we obtain

$$\mathscr{L}\{y'\} + \mathscr{L}\{y\} = \mathscr{L}\{\sin x\} \qquad \text{or} \qquad [sY(s) - 1] + Y(s) = \frac{1}{s^2+1}$$

Solving for $Y(s)$, we find

$$Y(s) = \frac{1}{(s+1)(s^2+1)} + \frac{1}{s+1}$$

Taking the inverse Laplace transform, and using the result of Problem 15.17, we obtain

$$y(x) = \mathscr{L}^{-1}\{Y(s)\} = \mathscr{L}^{-1}\left\{\frac{1}{(s+1)(s^2+1)}\right\} + \mathscr{L}^{-1}\left\{\frac{1}{s+1}\right\}$$
$$= \left(\frac{1}{2}e^{-x} - \frac{1}{2}\cos x + \frac{1}{2}\sin x\right) + e^{-x} = \frac{3}{2}e^{-x} - \frac{1}{2}\cos x + \frac{1}{2}\sin x$$

17.4. Solve $y'' + 4y = 0$; $y(0) = 2$, $y'(0) = 2$.

Taking Laplace transforms, we have $\mathscr{L}\{y''\} + 4\mathscr{L}\{y\} = \mathscr{L}\{0\}$. Then, using Eq. (*17.5*) with $c_0 = 2$ and $c_1 = 2$, we obtain

$$[s^2Y(s) - 2s - 2] + 4Y(s) = 0$$

or

$$Y(s) = \frac{2s+2}{s^2+4} = \frac{2s}{s^2+4} + \frac{2}{s^2+4}$$

Finally, taking the inverse Laplace transform, we obtain

$$y(x) = \mathscr{L}^{-1}\{Y(s)\} = 2\mathscr{L}^{-1}\left\{\frac{s}{s^2+4}\right\} + \mathscr{L}^{-1}\left\{\frac{2}{s^2+4}\right\} = 2\cos 2x + \sin 2x$$

17.5. Solve $y'' - 3y' + 4y = 0$; $y(0) = 1$, $y'(0) = 5$.

Taking Laplace transforms, we obtain $\mathscr{L}\{y''\} - 3\mathscr{L}\{y'\} + 4\mathscr{L}\{y\} = \mathscr{L}\{0\}$. Then, using *both* Eqs. *(17.4) and (17.5)* with $c_0 = 1$ and $c_1 = 5$, we have

$$[s^2Y(s) - s - 5] - 3[sY(s) - 1] + 4Y(s) = 0$$

or

$$Y(s) = \frac{s+2}{s^2 - 3s + 4}$$

Finally, taking the inverse Laplace transform and using the result of Problem 15.10, we obtain

$$y(x) = e^{(3/2)x}\cos\frac{\sqrt{7}}{2}x + \sqrt{7}\,e^{(3/2)x}\sin\frac{\sqrt{7}}{2}x$$

17.6. Solve $y'' - y' - 2y = 4x^2$; $y(0) = 1$, $y'(0) = 4$.

Taking Laplace transforms, we have $\mathscr{L}\{y''\} - \mathscr{L}\{y'\} - 2\mathscr{L}\{y\} = 4\mathscr{L}\{x^2\}$. Then, using *both* Eqs. *(17.4) and (17.5)* with $c_0 = 1$ and $c_1 = 4$, we obtain

$$[s^2Y(s) - s - 4] - [sY(s) - 1] - 2Y(s) = \frac{8}{s^3}$$

or, upon solving for $Y(s)$,

$$Y(s) = \frac{s+3}{s^2 - s - 2} + \frac{8}{s^3(s^2 - s - 2)}$$

Finally, taking the inverse Laplace transform and using the results of Problems 15.15 and 15.16 we obtain

$$y(x) = \left(\frac{5}{3}e^{2x} - \frac{2}{3}e^{-x}\right) + \left(-3 + 2x - 2x^2 + \frac{1}{3}e^{2x} + \frac{8}{3}e^{-x}\right)$$

$$= 2e^{2x} + 2e^{-x} - 2x^2 + 2x - 3$$

(See Problem 12.1.)

17.7. Solve $y'' + 4y' + 8y = \sin x$; $y(0) = 1$, $y'(0) = 0$.

Taking Laplace transforms, we obtain $\mathscr{L}\{y''\} + 4\mathscr{L}\{y'\} + 8\mathscr{L}\{y\} = \mathscr{L}\{\sin x\}$. Since $c_0 = 1$ and $c_1 = 0$, this becomes

$$[s^2Y(s) - s - 0] + 4[sY(s) - 1] + 8Y(s) = \frac{1}{s^2 + 1}$$

Thus,

$$Y(s) = \frac{s+4}{s^2 + 4s + 8} + \frac{1}{(s^2+1)(s^2 + 4s + 8)}$$

Finally, taking the inverse Laplace transform and using the results of Problems 15.9 and 15.18, we obtain

$$y(x) = (e^{-2x}\cos 2x + e^{-2x}\sin 2x)$$

$$+ \left(-\frac{4}{65}\cos x + \frac{7}{65}\sin x + \frac{4}{65}e^{-2x}\cos 2x + \frac{1}{130}e^{-2x}\sin 2x\right)$$

$$= e^{-2x}\left(\frac{69}{65}\cos 2x + \frac{131}{130}\sin 2x\right) + \frac{7}{65}\sin x - \frac{4}{65}\cos x$$

(See Problem 12.3.)

17.8. Solve $y'' - 2y' + y = f(x)$; $y(0) = 0$, $y'(0) = 0$.

In this equation $f(x)$ is unspecified. Taking Laplace transforms and designating $\mathscr{L}\{f(x)\}$ by $F(s)$, we

obtain

$$[s^2Y(s) - (0)s - 0] - 2[sY(s) - 0] + Y(s) = F(s) \quad \text{or} \quad Y(s) = \frac{F(s)}{(s-1)^2}$$

From Appendix A, entry 14, $\mathscr{L}^{-1}\{1/(s-1)^2\} = xe^x$. Thus, taking the inverse transform of $Y(s)$ and using convolutions, we conclude that

$$y(x) = xe^x * f(x) = \int_0^x te^t f(x-t)\,dt$$

17.9. Solve $y'' + y = f(x);\ y(0) = 0,\ y'(0) = 0$ if $f(x) = \begin{cases} 0 & x < 1 \\ 2 & x \geq 1 \end{cases}$.

Note that $f(x) = 2u(x-1)$. Taking Laplace transforms, we obtain

$$[s^2Y(s) - (0)s - 0] + Y(s) = \mathscr{L}\{f(x)\} = 2\mathscr{L}\{u(x-1)\} = 2e^{-s}/s$$

or

$$Y(s) = e^{-s}\frac{2}{s(s^2+1)}$$

Since

$$\mathscr{L}^{-1}\left\{\frac{2}{s(s^2+1)}\right\} = 2\mathscr{L}^{-1}\left\{\frac{1}{s}\right\} - 2\mathscr{L}^{-1}\left\{\frac{s}{s^2+1}\right\} = 2 - 2\cos x$$

it follows from Theorem 16.4 that

$$y(x) = \mathscr{L}^{-1}\left\{e^{-s}\frac{2}{s(s^2+1)}\right\} = [2 - 2\cos(x-1)]u(x-1)$$

17.10. Solve $y''' + y' = e^x;\ y(0) = y'(0) = y''(0) = 0$.

Taking Laplace transforms, we obtain $\mathscr{L}\{y'''\} + \mathscr{L}\{y'\} = \mathscr{L}\{e^x\}$. Then, using Eq. (*17.3*) with $n = 3$ and Eq. (*17.4*), we have

$$[s^3Y(s) - (0)s^2 - (0)s - 0] + [sY(s) - 0] = \frac{1}{s-1} \quad \text{or} \quad Y(s) = \frac{1}{(s-1)(s^3+s)}$$

Finally, using the method of partial fractions and taking the inverse transform, we obtain

$$y(x) = \mathscr{L}^{-1}\left\{-\frac{1}{s} + \frac{\frac{1}{2}}{s-1} + \frac{\frac{1}{2}s - \frac{1}{2}}{s^2+1}\right\} = -1 + \frac{1}{2}e^x + \frac{1}{2}\cos x - \frac{1}{2}\sin x$$

17.11. Solve $y' - 5y = 0$.

No initial conditions are specified. Taking the Laplace transform of both sides of the differential equation, we obtain

$$\mathscr{L}\{y'\} - 5\mathscr{L}\{y\} = \mathscr{L}\{0\}$$

Then, using Eq. (*17.4*) with $c_0 = y(0)$ kept arbitrary, we have

$$[sY(s) - c_0] - 5Y(s) = 0 \quad \text{or} \quad Y(s) = \frac{c_0}{s-5}$$

Taking the inverse Laplace transform, we find that

$$y(x) = \mathscr{L}^{-1}\{Y(s)\} = c_0\mathscr{L}^{-1}\left\{\frac{1}{s-5}\right\} = c_0e^{5x}$$

17.12. Solve $y'' - 3y' + 2y = e^{-x}$.

No initial conditions are specified. Taking Laplace transforms, we have $\mathscr{L}\{y''\} - 3\mathscr{L}\{y'\} + 2\mathscr{L}\{y\} = \mathscr{L}\{e^{-x}\}$, or

$$[s^2Y(s) - sc_0 - c_1] - 3[sY(s) - c_0] + 2[Y(s)] = 1/(s+1)$$

Here c_0 and c_1 must remain arbitrary, since they represent $y(0)$ and $y'(0)$, respectively, which are unknown. Thus,

$$Y(s) = c_0\frac{s-3}{s^2-3s+2} + c_1\frac{1}{s^2-3s+2} + \frac{1}{(s+1)(s^2-3s+2)}$$

Using the method of partial fractions and noting that $s^2 - 3s + 2 = (s-1)(s-2)$, we obtain

$$\begin{aligned} y(x) &= c_0\mathcal{L}^{-1}\left\{\frac{2}{s-1} + \frac{-1}{s-2}\right\} + c_1\mathcal{L}^{-1}\left\{\frac{-1}{s-1} + \frac{1}{s-2}\right\} + \mathcal{L}^{-1}\left\{\frac{1/6}{s+1} + \frac{-1/2}{s-1} + \frac{1/3}{s-2}\right\} \\ &= c_0(2e^x - e^{2x}) + c_1(-e^x + e^{2x}) + \left(\frac{1}{6}e^{-x} - \frac{1}{2}e^x + \frac{1}{3}e^{2x}\right) \\ &= \left(2c_0 - c_1 - \frac{1}{2}\right)e^x + \left(-c_0 + c_1 + \frac{1}{3}\right)e^{2x} + \frac{1}{6}e^{-x} \\ &= d_0e^x + d_1e^{2x} + \frac{1}{6}e^{-x} \end{aligned}$$

where $d_0 = 2c_0 - c_1 - \frac{1}{2}$ and $d_1 = -c_0 + c_1 + \frac{1}{3}$.

17.13. Solve $y'' - 3y' + 2y = e^{-x}$; $y(1) = 0$, $y'(1) = 0$.

The initial conditions are given at $x = 1$, not $x = 0$. Using the results of Problem 17.12, we have as the solution to just the differential equation

$$y = d_0e^x + d_1e^{2x} + \frac{1}{6}e^{-x}$$

Applying the initial conditions to this last equation, we find that $d_0 = -\frac{1}{2}e^{-2}$ and $d_1 = \frac{1}{3}e^{-3}$; hence,

$$y(x) = -\frac{1}{2}e^{x-2} + \frac{1}{3}e^{2x-3} + \frac{1}{6}e^{-x}$$

17.14. Solve $\dfrac{dN}{dt} = 0.05N$; $N(0) = 20{,}000$.

This is a differential equation for the unknown function $N(t)$ in the independent variable t. We set $N(s) = \mathcal{L}\{N(t)\}$. Taking Laplace transforms of the given differential equation and using *(17.4)* with N replacing y, we have

$$[sN(s) - N(0)] = 0.05N(s)$$
$$[sN(s) - 20{,}000] = 0.05N(s)$$

or, upon solving for $N(s)$,

$$N(s) = \frac{20{,}000}{s - 0.05}$$

Then from Appendix A, entry 7 with $a = 0.05$ and t replacing x, we obtain

$$N(t) = \mathcal{L}^{-1}\{N(s)\} = \mathcal{L}^{-1}\left\{\frac{20{,}000}{s-0.05}\right\} = 20{,}000\mathcal{L}^{-1}\left\{\frac{1}{s-0.05}\right\} = 20{,}000e^{0.05t}$$

Compare with (*2*) of Problem 6.1.

17.15. Solve $\dfrac{dI}{dt} + 50I = 5$; $I(0) = 0$.

This is a differential equation for the unknown function $I(t)$ in the independent variable t. We set $I(s) = \mathcal{L}\{I(t)\}$. Taking Laplace transforms of the given differential equation and using Eq. (*17.4*) with I

replacing y, we have

$$[sI(s) - I(0)] + 50I(s) = 5\left(\frac{1}{s}\right)$$

$$[sI(s) - 0] + 50I(s) = 5\left(\frac{1}{s}\right)$$

or, upon solving for $I(s)$,

$$I(s) = \frac{5}{s(s+50)}$$

Then using the method of partial fractions and Appendix A, with t replacing x, we obtain

$$I(t) = \mathscr{L}^{-1}\{I(s)\} = \mathscr{L}^{-1}\left\{\frac{5}{s(s+50)}\right\} = \mathscr{L}^{-1}\left\{\frac{1/10}{s} - \frac{1/10}{s+50}\right\}$$

$$= \frac{1}{10}\mathscr{L}^{-1}\left\{\frac{1}{s}\right\} - \frac{1}{10}\mathscr{L}^{-1}\left\{\frac{1}{s+50}\right\} = \frac{1}{10} - \frac{1}{10}e^{-50t}$$

Compare with *(1)* of Problem 6.19.

17.16. Solve $\ddot{x} + 16x = 2\sin 4t$; $x(0) = -\frac{1}{2}$, $\dot{x}(0) = 0$.

This is a differential equation for the unknown function $x(t)$ in the independent variable t. We set $X(s) = \mathscr{L}\{x(t)\}$. Taking Laplace transforms of the given differential equation and using Eq. *(17.5)* with x replacing y, we have

$$[s^2X(s) - sx(0) - \dot{x}(0)] + 16X(s) = 2\left(\frac{4}{s^2+16}\right)$$

$$\left[s^2X(s) - s\left(-\frac{1}{2}\right) - 0\right] + 16X(s) = \frac{8}{s^2+16}$$

$$(s^2+16)X(s) = \frac{8}{s^2+16} - \frac{s}{2}$$

or

$$X(s) = \frac{8}{(s^2+16)^2} - \frac{1}{2}\left(\frac{s}{s^2+16}\right)$$

Then using Appendix A, entries 17 and 9 with $a = 4$ and t replacing x, we obtain

$$x(t) = \mathscr{L}^{-1}\{X(s)\} = \mathscr{L}^{-1}\left\{\frac{8}{(s^2+16)^2} - \frac{1}{2}\left(\frac{s}{s^2+16}\right)\right\}$$

$$= \frac{1}{16}\mathscr{L}^{-1}\left\{\frac{128}{(s^2+16)^2}\right\} - \frac{1}{2}\mathscr{L}^{-1}\left\{\frac{s}{s^2+16}\right\}$$

$$= \frac{1}{16}(\sin 4t - 4t\cos 4t) - \frac{1}{2}\cos 4t$$

Compare with the results of Problem 13.10.

Supplementary Problems

Use Laplace transforms to solve the following problems.

17.17. $y' + 2y = 0$; $y(0) = 1$

17.18. $y' + 2y = 2$; $y(0) = 1$

17.19. $y' + 2y = e^x$; $y(0) = 1$

17.20. $y' + 2y = 0$; $y(1) = 1$

17.21. $y' + 5y = 0;\ y(1) = 0$

17.22. $y' - 5y = e^{5x};\ y(0) = 2$

17.23. $y' + y = xe^{-x};\ y(0) = -2$

17.24. $y' + y = \sin x$

17.25. $y' + 20y = 6\sin 2x;\ y(0) = 6$

17.26. $y'' - y = 0;\ y(0) = 1,\ y'(0) = 1$

17.27. $y'' - y = \sin x;\ y(0) = 0,\ y'(0) = 1$

17.28. $y'' - y = e^{x};\ y(0) = 1,\ y'(0) = 0$

17.29. $y'' + 2y' - 3y = \sin 2x;\ y(0) = y'(0) = 0$

17.30. $y'' + y = \sin x;\ y(0) = 0,\ y'(0) = 2$

17.31. $y'' + y' + y = 0;\ y(0) = 4,\ y'(0) = -3$

17.32. $y'' + 2y' + 5y = 3e^{-2x};\ y(0) = 1,\ y'(0) = 1$

17.33. $y'' + 5y' - 3y = u(x-4);\ y(0) = 0,\ y'(0) = 0$

17.34. $y'' + y = 0;\ y(\pi) = 0,\ y'(\pi) = -1$

17.35. $y''' - y = 5;\ y(0) = 0,\ y'(0) = 0,\ y''(0) = 0$

17.36. $y^{(4)} - y = 0;\ y(0) = 1,\ y'(0) = 0,\ y''(0) = 0,\ y'''(0) = 0$

17.37. $\dfrac{d^3y}{dx^3} - 3\dfrac{d^2y}{dx^2} + 3\dfrac{dy}{dx} - y = x^2e^x;\ y(0) = 1,\ y'(0) = 2,\ y''(0) = 3$

17.38. $\dfrac{dN}{dt} - 0.085N = 0;\ N(0) = 5000$

17.39. $\dfrac{dT}{dt} = 3T;\ T(0) = 100$

17.40. $\dfrac{dT}{dt} + 3T = 90;\ T(0) = 100$

17.41. $\dfrac{dv}{dt} + 2v = 32$

17.42. $\dfrac{dq}{dt} + q = 4\cos 2t;\ q(0) = 0$

17.43. $\ddot{x} + 9\dot{x} + 14x = 0;\ x(0) = 0,\ \dot{x}(0) = -1$

17.44. $\ddot{x} + 4\dot{x} + 4x = 0;\ x(0) = 2,\ \dot{x}(0) = -2$

17.45. $\dfrac{d^2x}{dt^2} + 8\dfrac{dx}{dt} + 25x = 0;\ x(\pi) = 0,\ \dot{x}(\pi) = 6$

17.46. $\dfrac{d^2q}{dt^2} + 9\dfrac{dq}{dt} + 14q = \dfrac{1}{2}\sin t;\ q(0) = 0,\ \dot{q}(0) = 1$

Chapter 18

Solutions of Linear Systems by Laplace Transforms

THE METHOD

Laplace transforms are useful for solving systems of linear differential equations; that is, sets of two or more differential equations with an equal number of unknown functions. If all of the coefficients are constants, then the method of solution is a straightforward generalization of the one given in Chapter 17. Laplace transforms are taken of each differential equation in the system; the transforms of the unknown functions are determined algebraically from the resulting set of simultaneous equations; inverse transforms for the unknown functions are calculated with the help of Appendix A.

Solved Problems

18.1. Solve the following system for the unknown functions $u(x)$ and $v(x)$:

$$u' + u - v = 0$$
$$v' - u + v = 2;$$
$$u(0) = 1, \qquad v(0) = 2$$

Denote $\mathscr{L}\{u(x)\}$ and $\mathscr{L}\{v(x)\}$ by $U(s)$ and $V(s)$, respectively. Taking Laplace transforms of both differential equations, we obtain

$$[sU(s) - 1] + U(s) - V(s) = 0$$
$$[sV(s) - 2] - U(s) + V(s) = \frac{2}{s}$$

or

$$(s + 1)U(s) - V(s) = 1$$
$$-U(s) + (s + 1)V(s) = \frac{2(s + 1)}{s}$$

The solution to this last set of simultaneous linear equations is

$$U(s) = \frac{s + 1}{s^2} \qquad V(s) = \frac{2s + 1}{s^2}$$

Taking inverse transforms, we obtain

$$u(x) = \mathscr{L}^{-1}\{U(s)\} = \mathscr{L}^{-1}\left\{\frac{s + 1}{s^2}\right\} = \mathscr{L}^{-1}\left\{\frac{1}{s} + \frac{1}{s^2}\right\} = 1 + x$$
$$v(x) = \mathscr{L}^{-1}\{V(s)\} = \mathscr{L}^{-1}\left\{\frac{2s + 1}{s^2}\right\} = \mathscr{L}^{-1}\left\{\frac{2}{s} + \frac{1}{s^2}\right\} = 2 + x$$

18.2. Solve the system

$$y' + z = x$$
$$z' + 4y = 0;$$
$$y(0) = 1, \qquad z(0) = -1$$

Denote $\mathscr{L}\{y(x)\}$ and $\mathscr{L}\{z(x)\}$ by $Y(s)$ and $Z(s)$, respectively. Then, taking Laplace transforms of both differential equations, we obtain

$$[sY(s) - 1] + Z(s) = \frac{1}{s^2} \qquad\qquad sY(s) + Z(s) = \frac{s^2+1}{s^2}$$

or

$$[sZ(s) + 1] + 4Y(s) = 0 \qquad\qquad 4Y(s) + sZ(s) = -1$$

The solution to this last set of simultaneous linear equations is

$$Y(s) = \frac{s^2+s+1}{s(s^2-4)} \qquad Z(s) = -\frac{s^3+4s^2+4}{s^2(s^2-4)}$$

Finally, using the method of partial fractions and taking inverse transforms, we obtain

$$y(x) = \mathscr{L}^{-1}\{Y(s)\} = \mathscr{L}^{-1}\left\{-\frac{1/4}{s} + \frac{7/8}{s-2} + \frac{3/8}{s+2}\right\}$$

$$= -\frac{1}{4} + \frac{7}{8}e^{2x} + \frac{3}{8}e^{-2x}$$

$$z(x) = \mathscr{L}^{-1}\{Z(s)\} = \mathscr{L}^{-1}\left\{\frac{1}{s^2} - \frac{7/4}{s-2} + \frac{3/4}{s+2}\right\}$$

$$= x - \frac{7}{4}e^{2x} + \frac{3}{4}e^{-2x}$$

18.3. Solve the system

$$w' + y = \sin x$$

$$y' - z = e^x$$

$$z' + w + y = 1;$$

$$w(0) = 0, \qquad y(0) = 1, \qquad z(0) = 1$$

Denote $\mathscr{L}\{w(x)\}$, $\mathscr{L}\{y(x)\}$, and $\mathscr{L}\{z(x)\}$ by $W(s)$, $Y(s)$, and $Z(s)$, respectively. Then, taking Laplace transforms of all three differential equations, we have

$$[sW(s) - 0] + Y(s) = \frac{1}{s^2+1} \qquad\qquad sW(s) + Y(s) = \frac{1}{s^2+1}$$

$$[sY(s) - 1] - Z(s) = \frac{1}{s-1} \qquad \text{or} \qquad sY(s) - Z(s) = \frac{s}{s-1}$$

$$[sZ(s) - 1] + W(s) + Y(s) = \frac{1}{s} \qquad\qquad W(s) + Y(s) + sZ(s) = \frac{s+1}{s}$$

The solution to this last system of simultaneous linear equations is

$$W(s) = \frac{-1}{s(s-1)} \qquad Y(s) = \frac{s^2+s}{(s-1)(s^2+1)} \qquad Z(s) = \frac{s}{s^2+1}$$

Using the method of partial fractions and then taking inverse transforms, we obtain

$$w(x) = \mathscr{L}^{-1}\{W(s)\} = \mathscr{L}^{-1}\left\{\frac{1}{s} - \frac{1}{s-1}\right\} = 1 - e^x$$

$$y(x) = \mathscr{L}^{-1}\{Y(s)\} = \mathscr{L}^{-1}\left\{\frac{1}{s-1} + \frac{1}{s^2+1}\right\} = e^x + \sin x$$

$$z(x) = \mathscr{L}^{-1}\{Z(s)\} = \mathscr{L}^{-1}\left\{\frac{s}{s^2+1}\right\} = \cos x$$

18.4. Solve the system

$$y'' + z + y = 0$$

$$z' + y' = 0;$$

$$y(0) = 0, \qquad y'(0) = 0, \qquad z(0) = 1$$

Taking Laplace transforms of both differential equations, we obtain

$$\begin{aligned} &[s^2Y(s) - (0)s - (0)] + Z(s) + Y(s) = 0 \\ &[sZ(s) - 1] + [sY(s) - 0] = 0 \end{aligned} \qquad \text{or} \qquad \begin{aligned} (s^2 + 1)Y(s) + Z(s) &= 0 \\ Y(s) + Z(s) &= \frac{1}{s} \end{aligned}$$

Solving this last system for $Y(s)$ and $Z(s)$, we find that

$$Y(s) = -\frac{1}{s^3} \qquad Z(s) = \frac{1}{s} + \frac{1}{s^3}$$

Thus, taking inverse transforms, we conclude that

$$y(x) = -\frac{1}{2}x^2 \qquad z(x) = 1 + \frac{1}{2}x^2$$

18.5. Solve the system

$$\begin{aligned} &z'' + y' = \cos x \\ &y'' - z = \sin x; \\ &z(0) = -1, \quad z'(0) = -1, \quad y(0) = 1, \quad y'(0) = 0 \end{aligned}$$

Taking Laplace transforms of both differential equations, we obtain

$$\begin{aligned} &[s^2Z(s) + s + 1] + [sY(s) - 1] = \frac{s}{s^2 + 1} \\ &[s^2Y(s) - s - 0] - Z(s) = \frac{1}{s^2 + 1} \end{aligned} \qquad \text{or} \qquad \begin{aligned} s^2Z(s) + sY(s) &= -\frac{s^3}{s^2 + 1} \\ -Z(s) + s^2Y(s) &= \frac{s^3 + s + 1}{s^2 + 1} \end{aligned}$$

Solving this last system for $Z(s)$ and $Y(s)$, we find that

$$Z(s) = -\frac{s + 1}{s^2 + 1} \qquad Y(s) = \frac{s}{s^2 + 1}$$

Finally, taking inverse transforms, we obtain

$$z(x) = -\cos x - \sin x \qquad y(x) = \cos x$$

18.6. Solve the system

$$\begin{aligned} &w'' - y + 2z = 3e^{-x} \\ &-2w' + 2y' + z = 0 \\ &2w' - 2y + z' + 2z'' = 0; \\ &w(0) = 1, \quad w'(0) = 1, \quad y(0) = 2, \quad z(0) = 2, \quad z'(0) = -2 \end{aligned}$$

Taking Laplace transforms of all three differential equations, we find that

$$\begin{gathered} [s^2W(s) - s - 1] - Y(s) + 2Z(s) = \frac{3}{s + 1} \\ -2[sW(s) - 1] + 2[sY(s) - 2] + Z(s) = 0 \\ 2[sW(s) - 1] - 2Y(s) + [sZ(s) - 2] + 2[s^2Z(s) - 2s + 2] = 0 \end{gathered}$$

or

$$\begin{aligned} s^2W(s) - Y(s) + 2Z(s) &= \frac{s^2 + 2s + 4}{s + 1} \\ -2sW(s) + 2sY(s) + Z(s) &= 2 \\ 2sW(s) - 2Y(s) + (2s^2 + s)Z(s) &= 4s \end{aligned}$$

The solution to this system is

$$W(s) = \frac{1}{s - 1} \qquad Y(s) = \frac{2s}{(s - 1)(s + 1)} \qquad Z(s) = \frac{2}{s + 1}$$

Hence,

$$w(x) = e^x \qquad y(x) = \mathscr{L}^{-1}\left\{\frac{1}{s - 1} + \frac{1}{s + 1}\right\} = e^x + e^{-x} \qquad z(x) = 2e^{-x}$$

Supplementary Problems

Use Laplace transforms to solve the following systems. All unknowns are functions of x.

18.7. $u' - 2v = 3$

$v' + v - u = -x^2;$

$u(0) = 0,\ v(0) = -1$

18.8. $u' + 4u - 6v = 0$

$v' + 3u - 5v = 0;$

$u(0) = 3,\ v(0) = 2$

18.9. $u' + 5u - 12v = 0$

$v' + 2u - 5v = 0;$

$u(0) = 8,\ v(0) = 3$

18.10. $y' + z = x$

$z' - y = 0;$

$y(0) = 1,\ z(0) = 0$

18.11. $y' - z = 0$

$y - z' = 0;$

$y(0) = 1,\ z(0) = 1$

18.12. $w' - w - 2y = 1$

$y' - 4w - 3y = -1;$

$w(0) = 1,\ y(0) = 2$

18.13. $w' - y = 0$

$w + y' + z = 1$

$w - y + z' = 2\sin x;$

$w(0) = 1,\ y(0) = 1,\ z(0) = 1$

18.14. $u'' + v = 0$

$u'' - v' = -2e^x;$

$u(0) = 0,\ u'(0) = -2,\ v(0) = 0,\ v'(0) = 2$

18.15. $u'' - 2v = 2$

$u + v' = 5e^{2x} + 1;$

$u(0) = 2,\ u'(0) = 2,\ v(0) = 1$

18.16. $w'' - 2z = 0$
$w' + y' - z = 2x$
$w' - 2y + z'' = 0;$
$w(0) = 0,\ w'(0) = 0,\ y(0) = 0,$
$z(0) = 1,\ z'(0) = 0$

18.17. $w'' + y + z = -1$
$w + y'' - z = 0$
$-w' - y' + z'' = 0;$
$w(0) = 0,\ w'(0) = 1,\ y(0) = 0,$
$y'(0) = 0,\ z(0) = -1,\ z'(0) = 1$

Chapter 19

Matrices

MATRICES AND VECTORS

A *matrix* (designated by an uppercase boldface letter) is a rectangular array of elements arranged in horizontal rows and vertical columns. In this book, the elements of matrices will always be numbers or functions of the variable t. If all the elements are numbers, then the matrix is called a *constant matrix.*

Example 19.1.

$$\begin{bmatrix} 1 & 2 \\ 3 & 4 \end{bmatrix}, \quad \begin{bmatrix} 1 & e^t & 2 \\ t & -1 & 1 \end{bmatrix}, \quad \text{and} \quad [1 \quad t^2 \quad \cos t]$$

are all matrices. In particular, the first matrix is a constant matrix, whereas the last two are not.

A general matrix $\mathbf{A}$ having p rows and n columns is given by

$$\mathbf{A} = [a_{ij}] = \begin{bmatrix} a_{11} & a_{12} & \cdots & a_{1n} \\ a_{21} & a_{22} & \cdots & a_{2n} \\ \vdots & \vdots & & \vdots \\ a_{p1} & a_{p2} & \cdots & a_{pn} \end{bmatrix}$$

where a_{ij} represents that element appearing in the ith row and jth column. A matrix is *square* if it has the same number of rows and columns.

A *vector* (designated by a lowercase boldface letter) is a matrix having only one column or one row. (The third matrix given in Example 19.1 is a vector.)

MATRIX ADDITION

The *sum* $\mathbf{A}+\mathbf{B}$ of two matrices $\mathbf{A}=[a_{ij}]$ and $\mathbf{B}=[b_{ij}]$ having the same number of rows and the same number of columns is the matrix obtained by adding the corresponding elements of $\mathbf{A}$ and $\mathbf{B}$. That is,

$$\mathbf{A} + \mathbf{B} = [a_{ij}] + [b_{ij}] = [a_{ij} + b_{ij}]$$

Matrix addition is both associative and commutative. Thus, $\mathbf{A}+(\mathbf{B}+\mathbf{C})=(\mathbf{A}+\mathbf{B})+\mathbf{C}$ and $\mathbf{A}+\mathbf{B}=\mathbf{B}+\mathbf{A}$.

SCALAR AND MATRIX MULTIPLICATION

If λ is either a number or a function of t, then $\lambda\mathbf{A}$ (or, equivalently, $\mathbf{A}\lambda$) is defined to be the matrix obtained by multiplying every element of $\mathbf{A}$ by λ. That is,

$$\lambda\mathbf{A} = \lambda[a_{ij}] = [\lambda a_{ij}]$$

Let $\mathbf{A}=[a_{ij}]$ and $\mathbf{B}=[b_{ij}]$ be two matrices such that $\mathbf{A}$ has r rows and n columns and $\mathbf{B}$ has n

rows and p columns. Then the *product* **AB** is defined to be the matrix $\mathbf{C} = [c_{ij}]$ given by

$$c_{ij} = \sum_{k=1}^{n} a_{ik}b_{kj} \qquad (i = 1, 2, \ldots, r; j = 1, 2, \ldots, p)$$

The element c_{ij} is obtained by multiplying the elements of the ith row of **A** with the corresponding elements of the jth column of **B** and summing the results.

Matrix multiplication is associative and distributes over addition; in general, however, it is *not* commutative. Thus,

$$\mathbf{A}(\mathbf{BC}) = (\mathbf{AB})\mathbf{C}, \qquad \mathbf{A}(\mathbf{B} + \mathbf{C}) = \mathbf{AB} + \mathbf{AC}, \qquad \text{and} \qquad (\mathbf{B} + \mathbf{C})\mathbf{A} = \mathbf{BA} + \mathbf{CA}$$

but, in general, $\mathbf{AB} \neq \mathbf{BA}$.

POWERS OF A SQUARE MATRIX

If n is a positive integer and **A** is a square matrix, then

$$\mathbf{A}^n = \underbrace{\mathbf{AA}\cdots\mathbf{A}}_{n \text{ times}}$$

In particular, $\mathbf{A}^2 = \mathbf{AA}$ and $\mathbf{A}^3 = \mathbf{AAA}$. By definition, $\mathbf{A}^0 = \mathbf{I}$, where

$$\mathbf{I} = \begin{bmatrix} 1 & 0 & 0 & \cdots & 0 & 0 \\ 0 & 1 & 0 & \cdots & 0 & 0 \\ 0 & 0 & 1 & \cdots & 0 & 0 \\ \vdots & & & \ddots & & \vdots \\ 0 & 0 & 0 & \cdots & 1 & 0 \\ 0 & 0 & 0 & \cdots & 0 & 1 \end{bmatrix}$$

is called an *identity matrix*. For any square matrix **A** and identity matrix **I** of the same size

$$\mathbf{AI} = \mathbf{IA} = \mathbf{A}$$

DIFFERENTIATION AND INTEGRATION OF MATRICES

The *derivative* of $\mathbf{A} = [a_{ij}]$ is the matrix obtained by differentiating each element of **A**; that is,

$$\frac{d\mathbf{A}}{dt} = \left[\frac{da_{ij}}{dt}\right]$$

Similarly, the *integral* of **A**, either definite or indefinite, is obtained by integrating each element of **A**. Thus,

$$\int_a^b \mathbf{A}\,dt = \left[\int_a^b a_{ij}\,dt\right] \qquad \text{and} \qquad \int \mathbf{A}\,dt = \left[\int a_{ij}\,dt\right]$$

THE CHARACTERISTIC EQUATION

The *characteristic equation* of a square matrix **A** is the polynomial equation in λ given by

$$\det(\mathbf{A} - \lambda\mathbf{I}) = 0 \tag{19.1}$$

where det( ) stands for "the determinant of." Those values of λ which satisfy (*19.1*), that is, the roots of (*19.1*), are the *eigenvalues* of **A**, a k-fold root being called an *eigenvalue of multiplicity k*.

Theorem 19.1 ***(Cayley-Hamilton theorem).*** Any square matrix satisfies its own characteristic equation. That is, if

$$\det(\mathbf{A} - \lambda\mathbf{I}) = b_n\lambda^n + b_{n-1}\lambda^{n-1} + \cdots + b_2\lambda^2 + b_1\lambda + b_0$$

then

$$b_n\mathbf{A}^n + b_{n-1}\mathbf{A}^{n-1} + \cdots + b_2\mathbf{A}^2 + b_1\mathbf{A} + b_0\mathbf{I} = \mathbf{0}$$

Solved Problems

19.1. Show that $\mathbf{A} + \mathbf{B} = \mathbf{B} + \mathbf{A}$ for

$$\mathbf{A} = \begin{bmatrix} 1 & 2 \\ 3 & 4 \end{bmatrix} \qquad \mathbf{B} = \begin{bmatrix} 5 & 6 \\ 7 & 8 \end{bmatrix}$$

$$\mathbf{A} + \mathbf{B} = \begin{bmatrix} 1 & 2 \\ 3 & 4 \end{bmatrix} + \begin{bmatrix} 5 & 6 \\ 7 & 8 \end{bmatrix} = \begin{bmatrix} 1+5 & 2+6 \\ 3+7 & 4+8 \end{bmatrix} = \begin{bmatrix} 6 & 8 \\ 10 & 12 \end{bmatrix}$$

$$\mathbf{B} + \mathbf{A} = \begin{bmatrix} 5 & 6 \\ 7 & 8 \end{bmatrix} + \begin{bmatrix} 1 & 2 \\ 3 & 4 \end{bmatrix} = \begin{bmatrix} 5+1 & 6+2 \\ 7+3 & 8+4 \end{bmatrix} = \begin{bmatrix} 6 & 8 \\ 10 & 12 \end{bmatrix}$$

Since the corresponding elements of the resulting matrices are equal, the desired equality follows.

19.2. Find $3\mathbf{A} - \frac{1}{2}\mathbf{B}$ for the matrices given in Problem 19.1.

$$3\mathbf{A} - \frac{1}{2}\mathbf{B} = 3\begin{bmatrix} 1 & 2 \\ 3 & 4 \end{bmatrix} + \left(-\frac{1}{2}\right)\begin{bmatrix} 5 & 6 \\ 7 & 8 \end{bmatrix}$$

$$= \begin{bmatrix} 3 & 6 \\ 9 & 12 \end{bmatrix} + \begin{bmatrix} -\frac{5}{2} & -3 \\ -\frac{7}{2} & -4 \end{bmatrix}$$

$$= \begin{bmatrix} 3 + \left(-\frac{5}{2}\right) & 6 + (-3) \\ 9 + \left(-\frac{7}{2}\right) & 12 + (-4) \end{bmatrix} = \begin{bmatrix} \frac{1}{2} & 3 \\ \frac{11}{2} & 8 \end{bmatrix}$$

19.3. Find **AB** and **BA** for the matrices given in Problem 19.1.

$$\mathbf{AB} = \begin{bmatrix} 1 & 2 \\ 3 & 4 \end{bmatrix}\begin{bmatrix} 5 & 6 \\ 7 & 8 \end{bmatrix} = \begin{bmatrix} 1(5) + 2(7) & 1(6) + 2(8) \\ 3(5) + 4(7) & 3(6) + 4(8) \end{bmatrix} = \begin{bmatrix} 19 & 22 \\ 43 & 50 \end{bmatrix}$$

$$\mathbf{BA} = \begin{bmatrix} 5 & 6 \\ 7 & 8 \end{bmatrix}\begin{bmatrix} 1 & 2 \\ 3 & 4 \end{bmatrix} = \begin{bmatrix} 5(1) + 6(3) & 5(2) + 6(4) \\ 7(1) + 8(3) & 7(2) + 8(4) \end{bmatrix} = \begin{bmatrix} 23 & 34 \\ 31 & 46 \end{bmatrix}$$

Note that for these matrices, $\mathbf{AB} \neq \mathbf{BA}$.

19.4. Find $(2\mathbf{A}-\mathbf{B})^2$ for the matrices given in Problem 19.1.

$$2\mathbf{A}-\mathbf{B} = 2\begin{bmatrix}1 & 2\\3 & 4\end{bmatrix} + (-1)\begin{bmatrix}5 & 6\\7 & 8\end{bmatrix} = \begin{bmatrix}2 & 4\\6 & 8\end{bmatrix} + \begin{bmatrix}-5 & -6\\-7 & -8\end{bmatrix} = \begin{bmatrix}-3 & -2\\-1 & 0\end{bmatrix}$$

and

$$(2\mathbf{A}-\mathbf{B})^2 = (2\mathbf{A}-\mathbf{B})(2\mathbf{A}-\mathbf{B}) = \begin{bmatrix}-3 & -2\\-1 & 0\end{bmatrix}\begin{bmatrix}-3 & -2\\-1 & 0\end{bmatrix}$$

$$= \begin{bmatrix}-3(-3)+(-2)(-1) & -3(-2)+(-2)(0)\\-1(-3)+0(-1) & -1(-2)+0(0)\end{bmatrix} = \begin{bmatrix}11 & 6\\3 & 2\end{bmatrix}$$

19.5. Find **AB** and **BA** for

$$\mathbf{A} = \begin{bmatrix}1 & 2 & 3\\4 & 5 & 6\end{bmatrix}, \qquad \mathbf{B} = \begin{bmatrix}7 & 0\\8 & -1\end{bmatrix}.$$

Since **A** has three columns and **B** has two rows, the product **AB** is not defined. But

$$\mathbf{BA} = \begin{bmatrix}7 & 0\\8 & -1\end{bmatrix}\begin{bmatrix}1 & 2 & 3\\4 & 5 & 6\end{bmatrix} = \begin{bmatrix}7(1)+(0)(4) & 7(2)+(0)(5) & 7(3)+(0)(6)\\8(1)+(-1)(4) & 8(2)+(-1)(5) & 8(3)+(-1)(6)\end{bmatrix}$$

$$= \begin{bmatrix}7 & 14 & 21\\4 & 11 & 18\end{bmatrix}$$

19.6. Find **AB** and **AC** if

$$\mathbf{A} = \begin{bmatrix}4 & 2 & 0\\2 & 1 & 0\\-2 & -1 & 1\end{bmatrix}, \qquad \mathbf{B} = \begin{bmatrix}2 & 3 & 1\\2 & -2 & -2\\-1 & 2 & 1\end{bmatrix}, \qquad \mathbf{C} = \begin{bmatrix}3 & 1 & -3\\0 & 2 & 6\\-1 & 2 & 1\end{bmatrix}$$

$$\mathbf{AB} = \begin{bmatrix}4(2)+2(2)+(0)(-1) & 4(3)+2(-2)+(0)(2) & 4(1)+2(-2)+(0)(1)\\2(2)+1(2)+(0)(-1) & 2(3)+1(-2)+(0)(2) & 2(1)+1(-2)+(0)(1)\\-2(2)+(-1)(2)+1(-1) & -2(3)+(-1)(-2)+1(2) & -2(1)+(-1)(-2)+1(1)\end{bmatrix}$$

$$= \begin{bmatrix}12 & 8 & 0\\6 & 4 & 0\\-7 & -2 & 1\end{bmatrix}$$

$$\mathbf{AC} = \begin{bmatrix}4(3)+2(0)+(0)(-1) & 4(1)+2(2)+(0)(2) & 4(-3)+2(6)+(0)(1)\\2(3)+1(0)+(0)(-1) & 2(1)+1(2)+(0)(2) & 2(-3)+1(6)+(0)(1)\\-2(3)+(-1)(0)+1(-1) & -2(1)+(-1)(2)+1(2) & -2(-3)+(-1)(6)+1(1)\end{bmatrix}$$

$$= \begin{bmatrix}12 & 8 & 0\\6 & 4 & 0\\-7 & -2 & 1\end{bmatrix}$$

Note that for these matrices $\mathbf{AB} = \mathbf{AC}$ and yet $\mathbf{B} \neq \mathbf{C}$. Therefore, the cancellation law is not valid for matrix multiplication.

19.7. Find $\mathbf{Ax}$ if

$$\mathbf{A} = \begin{bmatrix} 1 & 2 & 3 & 4 \\ 5 & 6 & 7 & 8 \end{bmatrix} \qquad \mathbf{x} = \begin{bmatrix} 9 \\ -1 \\ -2 \\ 0 \end{bmatrix}$$

$$\mathbf{Ax} = \begin{bmatrix} 1(9) + 2(-1) + 3(-2) + 4(0) \\ 5(9) + 6(-1) + 7(-2) + 8(0) \end{bmatrix} = \begin{bmatrix} 1 \\ 25 \end{bmatrix}$$

19.8. Find $\dfrac{d\mathbf{A}}{dt}$ if $\mathbf{A} = \begin{bmatrix} t^2+1 & e^{2t} \\ \sin t & 45 \end{bmatrix}$.

$$\frac{d\mathbf{A}}{dt} = \begin{bmatrix} \dfrac{d}{dt}(t^2+1) & \dfrac{d}{dt}(e^{2t}) \\ \dfrac{d}{dt}(\sin t) & \dfrac{d}{dt}(45) \end{bmatrix} = \begin{bmatrix} 2t & 2e^{2t} \\ \cos t & 0 \end{bmatrix}$$

19.9. Find $\dfrac{d\mathbf{x}}{dt}$ if $\mathbf{x} = \begin{bmatrix} x_1(t) \\ x_2(t) \\ x_3(t) \end{bmatrix}$.

$$\frac{d\mathbf{x}}{dt} = \begin{bmatrix} \dfrac{dx_1(t)}{dt} \\ \dfrac{dx_2(t)}{dt} \\ \dfrac{dx_3(t)}{dt} \end{bmatrix} = \begin{bmatrix} \dot{x}_1(t) \\ \dot{x}_2(t) \\ \dot{x}_3(t) \end{bmatrix}$$

19.10. Find $\int \mathbf{A}\,dt$ for $\mathbf{A}$ as given in Problem 19.8.

$$\int \mathbf{A}\,dt = \begin{bmatrix} \displaystyle\int (t^2+1)\,dt & \displaystyle\int e^{2t}\,dt \\ \displaystyle\int \sin t\,dt & \displaystyle\int 45\,dt \end{bmatrix} = \begin{bmatrix} \frac{1}{3}t^3 + t + c_1 & \frac{1}{2}e^{2t} + c_2 \\ -\cos t + c_3 & 45t + c_4 \end{bmatrix}$$

19.11. Find $\displaystyle\int_0^1 \mathbf{x}\,dt$ if $\mathbf{x} = \begin{bmatrix} 1 \\ e^t \\ 0 \end{bmatrix}$.

$$\int_0^1 \mathbf{x}\,dt = \begin{bmatrix} \displaystyle\int_0^1 1\,dt \\ \displaystyle\int_0^1 e^t\,dt \\ \displaystyle\int_0^1 0\,dt \end{bmatrix} = \begin{bmatrix} 1 \\ e-1 \\ 0 \end{bmatrix}$$

19.12. Find the eigenvalues of $\mathbf{A} = \begin{bmatrix} 1 & 3 \\ 4 & 2 \end{bmatrix}$.

We have

$$\mathbf{A} - \lambda\mathbf{I} = \begin{bmatrix} 1 & 3 \\ 4 & 2 \end{bmatrix} + (-\lambda)\begin{bmatrix} 1 & 0 \\ 0 & 1 \end{bmatrix}$$

$$= \begin{bmatrix} 1 & 3 \\ 4 & 2 \end{bmatrix} + \begin{bmatrix} -\lambda & 0 \\ 0 & -\lambda \end{bmatrix} = \begin{bmatrix} 1-\lambda & 3 \\ 4 & 2-\lambda \end{bmatrix}$$

Hence,

$$\det(\mathbf{A} - \lambda\mathbf{I}) = \det\begin{bmatrix} 1-\lambda & 3 \\ 4 & 2-\lambda \end{bmatrix}$$

$$= (1-\lambda)(2-\lambda) - (3)(4) = \lambda^2 - 3\lambda - 10$$

The characteristic equation of $\mathbf{A}$ is $\lambda^2 - 3\lambda - 10 = 0$, which can be factored into $(\lambda - 5)(\lambda + 2) = 0$. The roots of this equation are $\lambda_1 = 5$ and $\lambda_2 = -2$, which are the eigenvalues of $\mathbf{A}$.

19.13. Find the eigenvalues of $\mathbf{A}t$ if $\mathbf{A} = \begin{bmatrix} 2 & 5 \\ -1 & -2 \end{bmatrix}$.

$$\mathbf{A}t - \lambda\mathbf{I} = \begin{bmatrix} 2 & 5 \\ -1 & -2 \end{bmatrix}t + (-\lambda)\begin{bmatrix} 1 & 0 \\ 0 & 1 \end{bmatrix}$$

$$= \begin{bmatrix} 2t & 5t \\ -t & -2t \end{bmatrix} + \begin{bmatrix} -\lambda & 0 \\ 0 & -\lambda \end{bmatrix} = \begin{bmatrix} 2t-\lambda & 5t \\ -t & -2t-\lambda \end{bmatrix}$$

Then,

$$\det(\mathbf{A} - \lambda\mathbf{I}) = \det\begin{bmatrix} 2t-\lambda & 5t \\ -t & -2t-\lambda \end{bmatrix}$$

$$= (2t-\lambda)(-2t-\lambda) - (5t)(-t) = \lambda^2 + t^2$$

and the characteristic equation of $\mathbf{A}t$ is $\lambda^2 + t^2 = 0$. The roots of this equation, which are the eigenvalues of $\mathbf{A}t$, are $\lambda_1 = it$ and $\lambda_2 = -it$, where $i = \sqrt{-1}$.

19.14. Find the eigenvalues of $\mathbf{A} = \begin{bmatrix} 4 & 1 & 0 \\ -1 & 2 & 0 \\ 2 & 1 & -3 \end{bmatrix}$.

$$-\lambda\mathbf{I} = \begin{bmatrix} 4 & 1 & 0 \\ -1 & 2 & 0 \\ 2 & 1 & -3 \end{bmatrix} - \lambda\begin{bmatrix} 1 & 0 & 0 \\ 0 & 1 & 0 \\ 0 & 0 & 1 \end{bmatrix}$$

$$= \begin{bmatrix} 4-\lambda & 1 & 0 \\ -1 & 2-\lambda & 0 \\ 2 & 1 & -3-\lambda \end{bmatrix}$$

Thus,

$$\det(\mathbf{A} - \lambda\mathbf{I}) = \det\begin{bmatrix} 4-\lambda & 1 & 0 \\ -1 & 2-\lambda & 0 \\ 2 & 1 & -3-\lambda \end{bmatrix}$$

$$= (-3-\lambda)[(4-\lambda)(2-\lambda) - (1)(-1)]$$

$$= (-3-\lambda)(\lambda-3)(\lambda-3)$$

The characteristic equation of $\mathbf{A}$ is

$$(-3-\lambda)(\lambda-3)(\lambda-3)=0$$

Hence, the eigenvalues of $\mathbf{A}$ are $\lambda_1=-3$, $\lambda_2=3$, and $\lambda_3=3$. Here, $\lambda=3$ is an eigenvalue of multiplicity two, while $\lambda=-3$ is an eigenvalue of multiplicity one.

19.15. Find the eigenvalues of

$$\mathbf{A}=\begin{bmatrix}5 & 7 & 0 & 0\\ -3 & -5 & 0 & 0\\ 0 & 0 & -2 & 1\\ 0 & 0 & 0 & -2\end{bmatrix}$$

$$\mathbf{A}-\lambda\mathbf{I}=\begin{bmatrix}5-\lambda & 7 & 0 & 0\\ -3 & -5-\lambda & 0 & 0\\ 0 & 0 & -2-\lambda & 1\\ 0 & 0 & 0 & -2-\lambda\end{bmatrix}$$

and

$$\det(\mathbf{A}-\lambda\mathbf{I}) = [(5-\lambda)(-5-\lambda)-(-3)(7)](-2-\lambda)(-2-\lambda)$$
$$= (\lambda^2-4)(-2-\lambda)(-2-\lambda)$$

The characteristic equation of $\mathbf{A}$ is

$$(\lambda^2-4)(-2-\lambda)(-2-\lambda)=0$$

which has roots $\lambda_1=2$, $\lambda_2=-2$, $\lambda_3=-2$, and $\lambda_4=-2$. Thus, $\lambda=-2$ is an eigenvalue of multiplicity three, whereas $\lambda=2$ is an eigenvalue of multiplicity one.

19.16. Verify the Cayley-Hamilton theorem for $\mathbf{A}=\begin{bmatrix}2 & -7\\ 3 & 6\end{bmatrix}$.

For this matrix, we have $\det(\mathbf{A}-\lambda\mathbf{I})=\lambda^2-8\lambda+33$; hence

$$\mathbf{A}^2-8\mathbf{A}+33\mathbf{I}=\begin{bmatrix}2 & -7\\ 3 & 6\end{bmatrix}\begin{bmatrix}2 & -7\\ 3 & 6\end{bmatrix}-8\begin{bmatrix}2 & -7\\ 3 & 6\end{bmatrix}+33\begin{bmatrix}1 & 0\\ 0 & 1\end{bmatrix}$$

$$=\begin{bmatrix}-17 & -56\\ 24 & 15\end{bmatrix}-\begin{bmatrix}16 & -56\\ 24 & 48\end{bmatrix}+\begin{bmatrix}33 & 0\\ 0 & 33\end{bmatrix}$$

$$=\begin{bmatrix}0 & 0\\ 0 & 0\end{bmatrix}$$

19.17. Verify the Cayley-Hamilton theorem for the matrix of Problem 19.14.

For that matrix we found $\det(\mathbf{A}-\lambda\mathbf{I})=-(\lambda+3)(\lambda-3)^2$; hence

$$-(\mathbf{A}+3\mathbf{I})(\mathbf{A}-3\mathbf{I})^2=-\begin{bmatrix}7 & 1 & 0\\ -1 & 5 & 0\\ 2 & 1 & 0\end{bmatrix}\begin{bmatrix}1 & 1 & 0\\ -1 & -1 & 0\\ 2 & 1 & -6\end{bmatrix}^2$$

$$=-\begin{bmatrix}7 & 1 & 0\\ -1 & 5 & 0\\ 2 & 1 & 0\end{bmatrix}\begin{bmatrix}0 & 0 & 0\\ 0 & 0 & 0\\ -11 & -5 & 36\end{bmatrix}=\begin{bmatrix}0 & 0 & 0\\ 0 & 0 & 0\\ 0 & 0 & 0\end{bmatrix}$$

Supplementary Problems

In Problems 19.18 through 19.38, let

$$\mathbf{A} = \begin{bmatrix} 2 & 3 \\ -1 & -2 \end{bmatrix} \quad \mathbf{B} = \begin{bmatrix} 1 & -4 \\ 3 & 1 \end{bmatrix} \quad \mathbf{C} = \begin{bmatrix} 3 & 5 & 0 \\ -2 & -3 & 0 \\ 1 & 1 & 1 \end{bmatrix}$$

$$\mathbf{D} = \begin{bmatrix} 1 & 0 & 2 \\ 1 & 0 & 1 \\ 2 & 0 & 4 \end{bmatrix} \quad \mathbf{x} = \begin{bmatrix} 1 \\ -2 \end{bmatrix} \quad \mathbf{y} = \begin{bmatrix} 1 \\ 1 \\ 2 \end{bmatrix}$$

19.18. Find $\mathbf{A} + \mathbf{B}$.

19.19. Find $3\mathbf{A} - 2\mathbf{B}$.

19.20. Find $\mathbf{C} - \mathbf{D}$.

19.21. Find $2\mathbf{C} + 5\mathbf{D}$.

19.22. Find $\mathbf{A} + \mathbf{D}$.

19.23. Find $\mathbf{x} - 3\mathbf{y}$.

19.24. Find (*a*) $\mathbf{AB}$ and (*b*) $\mathbf{BA}$.

19.25. Find $\mathbf{A}^2$.

19.26. Find $\mathbf{A}^7$.

19.27. Find $\mathbf{B}^2$.

19.28. Find (*a*) $\mathbf{CD}$ and (*b*) $\mathbf{DC}$.

19.29. Find (*a*) $\mathbf{Ax}$ and (*b*) $\mathbf{xA}$.

19.30. Find $\mathbf{AC}$.

19.31. Find $(\mathbf{C} + \mathbf{D})\mathbf{y}$.

19.32. Find the characteristic equation and eigenvalues of $\mathbf{A}$.

19.33. Find the characteristic equation and eigenvalues of $\mathbf{B}$.

19.34. Find the characteristic equation and eigenvalues of $\mathbf{A} + \mathbf{B}$.

19.35. Find the characteristic equation and eigenvalues of $3\mathbf{A}$.

19.36. Find the characteristic equation and eigenvalues of $\mathbf{A} + 5\mathbf{I}$.

19.37. Find the characteristic equation and the eigenvalues of **C**. Determine the multiplicity of each eigenvalue.

19.38. Find the characteristic equation and the eigenvalues of **D**. Determine the multiplicity of each eigenvalue.

19.39. Find the characteristic equation and the eigenvalues of $\mathbf{A} = \begin{bmatrix} t & t^2 \\ 1 & 2t \end{bmatrix}$.

19.40. Find the characteristic equation and the eigenvalues of $\mathbf{A} = \begin{bmatrix} t & 6t & 0 \\ 4t & -t & 0 \\ 0 & 1 & 5t \end{bmatrix}$.

19.41. Find $\dfrac{d\mathbf{A}}{dt}$ for **A** as given in Problem 19.39.

19.42. Find $\dfrac{d\mathbf{A}}{dt}$ for $\mathbf{A} = \begin{bmatrix} \cos 2t \\ te^{3t^2} \end{bmatrix}$.

19.43. Find $\displaystyle\int_0^1 \mathbf{A}\, dt$ for **A** as given in Problem 19.42.

Chapter 20

$e^{\mathbf{A}t}$

DEFINITION

For a square matrix $\mathbf{A}$,

$$e^{\mathbf{A}t} \equiv \mathbf{I} + \frac{1}{1!}\mathbf{A}t + \frac{1}{2!}\mathbf{A}^2t^2 + \cdots = \sum_{n=0}^{\infty} \frac{1}{n!}\mathbf{A}^n t^n \tag{20.1}$$

The infinite series (*20.1*) converges for every $\mathbf{A}$ and t, so that $e^{\mathbf{A}t}$ is defined for all square matrices.

COMPUTATION OF $e^{\mathbf{A}t}$

For actually computing the elements of $e^{\mathbf{A}t}$, (*20.1*) is not generally useful. However, it follows (with some effort) from Theorem 19.1, applied to the matrix $\mathbf{A}t$, that the infinite series can be reduced to a polynomial in t. Thus:

Theorem 20.1. If $\mathbf{A}$ is a matrix having n rows and n columns, then

$$e^{\mathbf{A}t} = \alpha_{n-1}\mathbf{A}^{n-1}t^{n-1} + \alpha_{n-2}\mathbf{A}^{n-2}t^{n-2} + \cdots + \alpha_2\mathbf{A}^2t^2 + \alpha_1\mathbf{A}t + \alpha_0\mathbf{I} \tag{20.2}$$

where $\alpha_0, \alpha_1, \ldots, \alpha_{n-1}$ are functions of t which must be determined for each $\mathbf{A}$.

Example 20.1. When $\mathbf{A}$ has two rows and two columns, then $n = 2$ and

$$e^{\mathbf{A}t} = \alpha_1\mathbf{A}t + \alpha_0\mathbf{I} \tag{20.3}$$

When $\mathbf{A}$ has three rows and three columns, then $n = 3$ and

$$e^{\mathbf{A}t} = \alpha_2\mathbf{A}^2t^2 + \alpha_1\mathbf{A}t + \alpha_0\mathbf{I} \tag{20.4}$$

Theorem 20.2. Let $\mathbf{A}$ be as in Theorem 20.1, and define

$$r(\lambda) \equiv \alpha_{n-1}\lambda^{n-1} + \alpha_{n-2}\lambda^{n-2} + \cdots + \alpha_2\lambda^2 + \alpha_1\lambda + \alpha_0 \tag{20.5}$$

Then if λ_i is an eigenvalue of $\mathbf{A}t$,

$$e^{\lambda_i} = r(\lambda_i) \tag{20.6}$$

Furthermore, if λ_i is an eigenvalue of multiplicity k, $k > 1$, then the following equations are also valid:

$$\begin{aligned} e^{\lambda_i} &= \left.\frac{d}{d\lambda} r(\lambda)\right|_{\lambda=\lambda_i} \\ e^{\lambda_i} &= \left.\frac{d^2}{d\lambda^2} r(\lambda)\right|_{\lambda=\lambda_i} \\ &\cdots\cdots\cdots\cdots\cdots \\ e^{\lambda_i} &= \left.\frac{d^{k-1}}{d\lambda^{k-1}} r(\lambda)\right|_{\lambda=\lambda_i} \end{aligned} \tag{20.7}$$

Note that Theorem 20.2 involves the eigenvalues of $\mathbf{A}t$; these are t times the eigenvalues of $\mathbf{A}$. When computing the various derivatives in (*20.7*), one first calculates the appropriate derivatives of the expression (*20.5*) with respect to λ, and then substitutes $\lambda = \lambda_i$. The reverse procedure of first substituting $\lambda = \lambda_i$ (a function of t) into (20.5), and then calculating the derivatives with respect to t, can give erroneous results.

Example 20.2. Let $\mathbf{A}$ have four rows and four columns and let $\lambda = 5t$ and $\lambda = 2t$ be eigenvalues of $\mathbf{A}t$ of multiplicities three and one, respectively. Then $n = 4$ and

$$r(\lambda) = \alpha_3\lambda^3 + \alpha_2\lambda^2 + \alpha_1\lambda + \alpha_0$$
$$r'(\lambda) = 3\alpha_3\lambda^2 + 2\alpha_2\lambda + \alpha_1$$
$$r''(\lambda) = 6\alpha_3\lambda + 2\alpha_2$$

Since $\lambda = 5t$ is an eigenvalue of multiplicity three, it follows that $e^{5t} = r(5t)$, $e^{5t} = r'(5t)$, and $e^{5t} = r''(5t)$. Thus,

$$e^{5t} = \alpha_3(5t)^3 + \alpha_2(5t)^2 + \alpha_1(5t) + \alpha_0$$
$$e^{5t} = 3\alpha_3(5t)^2 + 2\alpha_2(5t) + \alpha_1$$
$$e^{5t} = 6\alpha_3(5t) + 2\alpha_2$$

Also, since $\lambda = 2t$ is an eigenvalue of multiplicity one, it follows that $e^{2t} = r(2t)$, or

$$e^{2t} = \alpha_3(2t)^3 + \alpha_2(2t)^2 + \alpha_1(2t) + \alpha_0$$

Notice that we now have four equations in the four unknown α's.

Method of computation: For each eigenvalue λ_i of $\mathbf{A}t$, apply Theorem 20.2 to obtain a set of linear equations. When this is done for each eigenvalue, the set of all equations so obtained can be solved for $\alpha_0, \alpha_1, \ldots, \alpha_{n-1}$. These values are then substituted into Eq. (20.2), which, in turn, is used to compute $e^{\mathbf{A}t}$.

Solved Problems

20.1. Find $e^{\mathbf{A}t}$ for $\mathbf{A} = \begin{bmatrix} 1 & 1 \\ 9 & 1 \end{bmatrix}$.

Here $n = 2$. From Eq. (20.3),

$$e^{\mathbf{A}t} = \alpha_1\mathbf{A}t + \alpha_0\mathbf{I} = \begin{bmatrix} \alpha_1 t + \alpha_0 & \alpha_1 t \\ 9\alpha_1 t & \alpha_1 t + \alpha_0 \end{bmatrix} \tag{1}$$

and from Eq. (20.5), $r(\lambda) = \alpha_1\lambda + \alpha_0$. The eigenvalues of $\mathbf{A}t$ are $\lambda_1 = 4t$ and $\lambda_2 = -2t$, which are both of multiplicity one. Substituting these values successively into Eq. (20.6), we obtain the two equations

$$e^{4t} = 4t\alpha_1 + \alpha_0$$
$$e^{-2t} = -2t\alpha_1 + \alpha_0$$

Solving these equations for α_1 and α_0, we find that

$$\alpha_1 = \frac{1}{6t}(e^{4t} - e^{-2t}) \quad \text{and} \quad \alpha_0 = \frac{1}{3}(e^{4t} + 2e^{-2t})$$

Substituting these values into (1) and simplifying, we have

$$e^{\mathbf{A}t} = \frac{1}{6}\begin{bmatrix} 3e^{4t} + 3e^{-2t} & e^{4t} - e^{-2t} \\ 9e^{4t} - 9e^{-2t} & 3e^{4t} + 3e^{-2t} \end{bmatrix}$$

20.2. Find $e^{\mathbf{A}t}$ for $\mathbf{A} = \begin{bmatrix} 0 & 1 \\ 8 & -2 \end{bmatrix}$.

Since $n = 2$, it follows from Eqs. (20.3) and (20.5) that

$$e^{\mathbf{A}t} = \alpha_1\mathbf{A}t + \alpha_0\mathbf{I} = \begin{bmatrix} \alpha_0 & \alpha_1 t \\ 8\alpha_1 t & -2\alpha_1 t + \alpha_0 \end{bmatrix} \tag{1}$$

and $r(\lambda) = \alpha_1\lambda + \alpha_0$. The eigenvalues of $\mathbf{A}t$ are $\lambda_1 = 2t$ and $\lambda_2 = -4t$, which are both of multiplicity one.

Substituting these values successively into (*20.6*), we obtain

$$e^{2t} = \alpha_1(2t) + \alpha_0 \qquad e^{-4t} = \alpha_1(-4t) + \alpha_0$$

Solving these equations for α_1 and α_0, we find that

$$\alpha_1 = \frac{1}{6t}(e^{2t} - e^{-4t}) \qquad \alpha_0 = \frac{1}{3}(2e^{2t} + e^{-4t})$$

Substituting these values into (*1*) and simplifying, we have

$$e^{\mathbf{A}t} = \frac{1}{6}\begin{bmatrix} 4e^{2t} + 2e^{-4t} & e^{2t} - e^{-4t} \\ 8e^{2t} - 8e^{-4t} & 2e^{2t} + 4e^{-4t} \end{bmatrix}$$

20.3. Find $e^{\mathbf{A}t}$ for $\mathbf{A} = \begin{bmatrix} 0 & 1 \\ -1 & 0 \end{bmatrix}$.

Here $n = 2$; hence,

$$e^{\mathbf{A}t} = \alpha_1\mathbf{A}t + \alpha_0\mathbf{I} = \begin{bmatrix} \alpha_0 & \alpha_1 t \\ -\alpha_1 t & \alpha_0 \end{bmatrix} \tag{1}$$

and $r(\lambda) = \alpha_1\lambda + \alpha_0$. The eigenvalues of $\mathbf{A}t$ are $\lambda_1 = it$ and $\lambda_2 = -it$, which are both of multiplicity one. Substituting these values successively into Eq. (*20.6*), we obtain

$$e^{it} = \alpha_1(it) + \alpha_0 \qquad e^{-it} = \alpha_1(-it) + \alpha_0$$

Solving these equations for α_1 and α_0 and using Euler's relations, we find that

$$\alpha_1 = \frac{1}{2it}(e^{it} - e^{-it}) = \frac{\sin t}{t}$$

$$\alpha_0 = \frac{1}{2}(e^{it} + e^{-it}) = \cos t$$

Substituting these values into (*1*), we obtain

$$e^{\mathbf{A}t} = \begin{bmatrix} \cos t & \sin t \\ -\sin t & \cos t \end{bmatrix}$$

20.4. Find $e^{\mathbf{A}t}$ for $\mathbf{A} = \begin{bmatrix} 0 & 1 \\ -9 & 6 \end{bmatrix}$.

Here $n = 2$. From Eq. (*20.3*),

$$e^{\mathbf{A}t} = \alpha_1\mathbf{A}t + \alpha_0\mathbf{I} = \begin{bmatrix} \alpha_0 & \alpha_1 t \\ -9\alpha_1 t & 6\alpha_1 t + \alpha_0 \end{bmatrix} \tag{1}$$

and from Eq. (*20.5*), $r(\lambda) = \alpha_1\lambda + \alpha_0$. Thus, $dr(\lambda)/d\lambda = \alpha_1$. The eigenvalues of $\mathbf{A}t$ are $\lambda_1 = \lambda_2 = 3t$, which is a single eigenvalue of multiplicity two. It follows from Theorem 20.2 that

$$e^{3t} = 3t\alpha_1 + \alpha_0$$
$$e^{3t} = \alpha_1$$

Solving these equations for α_1 and α_0, we find that

$$\alpha_1 = e^{3t} \qquad \text{and} \qquad \alpha_0 = e^{3t}(1 - 3t)$$

Substituting these values into (*1*) and simplifying, we have

$$e^{\mathbf{A}t} = e^{3t}\begin{bmatrix} 1-3t & t \\ -9t & 1+3t \end{bmatrix}$$

20.5. Find $e^{\mathbf{A}t}$ for $\mathbf{A} = \begin{bmatrix} 3 & 1 & 0 \\ 0 & 3 & 1 \\ 0 & 0 & 3 \end{bmatrix}$.

Here $n = 3$. From Eqs. (*20.4*) and (*20.5*) we have

$$e^{\mathbf{A}t} = \alpha_2\mathbf{A}^2t^2 + \alpha_1\mathbf{A}t + \alpha_0\mathbf{I}$$

$$= \alpha_2\begin{bmatrix} 9 & 6 & 1 \\ 0 & 9 & 6 \\ 0 & 0 & 9 \end{bmatrix}t^2 + \alpha_1\begin{bmatrix} 3 & 1 & 0 \\ 0 & 3 & 1 \\ 0 & 0 & 3 \end{bmatrix}t + \alpha_0\begin{bmatrix} 1 & 0 & 0 \\ 0 & 1 & 0 \\ 0 & 0 & 1 \end{bmatrix}$$

$$= \begin{bmatrix} 9\alpha_2t^2 + 3\alpha_1t + \alpha_0 & 6\alpha_2t^2 + \alpha_1t & \alpha_2t^2 \\ 0 & 9\alpha_2t^2 + 3\alpha_1t + \alpha_0 & 6\alpha_2t^2 + \alpha_1t \\ 0 & 0 & 9\alpha_2t^2 + 3\alpha_1t + \alpha_0 \end{bmatrix} \tag{1}$$

and $r(\lambda) = \alpha_2\lambda^2 + \alpha_1\lambda + \alpha_0$. Thus,

$$\frac{dr(\lambda)}{d\lambda} = 2\alpha_2\lambda + \alpha_1 \qquad \frac{d^2r(\lambda)}{d\lambda^2} = 2\alpha_2$$

Since the eigenvalues of $\mathbf{A}t$ are $\lambda_1 = \lambda_2 = \lambda_3 = 3t$, an eigenvalue of multiplicity three, it follows from Theorem 20.2 that

$$e^{3t} = \alpha_2 9t^2 + \alpha_1 3t + \alpha_0$$
$$e^{3t} = \alpha_2 6t + \alpha_1$$
$$e^{3t} = 2\alpha_2$$

The solution to this set of equations is

$$\alpha_2 = \frac{1}{2}e^{3t} \qquad \alpha_1 = (1-3t)e^{3t} \qquad \alpha_0 = \left(1 - 3t + \frac{9}{2}t^2\right)e^{3t}$$

Substituting these values into (*1*) and simplifying, we obtain

$$e^{\mathbf{A}t} = e^{3t}\begin{bmatrix} 1 & t & t^2/2 \\ 0 & 1 & t \\ 0 & 0 & 1 \end{bmatrix}$$

20.6. Find $e^{\mathbf{A}t}$ for $\mathbf{A} = \begin{bmatrix} 0 & 1 & 0 \\ 0 & 0 & 1 \\ 0 & -1 & 2 \end{bmatrix}$.

Here $n = 3$. From Eq. (*20.4*),

$$e^{\mathbf{A}t} = \alpha_2\mathbf{A}^2t^2 + \alpha_1\mathbf{A}t + \alpha_0\mathbf{I}$$

$$= \begin{bmatrix} \alpha_0 & \alpha_1t & \alpha_2t^2 \\ 0 & -\alpha_2t^2 + \alpha_0 & 2\alpha_2t^2 + \alpha_1t \\ 0 & -2\alpha_2t^2 - \alpha_1t & 3\alpha_2t^2 + 2\alpha_1t + \alpha_0 \end{bmatrix} \tag{1}$$

and from Eq. (*20.5*), $r(\lambda) = \alpha_2\lambda^2 + \alpha_1\lambda + \alpha_0$. The eigenvalues of $\mathbf{A}t$ are $\lambda_1 = 0$ and $\lambda_2 = \lambda_3 = t$; hence

$\lambda = t$ is an eigenvalue of multiplicity two, while $\lambda = 0$ is an eigenvalue of multiplicity one. It follows from Theorem 20.2 that $e^t = r(t)$, $e^t = r'(t)$, and $e^0 = r(0)$. Since $r'(\lambda) = 2\alpha_2\lambda + \alpha_1$, these equations become

$$\begin{aligned} e^t &= \alpha_2 t^2 + \alpha_1 t + \alpha_0 \\ e^t &= 2\alpha_2 t + \alpha_1 \\ e^0 &= \alpha_0 \end{aligned}$$

which have as their solution

$$\alpha_2 = \frac{te^t - e^t + 1}{t^2} \qquad \alpha_1 = \frac{-te^t + 2e^t - 2}{t} \qquad \alpha_0 = 1$$

Substituting these values into *(1)* and simplifying, we have

$$e^{\mathbf{A}t} = \begin{bmatrix} 1 & -te^t + 2e^t - 2 & te^t - e^t + 1 \\ 0 & -te^t + e^t & te^t \\ 0 & -te^t & te^t + e^t \end{bmatrix}$$

20.7. Find $e^{\mathbf{A}t}$ for $\mathbf{A} = \begin{bmatrix} 0 & 1 & 0 \\ 0 & -2 & -5 \\ 0 & 1 & 2 \end{bmatrix}$.

Here $n = 3$. From Eq. *(20.4)*,

$$\begin{aligned} e^{\mathbf{A}t} &= \alpha_2 \mathbf{A}^2 t^2 + \alpha_1 \mathbf{A}t + \alpha_0 \mathbf{I} \\ &= \begin{bmatrix} \alpha_0 & -2\alpha_2 t^2 + \alpha_1 t & -5\alpha_2 t^2 \\ 0 & -\alpha_2 t^2 - 2\alpha_1 t + \alpha_0 & -5\alpha_1 t \\ 0 & \alpha_1 t & -\alpha_2 t^2 + 2\alpha_1 t + \alpha_0 \end{bmatrix} \end{aligned} \tag{1}$$

and from Eq. *(20.5)*, $r(\lambda) = \alpha_2\lambda^2 + \alpha_1\lambda + \alpha_0$. The eigenvalues of $\mathbf{A}t$ are $\lambda_1 = 0$, $\lambda_2 = it$, and $\lambda_3 = -it$. Substituting these values successively into *(20.6)*, we obtain the three equations

$$\begin{aligned} e^0 &= \alpha_2(0)^2 + \alpha_1(0) + \alpha_0 \\ e^{it} &= \alpha_2(it)^2 + \alpha_1(it) + \alpha_0 \\ e^{-it} &= \alpha_2(-it)^2 + \alpha_1(-it) + \alpha_0 \end{aligned}$$

which have as their solution

$$\begin{aligned} \alpha_2 &= \frac{e^{it} + e^{-it} - 2}{-2t^2} = \frac{1 - \cos t}{t^2} \\ \alpha_1 &= \frac{e^{it} - e^{-it}}{2it} = \frac{\sin t}{t} \\ \alpha_0 &= 1 \end{aligned}$$

Substituting these values into *(1)* and simplifying, we have

$$e^{\mathbf{A}t} = \begin{bmatrix} 1 & -2 + 2\cos t + \sin t & -5 + 5\cos t \\ 0 & \cos t - 2\sin t & -5\sin t \\ 0 & \sin t & \cos t + 2\sin t \end{bmatrix}$$

20.8. Establish the necessary equations to find $e^{\mathbf{A}t}$ if

$$\mathbf{A} = \begin{bmatrix} 1 & 2 & 3 & 4 & 5 & 6 \\ 0 & 1 & 2 & 3 & 4 & 5 \\ 0 & 0 & 2 & 3 & 4 & 5 \\ 0 & 0 & 0 & 2 & 3 & 4 \\ 0 & 0 & 0 & 0 & 0 & 0 \\ 0 & 0 & 0 & 0 & 0 & 1 \end{bmatrix}$$

Here $n = 6$, so

$$e^{\mathbf{A}t} = \alpha_5\mathbf{A}^5t^5 + \alpha_4\mathbf{A}^4t^4 + \alpha_3\mathbf{A}^3t^3 + \alpha_2\mathbf{A}^2t^2 + \alpha_1\mathbf{A}t + \alpha_0\mathbf{I}$$

and

$$\begin{aligned} r(\lambda) &= \alpha_5\lambda^5 + \alpha_4\lambda^4 + \alpha_3\lambda^3 + \alpha_2\lambda^2 + \alpha_1\lambda + \alpha_0 \\ r'(\lambda) &= 5\alpha_5\lambda^4 + 4\alpha_4\lambda^3 + 3\alpha_3\lambda^2 + 2\alpha_2\lambda + \alpha_1 \\ r''(\lambda) &= 20\alpha_5\lambda^3 + 12\alpha_4\lambda^2 + 6\alpha_3\lambda + 2\alpha_2 \end{aligned}$$

The eigenvalues of $\mathbf{A}t$ are $\lambda_1 = \lambda_2 = \lambda_3 = t$, $\lambda_4 = \lambda_5 = 2t$, and $\lambda_6 = 0$. Hence, $\lambda = t$ is an eigenvalue of multiplicity three, $\lambda = 2t$ is an eigenvalue of multiplicity two, and $\lambda = 0$ is an eigenvalue of multiplicity one. It now follows from Theorem 20.2 that

$$\begin{aligned} e^{2t} &= r(2t) = \alpha_5(2t)^5 + \alpha_4(2t)^4 + \alpha_3(2t)^3 + \alpha_2(2t)^2 + \alpha_1(2t) + \alpha_0 \\ e^{2t} &= r'(2t) = 5\alpha_5(2t)^4 + 4\alpha_4(2t)^3 + 3\alpha_3(2t)^2 + 2\alpha_2(2t) + \alpha_1 \\ e^{2t} &= r''(2t) = 20\alpha_5(2t)^3 + 12\alpha_4(2t)^2 + 6\alpha_3(2t) + 2\alpha_2 \\ e^{t} &= r(t) = \alpha_5(t)^5 + \alpha_4(t)^4 + \alpha_3(t)^3 + \alpha_2(t)^2 + \alpha_1(t) + \alpha_0 \\ e^{t} &= r'(t) = 5\alpha_5(t)^4 + 4\alpha_4(t)^3 + 3\alpha_3(t)^2 + 2\alpha_2(t) + \alpha_1 \\ e^{0} &= r(0) = \alpha_5(0)^5 + \alpha_4(0)^4 + \alpha_3(0)^3 + \alpha_2(0)^2 + \alpha_1(0) + \alpha_0 \end{aligned}$$

or, more simply,

$$\begin{aligned} e^{2t} &= 32t^5\alpha_5 + 16t^4\alpha_4 + 8t^3\alpha_3 + 4t^2\alpha_2 + 2t\alpha_1 + \alpha_0 \\ e^{2t} &= 80t^4\alpha_5 + 32t^3\alpha_4 + 12t^2\alpha_3 + 4t\alpha_2 + \alpha_1 \\ e^{2t} &= 160t^3\alpha_5 + 48t^2\alpha_4 + 12t\alpha_3 + 2\alpha_2 \\ e^{t} &= t^5\alpha_5 + t^4\alpha_4 + t^3\alpha_3 + t^2\alpha_2 + t\alpha_1 + \alpha_0 \\ e^{t} &= 5t^4\alpha_5 + 4t^3\alpha_4 + 3t^2\alpha_3 + 2t\alpha_2 + \alpha_1 \\ 1 &= \alpha_0 \end{aligned}$$

20.9. Find $e^{\mathbf{A}t}e^{\mathbf{B}t}$ and $e^{(\mathbf{A}+\mathbf{B})t}$ for

$$\mathbf{A} = \begin{bmatrix} 0 & 1 \\ 0 & 0 \end{bmatrix} \quad \text{and} \quad \mathbf{B} = \begin{bmatrix} 0 & 0 \\ -1 & 0 \end{bmatrix}$$

and verify that, for these matrices, $e^{\mathbf{A}t}e^{\mathbf{B}t} \neq e^{(\mathbf{A}+\mathbf{B})t}$.

Here, $\mathbf{A} + \mathbf{B} = \begin{bmatrix} 0 & 1 \\ -1 & 0 \end{bmatrix}$. Using Theorem 20.1 and the result of Problem 20.3, we find that

$$e^{\mathbf{A}t} = \begin{bmatrix} 1 & t \\ 0 & 1 \end{bmatrix} \qquad e^{\mathbf{B}t} = \begin{bmatrix} 1 & 0 \\ -t & 1 \end{bmatrix} \qquad e^{(\mathbf{A}+\mathbf{B})t} = \begin{bmatrix} \cos t & \sin t \\ -\sin t & \cos t \end{bmatrix}$$

Thus,

$$e^{\mathbf{A}t}e^{\mathbf{B}t} = \begin{bmatrix} 1 & t \\ 0 & 1 \end{bmatrix}\begin{bmatrix} 1 & 0 \\ -t & 1 \end{bmatrix} = \begin{bmatrix} 1-t^2 & t \\ -t & 1 \end{bmatrix} \neq e^{(\mathbf{A}+\mathbf{B})t}$$

20.10. Prove that $e^{\mathbf{A}t}e^{\mathbf{B}t} = e^{(\mathbf{A}+\mathbf{B})t}$ if and only if the matrices $\mathbf{A}$ and $\mathbf{B}$ commute.

If $\mathbf{AB} = \mathbf{BA}$, and only then, we have

$$(\mathbf{A} + \mathbf{B})^2 = (\mathbf{A} + \mathbf{B})(\mathbf{A} + \mathbf{B}) = \mathbf{A}^2 + \mathbf{AB} + \mathbf{BA} + \mathbf{B}^2 = \mathbf{A}^2 + 2\mathbf{AB} + \mathbf{B}^2$$
$$= \sum_{k=0}^{2} \binom{2}{k} \mathbf{A}^{n-k}\mathbf{B}^k$$

and, in general,
$$(\mathbf{A} + \mathbf{B})^n = \sum_{k=0}^{n} \binom{n}{k} \mathbf{A}^{n-k}\mathbf{B}^k \tag{1}$$

where $\binom{n}{k} = \dfrac{n!}{k!\,(n-k)!}$ is the binomial coefficient ("n things taken k at a time").

Now, according to Eq. (*20.1*), we have for any $\mathbf{A}$ and $\mathbf{B}$:

$$e^{\mathbf{A}t}e^{\mathbf{B}t} = \left(\sum_{n=0}^{\infty} \frac{1}{n!}\mathbf{A}^n t^n\right)\left(\sum_{n=0}^{\infty} \frac{1}{n!}\mathbf{B}^n t^n\right) = \sum_{n=0}^{\infty}\sum_{k=0}^{n} \frac{\mathbf{A}^{n-k}t^{n-k}\,\mathbf{B}^k t^k}{(n-k)!\;k!}$$

$$= \sum_{n=0}^{\infty}\left[\sum_{k=0}^{n} \frac{\mathbf{A}^{n-k}\mathbf{B}^k}{(n-k)!\,k!}\right]t^n = \sum_{n=0}^{\infty}\left[\sum_{k=0}^{n}\binom{n}{k}\mathbf{A}^{n-k}\mathbf{B}^k\right]\frac{t^n}{n!} \tag{2}$$

and also
$$e^{(\mathbf{A}+\mathbf{B})t} = \sum_{n=0}^{\infty} \frac{1}{n!}(\mathbf{A} + \mathbf{B})^n t^n = \sum_{n=0}^{\infty} (\mathbf{A} + \mathbf{B})^n \frac{t^n}{n!} \tag{3}$$

We can equate the last series in (*3*) to the last series in (*2*) if and only if (*1*) holds; that is, if and only if $\mathbf{A}$ and $\mathbf{B}$ commute.

20.11. Prove that $e^{\mathbf{A}t}e^{-\mathbf{A}s} = e^{\mathbf{A}(t-s)}$.

Setting $t = 1$ in Problem 20.10, we conclude that $e^{\mathbf{A}}e^{\mathbf{B}} = e^{(\mathbf{A}+\mathbf{B})}$ if $\mathbf{A}$ and $\mathbf{B}$ commute. But the matrices $\mathbf{A}t$ and $-\mathbf{A}s$ commute, since

$$(\mathbf{A}t)(-\mathbf{A}s) = (\mathbf{AA})(-ts) = (\mathbf{AA})(-st) = (-\mathbf{A}s)(\mathbf{A}t)$$

Consequently, $e^{\mathbf{A}t}e^{-\mathbf{A}s} = e^{(\mathbf{A}t-\mathbf{A}s)} = e^{\mathbf{A}(t-s)}$.

20.12. Prove that $e^{\mathbf{0}} = \mathbf{I}$, where $\mathbf{0}$ denotes a square matrix all of whose elements are zero.

From the definition of matrix multiplication, $\mathbf{0}^n = \mathbf{0}$ for $n \geq 1$. Hence,

$$e^{\mathbf{0}} = e^{\mathbf{0}t} = \sum_{n=0}^{\infty} \frac{1}{n!}\mathbf{0}^n t^n = \mathbf{I} + \sum_{n=1}^{\infty} \frac{1}{n!}\mathbf{0}^n t^n = \mathbf{I} + \mathbf{0} = \mathbf{I}$$

Supplementary Problems

Find $e^{\mathbf{A}t}$ for the following matrices $\mathbf{A}$.

20.13. $\begin{bmatrix} 2 & 0 \\ 0 & -3 \end{bmatrix}$

20.14. $\begin{bmatrix} 3 & 2 \\ 4 & 1 \end{bmatrix}$

20.15. $\begin{bmatrix} 5 & 6 \\ -4 & -5 \end{bmatrix}$

20.16. $\begin{bmatrix} 0 & 1 \\ 8 & -2 \end{bmatrix}$

20.17. $\begin{bmatrix} 0 & 1 \\ -14 & -9 \end{bmatrix}$

20.18. $\begin{bmatrix} 2 & 0 \\ 0 & 2 \end{bmatrix}$

20.19. $\begin{bmatrix} 2 & 1 \\ 0 & 2 \end{bmatrix}$

20.20. $\begin{bmatrix} 4 & 5 \\ -4 & -4 \end{bmatrix}$

20.21. $\begin{bmatrix} 0 & 1 \\ -16 & 0 \end{bmatrix}$

20.22. $\begin{bmatrix} 0 & 1 \\ -64 & -16 \end{bmatrix}$

20.23. $\begin{bmatrix} 0 & 1 \\ -4 & -4 \end{bmatrix}$

20.24. $\begin{bmatrix} 0 & 1 \\ -36 & 0 \end{bmatrix}$

20.25. $\begin{bmatrix} 0 & 1 \\ -25 & -8 \end{bmatrix}$

20.26. $\begin{bmatrix} 4 & -2 \\ 8 & 2 \end{bmatrix}$

20.27. $\begin{bmatrix} 2 & 1 & 0 \\ 0 & 2 & 1 \\ 0 & 0 & 2 \end{bmatrix}$

20.28. $\begin{bmatrix} 2 & 0 & 0 \\ 0 & 2 & 1 \\ 0 & 0 & 2 \end{bmatrix}$

20.29. $\begin{bmatrix} -1 & 1 & 0 \\ 0 & 2 & 1 \\ 0 & 0 & 2 \end{bmatrix}$

20.30. $\begin{bmatrix} 0 & 0 & 0 \\ 0 & 0 & 0 \\ 0 & 0 & 0 \end{bmatrix}$

20.31. $\begin{bmatrix} 0 & 1 & 0 \\ 0 & 0 & 0 \\ 0 & 0 & 1 \end{bmatrix}$

20.32. $\begin{bmatrix} 0 & 0 & 0 \\ 1 & 0 & 0 \\ 1 & 0 & 1 \end{bmatrix}$

Chapter 21

Reduction of Linear Differential Equations to a First-Order System

REDUCTION OF ONE EQUATION

Every initial-value problem of the form

$$b_n(t)\frac{d^nx}{dt^n} + b_{n-1}(t)\frac{d^{n-1}x}{dt^{n-1}} + \cdots + b_1(t)\dot{x} + b_0(t)x = g(t); \tag{21.1}$$

$$x(t_0) = c_0, \qquad \dot{x}(t_0) = c_1, \ldots, x^{(n-1)}(t_0) = c_{n-1} \tag{21.2}$$

with $b_n(t) \neq 0$, can be reduced to the first-order matrix system

$$\begin{aligned} \dot{\mathbf{x}}(t) &= \mathbf{A}(t)\mathbf{x}(t) + \mathbf{f}(t) \\ \mathbf{x}(t_0) &= \mathbf{c} \end{aligned} \tag{21.3}$$

where $\mathbf{A}(t)$, $\mathbf{f}(t)$, $\mathbf{c}$, and the initial time t_0 are known. The method of reduction is as follows.

Step 1. Rewrite (*21.1*) so that d^nx/dt^n appears by itself. Thus,

$$\frac{d^nx}{dt^n} = a_{n-1}(t)\frac{d^{n-1}x}{dt^{n-1}} + \cdots + a_1(t)\dot{x} + a_0(t)x + f(t) \tag{21.4}$$

where $a_j(t) = -b_j(t)/b_n(t)$ $(j = 0, 1, \ldots, n-1)$ and $f(t) = g(t)/b_n(t)$.

Step 2. Define n new variables (the same number as the order of the original differential equation), $x_1(t), x_2(t), \ldots, x_n(t)$, by the equations

$$x_1(t) = x(t), \qquad x_2(t) = \frac{dx(t)}{dt}, \qquad x_3(t) = \frac{d^2x(t)}{dt^2}, \ldots, x_n(t) = \frac{d^{n-1}x(t)}{dt^{n-1}} \tag{21.5}$$

These new variables are interrelated by the equations

$$\begin{aligned} \dot{x}_1(t) &= x_2(t) \\ \dot{x}_2(t) &= x_3(t) \\ \dot{x}_3(t) &= x_4(t) \\ &\cdots\cdots\cdots\cdots \\ \dot{x}_{n-1}(t) &= x_n(t) \end{aligned} \tag{21.6}$$

Step 3. Express dx_n/dt in terms of the new variables. Proceed by first differentiating the last equation of (*21.5*) to obtain

$$\dot{x}_n(t) = \frac{d}{dt}\left[\frac{d^{n-1}x(t)}{dt^{n-1}}\right] = \frac{d^nx(t)}{dt^n}$$

Then, from Eqs. (*21.4*) and (*21.5*),

$$\begin{aligned} \dot{x}_n(t) &= a_{n-1}(t)\frac{d^{n-1}x(t)}{dt^{n-1}} + \cdots + a_1(t)\dot{x}(t) + a_0(t)x(t) + f(t) \\ &= a_{n-1}(t)x_n(t) + \cdots + a_1(t)x_2(t) + a_0(t)x_1(t) + f(t) \end{aligned}$$

For convenience, we rewrite this last equation so that $x_1(t)$ appears before $x_2(t)$, etc. Thus,

$$\dot{x}_n(t) = a_0(t)x_1(t) + a_1(t)x_2(t) + \cdots + a_{n-1}(t)x_n(t) + f(t) \tag{21.7}$$

Step 4. Equations (*21.6*) and (*21.7*) are a system of first-order linear differential equations in $x_1(t), x_2(t), \ldots, x_n(t)$. This system is equivalent to the single matrix equation $\dot{\mathbf{x}}(t) = \mathbf{A}(t)\mathbf{x}(t) + \mathbf{f}(t)$ if we define

$$\mathbf{x}(t) \equiv \begin{bmatrix} x_1(t) \\ x_2(t) \\ \vdots \\ x_n(t) \end{bmatrix} \tag{21.8}$$

$$\mathbf{f}(t) \equiv \begin{bmatrix} 0 \\ 0 \\ \vdots \\ 0 \\ f(t) \end{bmatrix} \tag{21.9}$$

$$\mathbf{A}(t) \equiv \begin{bmatrix} 0 & 1 & 0 & 0 & \cdots & 0 \\ 0 & 0 & 1 & 0 & \cdots & 0 \\ 0 & 0 & 0 & 1 & \cdots & 0 \\ \vdots & \vdots & \vdots & \vdots & & \vdots \\ 0 & 0 & 0 & 0 & \cdots & 1 \\ a_0(t) & a_1(t) & a_2(t) & a_3(t) & \cdots & a_{n-1}(t) \end{bmatrix} \tag{21.10}$$

Step 5. Define

$$\mathbf{c} \equiv \begin{bmatrix} c_0 \\ c_1 \\ \vdots \\ c_{n-1} \end{bmatrix}$$

Then the initial conditions (*21.2*) can be given by the matrix (vector) equation $\mathbf{x}(t_0) = \mathbf{c}$. This last equation is an immediate consequence of Eqs. (*21.8*), (*21.5*), and (*21.2*), since

$$\mathbf{x}(t_0) = \begin{bmatrix} x_1(t_0) \\ x_2(t_0) \\ \vdots \\ x_n(t_0) \end{bmatrix} = \begin{bmatrix} x(t_0) \\ \dot{x}(t_0) \\ \vdots \\ x^{(n-1)}(t_0) \end{bmatrix} = \begin{bmatrix} c_0 \\ c_1 \\ \vdots \\ c_{n-1} \end{bmatrix} \equiv \mathbf{c}$$

Observe that if no initial conditions are prescribed, Steps 1 through 4 by themselves reduce any linear differential equation (*21.1*) to the matrix equation $\dot{\mathbf{x}}(t) = \mathbf{A}(t)\mathbf{x}(t) + \mathbf{f}(t)$.

REDUCTION OF A SYSTEM

A set of linear differential equations with initial conditions also can be reduced to system (*21.3*). The procedure is nearly identical to the method for reducing a single equation to matrix form; only Step 2 changes. With a system of equations, Step 2 is generalized so that new variables are defined for *each* of the unknown functions in the set.

Solved Problems

21.1. Put the initial-value problem

$$\ddot{x} + 2\dot{x} - 8x = e^t; \qquad x(0) = 1, \qquad \dot{x}(0) = -4$$

into the form of System (*21.3*).

Following Step 1, we write $\ddot{x} = -2\dot{x} + 8x + e^t$; hence, $a_1(t) = -2$, $a_0(t) = 8$, and $f(t) = e^t$. Then, defining $x_1(t) = x$ and $x_2(t) = \dot{x}$ (the differential equation is second-order, so we need two new variables), we obtain $\dot{x}_1 = x_2$. Following Step 3, we find

$$\dot{x}_2 = \frac{d^2x}{dt^2} = -2\dot{x} + 8x + e^t = -2x_2 + 8x_1 + e^t$$

Thus,

$$\dot{x}_1 = 0x_1 + 1x_2 + 0$$
$$\dot{x}_2 = 8x_1 - 2x_2 + e^t$$

These equations are equivalent to the matrix equation $\dot{\mathbf{x}}(t) = \mathbf{A}(t)\mathbf{x}(t) + \mathbf{f}(t)$ if we define

$$\mathbf{x}(t) \equiv \begin{bmatrix} x_1(t) \\ x_2(t) \end{bmatrix} \qquad \mathbf{A}(t) \equiv \begin{bmatrix} 0 & 1 \\ 8 & -2 \end{bmatrix} \qquad \mathbf{f}(t) \equiv \begin{bmatrix} 0 \\ e^t \end{bmatrix}$$

Furthermore, if we also define $\mathbf{c} \equiv \begin{bmatrix} 1 \\ -4 \end{bmatrix}$, then the initial conditions can be given by $\mathbf{x}(t_0) = \mathbf{c}$, where $t_0 = 0$.

21.2. Put the initial-value problem

$$\ddot{x} + 2\dot{x} - 8x = 0; \qquad x(1) = 2, \qquad \dot{x}(1) = 3$$

into the form of System (*21.3*).

Proceeding as in Problem 21.1, with e^t replaced by zero, we define

$$\mathbf{x}(t) \equiv \begin{bmatrix} x_1(t) \\ x_2(t) \end{bmatrix} \qquad \mathbf{A}(t) \equiv \begin{bmatrix} 0 & 1 \\ 8 & -2 \end{bmatrix} \qquad \mathbf{f}(t) \equiv \begin{bmatrix} 0 \\ 0 \end{bmatrix}$$

The differential equation is then equivalent to the matrix equation $\dot{\mathbf{x}}(t) = \mathbf{A}(t)\mathbf{x}(t) + \mathbf{f}(t)$, or simply $\dot{\mathbf{x}}(t) = \mathbf{A}(t)\mathbf{x}(t)$, since $\mathbf{f}(t) = \mathbf{0}$. The initial conditions can be given by $\mathbf{x}(t_0) = \mathbf{c}$, if we define $t_0 = 1$ and $\mathbf{c} \equiv \begin{bmatrix} 2 \\ 3 \end{bmatrix}$.

21.3. Put the initial-value problem

$$\ddot{x} + x = 3; \qquad x(\pi) = 1, \qquad \dot{x}(\pi) = 2$$

into the form of System (*21.3*).

Following Step 1, we write $\ddot{x} = -x + 3$; hence, $a_1(t) = 0$, $a_0(t) = -1$, and $f(t) = 3$. Then defining $x_1(t) = x$ and $x_2(t) = \dot{x}$, we obtain $\dot{x}_1 = x_2$. Following Step 3, we find

$$\dot{x}_2 = \ddot{x} = -x + 3 = -x_1 + 3$$

Thus,

$$\dot{x}_1 = 0x_1 + 1x_2 + 0$$
$$\dot{x}_2 = -1x_1 + 0x_2 + 3$$

These equations are equivalent to the matrix equation $\dot{\mathbf{x}}(t) = \mathbf{A}(t)\mathbf{x}(t) + \mathbf{f}(t)$, if we define

$$\mathbf{x}(t) = \begin{bmatrix} x_1(t) \\ x_2(t) \end{bmatrix} \qquad \mathbf{A}(t) = \begin{bmatrix} 0 & 1 \\ -1 & 0 \end{bmatrix} \qquad \mathbf{f}(t) = \begin{bmatrix} 0 \\ 3 \end{bmatrix}$$

Furthermore, if we also define

$$\mathbf{c} = \begin{bmatrix} 1 \\ 2 \end{bmatrix}$$

then the initial conditions take the form $\mathbf{x}(t_0) = \mathbf{c}$, where $t_0 = \pi$.

21.4. Convert the differential equation $\ddot{x} - 6\dot{x} + 9x = t$ into the matrix equation

$$\dot{\mathbf{x}}(t) = \mathbf{A}(t)\mathbf{x}(t) + \mathbf{f}(t)$$

Here we omit Step 5, because the differential equation has no prescribed initial conditions. Following Step 1, we obtain

$$\ddot{x} = 6\dot{x} - 9x + t$$

Hence $a_1(t) = 6$, $a_0(t) = -9$, and $f(t) = t$. If we define two new variables, $x_1(t) = x$ and $x_2(t) = \dot{x}$, we have

$$\dot{x}_1 = x_2 \qquad \text{and} \qquad \dot{x}_2 = \ddot{x} = 6\dot{x} - 9x + t = 6x_2 - 9x_1 + t$$

Thus,

$$\dot{x}_1 = 0x_1 + 1x_2 + 0$$

$$\dot{x}_2 = -9x_1 + 6x_2 + t$$

These equations are equivalent to the matrix equation $\dot{\mathbf{x}}(t) = \mathbf{A}(t)\mathbf{x}(t) + \mathbf{f}(t)$ if we define

$$\mathbf{x}(t) \equiv \begin{bmatrix} x_1(t) \\ x_2(t) \end{bmatrix} \qquad \mathbf{A}(t) \equiv \begin{bmatrix} 0 & 1 \\ -9 & 6 \end{bmatrix} \qquad \mathbf{f}(t) \equiv \begin{bmatrix} 0 \\ t \end{bmatrix}$$

21.5. Convert the differential equation

$$\frac{d^3x}{dt^3} - 2\frac{d^2x}{dt^2} + \frac{dx}{dt} = 0$$

into the matrix equation $\dot{\mathbf{x}}(t) = \mathbf{A}(t)\mathbf{x}(t) + \mathbf{f}(t)$.

The given differential equation has no prescribed initial conditions, so Step 5 is omitted. Following Step 1, we obtain

$$\frac{d^3x}{dt^3} = 2\frac{d^2x}{dt^2} - \frac{dx}{dt}$$

Defining $x_1(t) = x$, $x_2(t) = \dot{x}$, and $x_3(t) = \ddot{x}$ (the differential equation is third-order, so we need three new variables), we have that $\dot{x}_1 = x_2$ and $\dot{x}_2 = x_3$. Following Step 3, we find

$$\dot{x}_3 = \frac{d^3x}{dt^3} = 2\ddot{x} - \dot{x} = 2x_3 - x_2$$

Thus,

$$\dot{x}_1 = 0x_1 + 1x_2 + 0x_3$$

$$\dot{x}_2 = 0x_1 + 0x_2 + 1x_3$$

$$\dot{x}_3 = 0x_1 - 1x_2 + 2x_3$$

We set

$$\mathbf{x}(t) = \begin{bmatrix} x_1(t) \\ x_2(t) \\ x_3(t) \end{bmatrix} \qquad \mathbf{A}(t) = \begin{bmatrix} 0 & 1 & 0 \\ 0 & 0 & 1 \\ 0 & -1 & 2 \end{bmatrix} \qquad \mathbf{f}(t) = \begin{bmatrix} 0 \\ 0 \\ 0 \end{bmatrix}$$

Then the original third-order differential equation is equivalent to the matrix equation $\dot{\mathbf{x}}(t) = \mathbf{A}(t)\mathbf{x}(t) + \mathbf{f}(t)$, or, more simply, $\dot{\mathbf{x}}(t) = \mathbf{A}(t)\mathbf{x}(t)$ because $\mathbf{f}(t) = \mathbf{0}$.

21.6. Put the initial-value problem

$$e^{-t}\frac{d^4x}{dt^4} - \frac{d^2x}{dt^2} + e^t t^2\frac{dx}{dt} = 5e^{-t};$$

$$x(1) = 2, \qquad \dot{x}(1) = 3, \qquad \ddot{x}(1) = 4, \qquad \dddot{x}(1) = 5$$

into the form of System (*21.3*).

Following Step 1, we obtain

$$\frac{d^4x}{dt^4} = e^t\frac{d^2x}{dt^2} - t^2e^{2t}\frac{dx}{dt} + 5$$

Hence, $a_3(t) = 0$, $a_2(t) = e^t$, $a_1(t) = -t^2e^{2t}$, $a_0(t) = 0$, and $f(t) = 5$. If we define four new variables,

$$x_1(t) = x \qquad x_2(t) = \frac{dx}{dt} \qquad x_3(t) = \frac{d^2x}{dt^2} \qquad x_4(t) = \frac{d^3x}{dt^3}$$

we obtain $\dot{x}_1 = x_2$, $\dot{x}_2 = x_3$, $\dot{x}_3 = x_4$, and, upon following Step 3,

$$\dot{x}_4 = \frac{d^4x}{dt^4} = e^t\ddot{x} - t^2e^{2t}\dot{x} + 5 = e^tx_3 - t^2e^{2t}x_2 + 5$$

Thus,

$$\begin{aligned}
\dot{x}_1 &= 0x_1 + 1x_2 + 0x_3 + 0x_4 + 0 \\
\dot{x}_2 &= 0x_1 + 0x_2 + 1x_3 + 0x_4 + 0 \\
\dot{x}_3 &= 0x_1 + 0x_2 + 0x_3 + 1x_4 + 0 \\
\dot{x}_4 &= 0x_1 - t^2e^{2t}x_2 + e^tx_3 + 0x_4 + 5
\end{aligned}$$

These equations are equivalent to the matrix equation $\dot{\mathbf{x}}(t) = \mathbf{A}(t)\mathbf{x}(t) + \mathbf{f}(t)$ if we define

$$\mathbf{x}(t) \equiv \begin{bmatrix} x_1(t) \\ x_2(t) \\ x_3(t) \\ x_4(t) \end{bmatrix} \qquad \mathbf{A}(t) \equiv \begin{bmatrix} 0 & 1 & 0 & 0 \\ 0 & 0 & 1 & 0 \\ 0 & 0 & 0 & 1 \\ 0 & -t^2e^{2t} & e^t & 0 \end{bmatrix} \qquad \mathbf{f}(t) \equiv \begin{bmatrix} 0 \\ 0 \\ 0 \\ 5 \end{bmatrix}$$

Furthermore, if we also define $\mathbf{c} \equiv \begin{bmatrix} 2 \\ 3 \\ 4 \\ 5 \end{bmatrix}$, then the initial conditions can be given by $\mathbf{x}(t_0) = \mathbf{c}$, where $t_0 = 1$.

21.7. Put the following system into the form of System (*21.3*):

$$\dddot{x} = t\ddot{x} + x - \dot{y} + t + 1$$

$$\ddot{y} = (\sin t)\dot{x} + x - y + t^2;$$

$$x(1) = 2, \qquad \dot{x}(1) = 3, \qquad \ddot{x}(1) = 4, \qquad y(1) = 5, \qquad \dot{y}(1) = 6$$

Since this system contains a third-order differential equation in x *and* a second-order differential equation in y, we will need three new x-variables and two new y-variables. Generalizing Step 2, we define

$$x_1(t) = x \qquad x_2(t) = \frac{dx}{dt} \qquad x_3(t) = \frac{d^2x}{dt^2}$$

$$y_1(t) = y \qquad y_2(t) = \frac{dy}{dt}$$

Thus,

$$\dot{x}_1 = x_2$$

$$\dot{x}_2 = x_3$$

$$\dot{x}_3 = \frac{d^3x}{dt^3} = t\ddot{x} + x - \dot{y} + t + 1 = tx_3 + x_1 - y_2 + t + 1$$

$$\dot{y}_1 = y_2$$

$$\dot{y}_2 = \frac{d^2y}{dt^2} = (\sin t)\dot{x} + x - y + t^2 = (\sin t)x_2 + x_1 - y_1 + t^2$$

or,

$$\dot{x}_1 = 0x_1 + 1x_2 + 0x_3 + 0y_1 + 0y_2 + 0$$

$$\dot{x}_2 = 0x_1 + 0x_2 + 1x_3 + 0y_1 + 0y_2 + 0$$

$$\dot{x}_3 = 1x_1 + 0x_2 + tx_3 + 0y_1 - 1y_2 + (t + 1)$$

$$\dot{y}_1 = 0x_1 + 0x_2 + 0x_3 + 0y_1 + 1y_2 + 0$$

$$\dot{y}_2 = 1x_1 + (\sin t)x_2 + 0x_3 - 1y_1 + 0y_2 + t^2$$

These equations are equivalent to the matrix equation $\dot{\mathbf{x}}(t) = \mathbf{A}(t)\mathbf{x}(t) + \mathbf{f}(t)$ if we define

$$\mathbf{x}(t) \equiv \begin{bmatrix} x_1(t) \\ x_2(t) \\ x_3(t) \\ y_1(t) \\ y_2(t) \end{bmatrix} \qquad \mathbf{A}(t) \equiv \begin{bmatrix} 0 & 1 & 0 & 0 & 0 \\ 0 & 0 & 1 & 0 & 0 \\ 1 & 0 & t & 0 & -1 \\ 0 & 0 & 0 & 0 & 1 \\ 1 & \sin t & 0 & -1 & 0 \end{bmatrix} \qquad \mathbf{f}(t) = \begin{bmatrix} 0 \\ 0 \\ t + 1 \\ 0 \\ t^2 \end{bmatrix}$$

Furthermore, if we define $\mathbf{c} \equiv \begin{bmatrix} 2 \\ 3 \\ 4 \\ 5 \\ 6 \end{bmatrix}$ and $t_0 = 1$, then the initial condition can be given by $\mathbf{x}(t_0) = \mathbf{c}$.

21.8. Put the following system into the form of System (*21.3*):

$$\ddot{x} = -2\dot{x} - 5y + 3$$

$$\dot{y} = \dot{x} + 2y;$$

$$x(0) = 0, \qquad \dot{x}(0) = 0, \qquad y(0) = 1$$

Since the system contains a second-order differential equation in x and a first-order differential equation in y, we define the three new variables

$$x_1(t) = x \qquad x_2(t) = \frac{dx}{dt} \qquad y_1(t) = y$$

Then,

$$\dot{x}_1 = x_2$$
$$\dot{x}_2 = \ddot{x} = -2\dot{x} - 5y + 3 = -2x_2 - 5y_1 + 3$$
$$\dot{y}_1 = \dot{y} = \dot{x} + 2y = x_2 + 2y_1$$

or,

$$\dot{x}_1 = 0x_1 + 1x_2 + 0y_1 + 0$$
$$\dot{x}_2 = 0x_1 - 2x_2 - 5y_1 + 3$$
$$\dot{y}_1 = 0x_1 + 1x_2 + 2y_1 + 0$$

These equations are equivalent to the matrix equation $\dot{\mathbf{x}}(t) = \mathbf{A}(t)\mathbf{x}(t) + \mathbf{f}(t)$ if we define

$$\mathbf{x}(t) \equiv \begin{bmatrix} x_1(t) \\ x_2(t) \\ y_1(t) \end{bmatrix} \qquad \mathbf{A}(t) = \begin{bmatrix} 0 & 1 & 0 \\ 0 & -2 & -5 \\ 0 & 1 & 2 \end{bmatrix} \qquad \mathbf{f}(t) = \begin{bmatrix} 0 \\ 3 \\ 0 \end{bmatrix}$$

If we also define $t_0 = 0$ and $\mathbf{c} \equiv \begin{bmatrix} 0 \\ 0 \\ 1 \end{bmatrix}$, then the initial conditions can be given by $\mathbf{x}(t_0) = \mathbf{c}$.

21.9. Put the following system into matrix form:

$$\dot{x} = x + y$$
$$\dot{y} = 9x + y$$

We proceed exactly as in Problems 21.7 and 21.8, except that now there are no initial conditions to consider. Since the system consists of two first-order differential equations, we define two new variables $x_1(t) = x$ and $y_1(t) = y$. Thus,

$$\dot{x}_1 = \dot{x} = x + y = x_1 + y_1 + 0$$
$$\dot{y}_1 = \dot{y} = 9x + y = 9x_1 + y_1 + 0$$

If we define

$$\mathbf{x}(t) \equiv \begin{bmatrix} x_1(t) \\ y_1(t) \end{bmatrix} \qquad \mathbf{A}(t) \equiv \begin{bmatrix} 1 & 1 \\ 9 & 1 \end{bmatrix} \qquad \mathbf{f}(t) \equiv \begin{bmatrix} 0 \\ 0 \end{bmatrix}$$

then this last set of equations is equivalent to the matrix equation $\dot{\mathbf{x}}(t) = \mathbf{A}(t)\mathbf{x}(t) + \mathbf{f}(t)$, or simply to $\dot{\mathbf{x}}(t) = \mathbf{A}(t)\mathbf{x}(t)$, since $\mathbf{f}(t) = \mathbf{0}$.

Supplementary Problems

Reduce each of the following systems to a first-order matrix system.

21.10. $\ddot{x} - 2\dot{x} + x = t + 1;\ x(1) = 1,\ \dot{x}(1) = 2$

21.11. $2\ddot{x} + x = 4e^t;\ x(0) = 1,\ \dot{x}(0) = 1$

21.12. $t\ddot{x} - 3\dot{x} - t^2x = \sin t;\ x(2) = 3,\ \dot{x}(2) = 4$

21.13. $\ddot{y} + 5\dot{y} - 2ty = t^2 + 1;\ y(0) = 11,\ \dot{y}(0) = 12$

21.14. $-\ddot{y} + 5\dot{y} + 6y = 0$

21.15. $e^t\dddot{x} - t\ddot{x} + \dot{x} - e^tx = 0;$
$x(-1) = 1,\ \dot{x}(-1) = 0,\ \ddot{x}(-1) = 1$

21.16. $2\dfrac{d^3y}{dt^3} + 3\dfrac{d^2y}{dt^2} - 4\dfrac{dy}{dt} + 5y = t^2 + 16t + 20;$
$y(\pi) = -1,\ y'(\pi) = -2,\ y''(\pi) = -3$

21.17. $\dddot{x} = t;\ x(0) = 0,\ \dot{x}(0) = 0,\ \ddot{x}(0) = 0$

21.18. $\ddot{x} = \dot{x} + \dot{y} - z + t$
$\ddot{y} = tx + \dot{y} - 2y + t^2 + 1$
$\dot{z} = x - y + \dot{y} + z;$
$x(1) = 1,\ \dot{x}(1) = 15,\ y(1) = 0,\ \dot{y}(1) = -7,\ z(1) = 4$

21.19. $\ddot{x} = 2\dot{x} + 5y + 3$
$\dot{y} = -\dot{x} - 2y;$
$x(0) = 0,\ \dot{x}(0) = 0,\ y(0) = 1$

21.20. $\dot{x} = x + 2y$
$\dot{y} = 4x + 3y;$
$x(7) = 2,\ y(7) = -3$

Chapter 22

Solutions of Linear Differential Equations with Constant Coefficients by Matrix Methods

SOLUTION OF THE INITIAL-VALUE PROBLEM

By the procedure of Chapter 21, any initial-value problem in which the differential equations are all linear *with constant coefficients*, can be reduced to the matrix system

$$\dot{\mathbf{x}}(t) = \mathbf{A}\mathbf{x}(t) + \mathbf{f}(t); \qquad \mathbf{x}(t_0) = \mathbf{c} \tag{22.1}$$

where $\mathbf{A}$ is a matrix of *constants*. The solution to Eq. (*22.1*) is

$$\mathbf{x}(t) = e^{\mathbf{A}(t-t_0)}\mathbf{c} + e^{\mathbf{A}t}\int_{t_0}^{t} e^{-\mathbf{A}s}\mathbf{f}(s)\,ds \tag{22.2}$$

or equivalently

$$\mathbf{x}(t) = e^{\mathbf{A}(t-t_0)}\mathbf{c} + \int_{t_0}^{t} e^{\mathbf{A}(t-s)}\mathbf{f}(s)\,ds \tag{22.3}$$

In particular, if the initial-value problem is *homogeneous* [i.e., $\mathbf{f}(t) = \mathbf{0}$], then both equations (*22.2*) and (*22.3*) reduce to

$$\mathbf{x}(t) = e^{\mathbf{A}(t-t_0)}\mathbf{c} \tag{22.4}$$

In the above solutions, the matrices $e^{\mathbf{A}(t-t_0)}$, $e^{-\mathbf{A}s}$, and $e^{\mathbf{A}(t-s)}$ are easily computed from $e^{\mathbf{A}t}$ by replacing the variable t by $t - t_0$, $-s$, and $t - s$, respectively. Usually $\mathbf{x}(t)$ is obtained quicker from (*22.3*) than from (*22.2*), since the former equation involves one less matrix multiplication. However, the integrals arising in (*22.3*) are generally more difficult to evaluate than those in (*22.2*).

SOLUTION WITH NO INITIAL CONDITIONS

If no initial conditions are prescribed, the solution of $\dot{\mathbf{x}}(t) = \mathbf{A}\mathbf{x}(t) + \mathbf{f}(t)$ is

$$\mathbf{x}(t) = e^{\mathbf{A}t}\mathbf{k} + e^{\mathbf{A}t}\int e^{-\mathbf{A}t}\mathbf{f}(t)\,dt \tag{22.5}$$

or, when $\mathbf{f}(t) = \mathbf{0}$,

$$\mathbf{x}(t) = e^{\mathbf{A}t}\mathbf{k} \tag{22.6}$$

where $\mathbf{k}$ is an arbitrary constant vector. All constants of integration can be disregarded when computing the integral in Eq. (*22.5*), since they are already included in $\mathbf{k}$.

Solved Problems

22.1. Solve $\ddot{x} + 2\dot{x} - 8x = 0$; $x(1) = 2$, $\dot{x}(1) = 3$.

From Problem 21.2, this initial-value problem is equivalent to Eq. (*22.1*) with

$$\mathbf{x}(t) = \begin{bmatrix} x_1(t) \\ x_2(t) \end{bmatrix} \qquad \mathbf{A} = \begin{bmatrix} 0 & 1 \\ 8 & -2 \end{bmatrix} \qquad \mathbf{f}(t) = \mathbf{0} \qquad \mathbf{c} = \begin{bmatrix} 2 \\ 3 \end{bmatrix} \qquad t_0 = 1$$

The solution to this system is given by Eq. (*22.4*). For this $\mathbf{A}$, $e^{\mathbf{A}t}$ is given in Problem 20.2; hence,

$$e^{\mathbf{A}(t-t_0)} = e^{\mathbf{A}(t-1)} = \frac{1}{6}\begin{bmatrix} 4e^{2(t-1)} + 2e^{-4(t-1)} & e^{2(t-1)} - e^{-4(t-1)} \\ 8e^{2(t-1)} - 8e^{-4(t-1)} & 2e^{2(t-1)} + 4e^{-4(t-1)} \end{bmatrix}$$

Therefore,

$$\begin{aligned}\mathbf{x}(t) &= e^{\mathbf{A}(t-1)}\mathbf{c} \\ &= \frac{1}{6}\begin{bmatrix} 4e^{2(t-1)} + 2e^{-4(t-1)} & e^{2(t-1)} - e^{-4(t-1)} \\ 8e^{2(t-1)} - 8e^{-4(t-1)} & 2e^{2(t-1)} + 4e^{-4(t-1)} \end{bmatrix}\begin{bmatrix} 2 \\ 3 \end{bmatrix} \\ &= \frac{1}{6}\begin{bmatrix} 2(4e^{2(t-1)} + 2e^{-4(t-1)}) + 3(e^{2(t-1)} - e^{-4(t-1)}) \\ 2(8e^{2(t-1)} - 8e^{-4(t-1)}) + 3(2e^{2(t-1)} + 4e^{-4(t-1)}) \end{bmatrix} \\ &= \begin{bmatrix} \frac{11}{6}e^{2(t-1)} + \frac{1}{6}e^{-4(t-1)} \\ \frac{22}{6}e^{2(t-1)} - \frac{4}{6}e^{-4(t-1)} \end{bmatrix}\end{aligned}$$

and the solution to the original initial-value problem is

$$x(t) = x_1(t) = \frac{11}{6}e^{2(t-1)} + \frac{1}{6}e^{-4(t-1)}$$

22.2. Solve $\ddot{x} + 2\dot{x} - 8x = e^t$; $x(0) = 1$, $\dot{x}(0) = -4$.

From Problem 21.1, this initial-value problem is equivalent to Eq. (*22.1*) with

$$\mathbf{x}(t) = \begin{bmatrix} x_1(t) \\ x_2(t) \end{bmatrix} \qquad \mathbf{A} = \begin{bmatrix} 0 & 1 \\ 8 & -2 \end{bmatrix} \qquad \mathbf{f}(t) = \begin{bmatrix} 0 \\ e^t \end{bmatrix} \qquad \mathbf{c} = \begin{bmatrix} 1 \\ -4 \end{bmatrix}$$

and $t_0 = 0$. The solution is given by either Eq. (*22.2*) or (*22.3*). Here, we use (*22.2*); the solution using (*22.3*) is found in Problem 22.3. For this $\mathbf{A}$, $e^{\mathbf{A}t}$ has already been calculated in Problem 20.2. Therefore,

$$e^{\mathbf{A}(t-t_0)}\mathbf{c} = e^{\mathbf{A}t}\mathbf{c} = \frac{1}{6}\begin{bmatrix} 4e^{2t} + 2e^{-4t} & e^{2t} - e^{-4t} \\ 8e^{2t} - 8e^{-4t} & 2e^{2t} + 4e^{-4t} \end{bmatrix}\begin{bmatrix} 1 \\ -4 \end{bmatrix} = \begin{bmatrix} e^{-4t} \\ -4e^{-4t} \end{bmatrix}$$

$$e^{-\mathbf{A}s}\mathbf{f}(s) = \frac{1}{6}\begin{bmatrix} 4e^{-2s} + 2e^{4s} & e^{-2s} - e^{4s} \\ 8e^{-2s} - 8e^{4s} & 2e^{-2s} + 4e^{4s} \end{bmatrix}\begin{bmatrix} 0 \\ e^s \end{bmatrix} = \begin{bmatrix} \frac{1}{6}e^{-s} - \frac{1}{6}e^{5s} \\ \frac{2}{6}e^{-s} + \frac{4}{6}e^{5s} \end{bmatrix}$$

$$\int_{t_0}^{t} e^{-\mathbf{A}s}\mathbf{f}(s)\,ds = \begin{bmatrix} \int_0^t \left(\frac{1}{6}e^{-s} - \frac{1}{6}e^{5s}\right) ds \\ \int_0^t \left(\frac{1}{3}e^{-s} + \frac{2}{3}e^{5s}\right) ds \end{bmatrix} = \frac{1}{30}\begin{bmatrix} -5e^{-t} - e^{5t} + 6 \\ -10e^{-t} + 4e^{5t} + 6 \end{bmatrix}$$

$$e^{\mathbf{A}t}\int_{t_0}^{t} e^{-\mathbf{A}s}\mathbf{f}(s)\,ds = \left(\frac{1}{6}\right)\left(\frac{1}{30}\right)\begin{bmatrix} 4e^{2t}+2e^{-4t} & e^{2t}-e^{-4t} \\ 8e^{2t}-8e^{-4t} & 2e^{2t}+4e^{-4t} \end{bmatrix}\begin{bmatrix} -5e^{-t}-e^{5t}+6 \\ -10e^{-t}+4e^{5t}+6 \end{bmatrix}$$

$$= \frac{1}{180}\begin{bmatrix} (4e^{2t}+2e^{-4t})(-5e^{-t}-e^{5t}+6)+(e^{2t}-e^{-4t})(-10e^{-t}+4e^{5t}+6) \\ (8e^{2t}-8e^{-4t})(-5e^{-t}-e^{5t}+6)+(2e^{2t}+4e^{-4t})(-10e^{-t}+4e^{5t}+6) \end{bmatrix}$$

$$= \frac{1}{30}\begin{bmatrix} -6e^{t}+5e^{2t}+e^{-4t} \\ -6e^{t}+10e^{2t}-4e^{-4t} \end{bmatrix}$$

Thus,

$$\mathbf{x}(t) = e^{\mathbf{A}(t-t_0)}\mathbf{c} + e^{\mathbf{A}t}\int_{t_0}^{t} e^{-\mathbf{A}s}\mathbf{f}(s)\,ds$$

$$= \begin{bmatrix} e^{-4t} \\ -4e^{-4t} \end{bmatrix} + \frac{1}{30}\begin{bmatrix} -6e^{t}+5e^{2t}+e^{-4t} \\ -6e^{t}+10e^{2t}-4e^{-4t} \end{bmatrix} = \begin{bmatrix} \frac{31}{30}e^{-4t}+\frac{1}{6}e^{2t}-\frac{1}{5}e^{t} \\ -\frac{62}{15}e^{-4t}+\frac{1}{3}e^{2t}-\frac{1}{5}e^{t} \end{bmatrix}$$

and
$$x(t) = x_1(t) = \frac{31}{30}e^{-4t}+\frac{1}{6}e^{2t}-\frac{1}{5}e^{t}$$

22.3. Use Eq. (*22.3*) to solve the initial-value problem of Problem 22.2.

The vector $e^{\mathbf{A}(t-t_0)}\mathbf{c}$ remains $\begin{bmatrix} e^{-4t} \\ -4e^{-4t} \end{bmatrix}$. Furthermore,

$$e^{\mathbf{A}(t-s)}\mathbf{f}(s) = \frac{1}{6}\begin{bmatrix} 4e^{2(t-s)}+2e^{-4(t-s)} & e^{2(t-s)}-e^{-4(t-s)} \\ 8e^{2(t-s)}-8e^{-4(t-s)} & 2e^{2(t-s)}+4e^{-4(t-s)} \end{bmatrix}\begin{bmatrix} 0 \\ e^{s} \end{bmatrix}$$

$$= \frac{1}{6}\begin{bmatrix} e^{(2t-s)}-e^{(-4t+5s)} \\ 2e^{(2t-s)}+4e^{(-4t+5s)} \end{bmatrix}$$

$$\int_{t_0}^{t} e^{\mathbf{A}(t-s)}\mathbf{f}(s)\,ds = \frac{1}{6}\begin{bmatrix} \int_0^t [e^{(2t-s)}-e^{(-4t+5s)}]\,ds \\ \int_0^t [2e^{(2t-s)}+4e^{(-4t+5s)}]\,ds \end{bmatrix}$$

$$= \frac{1}{6}\begin{bmatrix} \left[-e^{(2t-s)}-\frac{1}{5}e^{(-4t+5s)}\right]_{s=0}^{s=t} \\ \left[-2e^{(2t-s)}+\frac{4}{5}e^{(-4t+5s)}\right]_{s=0}^{s=t} \end{bmatrix} = \frac{1}{6}\begin{bmatrix} -\frac{6}{5}e^{t}+e^{2t}+\frac{1}{5}e^{-4t} \\ -\frac{6}{5}e^{t}+2e^{2t}-\frac{4}{5}e^{-4t} \end{bmatrix}$$

Thus,

$$\mathbf{x}(t) = e^{\mathbf{A}(t-t_0)}\mathbf{c} + \int_{t_0}^{t} e^{\mathbf{A}(t-s)}\mathbf{f}(s)\,ds$$

$$= \begin{bmatrix} e^{-4t} \\ -4e^{-4t} \end{bmatrix} + \frac{1}{6}\begin{bmatrix} -\frac{6}{5}e^{t}+e^{2t}+\frac{1}{5}e^{-4t} \\ -\frac{6}{5}e^{t}+2e^{2t}-\frac{4}{5}e^{-4t} \end{bmatrix} = \begin{bmatrix} \frac{31}{30}e^{-4t}+\frac{1}{6}e^{2t}-\frac{1}{5}e^{t} \\ -\frac{62}{15}e^{-4t}+\frac{1}{3}e^{2t}-\frac{1}{5}e^{t} \end{bmatrix}$$

as before.

22.4. Solve $\ddot{x} + x = 3$; $x(\pi) = 1$, $\dot{x}(\pi) = 2$.

From Problem 21.3, this initial-value problem is equivalent to Eq. (*22.1*) with

$$\mathbf{x}(t) = \begin{bmatrix} x_1(t) \\ x_2(t) \end{bmatrix} \qquad \mathbf{A} = \begin{bmatrix} 0 & 1 \\ -1 & 0 \end{bmatrix} \qquad \mathbf{f}(t) = \begin{bmatrix} 0 \\ 3 \end{bmatrix} \qquad \mathbf{c} = \begin{bmatrix} 1 \\ 2 \end{bmatrix}$$

and $t_0 = \pi$. Then, using Eq. (*22.3*) and the results of Problem 20.3, we find that

$$e^{\mathbf{A}(t-t_0)}\mathbf{c} = \begin{bmatrix} \cos(t-\pi) & \sin(t-\pi) \\ -\sin(t-\pi) & \cos(t-\pi) \end{bmatrix}\begin{bmatrix} 1 \\ 2 \end{bmatrix} = \begin{bmatrix} \cos(t-\pi) + 2\sin(t-\pi) \\ -\sin(t-\pi) + 2\cos(t-\pi) \end{bmatrix}$$

$$e^{\mathbf{A}(t-s)}\mathbf{f}(s) = \begin{bmatrix} \cos(t-s) & \sin(t-s) \\ -\sin(t-s) & \cos(t-s) \end{bmatrix}\begin{bmatrix} 0 \\ 3 \end{bmatrix} = \begin{bmatrix} 3\sin(t-s) \\ 3\cos(t-s) \end{bmatrix}$$

$$\int_{t_0}^{t} e^{\mathbf{A}(t-s)}\mathbf{f}(s)\,ds = \begin{bmatrix} \displaystyle\int_{\pi}^{t} 3\sin(t-s)\,ds \\ \displaystyle\int_{\pi}^{t} 3\cos(t-s)\,ds \end{bmatrix}$$

$$= \begin{bmatrix} 3\cos(t-s)\Big|_{s=\pi}^{s=t} \\ -3\sin(t-s)\Big|_{s=\pi}^{s=t} \end{bmatrix} = \begin{bmatrix} 3 - 3\cos(t-\pi) \\ 3\sin(t-\pi) \end{bmatrix}$$

Thus,
$$\mathbf{x}(t) = e^{\mathbf{A}(t-t_0)}\mathbf{c} + \int_{t_0}^{t} e^{\mathbf{A}(t-s)}\mathbf{f}(s)\,ds$$

$$= \begin{bmatrix} \cos(t-\pi) + 2\sin(t-\pi) \\ -\sin(t-\pi) + 2\cos(t-\pi) \end{bmatrix} + \begin{bmatrix} 3 - 3\cos(t-\pi) \\ 3\sin(t-\pi) \end{bmatrix}$$

$$= \begin{bmatrix} 3 - 2\cos(t-\pi) + 2\sin(t-\pi) \\ 2\cos(t-\pi) + 2\sin(t-\pi) \end{bmatrix}$$

and $x(t) = x_1(t) = 3 - 2\cos(t-\pi) + 2\sin(t-\pi)$.

Noting that $\cos(t-\pi) = -\cos t$ and $\sin(t-\pi) = -\sin t$, we also obtain

$$x(t) = 3 + 2\cos t - 2\sin t$$

22.5. Solve the differential equation $\ddot{x} - 6\dot{x} + 9x = t$.

This differential equation is equivalent to the standard matrix differential equation with

$$\mathbf{x}(t) = \begin{bmatrix} x_1(t) \\ x_2(t) \end{bmatrix} \qquad \mathbf{A} = \begin{bmatrix} 0 & 1 \\ -9 & 6 \end{bmatrix} \qquad \mathbf{f}(t) = \begin{bmatrix} 0 \\ t \end{bmatrix}$$

(See Problem 21.4). It follows from Problem 20.4 that

$$e^{\mathbf{A}t} = \begin{bmatrix} (1-3t)e^{3t} & te^{3t} \\ -9te^{3t} & (1+3t)e^{3t} \end{bmatrix} \quad \text{so} \quad e^{-\mathbf{A}t} = \begin{bmatrix} (1+3t)e^{-3t} & -te^{-3t} \\ 9te^{-3t} & (1-3t)e^{-3t} \end{bmatrix}$$

Then, using Eq. (*22.5*), we obtain

$$e^{\mathbf{A}t}\mathbf{k} = \begin{bmatrix} (1-3t)e^{3t} & te^{3t} \\ -9te^{3t} & (1+3t)e^{3t} \end{bmatrix}\begin{bmatrix} k_1 \\ k_2 \end{bmatrix} = \begin{bmatrix} [(-3k_1 + k_2)t + k_1]e^{3t} \\ [(-9k_1 + 3k_2)t + k_2]e^{3t} \end{bmatrix}$$

$$e^{-\mathbf{A}t}\mathbf{f}(t) = \begin{bmatrix} (1+3t)e^{-3t} & -te^{-3t} \\ 9te^{-3t} & (1-3t)e^{-3t} \end{bmatrix}\begin{bmatrix} 0 \\ t \end{bmatrix} = \begin{bmatrix} -t^2e^{-3t} \\ (t-3t^2)e^{-3t} \end{bmatrix}$$

$$\int e^{-\mathbf{A}t}\mathbf{f}(t)\,dt = \begin{bmatrix} -\int t^2e^{-3t}\,dt \\ \int (t-3t^2)e^{-3t}\,dt \end{bmatrix} = \begin{bmatrix} \left(\frac{1}{3}t^2+\frac{2}{9}t+\frac{2}{27}\right)e^{-3t} \\ \left(t^2+\frac{1}{3}t+\frac{1}{9}\right)e^{-3t} \end{bmatrix}$$

$$e^{\mathbf{A}t}\int e^{-\mathbf{A}t}\mathbf{f}(t)\,dt = \begin{bmatrix} (1-3t)e^{3t} & te^{3t} \\ -9te^{3t} & (1+3t)e^{3t} \end{bmatrix}\begin{bmatrix} \left(\frac{1}{3}t^2+\frac{2}{9}t+\frac{2}{27}\right)e^{-3t} \\ \left(t^2+\frac{1}{3}t+\frac{1}{9}\right)e^{-3t} \end{bmatrix} = \begin{bmatrix} \frac{1}{9}t+\frac{2}{27} \\ \frac{1}{9} \end{bmatrix}$$

and

$$\mathbf{x}(t) = e^{\mathbf{A}t}\mathbf{k} + e^{\mathbf{A}t}\int e^{-\mathbf{A}t}\mathbf{f}(t)\,dt$$

$$= \begin{bmatrix} [(-3k_1+k_2)t+k_1]e^{3t}+\frac{1}{9}t+\frac{2}{27} \\ [(-9k_1+3k_2)t+k_2]e^{3t}+\frac{1}{9} \end{bmatrix}$$

Thus,

$$x(t) = x_1(t) = [(-3k_1+k_2)t+k_1]e^{3t}+\frac{1}{9}t+\frac{2}{27} = (k_1+k_3t)e^{3t}+\frac{1}{9}t+\frac{2}{27}$$

where $k_3 = -3k_1 + k_2$.

22.6. Solve the differential equation $\dfrac{d^3x}{dt^3} - 2\dfrac{d^2x}{dt^2} + \dfrac{dx}{dt} = 0$.

Using the results of Problem 21.5, we reduce this homogeneous differential equation to the matrix equation $\dot{\mathbf{x}}(t) = \mathbf{A}\mathbf{x}(t)$ with

$$\mathbf{x}(t) = \begin{bmatrix} x_1(t) \\ x_2(t) \\ x_3(t) \end{bmatrix} \quad \text{and} \quad \mathbf{A} = \begin{bmatrix} 0 & 1 & 0 \\ 0 & 0 & 1 \\ 0 & -1 & 2 \end{bmatrix}$$

We have from Problem 20.6 that

$$e^{\mathbf{A}t} = \begin{bmatrix} 1 & -te^t+2e^t-2 & te^t-e^t+1 \\ 0 & -te^t+e^t & te^t \\ 0 & -te^t & te^t+e^t \end{bmatrix}$$

Then using Eq. (*22.6*), we calculate

$$e^{\mathbf{A}t}\mathbf{k} = \begin{bmatrix} 1 & -te^t+2e^t-2 & te^t-e^t+1 \\ 0 & -te^t+e^t & te^t \\ 0 & -te^t & te^t+e^t \end{bmatrix}\begin{bmatrix} k_1 \\ k_2 \\ k_3 \end{bmatrix}$$

$$= \begin{bmatrix} k_1+k_2(-te^t+2e^t-2)+k_3(te^t-e^t+1) \\ k_2(-te^t+e^t)+k_3(te^t) \\ k_2(-te^t)+k_3(te^t+e^t) \end{bmatrix}$$

Thus

$$\begin{aligned} x(t) = x_1(t) &= k_1+k_2(-te^t+2e^t-2)+k_3(te^t-e^t+1) \\ &= (k_1-2k_2+k_3)+(2k_2-k_3)e^t+(-k_2+k_3)te^t \\ &= k_4+k_5e^t+k_6te^t \end{aligned}$$

where $k_4 = k_1 - 2k_2 + k_3$, $k_5 = 2k_2 - k_3$, and $k_6 = -k_2 + k_3$.

22.7. Solve the system

$$\ddot{x} = -2\dot{x} - 5y + 3$$
$$\dot{y} = \dot{x} + 2y;$$
$$x(0) = 0, \qquad \dot{x}(0) = 0, \qquad y(0) = 1$$

This initial-value problem is equivalent to Eq. (*22.1*) with

$$\mathbf{x}(t) = \begin{bmatrix} x_1(t) \\ x_2(t) \\ y_1(t) \end{bmatrix} \qquad \mathbf{A} = \begin{bmatrix} 0 & 1 & 0 \\ 0 & -2 & -5 \\ 0 & 1 & 2 \end{bmatrix} \qquad \mathbf{f}(t) = \begin{bmatrix} 0 \\ 3 \\ 0 \end{bmatrix} \qquad \mathbf{c} = \begin{bmatrix} 0 \\ 0 \\ 1 \end{bmatrix}$$

and $t_0 = 0$. (See Problem 21.8.) For this $\mathbf{A}$, we have from Problem 20.7 that

$$e^{\mathbf{A}t} = \begin{bmatrix} 1 & -2 + 2\cos t + \sin t & -5 + 5\cos t \\ 0 & \cos t - 2\sin t & -5\sin t \\ 0 & \sin t & \cos t + 2\sin t \end{bmatrix}$$

Then, using Eq. (*22.3*), we calculate

$$e^{\mathbf{A}(t-t_0)}\mathbf{c} = \begin{bmatrix} 1 & -2 + 2\cos t + \sin t & -5 + 5\cos t \\ 0 & \cos t - 2\sin t & -5\sin t \\ 0 & \sin t & \cos t + 2\sin t \end{bmatrix} \begin{bmatrix} 0 \\ 0 \\ 1 \end{bmatrix} = \begin{bmatrix} -5 + 5\cos t \\ -5\sin t \\ \cos t + 2\sin t \end{bmatrix}$$

$$e^{\mathbf{A}(t-s)}\mathbf{f}(s) = \begin{bmatrix} 1 & -2 + 2\cos(t-s) + \sin(t-s) & -5 + 5\cos(t-s) \\ 0 & \cos(t-s) - 2\sin(t-s) & -5\sin(t-s) \\ 0 & \sin(t-s) & \cos(t-s) + 2\sin(t-s) \end{bmatrix} \begin{bmatrix} 0 \\ 3 \\ 0 \end{bmatrix}$$

$$= \begin{bmatrix} -6 + 6\cos(t-s) + 3\sin(t-s) \\ 3\cos(t-s) - 6\sin(t-s) \\ 3\sin(t-s) \end{bmatrix}$$

and

$$\int_{t_0}^{t} e^{\mathbf{A}(t-s)}\mathbf{f}(s)\,ds = \begin{bmatrix} \int_0^t [-6 + 6\cos(t-s) + 3\sin(t-s)]\,ds \\ \int_0^t [3\cos(t-s) - 6\sin(t-s)]\,ds \\ \int_0^t 3\sin(t-s)\,ds \end{bmatrix}$$

$$= \begin{bmatrix} [-6s - 6\sin(t-s) + 3\cos(t-s)]_{s=0}^{s=t} \\ [-3\sin(t-s) - 6\cos(t-s)]_{s=0}^{s=t} \\ 3\cos(t-s)\Big|_{s=0}^{s=t} \end{bmatrix}$$

$$= \begin{bmatrix} -6t + 3 + 6\sin t - 3\cos t \\ -6 + 3\sin t + 6\cos t \\ 3 - 3\cos t \end{bmatrix}$$

Therefore,
$$\mathbf{x}(t) = e^{\mathbf{A}(t-t_0)}\mathbf{c} + \int_{t_0}^{t} e^{\mathbf{A}(t-s)}\mathbf{f}(s)\,ds$$

$$= \begin{bmatrix} -5 + 5\cos t \\ -5\sin t \\ \cos t + 2\sin t \end{bmatrix} + \begin{bmatrix} -6t + 3 + 6\sin t - 3\cos t \\ -6 + 3\sin t + 6\cos t \\ 3 - 3\cos t \end{bmatrix}$$

$$= \begin{bmatrix} -2 - 6t + 2\cos t + 6\sin t \\ -6 + 6\cos t - 2\sin t \\ 3 - 2\cos t + 2\sin t \end{bmatrix}$$

Finally, $$x(t) = x_1(t) = 2\cos t + 6\sin t - 2 - 6t$$

$$y(t) = y_1(t) = -2\cos t + 2\sin t + 3$$

22.8. Solve the system of differential equations

$$\dot{x} = x + y$$

$$\dot{y} = 9x + y$$

This set of equations is equivalent to the matrix system $\dot{\mathbf{x}}(t) = \mathbf{A}\mathbf{x}(t)$ with

$$\mathbf{x}(t) = \begin{bmatrix} x_1(t) \\ y_1(t) \end{bmatrix} \qquad \mathbf{A} = \begin{bmatrix} 1 & 1 \\ 9 & 1 \end{bmatrix}$$

(See Problem 21.9.) The solution is given by Eq. (*22.6*). For this **A**, we have from Problem 20.1 that

$$e^{\mathbf{A}t} = \frac{1}{6}\begin{bmatrix} 3e^{4t} + 3e^{-2t} & e^{4t} - e^{-2t} \\ 9e^{4t} - 9e^{-2t} & 3e^{4t} + 3e^{-2t} \end{bmatrix}$$

hence, $$\mathbf{x}(t) = e^{\mathbf{A}t}\mathbf{k} = \frac{1}{6}\begin{bmatrix} 3e^{4t} + 3e^{-2t} & e^{4t} - e^{-2t} \\ 9e^{4t} - 9e^{-2t} & 3e^{4t} + 3e^{-2t} \end{bmatrix}\begin{bmatrix} k_1 \\ k_2 \end{bmatrix}$$

$$= \begin{bmatrix} \frac{1}{6}(3k_1 + k_2)e^{4t} + \frac{1}{6}(3k_1 - k_2)e^{-2t} \\ \frac{3}{6}(3k_1 + k_2)e^{4t} - \frac{3}{6}(3k_1 - k_2)e^{-2t} \end{bmatrix}$$

Thus, $$x(t) = x_1(t) = \frac{1}{6}(3k_1 + k_2)e^{4t} + \frac{1}{6}(3k_1 - k_2)e^{-2t}$$

$$y(t) = y_1(t) = \frac{3}{6}(3k_1 + k_2)e^{4t} - \frac{3}{6}(3k_1 - k_2)e^{-2t}$$

If we define two new arbitrary constants $k_3 = (3k_1 + k_2)/6$ and $k_4 = (3k_1 - k_2)/6$, then

$$x(t) = k_3e^{4t} + k_4e^{-2t} \quad \text{and} \quad y(t) = 3k_3e^{4t} - 3k_4e^{-2t}$$

Supplementary Problems

Solve each of the following systems by matrix methods. Note that $e^{\mathbf{A}t}$ for the first five problems is found in Problem 20.2, while $e^{\mathbf{A}t}$ for Problems 22.15 through 22.17 is given in Problem 20.3.

22.9. $\ddot{x} + 2\dot{x} - 8x = 0;\ x(1) = 1,\ \dot{x}(1) = 0$

22.10. $\ddot{x} + 2\dot{x} - 8x = 4;\ x(0) = 0,\ \dot{x}(0) = 0$

22.11. $\ddot{x} + 2\dot{x} - 8x = 4;\ x(1) = 0,\ \dot{x}(1) = 0$

22.12. $\ddot{x} + 2\dot{x} - 8x = 4$; $x(0) = 1$, $\dot{x}(0) = 2$

22.13. $\ddot{x} + 2\dot{x} - 8x = 9e^{-t}$; $x(0) = 0$, $\dot{x}(0) = 0$

22.14. The system of Problem 22.4, using Eq. (*22.2*)

22.15. $\ddot{x} + x = 0$

22.16. $\ddot{x} + x = 0$; $x(2) = 0$, $\dot{x}(2) = 0$

22.17. $\ddot{x} + x = t$; $x(1) = 0$, $\dot{x}(1) = 1$

22.18. $\ddot{y} - \dot{y} - 2y = 0$

22.19. $\ddot{y} - \dot{y} - 2y = 0$; $y(0) = 2$, $y'(0) = 1$

22.20. $\ddot{y} - \dot{y} - 2y = e^{3t}$; $y(0) = 2$, $y'(0) = 1$

22.21. $\ddot{y} - \dot{y} - 2y = e^{3t}$; $y(0) = 1$, $y'(0) = 2$

22.22. $\ddot{z} + 9\dot{z} + 14z = \frac{1}{2}\sin t$; $z(0) = 0$, $\dot{z}(0) = -1$

22.23. $\dot{x} = -4x + 6y$

$\dot{y} = -3x + 5y$;

$x(0) = 3$, $y(0) = 2$

22.24. $\dot{x} + 5x - 12y = 0$

$\dot{y} + 2x - 5y = 0$;

$x(0) = 8$, $y(0) = 3$

22.25. $\dot{x} - 2y = 3$

$\dot{y} + y - x = -t^2$;

$x(0) = 0$, $y(0) = -1$

22.26. $\dot{x} = x + 2y$

$\dot{y} = 4x + 3y$

22.27. $\dddot{x} = 6t$; $x(0) = 0$, $\dot{x}(0) = 0$, $\ddot{x}(0) = 12$

22.28. $\ddot{x} + y = 0$

$\dot{y} + x = 2e^{-t}$;

$x(0) = 0$, $\dot{x}(0) = -2$, $y(0) = 0$

22.29. $\ddot{x} = 2\dot{x} + 5y + 3$,

$\dot{y} = -\dot{x} - 2y$;

$x(0) = 0$, $\dot{x}(0) = 0$, $y(0) = 1$

Chapter 23

Linear Differential Equations with Variable Coefficients

SECOND-ORDER EQUATIONS

A *second-order* linear differential equation

$$b_2(x)y'' + b_1(x)y' + b_0(x)y = g(x) \tag{23.1}$$

has variable coefficients when $b_2(x)$, $b_1(x)$, and $b_0(x)$ are *not* all constants or constant multiples of one another. If $b_2(x)$ is not zero in a given interval, then we can divide by it and rewrite Eq. (*23.1*) as

$$y'' + P(x)y' + Q(x)y = \phi(x) \tag{23.2}$$

where $P(x) = b_1(x)/b_2(x)$, $Q(x) = b_0(x)/b_2(x)$, and $\phi(x) = g(x)/b_2(x)$. In this chapter and the next, we describe procedures for solving many equations in the form of (*23.1*) or (*23.2*). These procedures can be generalized in a straightforward manner to solve higher-order linear differential equations with variable coefficients.

ANALYTIC FUNCTIONS AND ORDINARY POINTS

A function $f(x)$ is *analytic* at x_0 if its Taylor series about x_0,

$$\sum_{n=0}^{\infty} \frac{f^{(n)}(x_0)(x - x_0)^n}{n!}$$

converges to $f(x)$ in some neighborhood of x_0.

Polynomials, $\sin x$, $\cos x$, and e^x are analytic everywhere; so too are sums, differences, and products of these functions. Quotients of any two of these functions are analytic at all points where the denominator is not zero.

The point x_0 is an *ordinary point* of the differential equation (*23.2*) if both $P(x)$ and $Q(x)$ are analytic at x_0. If either of these functions is not analytic at x_0, then x_0 is a *singular point* of (*23.2*).

SOLUTIONS AROUND THE ORIGIN OF HOMOGENEOUS EQUATIONS

Equation (*23.1*) is *homogeneous* when $g(x) \equiv 0$, in which case Eq. (*23.2*) specializes to

$$y'' + P(x)y' + Q(x)y = 0 \tag{23.3}$$

Theorem 23.1. If $x = 0$ is an ordinary point of Eq. (*23.3*), then the general solution in an interval containing this point has the form

$$y = \sum_{n=0}^{\infty} a_n x^n = a_0 y_1(x) + a_1 y_2(x) \tag{23.4}$$

where a_0 and a_1 are arbitrary constants and $y_1(x)$ and $y_2(x)$ are linearly independent functions analytic at $x = 0$.

To evaluate the coefficients a_n in the solution furnished by Theorem 23.1, use the following five-step procedure known as the *power series method.*

Step 1. Substitute into the left side of the homogeneous differential equation the power series

$$y = \sum_{n=0}^{\infty} a_n x^n = a_0 + a_1 x + a_2 x^2 + a_3 x^3 + a_4 x^4 + \cdots + a_n x^n + a_{n+1} x^{n+1} + a_{n+2} x^{n+2} + \cdots \quad (23.5)$$

together with the power series for

$$y' = a_1 + 2a_2 x + 3a_3 x^2 + 4a_4 x^3 + \cdots + na_n x^{n-1} + (n+1)a_{n+1} x^n + (n+2)a_{n+2} x^{n+1} + \cdots \quad (23.6)$$

and

$$y'' = 2a_2 + 6a_3 x + 12a_4 x^2 + \cdots + n(n-1)a_n x^{n-2} + (n+1)(n)a_{n+1} x^{n-1} + (n+2)(n+1)a_{n+2} x^n + \cdots \quad (23.7)$$

Step 2. Collect powers of x and set the coefficients of each power of x equal to zero.

Step 3. The equation obtained by setting the coefficient of x^n to zero in Step 2 will contain a_j terms for a finite number of j values. Solve this equation for the a_j term having the largest subscript. The resulting equation is known as the *recurrence formula* for the given differential equation.

Step 4. Use the recurrence formula to sequentially determine a_j $(j = 2, 3, 4, \ldots)$ in terms of a_0 and a_1.

Step 5. Substitute the coefficients determined in Step 4 into Eq. (*23.5*) and rewrite the solution in the form of Eq. (*23.4*).

The power series method is only applicable when $x = 0$ is an ordinary point. Although a differential equation must be in the form of Eq. (*23.2*) to determine whether $x = 0$ is an ordinary point, once this condition is verified, the power series method can be used on either form (*23.1*) or (*23.2*). If $P(x)$ or $Q(x)$ in (*23.2*) are quotients of polynomials, it is often simpler first to multiply through by the lowest common denominator, thereby clearing fractions, and then to apply the power series method to the resulting equation in the form of Eq. (*23.1*).

SOLUTIONS AROUND THE ORIGIN OF NONHOMOGENEOUS EQUATIONS

If $\phi(x)$ in Eq. (*23.2*) is analytic at $x = 0$, it has a Taylor series expansion around that point and the power series method given above can be modified to solve either Eq. (*23.1*) or (*23.2*). In Step 1, Eqs. (*23.5*) through (*23.7*) are substituted into the left side of the nonhomogeneous equation; the right side is written as a Taylor series around the origin. Steps 2 and 3 change so that the coefficients of each power of x on the left side of the equation resulting from Step 1 are set equal to their counterparts on the right side of that equation. The form of the solution in Step 5 becomes

$$y = a_0 y_1(x) + a_1 y_2(x) + y_3(x)$$

which has the form specified in Theorem 7.4. The first two terms comprise the general solution to the associated homogeneous differential equation while the last function is a particular solution to the nonhomogeneous equation.

INITIAL-VALUE PROBLEMS

Solutions to initial-value problems are obtained by first solving the given differential equation and then applying the specified initial conditions. An alternate technique that quickly generates the first few terms of the power series solution to an initial-value problem is described in Problem 23.23.

SOLUTIONS AROUND OTHER POINTS

When solutions are required around the ordinary point $x_0 \neq 0$, it generally simplifies the algebra if x_0 is translated to the origin by the change of variables $t = x - x_0$. The solution of the new differential equation that results can be obtained by the power series method about $t = 0$. Then the solution of the original equation is easily gotten by back-substitution.

Solved Problems

23.1. Determine whether $x = 0$ is an ordinary point of the differential equation

$$y'' - xy' + 2y = 0$$

Here $P(x) = -x$ and $Q(x) = 2$ are both polynomials; hence they are analytic everywhere. Therefore, every value of x, in particular $x = 0$, is an ordinary point.

23.2. Find a recurrence formula for the power series solution around $x = 0$ for the differential equation given in Problem 23.1.

It follows from Problem 23.1 that $x = 0$ is an ordinary point of the given equation, so Theorem 23.1 holds. Substituting Eqs. (*23.5*) through (*23.7*) into the left side of the differential equation, we find

$$\begin{aligned}[2a_2 + 6a_3x + 12a_4x^2 + \cdots + n(n-1)a_nx^{n-2} + (n+1)(n)a_{n+1}x^{n-1} + (n+2)(n+1)a_{n+2}x^n + \cdots] \\ - x[a_1 + 2a_2x + 3a_3x^2 + 4a_4x^3 + \cdots + na_nx^{n-1} + (n+1)a_{n+1}x^n + (n+2)a_{n+2}x^{n+1} + \cdots] \\ + 2[a_0 + a_1x + a_2x^2 + a_3x^3 + a_4x^4 + \cdots + a_nx^n + a_{n+1}x^{n+1} + a_{n+2}x^{n+2} + \cdots] = 0\end{aligned}$$

Combining terms that contain like powers of x, we have

$$\begin{aligned}&(2a_2 + 2a_0) + x(6a_3 + a_1) + x^2(12a_4) + x^3(20a_5 - a_3) \\ &\quad + \cdots + x^n[(n+2)(n+1)a_{n+2} - na_n + 2a_n] + \cdots \\ &= 0 + 0x + 0x^2 + 0x^3 + \cdots + 0x^n + \cdots\end{aligned}$$

The last equation holds if and only if each coefficient in the left-hand side is zero. Thus,

$$2a_2 + 2a_0 = 0, \qquad 6a_3 + a_1 = 0, \qquad 12a_4 = 0, \qquad 20a_5 - a_3 = 0, \qquad \ldots$$

In general, $$(n+2)(n+1)a_{n+2} - (n-2)a_n = 0, \text{ or,}$$

$$a_{n+2} = \frac{(n-2)}{(n+2)(n+1)}a_n$$

which is the recurrence formula for this problem.

23.3. Find the general solution near $x = 0$ of $y'' - xy' + 2y = 0$.

Successively evaluating the recurrence formula obtained in Problem 23.2 for $n = 0, 1, 2, \ldots,$ we calculate

$$
\begin{aligned}
a_2 &= -a_0 \\
a_3 &= -\frac{1}{6}a_1 \\
a_4 &= 0 \\
a_5 &= \frac{1}{20}a_3 = \frac{1}{20}\left(-\frac{1}{6}a_1\right) = -\frac{1}{120}a_1 \\
a_6 &= \frac{2}{30}a_4 = \frac{1}{15}(0) = 0 \\
a_7 &= \frac{3}{42}a_5 = \frac{1}{14}\left(-\frac{1}{120}\right)a_1 = -\frac{1}{1680}a_1 \\
a_8 &= \frac{4}{56}a_6 = \frac{1}{14}(0) = 0 \\
&\cdots\cdots\cdots\cdots\cdots\cdots
\end{aligned} \tag{1}
$$

Note that since $a_4 = 0$, it follows from the recurrence formula that all the even coefficients beyond a_4 are also zero. Substituting (*1*) into Eq. (*23.5*) we have

$$
\begin{aligned}
y &= a_0 + a_1 x - a_0 x^2 - \frac{1}{6}a_1 x^3 + 0x^4 - \frac{1}{120}a_1 x^5 + 0x^6 - \frac{1}{1680}a_1 x^7 - \cdots \\
&= a_0(1 - x^2) + a_1\left(x - \frac{1}{6}x^3 - \frac{1}{120}x^5 - \frac{1}{1680}x^7 - \cdots\right)
\end{aligned} \tag{2}
$$

If we define

$$
y_1(x) \equiv 1 - x^2 \qquad \text{and} \qquad y_2(x) \equiv x - \frac{1}{6}x^3 - \frac{1}{120}x^5 - \frac{1}{1680}x^7 - \cdots
$$

then the general solution (*2*) can be rewritten as $y = a_0 y_1(x) + a_1 y_2(x)$.

23.4. Determine whether $x = 0$ is an ordinary point of the differential equation

$$
y'' + y = 0
$$

Here $P(x) = 0$ and $Q(x) = 1$ are both constants; hence they are analytic everywhere. Therefore, every value of x, in particular $x = 0$, is an ordinary point.

23.5. Find a recurrence formula for the power series solution around $x = 0$ for the differential equation given in Problem 23.4.

It follows from Problem 23.4 that $x = 0$ is an ordinary point of the given equation, so Theorem 23.1 holds. Substituting Eqs. (*23.5*) through (*23.7*) into the left side of the differential equation, we find

$$
\begin{aligned}
&[2a_2 + 6a_3 x + 12a_4 x^2 + \cdots + n(n-1)a_n x^{n-2} + (n+1)na_{n+1}x^{n-1} + (n+2)(n+1)a_{n+2}x^n + \cdots] \\
&\quad + [a_0 + a_1 x + a_2 x^2 + a_3 x^3 + a_4 x^4 + \cdots + a_n x^n + a_{n+1}x^{n+1} + a_{n+2}x^{n+2} + \cdots] = 0
\end{aligned}
$$

or

$$
\begin{aligned}
&(2a_2 + a_0) + x(6a_3 + a_1) + x^2(12a_4 + a_2) + x^3(20a_5 + a_3) \\
&\quad + \cdots + x^n[(n+2)(n+1)a_{n+2} + a_n] + \cdots \\
&= 0 + 0x + 0x^2 + \cdots + 0x^n + \cdots
\end{aligned}
$$

Equating each coefficient to zero, we have

$$
2a_2 + a_0 = 0, \qquad 6a_3 + a_1 = 0, \qquad 12a_4 + a_2 = 0, \qquad 20a_5 + a_3 = 0, \qquad \ldots
$$

In general

$$(n+2)(n+1)a_{n+2} + a_n = 0,$$

which is equivalent to

$$a_{n+2} = \frac{-1}{(n+2)(n+1)}a_n$$

This equation is the recurrence formula for this problem.

23.6. Use the power series method to find the general solution near $x=0$ of $y''+y=0$.

Since this equation has constant coefficients, its solution is obtained easily by either the characteristic equation method, Laplace transforms, or matrix methods as $y = c_1 \cos x + c_2 \sin x$.

Solving by the power series method, we successively evaluate the recurrence formula found in Problem 23.5 for $n = 0, 1, 2, \ldots$, obtaining

$$a_2 = -\frac{1}{2}a_0 = -\frac{1}{2!}a_0$$

$$a_3 = -\frac{1}{6}a_1 = -\frac{1}{3!}a_1$$

$$a_4 = -\frac{1}{(4)(3)}a_2 = -\frac{1}{(4)(3)}\left(-\frac{1}{2!}a_0\right) = \frac{1}{4!}a_0$$

$$a_5 = -\frac{1}{(5)(4)}a_3 = -\frac{1}{(5)(4)}\left(-\frac{1}{3!}a_1\right) = \frac{1}{5!}a_1$$

$$a_6 = -\frac{1}{(6)(5)}a_4 = -\frac{1}{(6)(5)}\left(\frac{1}{4!}a_0\right) = -\frac{1}{6!}a_0$$

$$a_7 = -\frac{1}{(7)(6)}a_5 = -\frac{1}{(7)(6)}\left(\frac{1}{5!}a_1\right) = -\frac{1}{7!}a_1$$

. . . .

Recall that for a positive integer n, n factorial, which is denoted by $n!$, is defined by

$$n! = n(n-1)(n-2)\cdots(3)(2)(1)$$

and 0! is defined as one. Thus, $4! = (4)(3)(2)(1) = 24$ and $5! = (5)(4)(3)(2)(1) = 5(4!) = 120$. In general, $n! = n(n-1)!$.

Now substituting the above values for $a_2, a_3, a_4, \ldots$ into Eq. *(23.5)*, we have

$$y = a_0 + a_1 x - \frac{1}{2!}a_0x^2 - \frac{1}{3!}a_1x^3 + \frac{1}{4!}a_0x^4 + \frac{1}{5!}a_1x^5 - \frac{1}{6!}a_0x^6 - \frac{1}{7!}a_1x^7 + \cdots$$

$$= a_0\left(1 - \frac{1}{2!}x^2 + \frac{1}{4!}x^4 - \frac{1}{6!}x^6 + \cdots\right) + a_1\left(x - \frac{1}{3!}x^3 + \frac{1}{5!}x^5 - \frac{1}{7!}x^7 + \cdots\right) \tag{1}$$

But

$$\cos x = \sum_{n=0}^{\infty} \frac{(-1)^n x^{2n}}{(2n)!} = 1 - \frac{1}{2!}x^2 + \frac{1}{4!}x^4 - \frac{1}{6!}x^6 + \cdots$$

$$\sin x = \sum_{n=0}^{\infty} \frac{(-1)^n x^{2n+1}}{(2n+1)!} = x - \frac{1}{3!}x^3 + \frac{1}{5!}x^5 - \frac{1}{7!}x^7 + \cdots$$

Substituting these two results into *(1)* and letting $c_1 = a_0$ and $c_2 = a_1$, we obtain, as before,

$$y = c_1 \cos x + c_2 \sin x$$

23.7. Determine whether $x = 0$ is an ordinary point of the differential equation

$$2x^2y'' + 7x(x + 1)y' - 3y = 0$$

Dividing by $2x^2$, we have

$$P(x) = \frac{7(x + 1)}{2x} \qquad Q(x) = \frac{-3}{2x^2}$$

As neither function is analytic at $x = 0$ (both denominators are zero there), $x = 0$ is not an ordinary point but, rather, a singular point.

23.8. Determine whether $x = 0$ is an ordinary point of the differential equation

$$x^2y'' + 2y' + xy = 0$$

Here $P(x) = 2/x^2$ and $Q(x) = 1/x$. Neither of these functions is analytic at $x = 0$, so $x = 0$ is not an ordinary point but, rather, a singular point.

23.9. Find a recurrence formula for the power series solution around $t = 0$ for the differential equation

$$\frac{d^2y}{dt^2} + (t - 1)\frac{dy}{dt} + (2t - 3)y = 0$$

Both $P(t) = t - 1$ and $Q(t) = 2t - 3$ are polynomials; hence every point, in particular $t = 0$, is an ordinary point. Substituting Eqs. (*23.5*) through (*23.7*) into the left side of the differential equation, with t replacing x, we have

$$[2a_2 + 6a_3t + 12a_4t^2 + \cdots + n(n - 1)a_nt^{n-2} + (n + 1)na_{n+1}t^{n-1} + (n + 2)(n + 1)a_{n+2}t^n + \cdots]$$
$$+ (t - 1)[a_1 + 2a_2t + 3a_3t^2 + 4a_4t^3 + \cdots + na_nt^{n-1} + (n + 1)a_{n+1}t^n + (n + 2)a_{n+2}t^{n+1} + \cdots]$$
$$+ (2t - 3)[a_0 + a_1t + a_2t^2 + a_3t^3 + a_4t^4 + \cdots + a_nt^n + a_{n+1}t^{n+1} + a_{n+2}t^{n+2} + \cdots] = 0$$

or
$$(2a_2 - a_1 - 3a_0) + t(6a_3 + a_1 - 2a_2 + 2a_0 - 3a_1) + t^2(12a_4 + 2a_2 - 3a_3 + 2a_1 - 3a_2) + \cdots$$
$$+ t^n[(n + 2)(n + 1)a_{n+2} + na_n - (n + 1)a_{n+1} + 2a_{n-1} - 3a_n] + \cdots$$
$$= 0 + 0t + 0t^2 + \cdots + 0t^n + \cdots$$

Equating each coefficient to zero, we obtain

$$2a_2 - a_1 - 3a_0 = 0, \qquad 6a_3 - 2a_2 - 2a_1 + 2a_0 = 0, \qquad 12a_4 - 3a_3 - a_2 + 2a_1 = 0, \qquad \ldots(1)$$

In general,

$$(n + 2)(n + 1)a_{n+2} - (n + 1)a_{n+1} + (n - 3)a_n + 2a_{n-1} = 0$$

which is equivalent to

$$a_{n+2} = \frac{1}{n + 2}a_{n+1} - \frac{(n - 3)}{(n + 2)(n + 1)}a_n - \frac{2}{(n + 2)(n + 1)}a_{n-1} \qquad (2)$$

Equation (*2*) is the recurrence formula for this problem. Note, however, that it is not valid for $n = 0$, because a_{-1} is an undefined quantity. To obtain an equation for $n = 0$, we use the first equation in (*1*), which gives $a_2 = \frac{1}{2}a_1 + \frac{3}{2}a_0$.

23.10. Find the general solution near $t = 0$ for the differential equation given in Problem 23.9.

We have from Problem 23.9 that

$$a_2 = \frac{1}{2}a_1 + \frac{3}{2}a_0$$

Then evaluating recurrence formula (*2*) in Problem 23.9 for successive integer values of n beginning with $n = 1$, we find that

$$a_3 = \frac{1}{3}a_2 + \frac{1}{3}a_1 - \frac{1}{3}a_0 = \frac{1}{3}\left(\frac{1}{2}a_1 + \frac{3}{2}a_0\right) + \frac{1}{3}a_1 - \frac{1}{3}a_0 = \frac{1}{2}a_1 + \frac{1}{6}a_0$$

$$a_4 = \frac{1}{4}a_3 + \frac{1}{12}a_2 - \frac{1}{6}a_1 = \frac{1}{4}\left(\frac{1}{2}a_1 + \frac{1}{6}a_0\right) + \frac{1}{12}\left(\frac{1}{2}a_1 + \frac{3}{2}a_0\right) - \frac{1}{6}a_1 = \frac{1}{6}a_0$$

. . . .

Substituting these values into Eq. (*23.5*) with x replaced by t, we obtain as the general solution to the given differential equation

$$y = a_0 + a_1 t + \left(\frac{1}{2}a_1 + \frac{3}{2}a_0\right)t^2 + \left(\frac{1}{2}a_1 + \frac{1}{6}a_0\right)t^3 + \left(\frac{1}{6}a_0\right)t^4 + \cdots$$

$$= a_0\left(1 + \frac{3}{2}t^2 + \frac{1}{6}t^3 + \frac{1}{6}t^4 + \cdots\right) + a_1\left(t + \frac{1}{2}t^2 + \frac{1}{2}t^3 + 0t^4 + \cdots\right)$$

23.11. Determine whether $x = 0$ or $x = 1$ is an ordinary point of the differential equation

$$(1 - x^2)y'' - 2xy' + n(n + 1)y = 0$$

for any positive integer n.

We first transform the differential equation into the form of Eq. (*23.2*) by dividing by $x^2 - 1$. Then

$$P(x) = \frac{-2x}{x^2 - 1} \quad \text{and} \quad Q(x) = \frac{n(n + 1)}{x^2 - 1}$$

Both of these functions have Taylor series expansions around $x = 0$, so both are analytic there and $x = 0$ is an ordinary point. In contrast, the denominators of both functions are zero at $x = 1$, so neither function is defined there and, therefore, neither function is analytic there. Consequently, $x = 1$ is a singular point.

23.12. Find a recurrence formula for the power series solution around $x = 0$ for the differential equation given in Problem 23.11.

To avoid fractions, we work with the differential equation in its current form. Substituting Eqs. (*23.5*) through (*23.7*), with the dummy index n replaced by k, into the left side of this equation, we have that

$$(1 - x^2)[2a_2 + 6a_3x + 12a_4x^2 + \cdots + k(k - 1)a_kx^{k-2} + (k + 1)(k)a_{k+1}x^{k-1}$$
$$+ (k + 2)(k + 1)a_{k+2}x^k + \cdots] - 2x[a_1 + 2a_2x + 3a_3x^2 + \cdots + ka_kx^{k-1} + (k + 1)a_{k+1}x^k$$
$$+ (k + 2)a_{k+2}x^{k+1} + \cdots] + k(k + 1)[a_0 + a_1x + a_2x^2 + a_3x^3 + \cdots + a_kx^k$$
$$+ a_{k+1}x^{k+1} + a_{k+2}x^{k+2} + \cdots] = 0$$

Combining terms that contain like powers of x, we obtain

$$[2a_2 + (n^2 + n)a_0] + x[6a_3 + (n^2 + n - 2)a_1] + \cdots$$
$$+ x^k[(k + 2)(k + 1)a_{k+2} + (n^2 + n - k^2 - k)a_k] + \cdots = 0$$

Noting that $n^2 + n - k^2 - k = (n - k)(n + k + 1)$, we obtain the recurrence formula

$$a_{k+2} = -\frac{(n - k)(n + k + 1)}{(k + 2)(k + 1)}a_k \tag{1}$$

23.13. Show that whenever n is a positive integer, one solution near $x = 0$ of *Legendre's equation*

$$(1 - x^2)y'' - 2xy' + n(n + 1)y = 0$$

is a polynomial of degree n.

The recurrence formula for this equation is given by Eq. (*1*) in Problem 23.12. Because of the factor $n - k$, we find, upon letting $k = n$, that $a_{n+2} = 0$. It follows at once that $0 = a_{n+4} = a_{n+6} = a_{n+8} = \cdots$. Thus, if n is odd, all odd coefficients a_k $(k > n)$ are zero; whereas if n is even, all even coefficients a_k $(k > n)$ are zero. Therefore, either $y_1(x)$ or $y_2(x)$ in Eq. (*23.4*) (depending on whether n is even or odd, respectively) will contain only a finite number of nonzero terms up to and including a term in x^n; hence, it is a polynomial of degree n.

Since a_0 and a_1 are arbitrary, it is customary to choose them so that $y_1(x)$ or $y_2(x)$, whichever is the polynomial, will satisfy the condition $y(1) = 1$. The resulting polynomial, denoted by $P_n(x)$, is known as the *Legendre polynomial of degree n*. The first few of these are

$$P_0(x) = 1 \qquad P_1(x) = x \qquad P_2(x) = \frac{1}{2}(3x^2 - 1)$$

$$P_3(x) = \frac{1}{2}(5x^3 - 3x) \qquad P_4(x) = \frac{1}{8}(35x^4 - 30x^2 + 3)$$

23.14. Find a recurrence formula for the power series solution around $x = 0$ for the nonhomogeneous differential equation $(x^2 + 4)y'' + xy = x + 2$.

Dividing the given equation by $x^2 + 4$, we see that $x = 0$ is an ordinary point and that $\phi(x) = (x + 2)/(x^2 + 4)$ is analytic there. Hence, the power series method is applicable to the entire equation, which, furthermore, we may leave in the form originally given to simplify the algebra. Substituting Eqs. (*23.5*) through (*23.7*) into the given differential equation, we find that

$$(x^2 + 4)[2a_2 + 6a_3x + 12a_4x^2 + \cdots + n(n - 1)a_nx^{n-2}$$
$$+ (n + 1)na_{n+1}x^{n-1} + (n + 2)(n + 1)a_{n+2}x^n + \cdots]$$
$$+ x[a_0 + a_1x + a_2x^2 + a_3x^3 + \cdots + a_{n-1}x^{n-1} + \cdots] = x + 2$$

or

$$(8a_2) + x(24a_3 + a_0) + x^2(2a_2 + 48a_4 + a_1) + x^3(6a_3 + 80a_5 + a_2) + \cdots$$
$$+ x^n[n(n - 1)a_n + 4(n + 2)(n + 1)a_{n+2} + a_{n-1}] + \cdots$$
$$= 2 + (1)x + (0)x^2 + (0)x^3 + \cdots \tag{1}$$

Equating coefficients of like powers of x, we have

$$8a_2 = 2, \qquad 24a_3 + a_0 = 1, \qquad 2a_2 + 48a_4 + a_1 = 0, \qquad 6a_3 + 80a_5 + a_2 = 0, \ldots \tag{2}$$

In general,

$$n(n - 1)a_n + 4(n + 2)(n + 1)a_{n+2} + a_{n-1} = 0 \qquad (n = 2, 3, \ldots)$$

which is equivalent to

$$a_{n+2} = -\frac{n(n - 1)}{4(n + 2)(n + 1)}a_n - \frac{1}{4(n + 2)(n + 1)}a_{n-1} \tag{3}$$

$(n = 2, 3, \ldots)$. Note that the recurrence formula (*3*) is not valid for $n = 0$ or $n = 1$, since the coefficients of x^0 and x^1 on the right side of (*1*) are not zero. Instead, we use the first two equations in (*2*) to obtain

$$a_2 = \frac{1}{4} \qquad a_3 = \frac{1}{24} - \frac{1}{24}a_0 \tag{4}$$

23.15. Use the power series method to find the general solution near $x = 0$ of

$$(x^2 + 4)y'' + xy = x + 2$$

Using the results of Problem 23.14, we have that a_2 and a_3 are given by (*4*) and a_n for $(n = 4, 5, 6, \ldots)$ is given by (*3*). It follows from this recurrence formula that

$$a_4 = -\frac{1}{24}a_2 - \frac{1}{48}a_1 = -\frac{1}{24}\left(\frac{1}{4}\right) - \frac{1}{48}a_1 = -\frac{1}{96} - \frac{1}{48}a_1$$

$$a_5 = -\frac{3}{40}a_3 - \frac{1}{80}a_2 = -\frac{3}{40}\left(\frac{1}{24} - \frac{1}{24}a_0\right) - \frac{1}{80}\left(\frac{1}{4}\right) = \frac{-1}{160} + \frac{1}{320}a_0$$

$$\cdots\cdots\cdots\cdots\cdots\cdots\cdots\cdots\cdots\cdots\cdots$$

Thus,

$$y = a_0 + a_1 x + \frac{1}{4}x^2 + \left(\frac{1}{24} - \frac{1}{24}a_0\right)x^3 + \left(-\frac{1}{96} - \frac{1}{48}a_1\right)x^4 + \left(\frac{-1}{160} + \frac{1}{320}a_0\right)x^5 + \cdots$$

$$= a_0\left(1 - \frac{1}{24}x^3 + \frac{1}{320}x^5 + \cdots\right) + a_1\left(x - \frac{1}{48}x^4 + \cdots\right) + \left(\frac{1}{4}x^2 + \frac{1}{24}x^3 - \frac{1}{96}x^4 - \frac{1}{160}x^5 + \cdots\right)$$

The third series is the particular solution. The first and second series together represent the general solution of the associated homogeneous equation $(x^2 + 4)y'' + xy = 0$.

23.16. Find the recurrence formula for the power series solution around $t = 0$ for the nonhomogeneous differential equation $(d^2y/dt^2) + ty = e^{t+1}$.

Here $P(t) = 0$, $Q(t) = t$, and $\phi(t) = e^{t+1}$ are analytic everywhere, so $t = 0$ is an ordinary point. Substituting Eqs. (*23.5*) through (*23.7*), with t replacing x, into the given equation, we find that

$$[2a_2 + 6a_3 t + 12a_4 t^2 + \cdots + (n+2)(n+1)a_{n+2}t^n + \cdots]$$

$$+ t(a_0 + a_1 t + a_2 t^2 + \cdots + a_{n-1}t^{n-1} + \cdots) = e^{t+1}$$

Recall that e^{t+1} has the Taylor expansion $e^{t+1} = e\sum_{n=0}^{\infty} t^n/n!$ about $t = 0$. Thus, the last equation can be rewritten as

$$(2a_2) + t(6a_3 + a_0) + t^2(12a_4 + a_1) + \cdots + t^n[(n+2)(n+1)a_{n+2} + a_{n-1}] + \cdots$$

$$= \frac{e}{0!} + \frac{e}{1!}t + \frac{e}{2!}t^2 + \cdots + \frac{e}{n!}t^n + \cdots$$

Equating coefficients of like powers of t, we have

$$2a_2 = \frac{e}{0!}, \qquad 6a_3 + a_0 = \frac{e}{1!}, \qquad 12a_4 + a_1 = \frac{e}{2!}, \qquad \cdots \tag{1}$$

In general, $(n+2)(n+1)a_{n+2} + a_{n-1} = e/n!$ for $n = 1, 2, \ldots$, or,

$$a_{n+2} = -\frac{1}{(n+2)(n+1)}a_{n-1} + \frac{e}{(n+2)(n+1)n!} \tag{2}$$

which is the recurrence formula for $n = 1, 2, 3, \ldots$. Using the first equation in (*1*), we can solve for $a_2 = e/2$.

23.17. Use the power series method to find the general solution near $t = 0$ for the differential equation given in Problem 23.16.

Using the results of Problem 23.16, we have $a_2 = e/2$ and a recurrence formula given by Eq. (*2*).

Using this formula, we determine that

$$a_3 = -\frac{1}{6}a_0 + \frac{e}{6}$$

$$a_4 = -\frac{1}{12}a_1 + \frac{e}{24}$$

$$a_5 = -\frac{1}{20}a_2 + \frac{e}{120} = -\frac{1}{20}\left(\frac{e}{2}\right) + \frac{e}{120} = -\frac{e}{60}$$

. . . .

Substituting these results into Eq. *(23.5)*, with x replaced by t, we obtain the general solution

$$y = a_0 + a_1 t + \frac{e}{2}t^2 + \left(-\frac{1}{6}a_0 + \frac{e}{6}\right)t^3 + \left(-\frac{1}{12}a_1 + \frac{e}{24}\right)t^4 + \left(-\frac{e}{60}\right)t^5 + \cdots$$

$$= a_0\left(1 - \frac{1}{6}t^3 + \cdots\right) + a_1\left(t - \frac{1}{12}t^4 + \cdots\right) + e\left(\frac{1}{2}t^2 + \frac{1}{6}t^3 + \frac{1}{24}t^4 - \frac{1}{60}t^5 + \cdots\right)$$

23.18. Find the general solution near $x = 2$ of $y'' - (x - 2)y' + 2y = 0$.

To simplify the algebra, we first make the change of variables $t = x - 2$. From the chain rule we find the corresponding transformations of the derivatives of y:

$$\frac{dy}{dx} = \frac{dy}{dt}\frac{dt}{dx} = \frac{dy}{dt}(1) = \frac{dy}{dt}$$

$$\frac{d^2y}{dx^2} = \frac{d}{dx}\left(\frac{dy}{dx}\right) = \frac{d}{dx}\left(\frac{dy}{dt}\right) = \frac{d}{dt}\left(\frac{dy}{dt}\right)\frac{dt}{dx} = \frac{d^2y}{dt^2}(1) = \frac{d^2y}{dt^2}$$

Substituting these results into the differential equation, we obtain

$$\frac{d^2y}{dt^2} - t\frac{dy}{dt} + 2y = 0$$

and this equation is to be solved near $t = 0$. From Problem 23.3, with x replaced by t, we see that the solution is

$$y = a_0(1 - t^2) + a_1\left(t - \frac{1}{6}t^3 - \frac{1}{120}t^5 - \frac{1}{1680}t^7 - \cdots\right)$$

Substituting $t = x - 2$ into this last equation, we obtain the solution to the original problem as

$$y = a_0[1 - (x - 2)^2] + a_1\left[(x - 2) - \frac{1}{6}(x - 2)^3 - \frac{1}{120}(x - 2)^5 - \frac{1}{1680}(x - 2)^7 - \cdots\right] \tag{1}$$

23.19. Find the general solution near $x = -1$ of $y'' + xy' + (2x - 1)y = 0$.

To simplify the algebra, we make the substitution $t = x - (-1) = x + 1$. Then, as in Problem 23.18, $(dy/dx) = (dy/dt)$ and $(d^2y/dx^2) = (d^2y/dt^2)$. Substituting these results into the differential equation, we

obtain

$$\frac{d^2y}{dt^2} + (t-1)\frac{dy}{dt} + (2t-3)y = 0$$

The power series solution to this equation is found in Problems 23.9 and 23.10 as

$$y = a_0\left(1 + \frac{3}{2}t^2 + \frac{1}{6}t^3 + \frac{1}{6}t^4 + \cdots\right) + a_1\left(t + \frac{1}{2}t^2 + \frac{1}{2}t^3 + 0t^4 + \cdots\right)$$

Substituting back $t = x+1$, we obtain as the solution to the original problem

$$y = a_0\left[1 + \frac{3}{2}(x+1)^2 + \frac{1}{6}(x+1)^3 + \frac{1}{6}(x+1)^4 + \cdots\right]$$

$$+ a_1\left[(x+1) + \frac{1}{2}(x+1)^2 + \frac{1}{2}(x+1)^3 + 0(x+1)^4 + \cdots\right] \quad (1)$$

23.20. Find the general solution near $x = 1$ of $y'' + (x-1)y = e^x$.

We set $t = x - 1$, hence $x = t + 1$. As in Problem 23.18, $\frac{d^2y}{dx^2} = \frac{d^2y}{dt^2}$, so the given differential equation may be rewritten as

$$\frac{d^2y}{dt^2} + ty = e^{t+1}$$

Its solution is (see Problems 23.16 and 23.17)

$$y = a_0\left(1 - \frac{1}{6}t^3 + \cdots\right) + a_1\left(t - \frac{1}{12}t^4 + \cdots\right) + e\left(\frac{1}{2}t^2 + \frac{1}{6}t^3 + \frac{1}{24}t^4 - \frac{1}{60}t^5 + \cdots\right)$$

Substituting back $t = x - 1$, we obtain as the solution to the original problem

$$y = a_0\left[1 - \frac{1}{6}(x-1)^3 + \cdots\right] + a_1\left[(x-1) - \frac{1}{12}(x-1)^4 + \cdots\right]$$

$$+ e\left[\frac{1}{2}(x-1)^2 + \frac{1}{6}(x-1)^3 + \frac{1}{24}(x-1)^4 - \frac{1}{60}(x-1)^5 + \cdots\right]$$

23.21. Solve the initial-value problem

$$y'' - (x-2)y' + 2y = 0; \qquad y(2) = 5, \qquad y'(2) = 60$$

Since the initial conditions are prescribed at $x = 2$, they are most easily satisfied if the solution to the differential equation is obtained as a power series around this point. This has already been done in Eq. (*1*) of Problem 23.18. Applying the initial conditions directly to this solution, we find that $a_0 = 5$ and $a_1 = 60$. Thus, the solution is

$$y = 5[1 - (x-2)^2] + 60\left[(x-2) - \frac{1}{6}(x-2)^3 - \frac{1}{120}(x-2)^5 - \cdots\right]$$

$$= 5 + 60(x-2) - 5(x-2)^2 - 10(x-2)^3 - \frac{1}{2}(x-2)^5 - \cdots$$

23.22. Solve $y'' + xy' + (2x-1)y = 0$; $y(-1) = 2$, $y'(-1) = -2$.

Since the initial conditions are prescribed at $x = -1$, it is advantageous to obtain the general solution to the differential equation near $x = -1$. This has already been done in Eq. (*1*) of Problem

23.19. Applying the initial conditions, we find that $a_0 = 2$ and $a_1 = -2$. Thus, the solution is

$$y = 2\left[1 + \frac{3}{2}(x+1)^2 + \frac{1}{6}(x+1)^3 + \frac{1}{6}(x+1)^4 + \cdots\right]$$
$$- 2\left[(x+1) + \frac{1}{2}(x+1)^2 + \frac{1}{2}(x+1)^3 + 0(x+1)^4 + \cdots\right]$$
$$= 2 - 2(x+1) + 2(x+1)^2 - \frac{2}{3}(x+1)^3 + \frac{1}{3}(x+1)^4 + \cdots$$

23.23. Solve Problem 23.22 by another method.

TAYLOR SERIES METHOD. An alternative method for solving initial-value problems rests on the assumption that the solution can be expanded in a Taylor series about the initial point x_0; i.e.,

$$y = \sum_{n=0}^{\infty} \frac{y^{(n)}(x_0)}{n!}(x - x_0)^n$$
$$= \frac{y(x_0)}{0!} + \frac{y'(x_0)}{1!}(x - x_0) + \frac{y''(x_0)}{2!}(x - x_0)^2 + \cdots \tag{1}$$

The terms $y(x_0)$ and $y'(x_0)$ are given as initial conditions; the other terms $y^{(n)}(x_0)$ $(n = 2, 3, \ldots)$ can be obtained by successively differentiating the differential equation. For Problem 23.22 we have $x_0 = -1$, $y(x_0) = y(-1) = 2$, and $y'(x_0) = y'(-1) = -2$. Solving the differential equation of Problem 23.22 for y'', we find that

$$y'' = -xy' - (2x - 1)y \tag{2}$$

We obtain $y''(x_0) = y''(-1)$ by substituting $x_0 = -1$ into (2) and using the given initial conditions. Thus,

$$y''(-1) = -(-1)y'(-1) - [2(-1) - 1]y(-1) = 1(-2) - (-3)(2) = 4 \tag{3}$$

To obtain $y'''(-1)$, we differentiate (2) and then substitute $x_0 = -1$ into the resulting equation. Thus,

$$y'''(x) = -y' - xy'' - 2y - (2x - 1)y' \tag{4}$$

and

$$y'''(-1) = -y'(-1) - (-1)y''(-1) - 2y(-1) - [2(-1) - 1]y'(-1)$$
$$= -(-2) + 4 - 2(2) - (-3)(-2) = -4 \tag{5}$$

To obtain $y^{(4)}(-1)$, we differentiate (4) and then substitute $x_0 = -1$ into the resulting equation. Thus,

$$y^{(4)}(x) = -xy''' - (2x + 1)y'' - 4y' \tag{6}$$

and

$$y^{(4)}(-1) = -(-1)y'''(-1) - [2(-1) + 1]y''(-1) - 4y'(-1)$$
$$= -4 - (-1)(4) - 4(-2) = 8 \tag{7}$$

This process can be kept up indefinitely. Substituting Eqs. (3), (5), (7), and the initial conditions into (1), we obtain, as before,

$$y = 2 + \frac{-2}{1!}(x+1) + \frac{4}{2!}(x+1)^2 + \frac{-4}{3!}(x+1)^3 + \frac{8}{4!}(x+1)^4 + \cdots$$
$$= 2 - 2(x+1) + 2(x+1)^2 - \frac{2}{3}(x+1)^3 + \frac{1}{3}(x+1)^4 + \cdots$$

One advantage in using this alternative method, as compared to the usual method of first solving the differential equation and then applying the initial conditions, is that the Taylor series method is easier to apply when only the first few terms of the solution are required. One disadvantage is that the recurrence formula cannot be found by the Taylor series method, and, therefore, a general expression for the nth term of the solution cannot be obtained. Note that this alternative method is also useful in solving differential equations without initial conditions. In such cases, we set $y(x_0) = a_0$ and $y'(x_0) = a_1$, where a_0 and a_1 are unknown constants, and proceed as before.

23.24. Use the method outlined in Problem 23.23 to solve $y'' - 2xy = 0$; $y(2) = 1$, $y'(2) = 0$.

Using Eq. (*1*) of Problem 23.23, we assume a solution of the form

$$y(x) = \frac{y(2)}{0!} + \frac{y'(2)}{1!}(x-2) + \frac{y''(2)}{2!}(x-2)^2 + \frac{y'''(2)}{3!}(x-2)^3 + \cdots \qquad (1)$$

From the differential equation,

$$y''(x) = 2xy, \qquad y'''(x) = 2y + 2xy', \qquad y^{(4)}(x) = 4y' + 2xy'', \qquad \ldots$$

Substituting $x = 2$ into these equations and using the initial conditions, we find that

$$y''(2) = 2(2)y(2) = 4(1) = 4$$
$$y'''(2) = 2y(2) + 2(2)y'(2) = 2(1) + 4(0) = 2$$
$$y^{(4)}(2) = 4y'(2) + 2(2)y''(2) = 4(0) + 4(4) = 16$$
$$\cdots\cdots\cdots\cdots\cdots\cdots$$

Substituting these results into Eq. (*1*), we obtain the solution as

$$y = 1 + 2(x-2)^2 + \frac{1}{3}(x-2)^3 + \frac{2}{3}(x-2)^4 + \cdots$$

23.25. Show that the method of undetermined coefficients cannot be used to obtain a particular solution of $y'' + xy = 2$.

By the method of undetermined coefficients, we assume a particular solution of the form $y_p = A_0 x^m$, where m might be zero if the simple guess $y_p = A_0$ does not require modification (see Chapter 10). Substituting y_p into the differential equation, we find

$$m(m-1)A_0 x^{m-2} + A_0 x^{m+1} = 2 \qquad (1)$$

Regardless of the value of m, it is impossible to assign A_0 any *constant* value that will satisfy (*1*). It follows that the method of undetermined coefficients is not applicable.

One limitation on the method of undetermined coefficients is that it is only valid for linear equations with *constant coefficients.*

Supplementary Problems

In Problems 23.26 through 23.34, determine whether the given values of x are ordinary points or singular points of the given differential equations.

23.26. $x = 1$; $y'' + 3y' + 2xy = 0$

23.27. $x = 2$; $(x-2)y'' + 3(x^2 - 3x + 2)y' + (x-2)^2 y = 0$

23.28. $x = 0$; $(x+1)y'' + \frac{1}{x}y' + xy = 0$

23.29. $x = -1$; $(x+1)y'' + \frac{1}{x}y' + xy = 0$

23.30. $x = 0$; $x^3 y'' + y = 0$

23.31. $x = 0$; $x^3y'' + xy = 0$

23.32. $x = 0$; $e^xy'' + (\sin x)y' + xy = 0$

23.33. $x = -1$; $(x + 1)^3y'' + (x^2 - 1)(x + 1)y' + (x - 1)y = 0$

23.34. $x = 2$; $x^4(x^2 - 4)y'' + (x + 1)y' + (x^2 - 3x + 2)y = 0$

23.35. Find the general solution near $x = 0$ of $y'' - y' = 0$. Check your answer by solving the equation by the method of Chapter 8 and then expanding the result in a power series about $x = 0$.

In Problems 23.36 through 23.47, find (*a*) the recurrence formula and (*b*) the general solution of the given differential equation by the power series method around the given value of x.

23.36. $x = 0$; $y'' + xy = 0$

23.37. $x = 0$; $y'' - 2xy' - 2y = 0$

23.38. $x = 0$; $y'' + x^2y' + 2xy = 0$

23.39. $x = 0$; $y'' - x^2y' - y = 0$

23.40. $x = 0$; $y'' + 2x^2y = 0$

23.41. $x = 0$; $(x^2 - 1)y'' + xy' - y = 0$

23.42. $x = 0$; $y'' - xy = 0$

23.43. $x = 1$; $y'' - xy = 0$

23.44. $x = -2$; $y'' - x^2y' + (x + 2)y = 0$

23.45. $x = 0$; $(x^2 + 4)y'' + y = x$

23.46. $x = 1$; $y'' - (x - 1)y' = x^2 - 2x$

23.47. $x = 0$; $y'' - xy' = e^{-x}$

23.48. Use the Taylor series method described in Problem 23.23 to solve $y'' - 2xy' + x^2y = 0$; $y(0) = 1$, $y'(0) = -1$.

23.49. Use the Taylor series method described in Problem 23.23 to solve $y'' - 2xy = x^2$; $y(1) = 0$, $y'(1) = 2$.

Chapter 24

Regular Singular Points and the Method of Frobenius

REGULAR SINGULAR POINTS

The point x_0 is a *regular singular point* of the second-order homogeneous linear differential equation

$$y'' + P(x)y' + Q(x)y = 0 \tag{24.1}$$

if x_0 is not an ordinary point (see Chapter 23) but both $(x - x_0)P(x)$ and $(x - x_0)^2Q(x)$ are analytic at x_0. We only consider regular singular points at $x_0 = 0$; if this is not the case, then the change of variables $t = x - x_0$ will translate x_0 to the origin.

METHOD OF FROBENIUS

Theorem 24.1. If $x = 0$ is a regular singular point of *(24.1)*, then the equation has at least one solution of the form

$$y = x^\lambda \sum_{n=0}^{\infty} a_n x^n$$

where λ and a_n $(n = 0, 1, 2, \ldots)$ are constants. This solution is valid in an interval $0 < x < R$ for some real number R.

To evaluate the coefficients a_n and λ in Theorem 24.1, one proceeds as in the power series method of Chapter 23. The infinite series

$$y = x^\lambda \sum_{n=0}^{\infty} a_n x^n = \sum_{n=0}^{\infty} a_n x^{\lambda+n}$$

$$= a_0x^\lambda + a_1x^{\lambda+1} + a_2x^{\lambda+2} + \cdots + a_{n-1}x^{\lambda+n-1} + a_nx^{\lambda+n} + a_{n+1}x^{\lambda+n+1} + \cdots \tag{24.2}$$

with its derivatives

$$y' = \lambda a_0x^{\lambda-1} + (\lambda + 1)a_1x^\lambda + (\lambda + 2)a_2x^{\lambda+1} + \cdots$$
$$+ (\lambda + n - 1)a_{n-1}x^{\lambda+n-2} + (\lambda + n)a_nx^{\lambda+n-1} + (\lambda + n + 1)a_{n+1}x^{\lambda+n} + \cdots \tag{24.3}$$

and

$$y'' = \lambda(\lambda - 1)a_0x^{\lambda-2} + (\lambda + 1)(\lambda)a_1x^{\lambda-1} + (\lambda + 2)(\lambda + 1)a_2x^\lambda + \cdots$$
$$+ (\lambda + n - 1)(\lambda + n - 2)a_{n-1}x^{\lambda+n-3} + (\lambda + n)(\lambda + n - 1)a_nx^{\lambda+n-2}$$
$$+ (\lambda + n + 1)(\lambda + n)a_{n+1}x^{\lambda+n-1} + \cdots \tag{24.4}$$

are substituted into Eq. *(24.1)*. Terms with like powers of x are collected together and set equal to zero. When this is done for x^n the resulting equation is a recurrence formula. A quadratic equation in λ, called the *indicial equation*, arises when the coefficient of x^0 is set to zero and a_0 is left arbitrary.

The two roots of the indicial equation can be real or complex. If complex they will occur in a conjugate pair and the complex solutions that they produce can be combined (by using Euler's relations and the identity $x^{a \pm ib} = x^a e^{\pm ib \ln x}$) to form real solutions. In this book we shall, for simplicity, suppose that both roots of the indicial equation are real. Then, if λ is taken as the *larger* indicial root, $\lambda = \lambda_1 \geq \lambda_2$, the method of Frobenius always yields a solution

$$y_1(x) = x^{\lambda_1} \sum_{n=0}^{\infty} a_n(\lambda_1)x^n \tag{24.5}$$

to Eq. (*24.1*). [We have written $a_n(\lambda_1)$ to indicate the coefficients produced by the method when $\lambda = \lambda_1$.]

If $P(x)$ and $Q(x)$ are quotients of polynomials, it is usually easier first to multiply (*24.1*) by their lowest common denominator and then to apply the method of Frobenius to the resulting equation.

GENERAL SOLUTION

The method of Frobenius always yields one solution to (*24.1*) of the form (*24.5*). The general solution (see Theorem 7.2) has the form $y = c_1 y_1(x) + c_2 y_2(x)$ where c_1 and c_2 are arbitrary constants and $y_2(x)$ is a second solution of (*24.1*) that is linearly independent from $y_1(x)$. The method for obtaining this second solution depends on the relationship between the two roots of the indicial equation.

Case 1. If $\lambda_1 - \lambda_2$ is not an integer, then

$$y_2(x) = x^{\lambda_2} \sum_{n=0}^{\infty} a_n(\lambda_2)x^n \tag{24.6}$$

where $y_2(x)$ is obtained in an identical manner as $y_1(x)$ by the method of Frobenius, using λ_2 in place of λ_1.

Case 2. If $\lambda_1 = \lambda_2$, then

$$y_2(x) = y_1(x) \ln x + x^{\lambda_1} \sum_{n=0}^{\infty} b_n(\lambda_1)x^n \tag{24.7}$$

To generate this solution, keep the recurrence formula in terms of λ and use it to find the coefficients a_n $(n \geq 1)$ in terms of both λ and a_0, where the coefficient a_0 remains arbitrary. Substitute these a_n into Eq. (*24.2*) to obtain a function $y(\lambda, x)$ which depends on the variables λ and x. Then

$$y_2(x) = \left.\frac{\partial y(\lambda, x)}{\partial \lambda}\right|_{\lambda=\lambda_1} \tag{24.8}$$

Case 3. If $\lambda_1 - \lambda_2 = N$, a positive integer, then

$$y_2(x) = d_{-1} y_1(x) \ln x + x^{\lambda_2} \sum_{n=0}^{\infty} d_n(\lambda_2)x^n \tag{24.9}$$

To generate this solution, first try the method of Frobenius with λ_2. If it yields a second solution, then this solution is $y_2(x)$, having the form of (*24.9*) with $d_{-1} = 0$. Otherwise, proceed as in Case 2 to generate $y(\lambda, x)$, whence

$$y_2(x) = \frac{\partial}{\partial \lambda}[(\lambda - \lambda_2)y(\lambda, x)]|_{\lambda=\lambda_2} \tag{24.10}$$

Solved Problems

24.1. Determine whether $x = 0$ is a regular singular point of the differential equation

$$y'' - xy' + 2y = 0$$

As shown in Problem 23.1, $x = 0$ is an ordinary pont of this differential equation, so it cannot be a regular singular point.

24.2. Determine whether $x = 0$ is a regular singular point of the differential equation

$$2x^2y'' + 7x(x+1)y' - 3y = 0$$

Dividing by $2x^2$, we have

$$P(x) = \frac{7(x+1)}{2x} \quad \text{and} \quad Q(x) = \frac{-3}{2x^2}$$

As shown in Problem 23.7, $x = 0$ is a singular point. Furthermore, both

$$xP(x) = \frac{7}{2}(x+1) \quad \text{and} \quad x^2Q(x) = -\frac{3}{2}$$

are analytic everywhere: the first is a polynomial and the second a constant. Hence, both are analytic at $x = 0$, and this point is a regular singular point.

24.3. Determine whether $x = 0$ is a regular singular point of the differential equation

$$x^3y'' + 2x^2y' + y = 0$$

Dividing by x^3, we have

$$P(x) = \frac{2}{x} \quad \text{and} \quad Q(x) = \frac{1}{x^3}$$

Neither of these functions is defined at $x = 0$, so this point is a singular point. Here,

$$xP(x) = 2 \quad \text{and} \quad x^2Q(x) = \frac{1}{x}$$

The first of these terms is analytic everywhere, but the second is undefined at $x = 0$ and not analytic there. Therefore, $x = 0$ is *not* a regular singular point for the given differential equation.

24.4. Determine whether $x = 0$ is a regular singular point of the differential equation

$$8x^2y'' + 10xy' + (x-1)y = 0$$

Dividing by $8x^2$, we have

$$P(x) = \frac{5}{4x} \quad \text{and} \quad Q(x) = \frac{1}{8x} - \frac{1}{8x^2}$$

Neither of these functions is defined at $x = 0$, so this point is a singular point. Furthermore, both

$$xP(x) = \frac{5}{4} \quad \text{and} \quad x^2Q(x) = \frac{1}{8}(x-1)$$

are analytic everywhere: the first is a constant and the second a polynomial. Hence, both are analytic at $x = 0$, and this point is a regular singular point.

24.5. Find a recurrence formula and the indicial equation for an infinite series solution around $x = 0$ for the differential equation given in Problem 24.4.

It follows from Problem 24.4 that $x = 0$ is a regular singular point of the differential equation, so Theorem 24.1 holds. Substituting Eqs. *(24.2)* through *(24.4)* into the left side of the given differential equation and combining coefficients of like powers of x, we obtain

$$x^{\lambda}[8\lambda(\lambda-1)a_0+10\lambda a_0-a_0]+x^{\lambda+1}[8(\lambda+1)\lambda a_1+10(\lambda+1)a_1+a_0-a_1]+\cdots$$
$$+x^{\lambda+n}[8(\lambda+n)(\lambda+n-1)a_n+10(\lambda+n)a_n+a_{n-1}-a_n]+\cdots=0$$

Dividing by x^{λ} and simplifying, we have

$$[8\lambda^2+2\lambda-1]a_0+x[(8\lambda^2+18\lambda+9)a_1+a_0]+\cdots$$
$$+x^n\{[8(\lambda+n)^2+2(\lambda+n)-1]a_n+a_{n-1}\}+\cdots=0$$

Factoring the coefficient of a_n and equating the coefficient of each power of x to zero, we find

$$(8\lambda^2+2\lambda-1)a_0=0 \tag{1}$$

and, for $n \geq 1$,

$$[4(\lambda+n)-1][2(\lambda+n)+1]a_n+a_{n-1}=0$$

or,

$$a_n=\frac{-1}{[4(\lambda+n)-1][2(\lambda+n)+1]}a_{n-1} \tag{2}$$

Equation (2) is a recurrence formula for this differential equation.

From (1), either $a_0=0$ or

$$8\lambda^2+2\lambda-1=0 \tag{3}$$

It is convenient to keep a_0 arbitrary; therefore, we must choose λ to satisfy (3), which is the indicial equation.

24.6. Find the general solution near $x=0$ of $8x^2y''+10xy'+(x-1)y=0$.

The roots of the indicial equation given by (3) of Problem 24.5 are $\lambda_1=\frac{1}{4}$ and $\lambda_2=-\frac{1}{2}$. Since $\lambda_1-\lambda_2=\frac{3}{4}$, the solution is given by Eqs. (24.5) and (24.6). Substituting $\lambda=\frac{1}{4}$ into the recurrence formula (2) of Problem 24.5 and simplifying, we obtain

$$a_n=\frac{-1}{2n(4n+3)}a_{n-1} \qquad (n\geq 1)$$

Thus,

$$a_1=\frac{-1}{14}a_0, \qquad a_2=\frac{-1}{44}a_1=\frac{1}{616}a_0, \qquad \ldots$$

and

$$y_1(x)=a_0x^{1/4}\left(1-\frac{1}{14}x+\frac{1}{616}x^2+\cdots\right)$$

Substituting $\lambda=-\frac{1}{2}$ into recurrence formula (2) of Problem 24.5 and simplifying, we obtain

$$a_n=\frac{-1}{2n(4n-3)}a_{n-1}$$

Thus,

$$a_1=-\frac{1}{2}a_0, \qquad a_2=\frac{-1}{20}a_1=\frac{1}{40}a_0, \qquad \ldots$$

and

$$y_2(x)=a_0x^{-1/2}\left(1-\frac{1}{2}x+\frac{1}{40}x^2+\cdots\right)$$

The general solution is

$$y=c_1y_1(x)+c_2y_2(x)$$
$$=k_1x^{1/4}\left(1-\frac{1}{14}x+\frac{1}{616}x^2+\cdots\right)+k_2x^{-1/2}\left(1-\frac{1}{2}x+\frac{1}{40}x^2+\cdots\right)$$

where $k_1=c_1a_0$ and $k_2=c_2a_0$.

24.7. Find a recurrence formula and the indicial equation for an infinite series solution around $x = 0$ for the differential equation

$$2x^2y'' + 7x(x + 1)y' - 3y = 0$$

It follows from Problem 24.2 that $x = 0$ is a regular singular point of the differential equation, so Theorem 24.1 holds. Substituting Eqs. (*24.2*) through (*24.4*) into the left side of the given differential equation and combining coefficients of like powers of x, we obtain

$$x^\lambda[2\lambda(\lambda - 1)a_0 + 7\lambda a_0 - 3a_0] + x^{\lambda+1}[2(\lambda + 1)\lambda a_1 + 7\lambda a_0 + 7(\lambda + 1)a_1 - 3a_1] + \cdots$$
$$+ x^{\lambda+n}[2(\lambda + n)(\lambda + n - 1)a_n + 7(\lambda + n - 1)a_{n-1} + 7(\lambda + n)a_n - 3a_n] + \cdots = 0$$

Dividing by x^λ and simplifying, we have

$$(2\lambda^2 + 5\lambda - 3)a_0 + x[(2\lambda^2 + 9\lambda + 4)a_1 + 7\lambda a_0] + \cdots$$
$$+ x^n\{[2(\lambda + n)^2 + 5(\lambda + n) - 3]a_n + 7(\lambda + n - 1)a_{n-1}\} + \cdots = 0$$

Factoring the coefficient of a_n and equating each coefficient to zero, we find

$$(2\lambda^2 + 5\lambda - 3)a_0 = 0 \tag{1}$$

and, for $n \geq 1$,

$$[2(\lambda + n) - 1][(\lambda + n) + 3]a_n + 7(\lambda + n - 1)a_{n-1} = 0$$

or,

$$a_n = \frac{-7(\lambda + n - 1)}{[2(\lambda + n) - 1][(\lambda + n) + 3]}a_{n-1} \tag{2}$$

Equation (*2*) is a recurrence formula for this differential equation.

From (*1*), either $a_0 = 0$ or

$$2\lambda^2 + 5\lambda - 3 = 0 \tag{3}$$

It is convenient to keep a_0 arbitrary; therefore, we require λ to satisfy the indicial equation (*3*).

24.8. Find the general solution near $x = 0$ of $2x^2y'' + 7x(x + 1)y' - 3y = 0$.

The roots of the indicial equation given by (*3*) of Problem 24.7 are $\lambda_1 = \frac{1}{2}$ and $\lambda_2 = -3$. Since $\lambda_1 - \lambda_2 = \frac{7}{2}$, the solution is given by Eqs. (*24.5*) and (*24.6*). Substituting $\lambda = \frac{1}{2}$ into (*2*) of Problem 24.7 and simplifying, we obtain

$$a_n = \frac{-7(2n - 1)}{2n(2n + 7)}a_{n-1} \qquad (n \geq 1)$$

Thus,

$$a_1 = -\frac{7}{18}a_0, \qquad a_2 = -\frac{21}{44}a_1 = \frac{147}{792}a_0, \qquad \ldots$$

and

$$y_1(x) = a_0x^{1/2}\left(1 - \frac{7}{18}x + \frac{147}{792}x^2 + \cdots\right)$$

Substituting $\lambda = -3$ into (*2*) of Problem 24.7 and simplifying, we obtain

$$a_n = \frac{-7(n - 4)}{n(2n - 7)}a_{n-1} \qquad (n \geq 1)$$

Thus,

$$a_1 = -\frac{21}{5}a_0, \qquad a_2 = -\frac{7}{3}a_1 = \frac{49}{5}a_0, \qquad a_3 = -\frac{7}{3}a_2 = -\frac{343}{15}a_0, \qquad a_4 = 0$$

and, since $a_4 = 0$, $a_n = 0$ for $n \geq 4$. Thus,

$$y_2(x) = a_0x^{-3}\left(1 - \frac{21}{5}x + \frac{49}{5}x^2 - \frac{343}{15}x^3\right)$$

The general solution is

$$y = c_1y_1(x) + c_2y_2(x)$$
$$= k_1x^{1/2}\left(1 - \frac{7}{18}x + \frac{147}{792}x^2 + \cdots\right) + k_2x^{-3}\left(1 - \frac{21}{5}x + \frac{49}{5}x^2 - \frac{343}{15}x^3\right)$$

where $k_1 = c_1a_0$ and $k_2 = c_2a_0$.

24.9. Find the general solution near $x = 0$ of $3x^2y'' - xy' + y = 0$.

Here $P(x) = -1/(3x)$ and $Q(x) = 1/(3x^2)$; hence, $x = 0$ is a regular singular point and the method of Frobenius is applicable. Substituting Eqs. (*24.2*) through (*24.4*) into the differential equation and simplifying, we have

$$x^\lambda[3\lambda^2 - 4\lambda + 1]a_0 + x^{\lambda+1}[3\lambda^2 + 2\lambda]a_1 + \cdots + x^{\lambda+n}[3(\lambda + n)^2 - 4(\lambda + n) + 1]a_n + \cdots = 0$$

Dividing by x^λ and equating all coefficients to zero, we find

$$(3\lambda^2 - 4\lambda + 1)a_0 = 0 \tag{1}$$

and

$$[3(\lambda + n)^2 - 4(\lambda + n) + 1]a_n = 0 \qquad (n \geq 1) \tag{2}$$

From (*1*), we conclude that the indicial equation is $3\lambda^2 - 4\lambda + 1 = 0$, which has roots $\lambda_1 = 1$ and $\lambda_2 = \frac{1}{3}$. Since $\lambda_1 - \lambda_2 = \frac{2}{3}$, the solution is given by Eqs. (*24.5*) and (*24.6*). Note that for either value of λ, (*2*) is satisfied by simply choosing $a_n = 0$, $n \geq 1$. Thus,

$$y_1(x) = x^1 \sum_{n=0}^{\infty} a_n x^n = a_0 x \qquad y_2(x) = x^{1/3} \sum_{n=0}^{\infty} a_n x^n = a_0 x^{1/3}$$

and the general solution is

$$y = c_1 y_1(x) + c_2 y_2(x) = k_1 x + k_2 x^{1/3}$$

where $k_1 = c_1 a_0$ and $k_2 = c_2 a_0$.

24.10. Use the method of Frobenius to find one solution near $x = 0$ of $x^2y'' + xy' + x^2y = 0$.

Here $P(x) = 1/x$ and $Q(x) = 1$, so $x = 0$ is a regular singular point and the method of Frobenius is applicable. Substituting Eqs. (*24.2*) through (*24.4*) into the left side of the differential equation, as given, and combining coefficients of like powers of x, we obtain

$$x^\lambda[\lambda^2 a_0] + x^{\lambda+1}[(\lambda + 1)^2 a_1] + x^{\lambda+2}[(\lambda + 2)^2 a_2 + a_0] + \cdots + x^{\lambda+n}[(\lambda + n)^2 a_n + a_{n-2}] + \cdots = 0$$

Thus,

$$\lambda^2 a_0 = 0 \tag{1}$$

$$(\lambda + 1)^2 a_1 = 0 \tag{2}$$

and, for $n \geq 2$, $(\lambda + n)^2 a_n + a_{n-2} = 0$, or,

$$a_n = \frac{-1}{(\lambda + n)^2} a_{n-2} \qquad (n \geq 2) \tag{3}$$

The stipulation $n \geq 2$ is required in (*3*) because a_{n-2} is not defined for $n = 0$ or $n = 1$. From (*1*), the indicial equation is $\lambda^2 = 0$, which has roots $\lambda_1 = \lambda_2 = 0$. Thus, we will obtain only *one* solution of the form of (*24.5*); the second solution, $y_2(x)$, will have the form of (*24.7*).

Substituting $\lambda = 0$ into (*2*) and (*3*), we find that $a_1 = 0$ and $a_n = -(1/n^2)a_{n-2}$. Since $a_1 = 0$, it follows that $0 = a_3 = a_5 = a_7 = \cdots$. Furthermore,

$$a_2 = -\frac{1}{4}a_0 = -\frac{1}{2^2(1!)^2}a_0 \qquad a_4 = -\frac{1}{16}a_2 = \frac{1}{2^4(2!)^2}a_0$$

$$a_6 = -\frac{1}{36}a_4 = -\frac{1}{2^6(3!)^2}a_0 \qquad a_8 = -\frac{1}{64}a_6 = \frac{1}{2^8(4!)^2}a_0$$

and, in general, $a_{2k} = \dfrac{(-1)k}{2^{2k}(k!)^2}a_0$ $(k = 1, 2, 3, \ldots)$. Thus,

$$y_1(x) = a_0 x^0\left[1 - \frac{1}{2^2(1!)^2}x^2 + \frac{1}{2^4(2!)^2}x^4 + \cdots + \frac{(-1)^k}{2^{2k}(k!)^2}x^{2k} + \cdots\right]$$

$$= a_0 \sum_{n=0}^{\infty} \frac{(-1)^n}{2^{2n}(n!)^2}x^{2n} \tag{4}$$

24.11. Find the general solution near $x = 0$ to the differential equation given in Problem 24.10.

One solution is given by (*4*) in Problem 24.10. Because the roots of the indicial equation are equal, we use Eq. (*24.8*) to generate a second linearly independent solution. The recurrence formula is (*3*) of Problem 24.10, augmented by (*2*) of Problem 24.10 for the special case $n = 1$. From (*2*), $a_1 = 0$, which implies that $0 = a_3 = a_5 = a_7 = \cdots$. Then, from (*3*),

$$a_2 = \frac{-1}{(\lambda+2)^2}a_0, \qquad a_4 = \frac{-1}{(\lambda+4)^2}a_2 = \frac{1}{(\lambda+4)^2(\lambda+2)^2}a_0, \qquad \ldots$$

Substituting these values into Eq. (*24.2*), we have

$$y(\lambda, x) = a_0\left[x^\lambda - \frac{1}{(\lambda+2)^2}x^{\lambda+2} + \frac{1}{(\lambda+4)^2(\lambda+2)^2}x^{\lambda+4} + \cdots\right]$$

Recall that $\dfrac{\partial}{\partial\lambda}(x^{\lambda+k}) = x^{\lambda+k}\ln x$. (When differentiating with respect to λ, x can be thought of as a constant.) Thus,

$$\begin{aligned}\frac{\partial y(\lambda, x)}{\partial\lambda} = a_0\bigg[x^\lambda \ln x &+ \frac{2}{(\lambda+2)^3}x^{\lambda+2} - \frac{1}{(\lambda+2)^2}x^{\lambda+2}\ln x \\ &- \frac{2}{(\lambda+4)^3(\lambda+2)^2}x^{\lambda+4} - \frac{2}{(\lambda+4)^2(\lambda+2)^3}x^{\lambda+4} \\ &+ \frac{1}{(\lambda+4)^2(\lambda+2)^2}x^{\lambda+4}\ln x + \cdots\bigg]\end{aligned}$$

and

$$\begin{aligned}y_2(x) = \left.\frac{\partial y(\lambda, x)}{\partial\lambda}\right|_{\lambda=0} &= a_0\bigg(\ln x + \frac{2}{2^3}x^2 - \frac{1}{2^2}x^2\ln x \\ &\quad - \frac{2}{4^3 2^2}x^4 - \frac{2}{4^2 2^3}x^4 + \frac{1}{4^2 2^2}x^4 \ln x + \cdots\bigg) \\ &= (\ln x)a_0\left[1 - \frac{1}{2^2(1!)}x^2 + \frac{1}{2^4(2!)^2}x^4 + \cdots\right] \\ &\quad + a_0\left[\frac{x^2}{2^2(1!)^2}(1) - \frac{x^4}{2^4(2!)^2}\left(\frac{1}{2} + 1\right) + \cdots\right] \\ &= y_1(x)\ln x + a_0\left[\frac{x^2}{2^2(1!)^2}(1) - \frac{x^4}{2^4(2!)^2}\left(\frac{3}{2}\right) + \cdots\right] \qquad (1)\end{aligned}$$

which is the form claimed in Eq. (*24.7*). The general solution is $y = c_1y_1(x) + c_2y_2(x)$.

24.12. Use the method of Frobenius to find one solution near $x = 0$ of $x^2y'' - xy' + y = 0$.

Here $P(x) = -1/x$ and $Q(x) = 1/x^2$, so $x = 0$ is a regular singular point and the method of Frobenius is applicable. Substituting Eqs. (*24.2*) through (*24.4*) into the left side of the differential equation, as given, and combining coefficients of like powers of x, we obtain

$$x^\lambda(\lambda - 1)^2a_0 + x^{\lambda+1}[\lambda^2a_1] + \cdots + x^{\lambda+n}[(\lambda+n)^2 - 2(\lambda+n) + 1]a_n + \cdots = 0$$

Thus,
$$(\lambda - 1)^2a_0 = 0 \qquad (1)$$

and, in general,
$$[(\lambda+n)^2 - 2(\lambda+n) + 1]a_n = 0 \qquad (2)$$

From (*1*), the indicial equation is $(\lambda - 1)^2 = 0$, which has roots $\lambda_1 = \lambda_2 = 1$. Substituting $\lambda = 1$ into (*2*), we obtain $n^2a_n = 0$, which implies that $a_n = 0$, $n \geq 1$. Thus, $y_1(x) = a_0x$.

24.13. Find the general solution near $x = 0$ to the differential equation given in Problem 24.12.

One solution is given in Problem 24.12. Because the roots of the indicial equation are equal, we use Eq. (*24.8*) to generate a second linearly independent solution. The recurrence formula is (*2*) of Problem

24.12. Solving it for a_n in terms of λ, we find that $a_n = 0$ $(n \geq 1)$, and when these values are substituted into Eq. (*24.2*), we have $y(\lambda, x) = a_0 x^\lambda$. Thus,

$$\frac{\partial y(\lambda, x)}{\partial \lambda} = a_0 x^\lambda \ln x$$

and

$$y_2(x) = \left.\frac{\partial y(\lambda, x)}{\partial \lambda}\right|_{\lambda=1} = a_0 x \ln x = y_1(x) \ln x$$

which is precisely the form of Eq. (*20.5*), where, for this particular differential equation, $b_n(\lambda_1) = 0$ $(n = 0, 1, 2, \ldots)$. The general solution is

$$y = c_1 y_1(x) + c_2 y_2(x) = k_1 x + k_2 x \ln x$$

where $k_1 = c_1 a_0$ and $k_2 = c_2 a_0$.

24.14. Use the method of Frobenius to find one solution near $x = 0$ of $x^2 y'' + (x^2 - 2x)y' + 2y = 0$.

Here

$$P(x) = 1 - \frac{2}{x} \quad \text{and} \quad Q(x) = \frac{2}{x^2}$$

so $x = 0$ is a regular singular point and the method of Frobenius is applicable. Substituting Eqs. (*24.2*) through (*24.4*) into the left side of the differential equation, as given, and combining coefficients of like powers of x, we obtain

$$x^\lambda[(\lambda^2 - 3\lambda + 2)a_0] + x^{\lambda+1}[(\lambda^2 - \lambda)a_1 + \lambda a_0] + \cdots$$
$$+ x^{\lambda+n}\{[(\lambda + n)^2 - 3(\lambda + n) + 2]a_n + (\lambda + n - 1)a_{n-1}\} + \cdots = 0$$

Dividing by x^λ, factoring the coefficient of a_n, and equating the coefficient of each power of x to zero, we obtain

$$(\lambda^2 - 3\lambda + 2)a_0 = 0 \tag{1}$$

and, in general, $[(\lambda + n) - 2][(\lambda + n) - 1]a_n + (\lambda + n - 1)a_{n-1} = 0$, or,

$$a_n = -\frac{1}{\lambda + n - 2} a_{n-1} \qquad (n \geq 1) \tag{2}$$

From (*1*), the indicial equation is $\lambda^2 - 3\lambda + 2 = 0$, which has roots $\lambda_1 = 2$ and $\lambda_2 = 1$. Since $\lambda_1 - \lambda_2 = 1$, a positive integer, the solution is given by Eqs. (*24.5*) and (*24.9*). Substituting $\lambda = 2$ into (*2*), we have $a_n = -(1/n)a_{n-1}$, from which we obtain

$$a_1 = -a_0$$
$$a_2 = -\frac{1}{2}a_1 = \frac{1}{2!}a_0$$
$$a_3 = -\frac{1}{3}a_2 = -\frac{1}{3}\frac{1}{2!}a_0 = -\frac{1}{3!}a_0$$

and, in general, $a_k = \dfrac{(-1)^k}{k!} a_0$. Thus,

$$y_1(x) = a_0 x^2 \sum_{n=0}^{\infty} \frac{(-1)^n}{n!} x^n = a_0 x^2 e^{-x} \tag{3}$$

24.15. Find the general solution near $x = 0$ to the differential equation given in Problem 24.14.

One solution is given by (*3*) in Problem 24.14 for the indicial root $\lambda_1 = 2$. If we try the method of Frobenius with the indicial root $\lambda_2 = 1$, recurrence formula (*2*) of Problem 24.14 becomes

$$a_n = -\frac{1}{n - 1} a_{n-1}$$

which leaves a_1 undefined because the denominator is zero when $n = 1$. Instead, we must use (*24.10*) to generate a second linearly independent solution. Using the recurrence formula (*2*) of Problem 24.14 to solve sequentially for a_n $(n = 1, 2, 3, \ldots)$ in terms of λ, we find

$$a_1 = -\frac{1}{\lambda - 1}a_0, \qquad a_2 = -\frac{1}{\lambda}a_1 = \frac{1}{\lambda(\lambda - 1)}a_0, \qquad a_3 = -\frac{1}{\lambda + 1}a_2 = \frac{-1}{(\lambda + 1)\lambda(\lambda - 1)}a_0, \qquad \ldots$$

Substituting these values into Eq. (*24.2*), we obtain

$$y(\lambda, x) = a_0\left[x^\lambda - \frac{1}{(\lambda - 1)}x^{\lambda+1} + \frac{1}{\lambda(\lambda - 1)}x^{\lambda+2} - \frac{1}{(\lambda + 1)\lambda(\lambda - 1)}x^{\lambda+3} + \cdots\right]$$

and, since $\lambda - \lambda_2 = \lambda - 1$,

$$(\lambda - \lambda_2)y(\lambda, x) = a_0\left[(\lambda - 1)x^\lambda - x^{\lambda+1} + \frac{1}{\lambda}x^{\lambda+2} - \frac{1}{\lambda(\lambda + 1)}x^{\lambda+3} + \cdots\right]$$

Then

$$\frac{\partial}{\partial\lambda}[(\lambda - \lambda_2)y(\lambda, x)] = a_0\left[x^\lambda + (\lambda - 1)x^\lambda \ln x - x^{\lambda+1}\ln x - \frac{1}{\lambda^2}x^{\lambda+2} + \frac{1}{\lambda}x^{\lambda+2}\ln x \right.$$

$$\left. + \frac{1}{\lambda^2(\lambda + 1)}x^{\lambda+3} + \frac{1}{\lambda(\lambda + 1)^2}x^{\lambda+3} - \frac{1}{\lambda(\lambda + 1)}x^{\lambda+3}\ln x + \cdots\right]$$

and

$$y_2(x) = \frac{\partial}{\partial\lambda}[(\lambda - \lambda_2)y(\lambda, x)]\bigg|_{\lambda=\lambda_2=1}$$

$$= a_0\left(x + 0 - x^2\ln x - x^3 + x^3\ln x + \frac{1}{2}x^4 + \frac{1}{4}x^4 - \frac{1}{2}x^4\ln x + \cdots\right)$$

$$= (-\ln x)a_0\left(x^2 - x^3 + \frac{1}{2}x^4 + \cdots\right) + a_0\left(x - x^3 + \frac{3}{4}x^4 + \cdots\right)$$

$$= -y_1(x)\ln x + a_0 x\left(1 - x^2 + \frac{3}{4}x^3 + \cdots\right)$$

This is the form claimed in Eq. (*24.9*), with $d_{-1} = -1$, $d_0 = a_0$, $d_1 = 0$, $d_2 = -a_0$, $d_3 = \frac{3}{4}a_0, \ldots$. The general solution is $y = c_1 y_1(x) + c_2 y_2(x)$.

24.16. Use the method of Frobenius to find one solution near $x = 0$ of $x^2y'' + xy' + (x^2 - 1)y = 0$.

Here

$$P(x) = \frac{1}{x} \qquad \text{and} \qquad Q(x) = 1 - \frac{1}{x^2}$$

so $x = 0$ is a regular singular point and the method of Frobenius is applicable. Substituting Eqs. (*24.2*) through (*24.4*) into the left side of the differential equation, as given, and combining coefficients of like powers of x, we obtain

$$x^\lambda[(\lambda^2 - 1)a_0] + x^{\lambda+1}[(\lambda + 1)^2 - 1]a_1 + x^{\lambda+2}\{[(\lambda + 2)^2 - 1]a_2 + a_0\} + \cdots$$
$$+ x^{\lambda+n}\{[(\lambda + n)^2 - 1]a_n + a_{n-2}\} + \cdots = 0$$

Thus,

$$(\lambda^2 - 1)a_0 = 0 \tag{1}$$

$$[(\lambda + 1)^2 - 1]a_1 = 0 \tag{2}$$

and, for $n \geq 2$, $[(\lambda + n)^2 - 1]a_n + a_{n-2} = 0$, or,

$$a_n = \frac{-1}{(\lambda + n)^2 - 1}a_{n-2} \qquad (n \geq 2) \tag{3}$$

From (*1*), the indicial equation is $\lambda^2 - 1 = 0$, which has roots $\lambda_1 = 1$ and $\lambda_2 = -1$. Since $\lambda_1 - \lambda_2 = 2$, a positive integer, the solution is given by (*24.5*) and (*24.9*). Substituting $\lambda = 1$ into (*2*) and (*3*), we obtain

$a_1 = 0$ and

$$a_n = \frac{-1}{n(n+2)} a_{n-2} \qquad (n \geq 2)$$

Since $a_1 = 0$, it follows that $0 = a_3 = a_5 = a_7 = \cdots$. Furthermore,

$$a_2 = \frac{-1}{2(4)} a_0 = \frac{-1}{2^2 1!\, 2!} a_0, \qquad a_4 = \frac{-1}{4(6)} a_2 = \frac{1}{2^4 2!\, 3!} a_0, \qquad a_6 = \frac{-1}{6(8)} a_4 = \frac{-1}{2^6 3!\, 4!} a_0$$

and, in general,

$$a_{2k} = \frac{(-1)^k}{2^{2k} k!\, (k+1)!} a_0 \qquad (k = 1, 2, 3, \ldots)$$

Thus,

$$y_1(x) = a_0 x \sum_{n=0}^{\infty} \frac{(-1)^n}{2^{2n} n!\, (n+1)!} x^{2n} \tag{4}$$

24.17. Find the general solution near $x = 0$ to the differential equation given in Problem 24.16.

One solution is given by (4) in Problem 24.16 for the indicial root $\lambda_1 = 1$. If we try the method of Frobenius with the indicial root $\lambda_2 = -1$, recurrence formula (3) of Problem 24.16 becomes

$$a_n = -\frac{1}{n(n-2)} a_{n-2}$$

which fails to define a_2 because the denominator is zero when $n = 2$. Instead, we must use Eq. (24.10) to generate a second linearly independent solution. Using Eqs. (2) and (3) of Problem 24.16 to solve sequentially for a_n $(n = 1, 2, 3, \ldots)$ in terms of λ, we find $0 = a_1 = a_3 = a_5 = \cdots$ and

$$a_2 = \frac{-1}{(\lambda+3)(\lambda+1)} a_0, \qquad a_4 = \frac{1}{(\lambda+5)(\lambda+3)^2(\lambda+1)} a_0, \qquad \ldots$$

Thus,

$$y(\lambda, x) = a_0\left[x^\lambda - \frac{1}{(\lambda+3)(\lambda+1)} x^{\lambda+2} + \frac{1}{(\lambda+5)(\lambda+3)^2(\lambda+1)} x^{\lambda+4} + \cdots\right]$$

Since $\lambda - \lambda_2 = \lambda + 1$,

$$(\lambda - \lambda_2) y(\lambda, x) = a_0\left[(\lambda+1)x^\lambda - \frac{1}{(\lambda+3)} x^{\lambda+2} + \frac{1}{(\lambda+5)(\lambda+3)^2} x^{\lambda+4} + \cdots\right]$$

and

$$\begin{aligned}\frac{\partial}{\partial\lambda}[(\lambda - \lambda_2) y(\lambda, x)] = a_0\bigg[&x^\lambda + (\lambda+1)x^\lambda \ln x + \frac{1}{(\lambda+3)^2} x^{\lambda+2} \\ &- \frac{1}{(\lambda+3)} x^{\lambda+2} \ln x - \frac{1}{(\lambda+5)^2(\lambda+3)^2} x^{\lambda+4} \\ &- \frac{2}{(\lambda+5)(\lambda+3)^3} x^{\lambda+4} + \frac{1}{(\lambda+5)(\lambda+3)^2} x^{\lambda+4} \ln x + \cdots\bigg]\end{aligned}$$

Then

$$\begin{aligned} y_2(x) &= \frac{\partial}{\partial\lambda}[(\lambda - \lambda_2) y(\lambda, x)]\bigg|_{\lambda=\lambda_2=-1} \\ &= a_0\left(x^{-1} + 0 + \frac{1}{4}x - \frac{1}{2}x \ln x - \frac{1}{64}x^3 - \frac{2}{32}x^3 + \frac{1}{16}x^3 \ln x + \cdots\right) \\ &= -\frac{1}{2}(\ln x) a_0 x\left(1 - \frac{1}{8}x^2 + \cdots\right) + a_0\left(x^{-1} + \frac{1}{4}x - \frac{5}{64}x^3 + \cdots\right) \\ &= -\frac{1}{2}(\ln x) y_1(x) + a_0 x^{-1}\left(1 + \frac{1}{4}x^2 - \frac{5}{64}x^4 + \cdots\right) \end{aligned} \tag{1}$$

This is in the form of (24.9), with $d_{-1} = -\frac{1}{2}$, $d_0 = a_0$, $d_1 = 0$, $d_2 = \frac{1}{4}a_0$, $d_3 = 0$, $d_4 = \frac{-5}{64}a_0, \ldots$. The general solution is $y = c_1 y_1(x) + c_2 y_2(x)$.

24.18. Use the method of Frobenius to find one solution near $x=0$ of $x^2y''+(x^2+2x)y'-2y=0$.

Here

$$P(x) = 1 + \frac{2}{x} \quad \text{and} \quad Q(x) = -\frac{2}{x^2}$$

so $x=0$ is a regular singular point and the method of Frobenius is applicable. Substituting Eqs. (*24.2*) through (*24.4*) into the left side of the differential equation, as given, and combining coefficients of like powers of x, we obtain

$$x^{\lambda}[(\lambda^2+\lambda-2)a_0] + x^{\lambda+1}[(\lambda^2+3\lambda)a_1+\lambda a_0] + \cdots$$
$$+ x^{\lambda+n}\{[(\lambda+n)^2+(\lambda+n)-2]a_n + (\lambda+n-1)a_{n-1}\} + \cdots = 0$$

Dividing by x^{λ}, factoring the coefficient of a_n, and equating to zero the coefficient of each power of x, we obtain

$$(\lambda^2+\lambda-2)a_0 = 0 \tag{1}$$

and, for $n \geq 1$,

$$[(\lambda+n)+2][(\lambda+n)-1]a_n + (\lambda+n-1)a_{n-1} = 0$$

which is equivalent to

$$a_n = -\frac{1}{\lambda+n+2}a_{n-1} \qquad (n \geq 1) \tag{2}$$

From (*1*), the indicial equation is $\lambda^2+\lambda-2=0$, which has roots $\lambda_1=1$ and $\lambda_2=-2$. Since $\lambda_1-\lambda_2=3$, a positive integer, the solution is given by Eqs. (*24.5*) and (*24.9*). Substituting $\lambda=1$ into (*2*), we obtain $a_n=[-1/(n+3)]a_{n-1}$, which in turn yields

$$a_1 = -\frac{1}{4}a_0 = -\frac{3!}{4!}a_0$$

$$a_2 = -\frac{1}{5}a_1 = \left(-\frac{1}{5}\right)\left(-\frac{3!}{4!}\right)a_0 = \frac{3!}{5!}a_0$$

$$a_3 = -\frac{1}{6}a_2 = -\frac{3!}{6!}a_0$$

and, in general,

$$a_k = \frac{(-1)^k 3!}{(k+3)!}a_0$$

Hence,
$$y_1(x) = a_0x\left[1 + 3!\sum_{n=1}^{\infty}\frac{(-1)^n x^n}{(n+3)!}\right] = a_0x\sum_{n=0}^{\infty}\frac{(-1)^n 3!\, x^n}{(n+3)!}$$

which can be simplified to

$$y_1(x) = \frac{3a_0}{x^2}(2-2x+x^2-2e^{-x}) \tag{3}$$

24.19. Find the general solution near $x=0$ to the differential equation given in Problem 24.18.

One solution is given by (*3*) in Problem 24.18 for the indicial root $\lambda_1=1$. If we try the method of

Frobenius with the indicial root $\lambda_2 = -2$, recurrence formula (*2*) of Problem 24.18 becomes

$$a_n = -\frac{1}{n} a_{n-1} \tag{1}$$

which does define all a_n $(n \geq 1)$. Solving sequentially, we obtain

$$a_1 = -a_0 = -\frac{1}{1!} a_0 \qquad a_2 = -\frac{1}{2} a_2 = \frac{1}{2!} a_0$$

and, in general, $a_k = (-1)^k a_0 / k!$. Therefore,

$$y_2(x) = a_0 x^{-2}\left[1 - \frac{1}{1!}x + \frac{1}{2!}x^2 + \cdots + \frac{(-1)^k}{k!}x^k + \cdots\right]$$

$$= a_0 x^{-2} \sum_{n=0}^{\infty} \frac{(-1)^n x^n}{n!} = a_0 x^{-2} e^{-x}$$

This is precisely in the form of (*24.9*), with $d_{-1} = 0$ and $d_n = (-1)^n a_0 / n!$. The general solution is

$$y = c_1 y_1(x) + c_2 y_2(x)$$

24.20. Find a general expression for the indicial equation of (*24.1*).

Since $x = 0$ is a regular singular point, $xP(x)$ and $x^2Q(x)$ are analytic near the origin and can be expanded in Taylor series there. Thus,

$$xP(x) = \sum_{n=0}^{\infty} p_n x^n = p_0 + p_1 x + p_2 x^2 + \cdots$$

$$x^2Q(x) = \sum_{n=0}^{\infty} q_n x^n = q_0 + q_1 x + q_2 x^2 + \cdots$$

Dividing by x and x^2, respectively, we have

$$P(x) = p_0 x^{-1} + p_1 + p_2 x + \cdots \qquad Q(x) = q_0 x^{-2} + q_1 x^{-1} + q_2 + \cdots$$

Substituting these two results with Eqs. (*24.2*) through (*24.4*) into (*24.1*) and combining, we obtain

$$x^{\lambda-2}[\lambda(\lambda - 1)a_0 + \lambda a_0 p_0 + a_0 q_0] + \cdots = 0$$

which can hold only if

$$a_0[\lambda^2 + (p_0 - 1)\lambda + q_0] = 0$$

Since $a_0 \neq 0$ (a_0 is an arbitrary constant, hence can be chosen nonzero), the indicial equation is

$$\lambda^2 + (p_0 - 1)\lambda + q_0 = 0 \tag{1}$$

24.21. Find the indicial equation of $x^2y'' + xe^xy' + (x^3 - 1)y = 0$ if the solution is required near $x = 0$.

Here

$$P(x) = \frac{e^x}{x} \qquad \text{and} \qquad Q(x) = x - \frac{1}{x^2}$$

and we have

$$xP(x) = e^x = 1 + x + \frac{x^2}{2!} + \cdots$$

$$x^2Q(x) = x^3 - 1 = -1 + 0x + 0x^2 + 1x^3 + 0x^4 + \cdots$$

from which $p_0 = 1$ and $q_0 = -1$. Using (*1*) of Problem 24.20, we obtain the indicial equation as $\lambda^2 - 1 = 0$.

24.22. Solve Problem 24.9 by an alternative method.

The given differential equation, $3x^2y'' - xy' + y = 0$, is a special case of *Euler's equation*

$$b_nx^ny^{(n)} + b_{n-1}x^{n-1}y^{(n-1)} + \cdots + b_2x^2y'' + b_1xy' + b_0y = \phi(x) \tag{1}$$

where b_j $(j = 0, 1, \ldots, n)$ is a constant. Euler's equation can always be transformed into a linear differential equation with *constant coefficients* by the change of variables

$$z = \ln x \qquad \text{or} \qquad x = e^z \tag{2}$$

It follows from (2) and from the chain rule and the product rule of differentiation that

$$\frac{dy}{dx} = \frac{dy}{dz}\frac{dz}{dx} = \frac{1}{x}\frac{dy}{dz} = e^{-z}\frac{dy}{dz} \tag{3}$$

$$\frac{d^2y}{dx^2} = \frac{d}{dx}\left(\frac{dy}{dx}\right) = \frac{d}{dx}\left(e^{-z}\frac{dy}{dz}\right) = \left[\frac{d}{dz}\left(e^{-z}\frac{dy}{dz}\right)\right]\frac{dz}{dx}$$

$$= \left[-e^{-z}\left(\frac{dy}{dz}\right) + e^{-z}\left(\frac{d^2y}{dz^2}\right)\right]e^{-z} = e^{-2z}\left(\frac{d^2y}{dz^2}\right) - e^{-2z}\left(\frac{dy}{dz}\right) \tag{4}$$

Substituting Eqs. (2), (3), and (4) into the given differential equation and simplifying, we obtain

$$\frac{d^2y}{dz^2} - \frac{4}{3}\frac{dy}{dz} + \frac{1}{3}y = 0$$

Using the method of Chapter 8 we find that the solution of this last equation is $y = c_1e^z + c_2e^{(1/3)z}$. Then using (2) and noting that $e^{(1/3)z} = (e^z)^{1/3}$, we have, as before,

$$y = c_1x + c_2x^{1/3}$$

24.23. Solve the differential equation given in Problem 24.12 by an alternative method.

The given differential equation, $x^2y'' - xy' + y = 0$, is a special case of Euler's equation, (1) of Problem 24.22. Using the transformations (2), (3), and (4) of Problem 24.22, we reduce the given equation to

$$\frac{d^2y}{dz^2} - 2\frac{dy}{dz} + y = 0$$

The solution to this equation is (see Chapter 8) $y = c_1e^z + c_2ze^z$. Then, using (2) of Problem 24.22, we have for the solution of the original differential equation

$$y = c_1x + c_2x\ln x$$

as before.

24.24. Find the general solution near $x = 0$ of the *hypergeometric equation*

$$x(1 - x)y'' + [C - (A + B + 1)x]y' - ABy = 0$$

where A and B are any real numbers, and C is any real nonintegral number.

Since $x = 0$ is a regular singular point, the method of Frobenius is applicable. Substituting Eqs. (24.2) through (24.4) into the differential equation, simplifying and equating the coefficient of each power of x to zero, we obtain

$$\lambda^2 + (C - 1)\lambda = 0 \tag{1}$$

as the indicial equation and

$$a_{n+1} = \frac{(\lambda + n)(\lambda + n + A + B) + AB}{(\lambda + n + 1)(\lambda + n + C)}a_n \tag{2}$$

as the recurrence formula. The roots of (1) are $\lambda_1 = 0$ and $\lambda_2 = 1 - C$; hence, $\lambda_1 - \lambda_2 = C - 1$. Since C is not an integer, the solution of the hypergeometric equation is given by Eqs. (24.5) and (24.6).

Substituting $\lambda = 0$ into (2), we have

$$a_{n+1} = \frac{n(n+A+B)+AB}{(n+1)(n+C)}a_n$$

which is equivalent to

$$a_{n+1} = \frac{(A+n)(B+n)}{(n+1)(n+C)}a_n$$

Thus

$$a_1 = \frac{AB}{C}a_0 = \frac{AB}{1!\,C}a_0$$

$$a_2 = \frac{(A+1)(B+1)}{2(C+1)}a_1 = \frac{A(A+1)B(B+1)}{2!\,C(C+1)}a_0$$

$$a_3 = \frac{(A+2)(B+2)}{3(C+2)}a_2 = \frac{A(A+1)(A+2)B(B+1)(B+2)}{3!\,C(C+1)(C+2)}a_0$$

....

and $y_1(x) = a_0 F(A, B; C; x)$, where

$$F(A, B; C; x) = 1 + \frac{AB}{1!\,C}x + \frac{A(A+1)B(B+1)}{2!\,C(C+1)}x^2 + \frac{A(A+1)(A+2)B(B+1)(B+2)}{3!\,C(C+1)(C+2)}x^3 + \cdots$$

The series $F(A, B; C; x)$ is known as the *hypergeometric series*; it can be shown that this series converges for $-1 < x < 1$. It is customary to assign the arbitrary constant a_0 the value 1. Then $y_1(x) = F(A, B; C; x)$ and the hypergeometric series is a solution of the hypergeometric equation.

To find $y_2(x)$, we substitute $\lambda = 1 - C$ into (2) and obtain

$$a_{n+1} = \frac{(n+1-C)(n+1+A+B-C)+AB}{(n+2-C)(n+1)}a_n$$

or

$$a_{n+1} = \frac{(A-C+n+1)(B-C+n+1)}{(n+2-C)(n+1)}a_n$$

Solving for a_n in terms of a_0, and again setting $a_0 = 1$, it follows that

$$y_2(x) = x^{1-C}F(A-C+1, B-C+1; 2-C; x)$$

The general solution is $y = c_1 y_1(x) + c_2 y_2(x)$.

Supplementary Problems

In Problems 24.25 through 24.33, find two linearly independent solutions to the given differential equations.

24.25. $2x^2y'' - xy' + (1-x)y = 0$

24.26. $2x^2y'' + (x^2 - x)y' + y = 0$

24.27. $3x^2y'' - 2xy' - (2 + x^2)y = 0$

24.28. $xy'' + y' - y = 0$

24.29. $x^2y'' + xy' + x^3y = 0$

24.30. $x^2y'' + (x - x^2)y' - y = 0$

24.31. $xy'' - (x+1)y' - y = 0$

24.32. $4x^2y'' + (4x + 2x^2)y' + (3x - 1)y = 0$

24.33. $x^2y'' + (x^2 - 3x)y' - (x - 4)y = 0$

In Problems 24.34 through 24.38, find the general solution to the given equations using the method described in Problem 24.22.

24.34. $4x^2y'' + 4xy' - y = 0$

24.35. $x^2y'' - 3xy' + 4y = 0$

24.36. $2x^2y'' + 11xy' + 4y = 0$

24.37. $x^2y'' - 2y = 0$

24.38. $x^2y'' - 6xy' = 0$

Chapter 25

Gamma and Bessel Functions

GAMMA FUNCTION

The *gamma function*, $\Gamma(p)$, is defined for any positive real number p by

$$\Gamma(p) = \int_0^{\infty} x^{p-1}e^{-x}\,dx \tag{25.1}$$

Consequently, $\Gamma(1) = 1$ and for any positive real number p,

$$\Gamma(p+1) = p\Gamma(p) \tag{25.2}$$

Furthermore, when $p = n$, a positive integer,

$$\Gamma(n+1) = n! \tag{25.3}$$

Thus, the gamma function (which is defined on all positive real numbers) is an extension of the factorial function (which is defined only on the nonnegative integers).

Equation (*25.2*) may be rewritten as

$$\Gamma(p) = \frac{1}{p}\Gamma(p+1) \tag{25.4}$$

which defines the gamma function iteratively for all nonintegral negative values of p. $\Gamma(0)$ remains undefined, because

$$\lim_{p\to 0^+}\Gamma(p) = \lim_{p\to 0^+}\frac{\Gamma(p+1)}{p} = \infty \qquad \text{and} \qquad \lim_{p\to 0^-}\Gamma(p) = \lim_{p\to 0^-}\frac{\Gamma(p+1)}{p} = -\infty$$

It then follows from Eq. (*25.4*) that $\Gamma(p)$ is undefined for negative integer values of p.

Table 25-1 lists values of the gamma function in the interval $1 \le p < 2$. These tabular values are used with Eqs. (*25.2*) and (*25.4*) to generate values of $\Gamma(p)$ in other intervals.

BESSEL FUNCTIONS

Let p represent any real number. The *Bessel function of the first kind of order* p, $J_p(x)$, is

$$J_p(x) = \sum_{k=0}^{\infty} \frac{(-1)^k x^{2k+p}}{2^{2k+p}k!\,\Gamma(p+k+1)} \tag{25.5}$$

The function $J_p(x)$ is a solution near the regular singular point $x = 0$ of *Bessel's differential equation of order* p:

$$x^2y'' + xy' + (x^2 - p^2)y = 0 \tag{25.6}$$

In fact, $J_p(x)$ is that solution of Eq. (*25.6*) guaranteed by Theorem 24.1.

ALGEBRAIC OPERATIONS ON INFINITE SERIES

Changing the dummy index. The dummy index in an infinite series can be changed at will without altering the series. For example,

$$\sum_{k=0}^{\infty}\frac{1}{(k+1)!} = \sum_{n=0}^{\infty}\frac{1}{(n+1)!} = \sum_{p=0}^{\infty}\frac{1}{(p+1)!} = \frac{1}{1!} + \frac{1}{2!} + \frac{1}{3!} + \frac{1}{4!} + \frac{1}{5!} + \cdots$$

Table 25-1 The Gamma Function $(1.00 \le x \le 1.99)$

x	$\Gamma(x)$	x	$\Gamma(x)$	x	$\Gamma(x)$	x	$\Gamma(x)$
1.00	1.0000 0000	1.25	0.9064 0248	1.50	0.8862 2693	1.75	0.9190 6253
1.01	0.9943 2585	1.26	0.9043 9712	1.51	0.8865 9169	1.76	0.9213 7488
1.02	0.9888 4420	1.27	0.9025 0306	1.52	0.8870 3878	1.77	0.9237 6313
1.03	0.9835 4995	1.28	0.9007 1848	1.53	0.8875 6763	1.78	0.9262 2731
1.04	0.9784 3820	1.29	0.8990 4159	1.54	0.8881 7766	1.79	0.9287 6749
1.05	0.9735 0427	1.30	0.8974 7070	1.55	0.8888 6835	1.80	0.9313 8377
1.06	0.9687 4365	1.31	0.8960 0418	1.56	0.8896 3920	1.81	0.9340 7626
1.07	0.9641 5204	1.32	0.8946 4046	1.57	0.8904 8975	1.82	0.9368 4508
1.08	0.9597 2531	1.33	0.8933 7805	1.58	0.8914 1955	1.83	0.9396 9040
1.09	0.9554 5949	1.34	0.8922 1551	1.59	0.8924 2821	1.84	0.9426 1236
1.10	0.9513 5077	1.35	0.8911 5144	1.60	0.8935 1535	1.85	0.9456 1118
1.11	0.9473 9550	1.36	0.8901 8453	1.61	0.8946 8061	1.86	0.9486 8704
1.12	0.9435 9019	1.37	0.8893 1351	1.62	0.8959 2367	1.87	0.9518 4019
1.13	0.9399 3145	1.38	0.8885 3715	1.63	0.8972 4423	1.88	0.9550 7085
1.14	0.9364 1607	1.39	0.8878 5429	1.64	0.8986 4203	1.89	0.9583 7931
1.15	0.9330 4093	1.40	0.8872 6382	1.65	0.9001 1682	1.90	0.9617 6583
1.16	0.9298 0307	1.41	0.8867 6466	1.66	0.9016 6837	1.91	0.9652 3073
1.17	0.9266 9961	1.42	0.8863 5579	1.67	0.9032 9650	1.92	0.9787 7431
1.18	0.9237 2781	1.43	0.8860 3624	1.68	0.9050 0103	1.93	0.9723 9692
1.19	0.9208 8504	1.44	0.8858 0506	1.69	0.9067 8182	1.94	0.9760 9891
1.20	0.9181 6874	1.45	0.8856 6138	1.70	0.9086 3873	1.95	0.9798 8065
1.21	0.9155 7649	1.46	0.8856 0434	1.71	0.9105 7168	1.96	0.9837 4254
1.22	0.9131 0595	1.47	0.8856 3312	1.72	0.9125 8058	1.97	0.9876 8498
1.23	0.9107 5486	1.48	0.8857 4696	1.73	0.9146 6537	1.98	0.9917 0841
1.24	0.9085 2106	1.49	0.8859 4513	1.74	0.9168 2603	1.99	0.9958 1326

Change of variables. Consider the infinite series $\sum_{k=0}^{\infty} \frac{1}{(k+1)!}$. If we make the change of variables $j = k + 1$, or $k = j - 1$, then

$$\sum_{k=0}^{\infty} \frac{1}{(k+1)!} = \sum_{j=1}^{\infty} \frac{1}{j!}$$

Note that a change of variables generally changes the limits on the summation. For instance, if $j = k + 1$, it follows that $j = 1$ when $k = 0$, $j = \infty$ when $k = \infty$, and, as k runs from 0 to ∞, j runs from 1 to ∞.

The two operations given above are often used in concert. For example,

$$\sum_{k=0}^{\infty} \frac{1}{(k+1)!} = \sum_{j=2}^{\infty} \frac{1}{(j-1)!} = \sum_{k=2}^{\infty} \frac{1}{(k-1)!}$$

Here, the second series results from the change of variables $j = k + 2$ in the first series, while the third series is the result of simply changing the dummy index in the second series from j to k. Note that all three series equal

$$\frac{1}{1!} + \frac{1}{2!} + \frac{1}{3!} + \frac{1}{4!} + \cdots = e - 1$$

Solved Problems

25.1. Determine $\Gamma(3.5)$.

It follows from Table 25-1 that $\Gamma(1.5) = 0.8862$, rounded to four decimal places. Using Eq. (*25.2*) with $p = 2.5$, we obtain $\Gamma(3.5) = (2.5)\Gamma(2.5)$. But also from Eq. (*25.2*), with $p = 1.5$, we have $\Gamma(2.5) = (1.5)\Gamma(1.5)$. Thus, $\Gamma(3.5) = (2.5)(1.5)\Gamma(1.5) = (3.75)(0.8862) = 3.3233$.

25.2. Determine $\Gamma(-0.5)$.

It follows from Table 25-1 that $\Gamma(1.5) = 0.8862$, rounded to four decimal places. Using Eq. (*25.4*) with $p = 0.5$, we obtain $\Gamma(0.5) = 2\Gamma(1.5)$. But also from Eq. (*25.4*), with $p = -0.5$, we have $\Gamma(-0.5) = -2\Gamma(0.5)$. Thus, $\Gamma(-0.5) = (-2)(2)\Gamma(1.5) = -4(0.8862) = -3.5448$.

25.3. Determine $\Gamma(-1.42)$.

It follows repeatedly from Eq. (*25.4*) that

$$\Gamma(-1.42) = \frac{1}{-1.42}\Gamma(-0.42) = \frac{1}{-1.42}\left(\frac{1}{-0.42}\Gamma(0.58)\right) = \frac{1}{1.42(0.42)}\left(\frac{1}{0.58}\Gamma(1.58)\right)$$

From Table 25-1, we have $\Gamma(1.58) = 0.8914$, rounded to four decimal places; hence

$$\Gamma(-1.42) = \frac{0.8914}{1.42(0.42)(0.58)} = 2.5770$$

25.4. Prove that $\Gamma(p+1) = p\Gamma(p)$, $p > 0$.

Using (*25.1*) and integration by parts, we have

$$\Gamma(p+1) = \int_0^\infty x^{(p+1)-1}e^{-x}\,dx = \lim_{r\to\infty}\int_0^r x^p e^{-x}\,dx$$

$$= \lim_{r\to\infty}\left[-x^p e^{-x}\Big|_0^r + \int_0^r p x^{p-1}e^{-x}\,dx\right]$$

$$= \lim_{r\to\infty}(-r^p e^{-r} + 0) + p\int_0^\infty x^{p-1}e^{-x}\,dx = p\Gamma(p)$$

The result $\lim_{r\to\infty} r^p e^{-r} = 0$ is easily obtained by first writing $r^p e^{-r}$ as r^p/e^r and then using L'Hôpital's rule.

25.5. Prove that $\Gamma(1) = 1$.

Using Eq. (*25.1*), we find that

$$\Gamma(1) = \int_0^\infty x^{1-1}e^{-x}\,dx = \lim_{r\to\infty}\int_0^r e^{-x}\,dx$$

$$= \lim_{r\to\infty} - e^{-x}\Big|_0^r = \lim_{r\to\infty}(-e^{-r} + 1) = 1$$

25.6. Prove that if $p = n$, a positive integer, then $\Gamma(n+1) = n!$.

The proof is by induction. First we consider $n = 1$. Using Problem 25.4 with $p = 1$ and then Problem

25.5, we have

$$\Gamma(1+1) = 1\Gamma(1) = 1(1) = 1 = 1!$$

Next we assume that $\Gamma(n+1) = n!$ holds for $n = k$ and then try to prove its validity for $n = k+1$:

$$\begin{aligned} \Gamma[(k+1)+1] &= (k+1)\Gamma(k+1) && \text{(Problem 25.4 with } p = k+1\text{)} \\ &= (k+1)(k!) && \text{(from the induction hypothesis)} \\ &= (k+1)! \end{aligned}$$

Thus, $\Gamma(n+1) = n!$ is true by induction.

Note that we can now use this equality to define $0!$; that is,

$$0! = \Gamma(0+1) = \Gamma(1) = 1$$

25.7. Prove that $\Gamma(p+k+1) = (p+k)(p+k-1)\cdots(p+2)(p+1)\Gamma(p+1)$.

Using Problem 25.4 repeatedly, where first p is replaced by $p+k$, then by $p+k-1$, etc., we obtain

$$\begin{aligned} \Gamma(p+k+1) &= \Gamma[(p+k)+1] = (p+k)\Gamma(p+k) \\ &= (p+k)\Gamma[(p+k-1)+1] = (p+k)(p+k-1)\Gamma(p+k-1) \\ &= \cdots = (p+k)(p+k-1)\cdots(p+2)(p+1)\Gamma(p+1) \end{aligned}$$

25.8. Express $\int_0^\infty e^{-x^2}\,dx$ as a gamma function.

Let $z = x^2$; hence $x = z^{1/2}$ and $dx = \frac{1}{2}z^{-1/2}\,dz$. Substituting these values into the integral and noting that as x goes from 0 to ∞ so does z, we have

$$\int_0^\infty e^{-x^2}\,dx = \int_0^\infty e^{-z}\left(\frac{1}{2}z^{-1/2}\right)dz = \frac{1}{2}\int_0^\infty z^{(1/2)-1}e^{-z}\,dz = \frac{1}{2}\Gamma\left(\frac{1}{2}\right)$$

The last equality follows from Eq. (*25.1*), with the dummy variable x replaced by z and with $p = \frac{1}{2}$.

25.9. Use the method of Frobenius to find one solution of Bessel's equation of order p:

$$x^2y'' + xy' + (x^2 - p^2)y = 0$$

Substituting Eqs. (*24.2*) through (*24.4*) into Bessel's equation and simplifying, we find that

$$\begin{aligned} x^\lambda(\lambda^2 - p^2)a_0 &+ x^{\lambda+1}[(\lambda+1)^2 - p^2]a_1 + x^{\lambda+2}\{[(\lambda+2)^2 - p^2]a_2 + a_0\} + \cdots \\ &+ x^{\lambda+n}\{[(\lambda+n)^2 - p^2]a_n + a_{n-2}\} + \cdots = 0 \end{aligned}$$

Thus,

$$(\lambda^2 - p^2)a_0 = 0 \qquad [(\lambda+1)^2 - p^2]a_1 = 0 \tag{1}$$

and, in general, $[(\lambda+n)^2 - p^2]a_n + a_{n-2} = 0$, or,

$$a_n = -\frac{1}{(\lambda+n)^2 - p^2}a_{n-2} \qquad (n \geq 2) \tag{2}$$

The indicial equation is $\lambda^2 - p^2 = 0$, which has the roots $\lambda_1 = p$ and $\lambda_2 = -p$ (p nonnegative).

Substituting $\lambda = p$ into (1) and (2) and simplifying, we find that $a_1 = 0$ and

$$a_n = -\frac{1}{n(2p+n)}a_{n-2} \qquad (n \geq 2)$$

Hence, $0 = a_1 = a_3 = a_5 = a_7 = \cdots$ and

$$a_2 = \frac{-1}{2^2 1!\,(p+1)}a_0$$

$$a_4 = -\frac{1}{2^2 2(p+2)}a_2 = \frac{1}{2^4 2!\,(p+2)(p+1)}a_0$$

$$a_6 = -\frac{1}{2^2 3(p+3)}a_4 = \frac{-1}{2^6 3!\,(p+3)(p+2)(p+1)}a_0$$

and, in general,

$$a_{2k} = \frac{(-1)^k}{2^{2k}k!\,(p+k)(p+k-1)\cdots(p+2)(p+1)}a_0 \qquad (k \geq 1)$$

Thus,

$$y_1(x) = x^\lambda \sum_{n=0}^{\infty} a_n x^n = x^p\left[a_0 + \sum_{k=1}^{\infty} a_{2k}x^{2k}\right]$$

$$= a_0 x^p\left[1 + \sum_{k=1}^{\infty} \frac{(-1)^k x^{2k}}{2^{2k}k!\,(p+k)(p+k-1)\cdots(p+2)(p+1)}\right] \qquad (3)$$

It is customary to choose the arbitrary constant a_0 as $a_0 = \dfrac{1}{2^p\Gamma(p+1)}$. Then bringing a_0x^p inside the brackets and summation in (3), combining, and finally using Problem 25.4, we obtain

$$y_1(x) = \frac{1}{2^p\Gamma(p+1)}x^p + \sum_{k=1}^{\infty} \frac{(-1)^k x^{2k+p}}{2^{2k+p}k!\,\Gamma(p+k+1)}$$

$$= \sum_{k=0}^{\infty} \frac{(-1)^k x^{2k+p}}{2^{2k+p}k!\,\Gamma(p+k+1)} \equiv J_p(x)$$

25.10. Find the general solution to Bessel's equation of order zero.

For $p = 0$, the equation is $x^2y'' + xy' + x^2y = 0$, which was solved in Chapter 24. By (4) of Problem 24.10, one solution is

$$y_1(x) = a_0 \sum_{n=0}^{\infty} \frac{(-1)^n x^{2n}}{2^{2n}(n!)^2}$$

Changing n to k, using Problem 25.6, and letting $a_0 = \dfrac{1}{2^0\Gamma(0+1)} = 1$ as indicated in Problem 25.9, it follows that $y_1(x) = J_0(x)$. A second solution is [see (1) of Problem 24.11, with a_0 again chosen to be 1]

$$y_2(x) = J_0(x)\ln x + \left[\frac{x^2}{2^2(1!)^2}(1) - \frac{x^4}{2^4(2!)^2}\left(1 + \frac{1}{2}\right) + \frac{x^6}{2^6(3!)^2}\left(1 + \frac{1}{2} + \frac{1}{3}\right) - \cdots\right]$$

which is usually designated by $N_0(x)$. Thus, the general solution to Bessel's equation of order zero is $y = c_1J_0(x) + c_2N_0(x)$.

Another common form of the general solution is obtained when the second linearly independent solution is not taken to be $N_0(x)$, but a combination of $N_0(x)$ and $J_0(x)$. In particular, if we define

$$Y_0(x) = \frac{2}{\pi}[N_0(x) + (\gamma - \ln 2)J_0(x)] \qquad (1)$$

where γ is the *Euler constant* defined by

$$\gamma = \lim_{k\to\infty}\left(1 + \frac{1}{2} + \frac{1}{3} + \cdots + \frac{1}{k} - \ln k\right) \simeq 0.57721566$$

then the general solution to Bessel's equation of order zero can be given as $y = c_1J_0(x) + c_2Y_0(x)$.

25.11. Prove that

$$\sum_{k=0}^{\infty} \frac{(-1)^k (2k) x^{2k-1}}{2^{2k+p} k!\, \Gamma(p+k+1)} = -\sum_{k=0}^{\infty} \frac{(-1)^k x^{2k+1}}{2^{2k+p+1} k!\, \Gamma(p+k+2)}$$

Writing the $k=0$ term separately, we have

$$\sum_{k=0}^{\infty} \frac{(-1)^k (2k) x^{2k-1}}{2^{2k+p} k!\, \Gamma(p+k+1)} = 0 + \sum_{k=1}^{\infty} \frac{(-1)^k (2k) x^{2k-1}}{2^{2k+p} k!\, \Gamma(p+k+1)}$$

which, under the change of variables $j = k - 1$, becomes

$$\begin{aligned}
\sum_{j=0}^{\infty} \frac{(-1)^{j+1} 2(j+1) x^{2(j+1)-1}}{2^{2(j+1)+p} (j+1)!\, \Gamma(p+j+1+1)} &= \sum_{j=0}^{\infty} \frac{(-1)(-1)^j 2(j+1) x^{2j+1}}{2^{2j+p+2} (j+1)!\, \Gamma(p+j+2)} \\
&= -\sum_{j=0}^{\infty} \frac{(-1)^j 2(j+1) x^{2j+1}}{2^{2j+p+1}(2)(j+1)(j!)\Gamma(p+j+2)} \\
&= -\sum_{j=0}^{\infty} \frac{(-1)^j x^{2j+1}}{2^{2j+p+1} j!\, \Gamma(p+j+2)}
\end{aligned}$$

The desired result follows by changing the dummy variable in the last summation from j to k.

25.12. Prove that

$$-\sum_{k=0}^{\infty} \frac{(-1)^k x^{2k+p+2}}{2^{2k+p+1} k!\, \Gamma(p+k+2)} = \sum_{k=0}^{\infty} \frac{(-1)^k (2k) x^{2k+p}}{2^{2k+p} k!\, \Gamma(p+k+1)}$$

Make the change of variables $j = k + 1$:

$$\begin{aligned}
-\sum_{k=0}^{\infty} \frac{(-1)^k x^{2k+p+2}}{2^{2k+p+1} k!\, \Gamma(p+k+2)} &= -\sum_{j=1}^{\infty} \frac{(-1)^{j-1} x^{2(j-1)+p+2}}{2^{2(j-1)+p+1} (j-1)!\, \Gamma(p+j-1+2)} \\
&= \sum_{j=1}^{\infty} \frac{(-1)^j x^{2j+p}}{2^{2j+p-1} (j-1)!\, \Gamma(p+j+1)}
\end{aligned}$$

Now, multiply the numerator and denominator in the last summation by $2j$, noting that $j(j-1)! = j!$ and $2^{2j+p-1}(2) = 2^{2j+p}$. The result is

$$\sum_{j=1}^{\infty} \frac{(-1)^j (2j) x^{2j+p}}{2^{2j+p} j!\, \Gamma(p+j+1)}$$

Owing to the factor j in the numerator, the last infinite series is not altered if the lower limit in the sum is changed from $j = 1$ to $j = 0$. Once this is done, the desired result is achieved by simply changing the dummy index from j to k.

25.13. Prove that $\frac{d}{dx}[x^{p+1} J_{p+1}(x)] = x^{p+1} J_p(x)$.

We may differentiate the series for the Bessel function term by term. Thus,

$$\begin{aligned}
\frac{d}{dx}[x^{p+1} J_{p+1}(x)] &= \frac{d}{dx}\left[x^{p+1} \sum_{k=0}^{\infty} \frac{(-1)^k x^{2k+p+1}}{2^{2k+p+1} k!\, \Gamma(k+p+1+1)}\right] \\
&= \frac{d}{dx}\left[\sum_{k=0}^{\infty} \frac{(-1)^k x^{2k+2p+2}}{2^{2k+p}(2) k!\, \Gamma(k+p+2)}\right] \\
&= \sum_{k=0}^{\infty} \frac{(-1)^k (2k+2p+2) x^{2k+2p+1}}{2^{2k+p} k!\, 2\Gamma(k+p+2)}
\end{aligned}$$

Noting that $2\Gamma(k+p+2) = 2(k+p+1)\Gamma(k+p+1)$ and that the factor $2(k+p+1)$ cancels, we have

$$\frac{d}{dx}[x^{p+1} J_{p+1}(x)] = \sum_{k=0}^{\infty} \frac{(-1)^k x^{2k+2p+1}}{2^{2k+p} k!\, \Gamma(k+p+1)} = x^{p+1} J_p(x)$$

For the particular case $p = 0$, it follows that

$$\frac{d}{dx}[x J_1(x)] = x J_0(x) \tag{1}$$

25.14. Prove that $xJ'_p(x) = pJ_p(x) - xJ_{p+1}(x)$.

We have

$$pJ_p(x) - xJ_{p+1}(x) = p\sum_{k=0}^{\infty} \frac{(-1)^k x^{2k+p}}{2^{2k+p}k!\,\Gamma(p+k+1)} - x\sum_{k=0}^{\infty} \frac{(-1)^k x^{2k+p+1}}{2^{2k+p+1}k!\,\Gamma(p+k+2)}$$

$$= \sum_{k=0}^{\infty} \frac{(-1)^k p x^{2k+p}}{2^{2k+p}k!\,\Gamma(p+k+1)} - \sum_{k=0}^{\infty} \frac{(-1)^k x^{2k+p+2}}{2^{2k+p+1}k!\,\Gamma(p+k+2)}$$

Using Problem 25.12 on the last summation, we find

$$pJ_p(x) - xJ_{p+1}(x) = \sum_{k=0}^{\infty} \frac{(-1)^k p x^{2k+p}}{2^{2k+p}k!\,\Gamma(p+k+1)} + \sum_{k=0}^{\infty} \frac{(-1)^k (2k) x^{2k+p}}{2^{2k+p}k!\,\Gamma(p+k+1)}$$

$$= \sum_{k=0}^{\infty} \frac{(-1)^k (p+2k) x^{2k+p}}{2^{2k+p}k!\,\Gamma(p+k+1)} = xJ'_p(x)$$

For the particular case $p = 0$, it follows that $xJ'_0(x) = -xJ_1(x)$, or

$$J'_0(x) = -J_1(x) \tag{1}$$

25.15. Prove that $xJ'_p(x) = -pJ_p(x) + xJ_{p-1}(x)$.

$$-pJ_p(x) + xJ_{p-1}(x) = -p\sum_{k=0}^{\infty} \frac{(-1)^k x^{2k+p}}{2^{2k+p}k!\,\Gamma(p+k+1)} + x\sum_{k=0}^{\infty} \frac{(-1)^k x^{2k+p-1}}{2^{2k+p-1}k!\,\Gamma(p+k)}$$

Multiplying the numerator and denominator in the second summation by $2(p+k)$ and noting that $(p+k)\Gamma(p+k) = \Gamma(p+k+1)$, we find

$$-pJ_p(x) + xJ_{p-1}(x) = \sum_{k=0}^{\infty} \frac{(-1)^k (-p) x^{2k+p}}{2^{2k+p}k!\,\Gamma(p+k+1)} + \sum_{k=0}^{\infty} \frac{(-1)^k 2(p+k) x^{2k+p}}{2^{2k+p}k!\,\Gamma(p+k+1)}$$

$$= \sum_{k=0}^{\infty} \frac{(-1)^k [-p + 2(p+k)] x^{2k+p}}{2^{2k+p}k!\,\Gamma(p+k+1)}$$

$$= \sum_{k=0}^{\infty} \frac{(-1)^k (2k+p) x^{2k+p}}{2^{2k+p}k!\,\Gamma(p+k+1)} = xJ'_p(x)$$

25.16. Use Problems 25.14 and 25.15 to derive the recurrence formula

$$J_{p+1}(x) = \frac{2p}{x} J_p(x) - J_{p-1}(x)$$

Subtracting the results of Problem 25.15 from the results of Problem 25.14, we find that

$$0 = 2pJ_p(x) - xJ_{p-1}(x) - xJ_{p+1}(x)$$

Upon solving for $J_{p+1}(x)$, we obtain the desired result.

25.17. Show that $y = xJ_1(x)$ is a solution of $xy'' - y' - x^2J'_0(x) = 0$.

First note that $J_1(x)$ is a solution of Bessel's equation of order one:

$$x^2J''_1(x) + xJ'_1(x) + (x^2 - 1)J_1(x) = 0 \tag{1}$$

Now substitute $y = xJ_1(x)$ into the left side of the given differential equation:

$$x[xJ_1(x)]'' - [xJ_1(x)]' - x^2J'_0(x) = x[2J'_1(x) + xJ''_1(x)] - [J_1(x) + xJ'_1(x)] - x^2J'_0(x)$$

But $J'_0(x) = -J_1(x)$ [by (*1*) of Problem 25.14], so that the right-hand side becomes

$$x^2J''_1(x) + 2xJ'_1(x) - J_1(x) - xJ'_1(x) + x^2J_1(x) = x^2J''_1(x) + xJ'_1(x) + (x^2 - 1)J_1(x) = 0$$

the last equality following from (*1*).

25.18. Show that $y = \sqrt{x}\,J_{3/2}(x)$ is a solution of $x^2y'' + (x^2 - 2)y = 0$.

Observe that $J_{3/2}(x)$ is a solution of Bessel's equation of order $\frac{3}{2}$:

$$x^2J''_{3/2}(x) + xJ'_{3/2}(x) + \left(x^2 - \frac{9}{4}\right)J_{3/2}(x) = 0 \tag{1}$$

Now substitute $y = \sqrt{x}\,J_{3/2}(x)$ into the left side of the given differential equation, obtaining

$$x^2[\sqrt{x}\,J_{3/2}(x)]'' + (x^2 - 2)\sqrt{x}\,J_{3/2}(x)$$

$$= x^2\left[-\frac{1}{4}x^{-3/2}J_{3/2}(x) + x^{-1/2}J'_{3/2}(x) + x^{1/2}J''_{3/2}(x)\right] + (x^2 - 2)x^{1/2}J_{3/2}(x)$$

$$= \sqrt{x}\left[x^2J''_{3/2}(x) + xJ'_{3/2}(x) + \left(x^2 - \frac{9}{4}\right)J_{3/2}(x)\right] = 0$$

the last equality following from (*1*). Thus $\sqrt{x}\,J_{3/2}(x)$ satisfies the given differential equation.

Supplementary Problems

25.19. Find $\Gamma(2.6)$.

25.20. Find $\Gamma(-1.4)$.

25.21. Find $\Gamma(4.14)$.

25.22. Find $\Gamma(-2.6)$.

25.23. Find $\Gamma(-1.33)$.

25.24. Express $\int_0^\infty e^{-x^3}\,dx$ as a gamma function.

25.25. Evaluate $\int_0^\infty x^3e^{-x^2}\,dx$.

25.26. Prove that $\sum_{k=0}^{\infty} \frac{(-1)^k(2k)x^{2k-1}}{2^{2k-1}k!\,\Gamma(p+k)} = -\sum_{k=0}^{\infty} \frac{(-1)^kx^{2k+1}}{2^{2k}k!\,\Gamma(p+k+1)}$.

25.27. Prove that $\frac{d}{dx}[x^{-p}J_p(x)] = -x^{-p}J_{p+1}(x)$.
Hint: Use Problem 25.11.

25.28. Prove that $J_{p-1}(x) - J_{p+1}(x) = 2J'_p(x)$.

25.29. (*a*) Prove that the derivative of $(\frac{1}{2}x^2)[J_0^2(x) + J_1^2(x)]$ is $xJ_0^2(x)$.
Hint: Use (*1*) of Problem 25.13 and (*1*) of Problem 25.14.

(*b*) Evaluate $\int_0^1 xJ_0^2(x)\,dx$ in terms of Bessel functions.

25.30. Show that $y = xJ_n(x)$ is a solution of $x^2y'' - xy' + (1 + x^2 - n^2)y = 0$.

25.31. Show that $y = x^2J_2(x)$ is a solution of $xy'' - 3y' + xy = 0$.

Chapter 26

Graphical Methods for Solving First-Order Differential Equations

DIRECTION FIELDS

Graphical methods produce plots of solutions to first-order differential equations of the form

$$y' = f(x, y) \tag{26.1}$$

where the derivative appears only on the left side of the equation.

Example 26.1. (*a*) For the problem $y' = -y + x + 2$, we have $f(x, y) = -y + x + 2$. (*b*) For the problem $y' = y^2 + 1$, we have $f(x, y) = y^2 + 1$. (*c*) For the problem $y' = 3$, we have $f(x, y) \equiv 3$. Observe that in a particular problem, $f(x, y)$ may be independent of x, of y, or of x and y.

Equation (*26.1*) defines the slope of the solution curve $y(x)$ at any point (x, y) in the plane. A *line element* is a short line segment that begins at the point (x, y) and has a slope specified by (*26.1*); it represents an approximation to the solution curve through that point. A collection of line elements is a *direction field*. The graphs of solutions to (*26.1*) are generated from direction fields by drawing curves that pass through the points at which line elements are drawn and also are tangent to those line elements.

If the left side of Eq. (*26.1*) is set equal to a constant, the graph of the resulting equation is called an *isocline*. Different constants define different isoclines, and each isocline has the property that all line elements emanating from points on that isocline have the same slope, a slope equal to the constant that generated the isocline. When they are simple to draw, isoclines yield many line elements at once which is useful for constructing direction fields.

EULER'S METHOD

If an initial condition of the form

$$y(x_0) = y_0 \tag{26.2}$$

is also specified, then the only solution curve of Eq. (*26.1*) of interest is the one that passes through the initial point (x_0, y_0).

To obtain a graphical approximation to the solution curve of Eqs. (*26.1*) and (*26.2*), begin by constructing a line element at the inital point (x_0, y_0) and then continuing it for a short distance. Denote the terminal point of this line element as (x_1, y_1). Then construct a second line element at (x_1, y_1) and continue it a short distance. Denote the terminal point of this second line element as (x_2, y_2). Follow with a third line element constructed at (x_2, y_2) and continue it a short distance. The process proceeds iteratively and concludes when enough of the solution curve has been drawn to meet the needs of those concerned with the problem.

If the difference between successive x values are equal, that is, if for a specified constant h, $h = x_1 - x_0 = x_2 - x_1 = x_3 - x_2 = \dots$, then the graphical method given above for a first-order initial-value problem is known as Euler's method. It satisfies the formula

$$y_{n+1} = y_n + hf(x_n, y_n) \tag{26.3}$$

for $n = 1, 2, 3, \dots$. This formula is often written as

$$y_{n+1} = y_n + hy'_n \tag{26.4}$$

where
$$y'_n = f(x_n, y_n) \tag{26.5}$$
as required by Eq. (*26.1*).

STABILITY

The constant h in Eqs. (*26.3*) and (*26.4*) is called the *step-size,* and its value is arbitrary. In general, the smaller the step-size, the more accurate the approximate solution becomes at the price of more work to obtain that solution. Thus, the final choice of h may be a compromise between accuracy and effort. If h is chosen too large, then the approximate solution may not resemble the real solution at all, a condition known as *numerical instability.* To avoid numerical instability, Euler's method is repeated, each time with a step-size one half its previous value, until two successive approximations are close enough to each other to satisfy the needs of the solver.

Solved Problems

26.1. Construct a direction field for the differential equation $y' = 2y - x$.

Here $f(x, y) = 2y - x$.
At $x = 1$, $y = 1$, $f(1, 1) = 2(1) - 1 = 1$, equivalent to an angle of 45°.
At $x = 1$, $y = 2$, $f(1, 2) = 2(2) - 1 = 3$, equivalent to an angle of 71.6°.
At $x = 2$, $y = 1$, $f(2, 1) = 2(1) - 2 = 0$, equivalent to an angle of 0°.
At $x = 2$, $y = 2$, $f(2, 2) = 2(2) - 2 = 2$, equivalent to an angle of 63.4°.
At $x = 1$, $y = -1$, $f(1, -1) = 2(-1) - 1 = -3$, equivalent to an angle of −71.6°.
At $x = -2$, $y = -1$, $f(-2, -1) = 2(-1) - (-2) = 0$, equivalent to an angle of 0°.

Line elements at these points with their respective slopes are graphed in Fig. 26-1. Continuing in this manner we generate the more complete direction field shown in Fig. 26-2. To avoid confusion between line elements associated with the differential equation and axis markings, we deleted the axes in Fig. 26-2. The origin is at the center of the graph.

26.2. Describe the isoclines associated with the differential equation defined in Problem 26.1.

Isoclines are defined by setting $y' = c$, a constant. For the differential equation in Problem 26.1, we obtain
$$c = 2y - x \qquad \text{or} \qquad y = \tfrac{1}{2}x + \tfrac{1}{2}c$$
which is the equation for a straight line. Three such isoclines, corresponding to $c = 1$, $c = 0$, and $c = -1$, are graphed in Fig. 26-3. On the isocline corresponding to $c = 1$, every line element beginning on the isocline will have a slope of unity. On the isocline corresponding to $c = 0$, every line element beginning on the isocline will have a slope of zero. On the isocline corresponding to $c = -1$, every line element beginning on the isocline will have a slope of negative one. Some of these line elements are also drawn in Fig. 26-3.

26.3. Draw two solution curves to the differential equation given in Problem 26.1.

A direction field for this equation is given by Fig. 26-2. Two solution curves are shown in Fig. 26-4, one that passes through the point $(0, 0)$ and a second that passes through the point $(0, 2)$. Observe that each solution curve follows the flow of the line elements in the direction field.

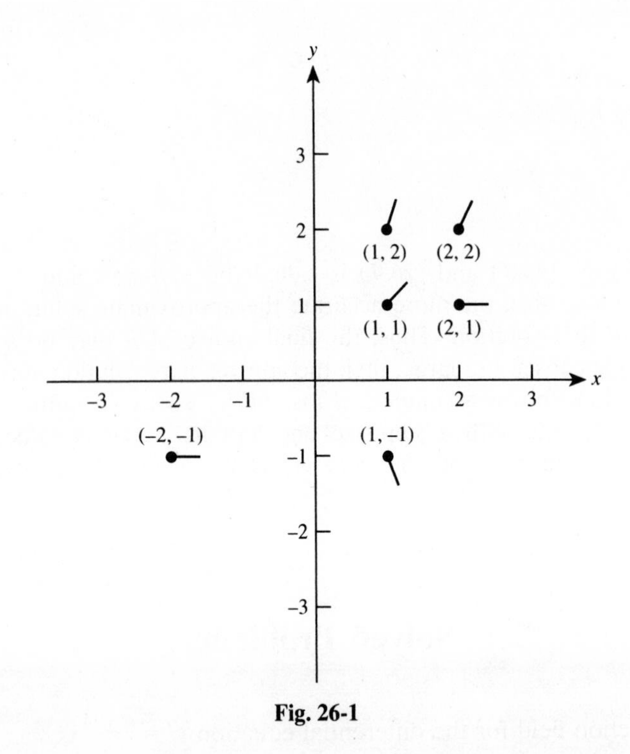

Fig. 26-1

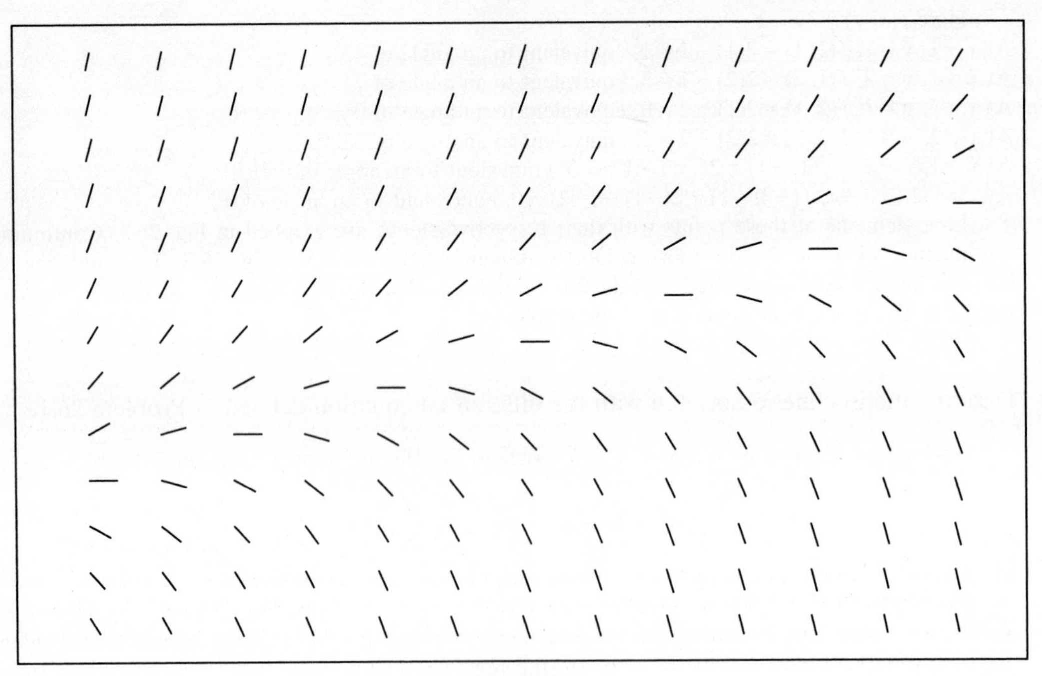

Fig. 26-2

26.4. Construct a direction field for the differential equation $y' = x^2 + y^2 - 1$.

Here $f(x, y) = x^2 + y^2 - 1$.
At $x = 0$, $y = 0$, $f(0, 0) = (0)^2 + (0)^2 - 1 = -1$, equivalent to an angle of $-45°$.
At $x = 1$, $y = 2$, $f(1, 2) = (1)^2 + (2)^2 - 1 = 4$, equivalent to an angle of $76.0°$.
At $x = -1$, $y = 2$, $f(-1, 2) = (-1)^2 + (2)^2 - 1 = 4$, equivalent to an angle of $76.0°$.

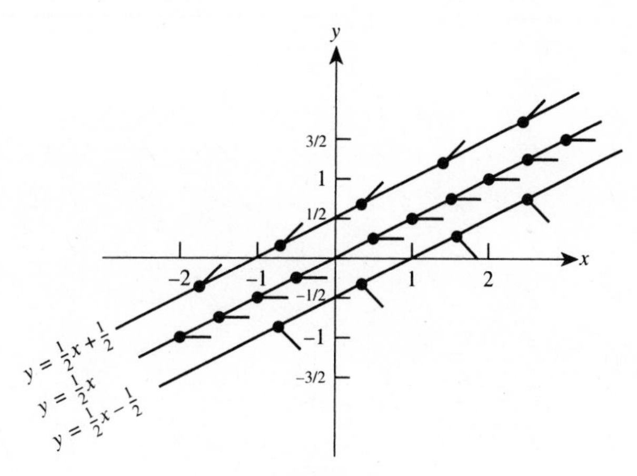

Fig. 26-3

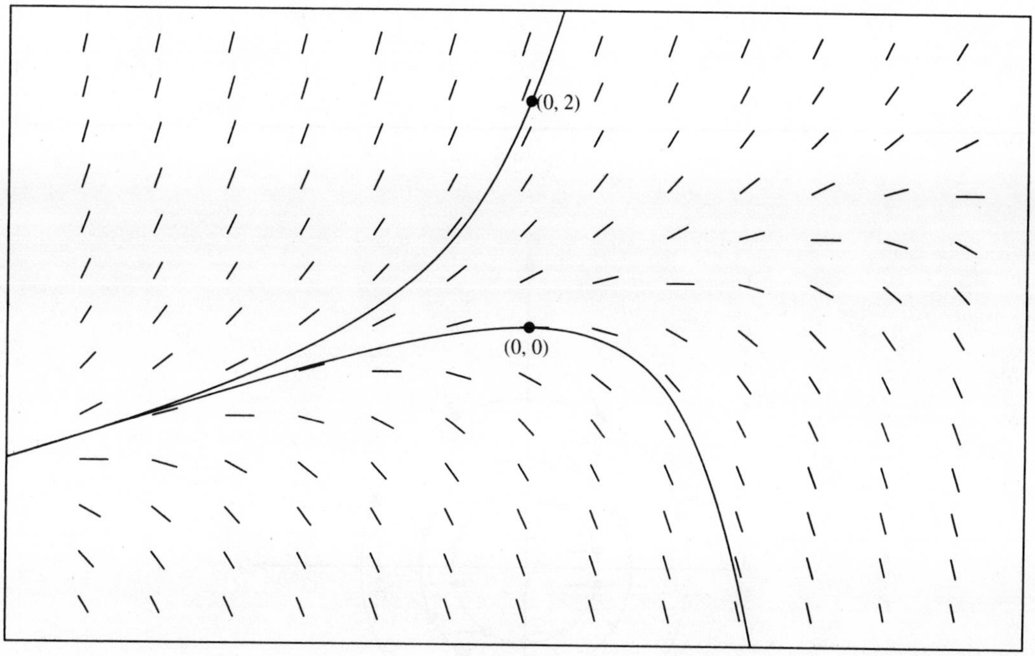

(0, 2)

(0, 0)

Fig. 26-4

At $x = 0.25$, $y = 0.5$, $f(0.25, 0.5) = (0.25)^2 + (0.5)^2 - 1 = -0.6875$, equivalent to an angle of $-34.5°$.
At $x = -0.3$, $y = -0.1$, $f(-0.3, -0.1) = (-0.3)^2 + (-0.1)^2 - 1 = -0.9$, equivalent to an angle of $-42.0°$.

Continuing in this manner, we generate Fig. 26-5. At each point, we graph a short line segment emanating from the point at the specified angle from the horizontal. To avoid confusion between line elements associated with the differential equation and axis markings, we deleted the axes in Fig. 26-5. The origin is at the center of the graph.

26.5. Describe the isoclines associated with the differential equation defined in Problem 26.4.

Isoclines are defined by setting $y' = c$, a constant. For the differential equation in Problem 26.4, we obtain $c = x^2 + y^2 - 1$ or $x^2 + y^2 = c + 1$, which is the equation for a circle centered at the origin. Three such isoclines, corresponding to $c = 4$, $c = 1$, and $c = 0$, are graphed in Fig. 26-6. On the isocline corresponding to $c = 4$, every line element beginning on the isocline will have a slope of four. On the

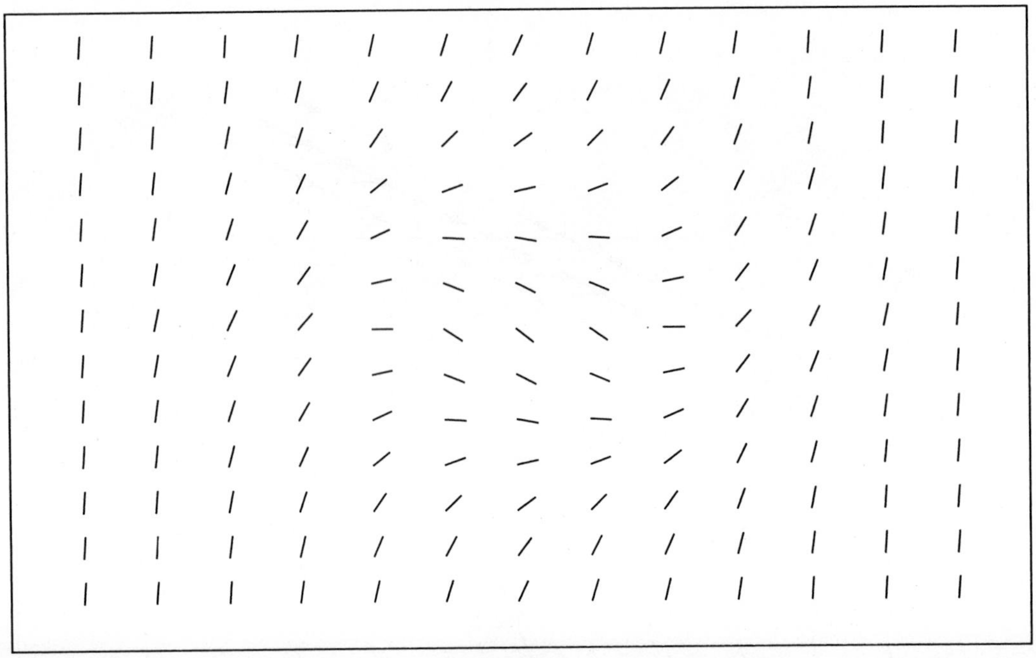

Fig. 26-5

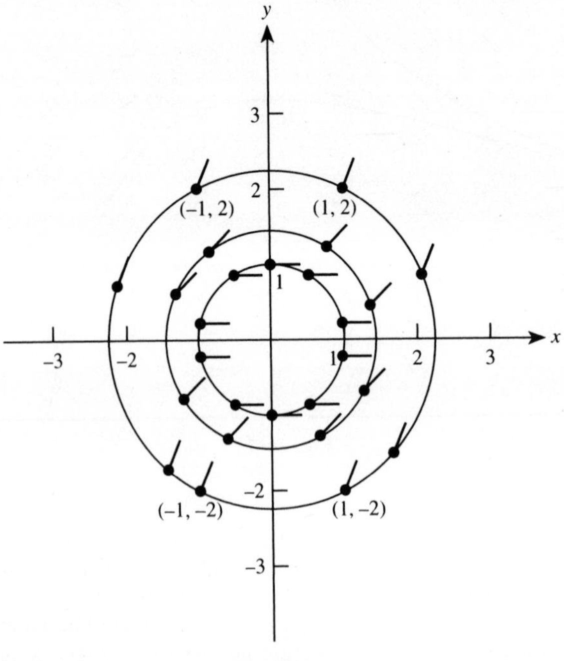

Fig. 26-6

isocline corresponding to $c = 1$, every line element beginning on the isocline will have a slope of unity. On the isocline corresponding to $c = 0$, every line element beginning on the isocline will have a slope of zero. Some of these line elements are also drawn in Fig. 26-6.

26.6. Draw three solution curves to the differential equation given in Problem 26.4.

A direction field for this equation is given by Fig. 26-5. Three solution curves are shown in Fig. 26-7, the top one passes through $(0, 1)$, the middle curve passes through $(0, 0)$, and the bottom curve passes through $(0, -1)$. Observe that each solution curve follows the flow of the line elements in the direction field.

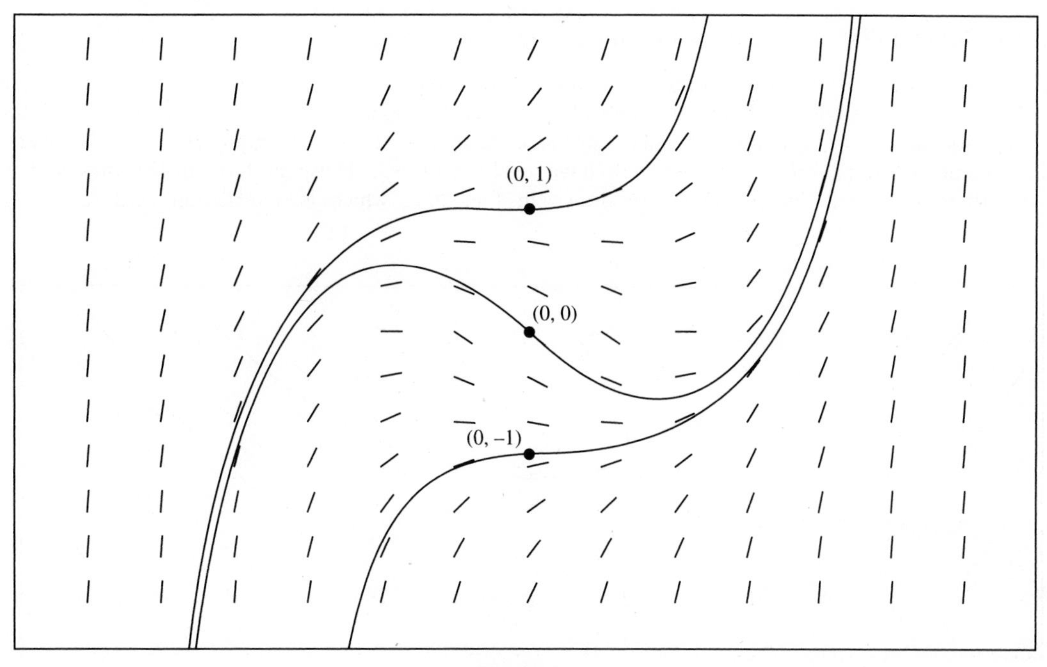

Fig. 26-7

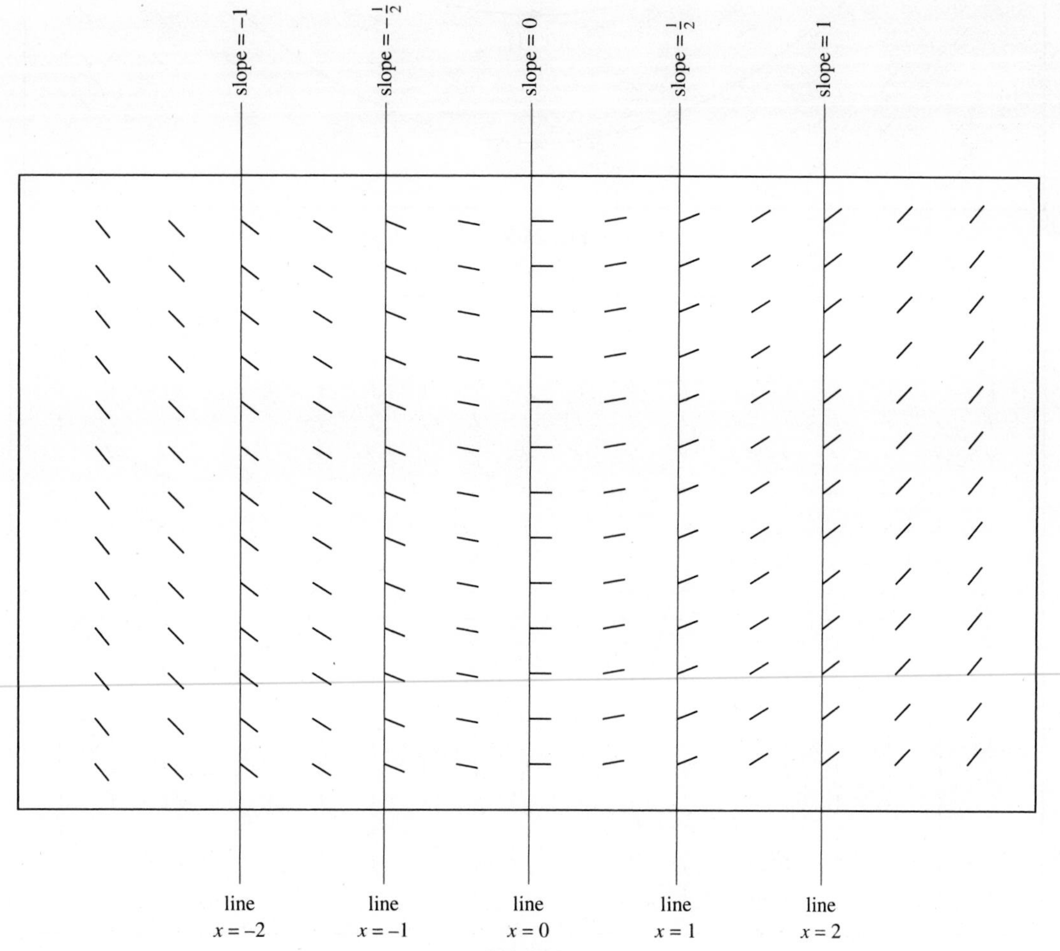

Fig. 26-8

26.7. Construct a direction field for the differential equation $y' = x/2$.

Isoclines are defined by setting $y' = c$, a constant. Doing so, we obtain $x = 2c$ which is the equation for a vertical straight line. On the isocline $x = 2$, corresponding to $c = 1$, every line element beginning on the isocline will have a slope of unity. On the isocline $x = -1$, corresponding to $c = -1/2$, every line element beginning on the isocline will have a slope of $-\frac{1}{2}$. These and other isoclines with some of their associated line elements are drawn in Fig. 26-8, which is a direction field for the given differential equation.

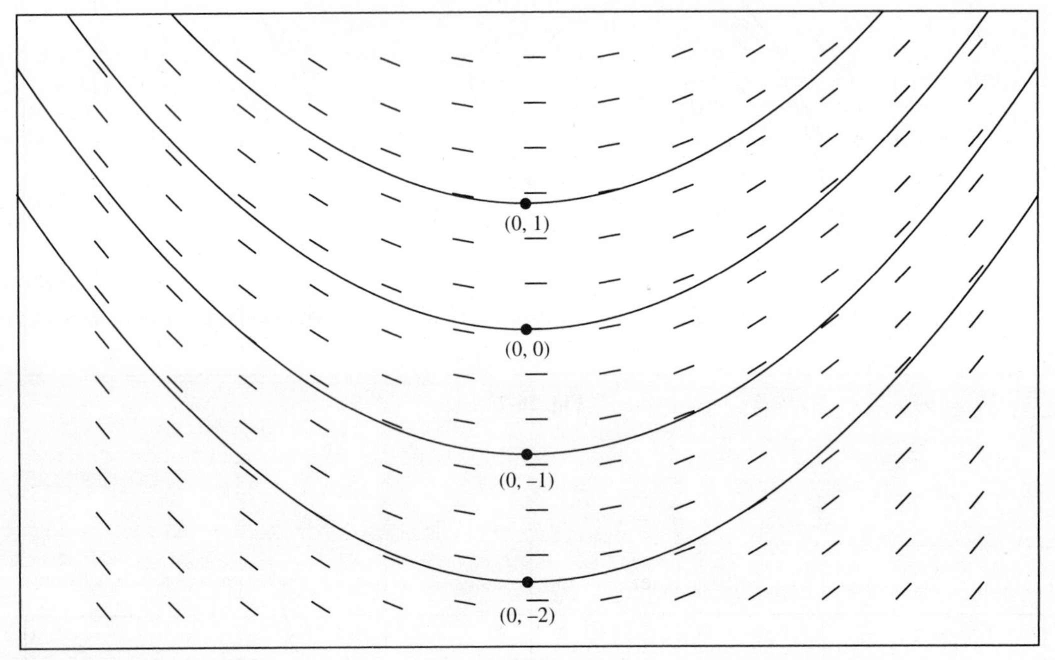

Fig. 26-9

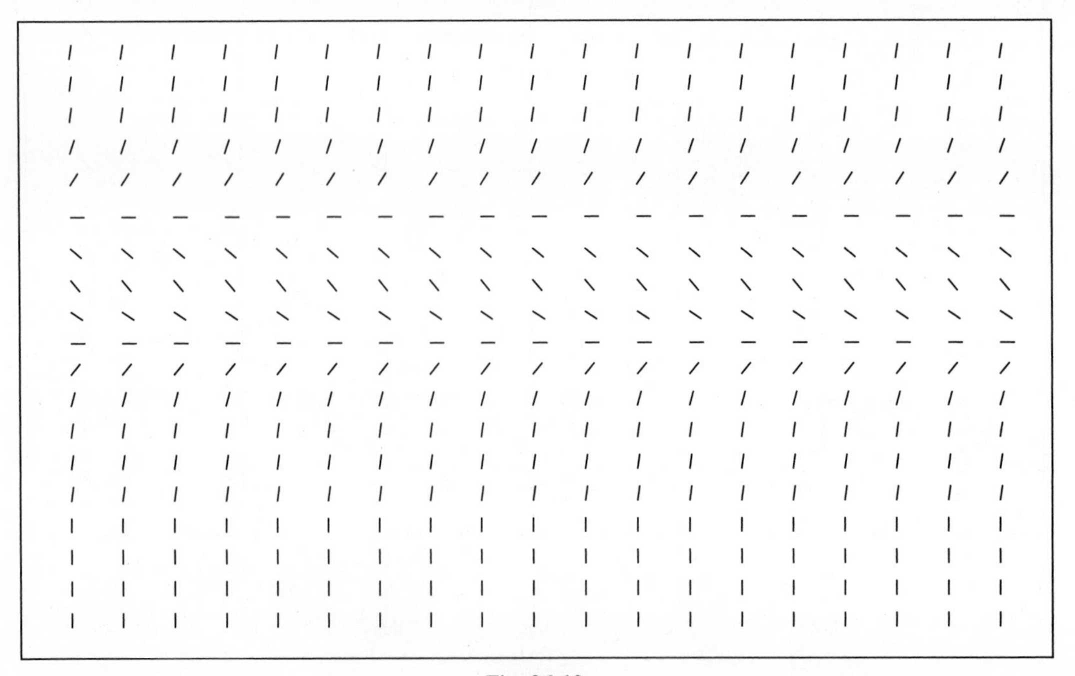

Fig. 26-10

26.8. Draw four solution curves to the differential equation given in Problem 26.7.

A direction field for this equation is given by Fig. 26-8. Four solution curves are drawn in Fig. 26-9, which from top to bottom pass through the points $(0, 1)$, $(0, 0)$, $(0, -1)$, and $(0, -2)$, respectively. Note that the differential equation is solved easily by direct integration. Its solution, $y = x^2/4 + k$, where k is a constant of integration, is a family of parabolas, one for each value of k.

26.9. Draw solution curves to the differential equation $y' = 5y(y - 1)$.

A direction field for this equation is given by Fig. 26-10. Two isoclines with line elements having zero slopes are the horizontal straight lines $y = 0$ and $y = 1$. Observe that solution curves have different shapes depending on whether they are above both of these isoclines, between them, or below them. A representative solution curve of each type is drawn in Fig 26-11(a) through (c).

26.10. Give a geometric derivation of Euler's method.

Assume that $y_n = y(x_n)$ has already been computed, so that y'_n is also known, via Eq. (*26.5*). Draw a straight line $l(x)$ emanating from (x_n, y_n) and having slope y'_n, and use $l(x)$ to approximate $y(x)$ on the interval $[x_n, x_{n+1}]$ (see Fig. 26-12). The value $l(x_{n+1})$ is taken to be y_{n+1}. Thus

$$l(x) = (y'_n)x + [y_n - (y'_n)x_n]$$

and

$$l(x_{n+1}) = (y'_n)x_{n+1} + [y_n - (y'_n)x_n]$$

$$= y_n + (y'_n)(x_{n+1} - x_n) = y_n + hy'_n$$

Hence, $y_{n+1} = y_n + hy'_n$, which is Euler's method.

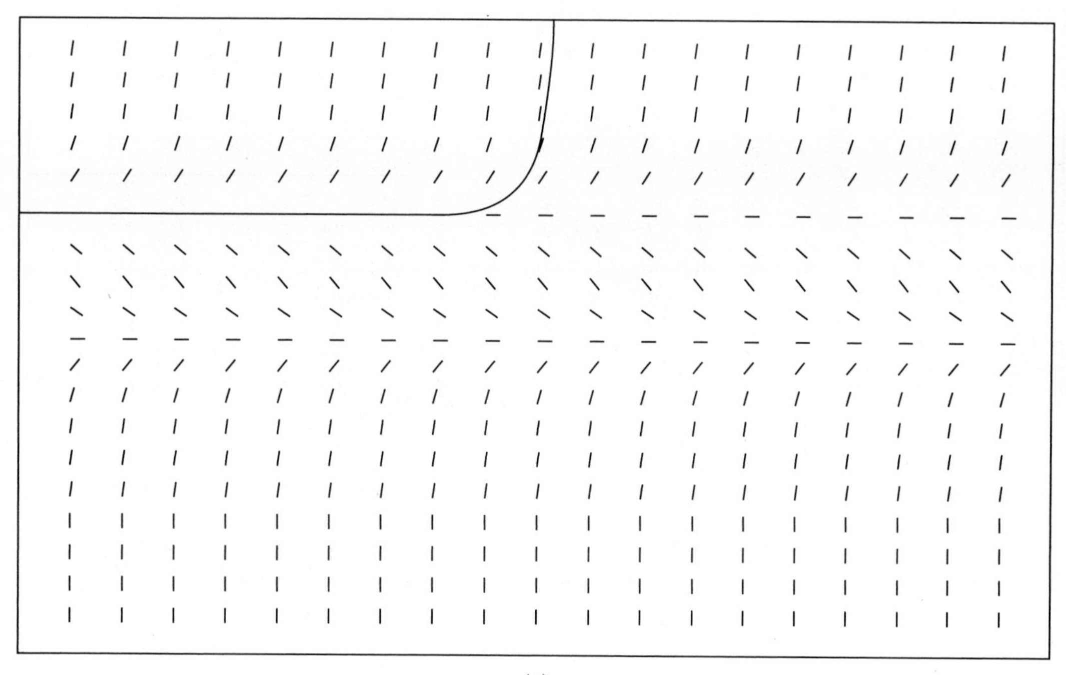

(a)

Fig. 26-11

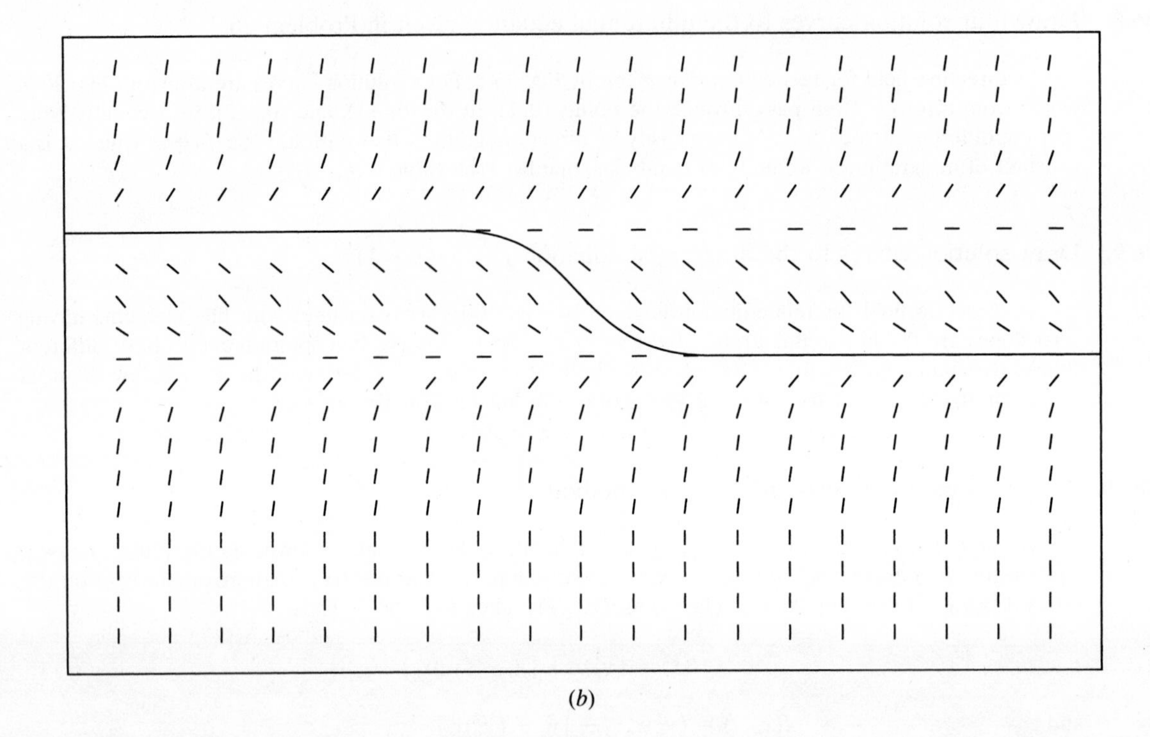

(*b*)

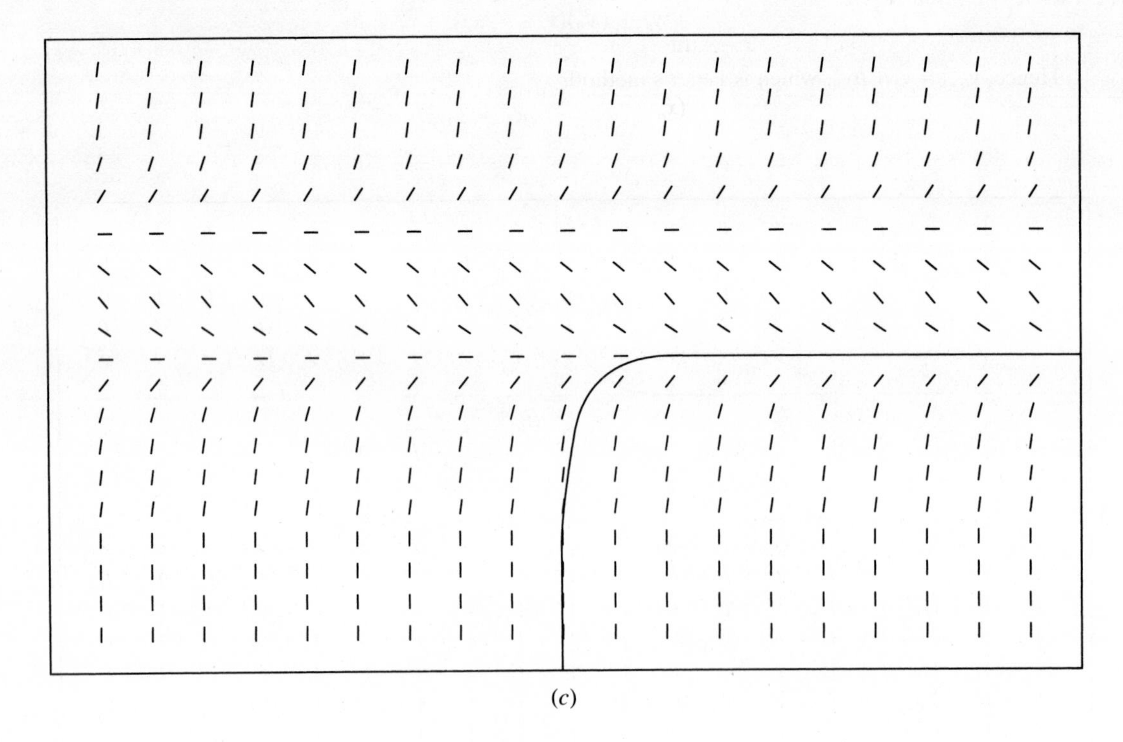

(*c*)

Fig. 26-11 (*cont.*)

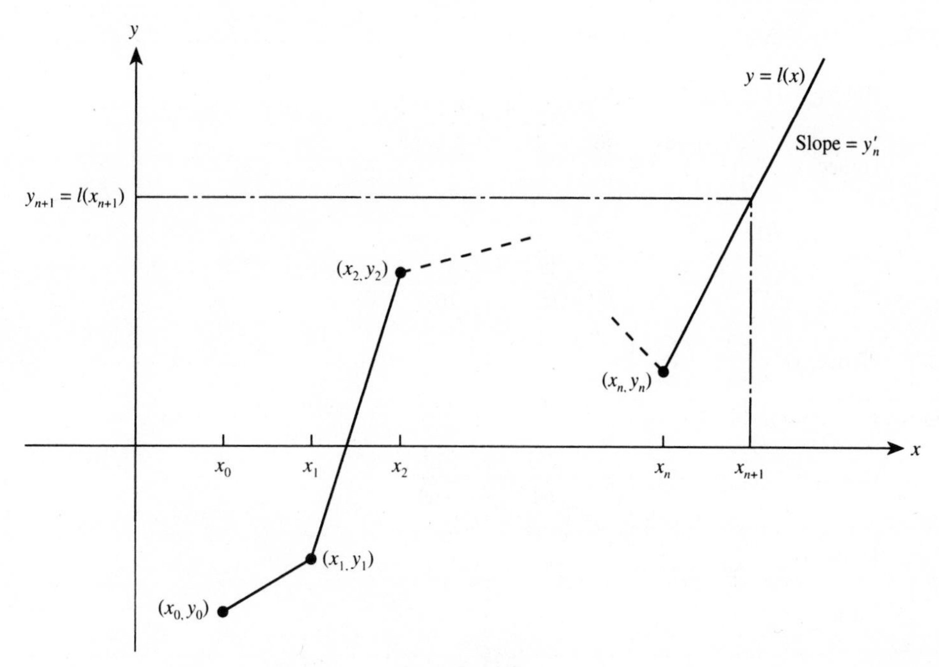

Fig. 26-12

26.11. Give an analytic derivation of Euler's method.

Let $Y(x)$ represent the true solution. Then, using the definition of the derivative, we have

$$Y'(x_n) = \lim_{\Delta x \to 0} \frac{Y(x_n + \Delta x) - Y(x_n)}{\Delta x}$$

If Δx is small, then

$$Y'(x_n) \simeq \frac{Y(x_n + \Delta x) - Y(x_n)}{\Delta x}$$

Setting $\Delta x = h$ and solving for $Y(x_n + \Delta x) = Y(x_{n+1})$, we obtain

$$Y(x_{n+1}) \simeq Y(x_n) + hY'(x_n) \tag{1}$$

Finally, if we use y_n and y'_n to approximate $Y(x_n)$ and $Y'(x_n)$, respectively, the right side of (1) can be used to approximate $Y(x_{n+1})$. Thus,

$$y_{n+1} = y_n + hy'_n$$

which is Euler's method.

26.12. Find $y(1)$ for $y' = y - x$; $y(0) = 2$, using Euler's method with $h = \frac{1}{4}$.

For this problem, $x_0 = 0$, $y_0 = 2$, and $f(x, y) = y - x$; so Eq. (26.5) becomes $y'_n = y_n - x_n$. Because $h = \frac{1}{4}$,

$$x_1 = x_0 + h = \frac{1}{4} \qquad x_2 = x_1 + h = \frac{1}{2} \qquad x_3 = x_2 + h = \frac{3}{4} \qquad x_4 = x_3 + h = 1$$

Using Eq. (26.4) with $n = 0, 1, 2, 3$ successively, we now compute the corresponding y-values.

$n = 0$: $y_1 = y_0 + hy'_0$

But $y'_0 = f(x_0, y_0) = y_0 - x_0 = 2 - 0 = 2$

Hence, $y_1 = 2 + \frac{1}{4}(2) = \frac{5}{2}$

$n = 1$: $y_2 = y_1 + hy'_1$

But $y'_1 = f(x_1, y_1) = y_1 - x_1 = \frac{5}{2} - \frac{1}{4} = \frac{9}{4}$

Hence, $y_2 = \frac{5}{2} + \frac{1}{4}\left(\frac{9}{4}\right) = \frac{49}{16}$

$n = 2$: $y_3 = y_2 + hy'_2$

But $y'_2 = f(x_2, y_2) = y_2 - x_2 = \frac{49}{16} - \frac{1}{2} = \frac{41}{16}$

Hence, $y_3 = \frac{49}{16} + \frac{1}{4}\left(\frac{41}{16}\right) = \frac{237}{64}$

$n = 3$: $y_4 = y_3 + hy'_3$

But $y'_3 = f(x_3, y_3) = y_3 - x_3 = \frac{237}{64} - \frac{3}{4} = \frac{189}{64}$

Hence, $y_4 = \frac{237}{64} + \frac{1}{4}\left(\frac{189}{64}\right) = \frac{1137}{256}$

Thus,

$$y(1) = y_4 = \frac{1137}{256} = 4.441$$

Note that the true solution is $Y(x) = e^x + x + 1$, so that $Y(1) = 4.718$. If we plot (x_n, y_n) for $n = 0, 1, 2, 3$, and 4, and then connect successive points with straight line segments, as done in Fig. 26-13, we have an approximation to the solution curve on $[0, 1]$ for this initial-value problem.

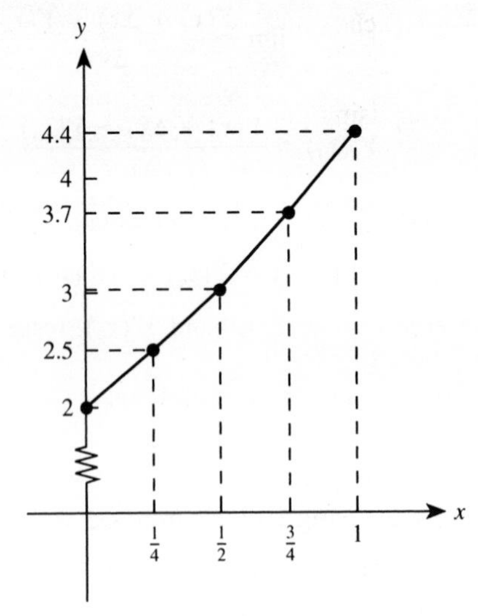

Fig. 26-13

26.13. Solve Problem 26.12 with $h = 0.1$.

With $h = 0.1$, $y(1) = y_{10}$. As before, $y'_n = y_n - x_n$. Then, using Eq. (*26.4*) with $n = 0, 1, \ldots, 9$ successively, we obtain

$n = 0$: $x_0 = 0$, $\quad y_0 = 2$, $\quad y'_0 = y_0 - x_0 = 2 - 0 = 2$

$y_1 = y_0 + hy'_0 = 2 + (0.1)(2) = 2.2$

$n = 1$: $x_1 = 0.1, \quad y_1 = 2.2, \quad y_1' = y_1 - x_1 = 2.2 - 0.1 = 2.1$
$y_2 = y_1 + hy_1' = 2.2 + (0.1)(2.1) = 2.41$

$n = 2$: $x_2 = 0.2, \quad y_2 = 2.41, \quad y_2' = y_2 - x_2 = 2.41 - 0.2 = 2.21$
$y_3 = y_2 + hy_2' = 2.41 + (0.1)(2.21) = 2.631$

$n = 3$: $x_3 = 0.3, \quad y_3 = 2.631, \quad y_3' = y_3 - x_3 = 2.631 - 0.3 = 2.331$
$y_4 = y_3 + hy_3' = 2.631 + (0.1)(2.331) = 2.864$

$n = 4$: $x_4 = 0.4, \quad y_4 = 2.864, \quad y_4' = y_4 - x_4 = 2.864 - 0.4 = 2.464$
$y_5 = y_4 + hy_4' = 2.864 + (0.1)(2.464) = 3.110$

$n = 5$: $x_5 = 0.5, \quad y_5 = 3.110, \quad y_5' = y_5 - x_5 = 3.110 - 0.5 = 2.610$
$y_6 = y_5 + hy_5' = 3.110 + (0.1)(2.610) = 3.371$

$n = 6$: $x_6 = 0.6, \quad y_6 = 3.371, \quad y_6' = y_6 - x_6 = 3.371 - 0.6 = 2.771$
$y_7 = y_6 + hy_6' = 3.371 + (0.1)(2.771) = 3.648$

$n = 7$: $x_7 = 0.7, \quad y_7 = 3.648, \quad y_7' = y_7 - x_7 = 3.648 - 0.7 = 2.948$
$y_8 = y_7 + hy_7' = 3.648 + (0.1)(2.948) = 3.943$

$n = 8$: $x_8 = 0.8, \quad y_8 = 3.943, \quad y_8' = y_8 - x_8 = 3.943 - 0.8 = 3.143$
$y_9 = y_8 + hy_8' = 3.943 + (0.1)(3.143) = 4.257$

$n = 9$: $x_9 = 0.9, \quad y_9 = 4.257, \quad y_9' = y_9 - x_9 = 4.257 - 0.9 = 3.357$
$y_{10} = y_9 + hy_9' = 4.257 + (0.1)(3.357) = 4.593$

The above results are displayed in Table 26-1: For comparison, Table 26-1 also contains results for $h = 0.05$, $h = 0.01$, and $h = 0.005$, with all computations rounded to four decimal places. Note that more accurate results are obtained when smaller values of h are used.

If we plot (x_n, y_n) for integer values of n between 0 and 10, inclusively, and then connect successive points with straight line segments, we would generate a graph almost indistinguishable from Fig. 26-13, because graphical accuracy with the chosen scales on the axes is limited to one decimal place.

26.14. Find $y(0.5)$ for $y' = y$; $y(0) = 1$, using Euler's method with $h = 0.1$.

For this problem, $f(x, y) = y$, $x_0 = 0$, and $y_0 = 1$; hence, from Eq. (*26.5*), $y_n' = f(x_n, y_n) = y_n$. With $h = 0.1$, $y(0.5) = y_5$. Then, using Eq. (*26.4*) with $n = 0, 1, 2, 3, 4$ successively, we obtain

$n = 0$: $x_0 = 0, \quad y_0 = 1, \quad y_0' = y_0 = 1$
$y_1 = y_0 + hy_0' = 1 + (0.1)(1) = 1.1$

$n = 1$: $x_1 = 0.1, \quad y_1 = 1.1, \quad y_1' = y_1 = 1.1$
$y_2 = y_1 + hy_1' = 1.1 + (0.1)(1.1) = 1.21$

$n = 2$: $x_2 = 0.2, \quad y_2 = 1.21, \quad y_2' = y_2 = 1.21$
$y_3 = y_2 + hy_2' = 1.21 + (0.1)(1.21) = 1.331$

$n = 3$: $x_3 = 0.3, \quad y_3 = 1.331, \quad y_3' = y_3 = 1.331$
$y_4 = y_3 + hy_3' = 1.331 + (0.1)(1.331) = 1.464$

$n = 4$: $x_4 = 0.4, \quad y_4 = 1.464, \quad y_4' = y_4 = 1.464$
$y_5 = y_4 + hy_4' = 1.464 + (0.1)(1.464) = 1.610$

Thus, $y(0.5) = y_5 = 1.610$. Note that since the true solution is $Y(x) = e^x$, $Y(0.5) = e^{0.5} = 1.649$.

26.15. Find $y(1)$ for $y' = y$; $y(0) = 1$, using Euler's method with $h = 0.1$.

We proceed exactly as in Problem 26.14, except that we now calculate through $n = 9$. The results of these computations are given in Table 26-2. For comparison, Table 26-2 also contains results for $h = 0.05$, $h = 0.001$, and $h = 0.005$, with all calculations rounded to four decimal places.

Table 26-1

Method: EULER'S METHOD					
Problem: $y' = y - x;\ y(0) = 2$					
x_n	y_n				True solution $Y(x) = e^x + x + 1$
	$h = 0.1$	$h = 0.05$	$h = 0.01$	$h = 0.005$	
0.0	2.0000	2.0000	2.0000	2.0000	2.0000
0.1	2.2000	2.2025	2.2046	2.2049	2.2052
0.2	2.4100	2.4155	2.4202	2.4208	2.4214
0.3	2.6310	2.6401	2.6478	2.6489	2.6499
0.4	2.8641	2.8775	2.8889	2.8903	2.8918
0.5	3.1105	3.1289	3.1446	3.1467	3.1487
0.6	3.3716	3.3959	3.4167	3.4194	3.4221
0.7	3.6487	3.6799	3.7068	3.7102	3.7138
0.8	3.9436	3.9829	4.0167	4.0211	4.0255
0.9	4.2579	4.3066	4.3486	4.3541	4.3596
1.0	4.5937	4.6533	4.7048	4.7115	4.7183

Table 26-2

Method: EULER'S METHOD					
Problem: $y' = y;\ y(0) = 1$					
x_n	y_n				True solution $Y(x) = e^x$
	$h = 0.1$	$h = 0.05$	$h = 0.01$	$h = 0.005$	
0.0	1.0000	1.0000	1.0000	1.0000	1.0000
0.1	1.1000	1.1025	1.1046	1.1049	1.1052
0.2	1.2100	1.2155	1.2202	1.2208	1.2214
0.3	1.3310	1.3401	1.3478	1.3489	1.3499
0.4	1.4641	1.4775	1.4889	1.4903	1.4918
0.5	1.6105	1.6289	1.6446	1.6467	1.6487
0.6	1.7716	1.7959	1.8167	1.8194	1.8221
0.7	1.9487	1.9799	2.0068	2.0102	2.0138
0.8	2.1436	2.1829	2.2167	2.2211	2.2255
0.9	2.3579	2.4066	2.4486	2.4541	2.4596
1.0	2.5937	2.6533	2.7048	2.7115	2.7183

26.16. Find $y(1)$ for $y' = y^2 + 1$; $y(0) = 0$, using Euler's method with $h = 0.1$.

Here, $f(x, y) = y^2 + 1$, $x_0 = 0$, and $y_0 = 0$; hence, from Eq. (*26.5*), $y'_n = f(x_n, y_n) = (y_n)^2 + 1$. With $h = 0.1$, $y(1) = y_{10}$. Then, using Eq. (*26.4*) with $n = 0, 1, \ldots, 9$ successively, we obtain

n = 0: $x_0 = 0, \quad y_0 = 0, \quad y'_0 = (y_0)^2 + 1 = (0)^2 + 1 = 1$
$y_1 = y_0 + hy'_0 = 0 + (0.1)(1) = 0.1$

n = 1: $x_1 = 0.1, \quad y_1 = 0.1, \quad y'_1 = (y_1)^2 + 1 = (0.1)^2 + 1 = 1.01$
$y_2 = y_1 + hy'_1 = 0.1 + (0.1)(1.01) = 0.201$

n = 2: $x_2 = 0.2, \quad y_2 = 0.201$
$y'_2 = (y_2)^2 + 1 = (0.201)^2 + 1 = 1.040$
$y_3 = y_2 + hy'_2 = 0.201 + (0.1)(1.040) = 0.305$

n = 3: $x_3 = 0.3, \quad y_3 = 0.305$
$y'_3 = (y_3)^2 + 1 = (0.305)^2 + 1 = 1.093$
$y_4 = y_3 + hy'_3 = 0.305 + (0.1)(1.093) = 0.414$

n = 4: $x_4 = 0.4, \quad y_4 = 0.414$
$y'_4 = (y_4)^2 + 1 = (0.414)^2 + 1 = 1.171$
$y_5 = y_4 + hy'_4 = 0.414 + (0.1)(1.171) = 0.531$

Continuing in this manner, we find that $y_{10} = 1.396$.

The calculations are found in Table 26-3. For comparison, Table 26-3 also contains results for $h = 0.05$, $h = 0.01$, and $h = 0.005$, with all computations rounded to four decimal places. The true solution to this problem is $Y(x) = \tan x$, hence $Y(1) = 1.557$.

Table 26-3

Method: EULER'S METHOD					
Problem: $y' = y^2 + 1$; $y(0) = 0$					
x_n	y_n				True solution
	$h = 0.1$	$h = 0.05$	$h = 0.01$	$h = 0.005$	$Y(x) = \tan x$
0.0	0.0000	0.0000	0.0000	0.0000	0.0000
0.1	0.1000	0.1001	0.1003	0.1003	0.1003
0.2	0.2010	0.2018	0.2025	0.2026	0.2027
0.3	0.3050	0.3070	0.3088	0.3091	0.3093
0.4	0.4143	0.4183	0.4218	0.4223	0.4228
0.5	0.5315	0.5384	0.5446	0.5455	0.5463
0.6	0.6598	0.6711	0.6814	0.6827	0.6841
0.7	0.8033	0.8212	0.8378	0.8400	0.8423
0.8	0.9678	0.9959	1.0223	1.0260	1.0296
0.9	1.1615	1.2055	1.2482	1.2541	1.2602
1.0	1.3964	1.4663	1.5370	1.5470	1.5574

Supplementary Problems

Direction fields are provided in Problems 26.17 through 26.22. Sketch some of the solution curves.

26.17. See Fig. 26-14.

26.18. See Fig. 26-15.

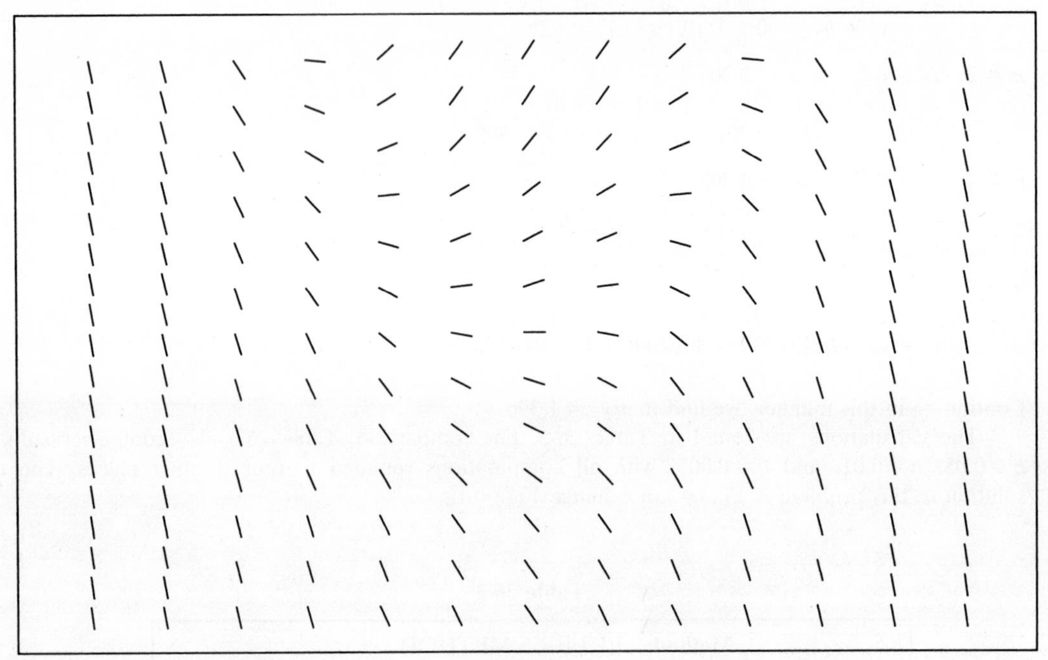

Fig. 26-14

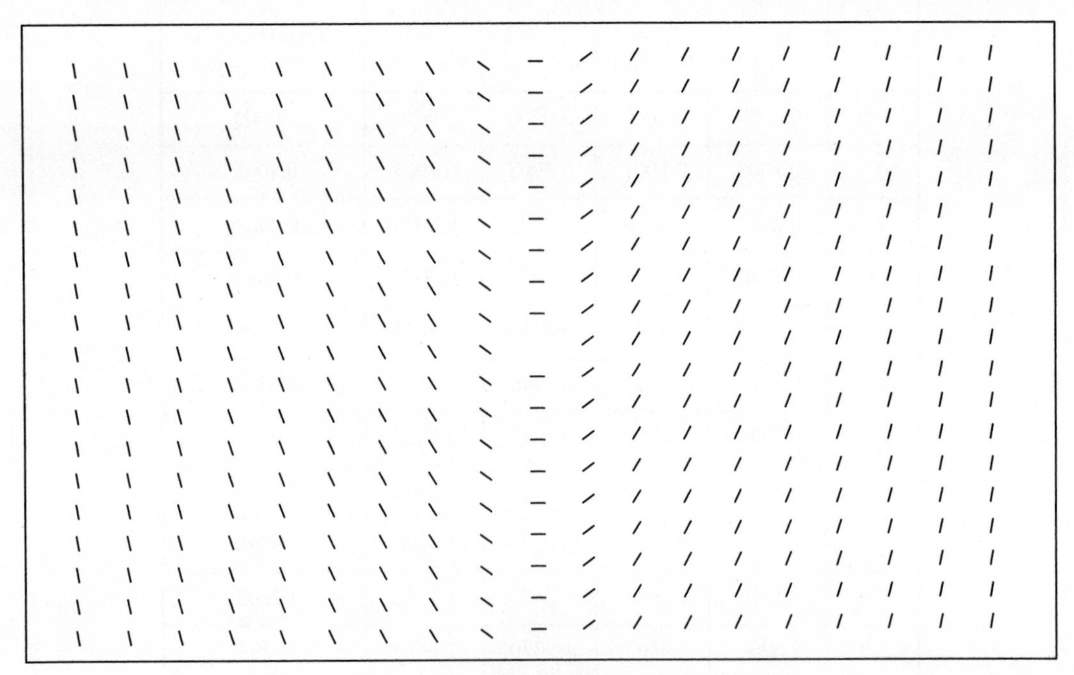

Fig. 26-15

26.19. See Fig. 26-16.

26.20. See Fig. 26-17.

26.21. See Fig. 26-18.

26.22. See Fig. 26-19.

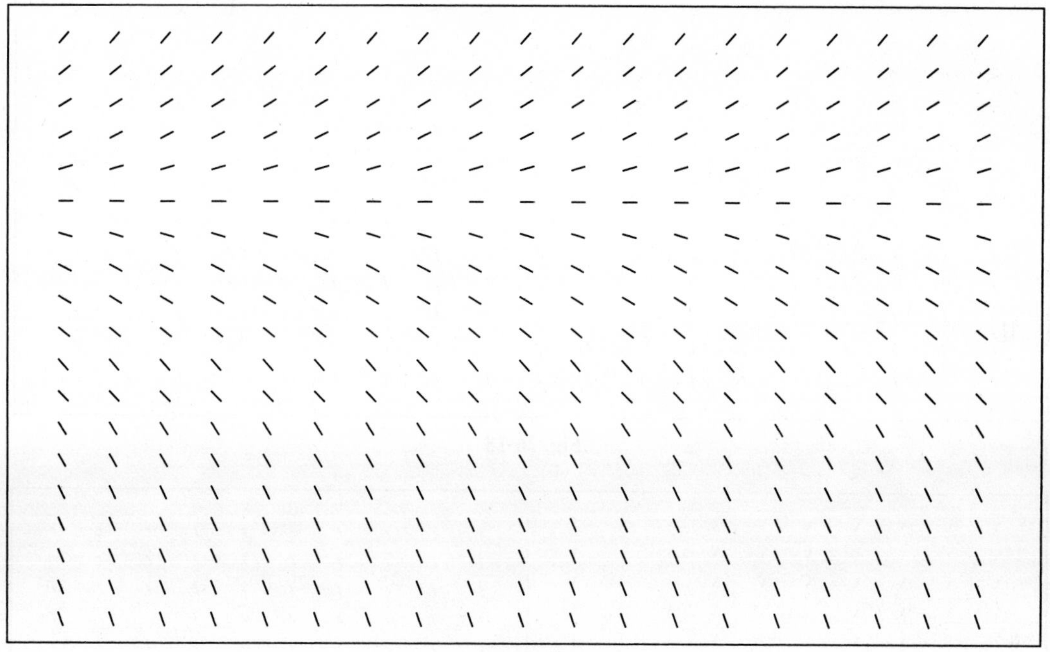

Fig. 26-16

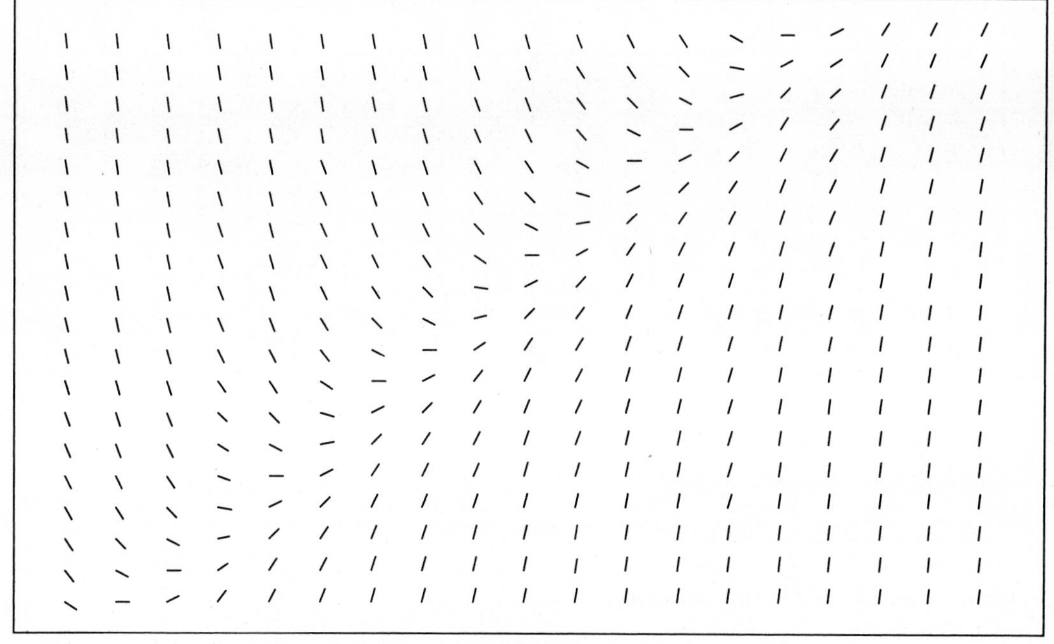

Fig. 26-17

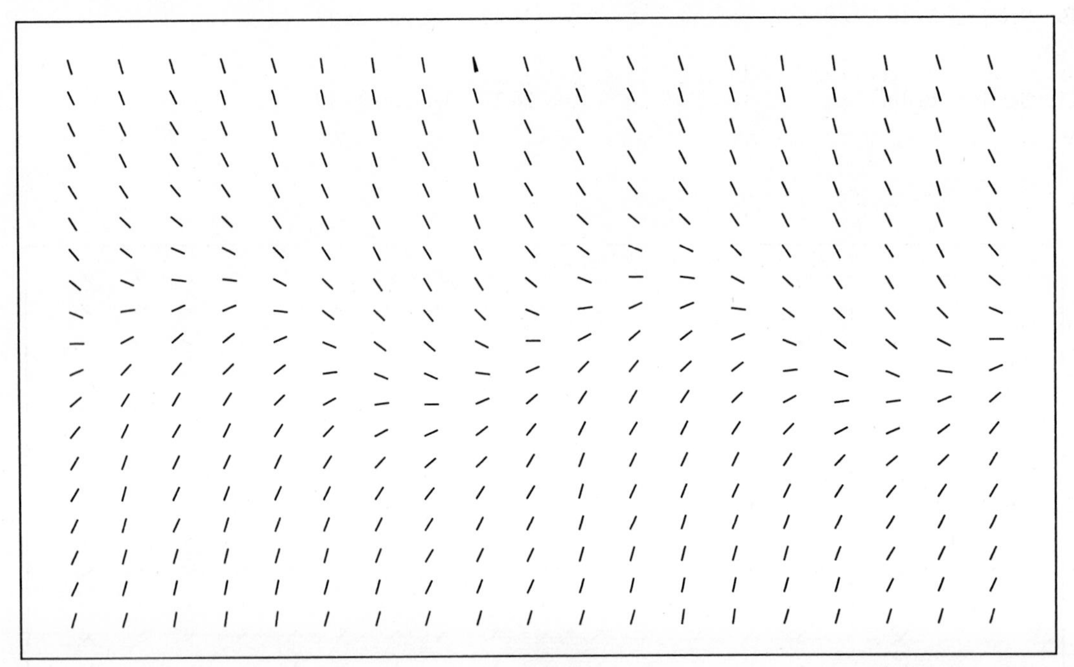

Fig. 26-18

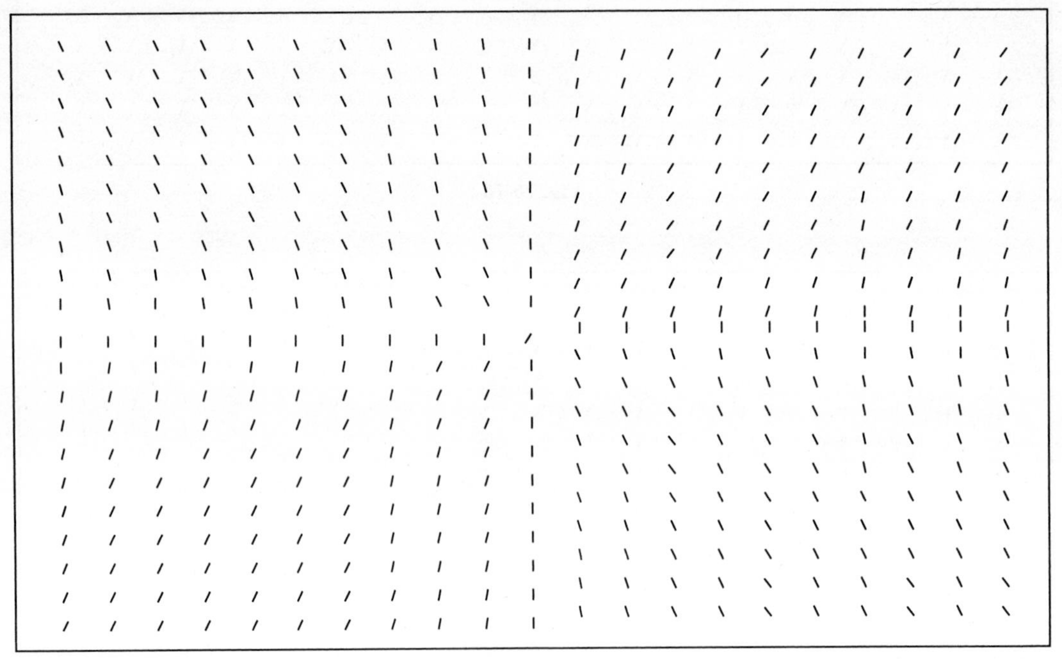

Fig. 26-19

26.23. Draw a direction field for the equation $y' = x - y + 1$.

26.24. Describe the isoclines for the equation in Problem 26.23.

26.25. Draw a direction field for the equation $y' = 2x$.

26.26. Describe the isoclines for the equation in Problem 26.25.

26.27. Draw a direction field for the equation $y' = y - 1$.

26.28. Describe the isoclines for the equation in Problem 26.27.

26.29. Draw a direction field for the equation $y' = y - x^2$.

26.30. Describe the isoclines for the equation in Problem 26.29.

26.31. Draw a direction field for the equation $y' = \sin x - y$.

26.32. Describe the isoclines for the equation in Problem 26.31.

26.33. Find $y(1.0)$ for $y' = -y$; $y(0) = 1$, using Euler's method with $h = 0.1$.

26.34. Find $y(0.5)$ for $y' = 2x$; $y(0) = 0$, using Euler's method with $h = 0.1$.

26.35. Find $y(0.5)$ for $y' = -y + x + 2$; $y(0) = 2$, using Euler's method with $h = 0.1$.

26.36. Find $y(0.5)$ for $y' = 4x^3$; $y(0) = 0$, using Euler's method with $h = 0.1$.

Chapter 27

Numerical Methods for Solving First-Order Differential Equations

GENERAL REMARKS

A *numerical method* for solving an initial-value problem is a procedure that produces approximate solutions at particular points using only the operations of addition, subtraction, multiplication, division, and functional evaluations. In this chapter, we consider only first-order initial-value problems of the form

$$y' = f(x, y); \qquad y(x_0) = y_0 \tag{27.1}$$

Generalizations to higher-order problems are given in Chapter 28. Each numerical method will produce approximate solutions at the points $x_0, x_1, x_2, \ldots,$ where the difference between any two successive x-values is a constant step-size h; that is, $x_{n+1} - x_n = h$ $(n = 0, 1, 2, \ldots)$. Remarks made in Chapter 26 on the step-size remain valid for all the numerical methods presented below.

The approximate solution at x_n will be designated by $y(x_n)$, or simply y_n. The true solution at x_n will be denoted by either $Y(x_n)$ or Y_n. Note that once y_n is known, Eq. (*27.1*) can be used to obtain y'_n as

$$y'_n = f(x_n, y_n) \tag{27.2}$$

The simplest numerical method is Euler's method, described in Chapter 27.

A *predictor-corrector* method is a set of two equations for y_{n+1}. The first equation, called the *predictor,* is used to predict (obtain a first approximation to) y_{n+1}; the second equation, called the *corrector,* is then used to obtain a corrected value (second approximation) to y_{n+1}. In general, the corrector depends on the predicted value.

MODIFIED EULER'S METHOD

This is a simple predictor-corrector method that uses Euler's method (see Chapter 26) as the predictor and then uses the average value of y' at both the left and right end points of the interval $[x_n, x_{n+1}]$ $(n = 0, 1, 2, \ldots)$ as the slope of the line element approximation to the solution over that interval. The resulting equations are:

$$\text{predictor:} \qquad y_{n+1} = y_n + hy'_n$$

$$\text{corrector:} \qquad y_{n+1} = y_n + \frac{h}{2}(y'_{n+1} + y'_n)$$

For notational convenience, we designate the predicted value of y_{n+1} by py_{n+1}. It then follows from Eq. (*27.2*) that

$$py'_{n+1} = f(x_{n+1}, py_{n+1}) \tag{27.3}$$

The modified Euler's method becomes

$$\begin{aligned} &\text{predictor:} \qquad py_{n+1} = y_n + hy'_n \\ &\text{corrector:} \qquad y_{n+1} = y_n + \frac{h}{2}(py'_{n+1} + y'_n) \end{aligned} \tag{27.4}$$

RUNGE–KUTTA METHOD

$$y_{n+1} = y_n + \frac{1}{6}(k_1 + 2k_2 + 2k_3 + k_4) \tag{27.5}$$

where

$$k_1 = hf(x_n, y_n)$$

$$k_2 = hf\left(x_n + \frac{1}{2}h, y_n + \frac{1}{2}k_1\right)$$

$$k_3 = hf\left(x_n + \frac{1}{2}h, y_n + \frac{1}{2}k_2\right)$$

$$k_4 = hf(x_n + h, y_n + k_3)$$

This is *not* a predictor-corrector method.

ADAMS–BASHFORTH–MOULTON METHOD

$$\begin{aligned} \text{predictor:} \quad & py_{n+1} = y_n + \frac{h}{24}(55y'_n - 59y'_{n-1} + 37y'_{n-2} - 9y'_{n-3}) \\ \text{corrector:} \quad & y_{n+1} = y_n + \frac{h}{24}(9py'_{n+1} + 19y'_n - 5y'_{n-1} + y'_{n-2}) \end{aligned} \tag{27.6}$$

MILNE'S METHOD

$$\begin{aligned} \text{predictor:} \quad & py_{n+1} = y_{n-3} + \frac{4h}{3}(2y'_n - y'_{n-1} + 2y'_{n-2}) \\ \text{corrector:} \quad & y_{n+1} = y_{n-1} + \frac{h}{3}(py'_{n+1} + 4y'_n + y'_{n-1}) \end{aligned} \tag{27.7}$$

STARTING VALUES

The Adams–Bashforth–Moulton method and Milne's method require information at y_0, y_1, y_2, and y_3 to start. The first of these values is given by the initial condition in Eq. (*27.1*). The other three starting values are gotten by the Runge–Kutta method.

ORDER OF A NUMERICAL METHOD

A numerical method is of *order n*, where n is a positive integer, if the method is exact for polynomials of degree n or less. In other words, if the true solution of an initial-value problem is a polynomial of degree n or less, then the approximate solution and the true solution will be identical for a method of order n.

In general, the higher the order, the more accurate the method. Euler's method, Eq. (*26.4*), is of order one, the modified Euler's method, Eq. (*27.4*), is of order two, while the other three, Eqs. (*27.5*) through (*27.7*), are fourth-order methods.

Solved Problems

27.1. Use the modified Euler's method to solve $y' = y - x$; $y(0) = 2$ on the interval $[0, 1]$ with $h = 0.1$.

Here $f(x, y) = y - x$, $x_0 = 0$, and $y_0 = 2$. From Eq. (27.2) we have $y'_0 = f(0, 2) = 2 - 0 = 2$. Then using Eqs. (27.4) and (27.3), we compute

$n = 0$: $x_1 = 0.1$

$$py_1 = y_0 + hy'_0 = 2 + 0.1(2) = 2.2$$
$$py'_1 = f(x_1, py_1) = f(0.1, 2.2) = 2.2 - 0.1 = 2.1$$
$$y_1 = y_0 + \frac{h}{2}(py'_1 + y'_0) = 2 + 0.05(2.1 + 2) = 2.205$$
$$y'_1 = f(x_1, y_1) = f(0.1, 2.205) = 2.205 - 0.1 = 2.105$$

$n = 1$: $x_2 = 0.2$

$$py_2 = y_1 + hy'_1 = 2.205 + 0.1(2.105) = 2.4155$$
$$py'_2 = f(x_2, py_2) = f(0.2, 2.4155) = 2.4155 - 0.2 = 2.2155$$
$$y_2 = y_1 + \frac{h}{2}(py'_2 + y'_1) = 2.205 + 0.05(2.2155 + 2.105) = 2.421025$$
$$y'_2 = f(x_2, y_2) = f(0.2, 2.421025) = 2.421025 - 0.2 = 2.221025$$

$n = 2$: $x_3 = 0.3$

$$py_3 = y_2 + hy'_2 = 2.421025 + 0.1(2.221025) = 2.6431275$$
$$py'_3 = f(x_3, py_3) = f(0.3, 2.6431275) = 2.6431275 - 0.3 = 2.3431275$$
$$y_3 = y_2 + \frac{h}{2}(py'_3 + y'_2) = 2.421025 + 0.05(2.3431275 + 2.221025) = 2.6492326$$
$$y'_3 = f(x_3, y_3) = f(0.3, 2.6492326) = 2.6492326 - 0.3 = 2.3492326$$

Continuing in this manner, we generate Table 27-1. Compare it to Table 26-1.

Table 27-1

Method: MODIFIED EULER'S METHOD			
Problem: $y' = y - x;\ y(0) = 2$			
x_n	$h = 0.1$		True solution
	py_n	y_n	$Y(x) = e^x + x + 1$
0.0	—	2.0000000	2.0000000
0.1	2.2000000	2.2050000	2.2051709
0.2	2.4155000	2.4210250	2.4214028
0.3	2.6431275	2.6492326	2.6498588
0.4	2.8841559	2.8909021	2.8918247
0.5	3.1399923	3.1474468	3.1487213
0.6	3.4121914	3.4204287	3.4221188
0.7	3.7024715	3.7115737	3.7137527
0.8	4.0127311	4.0227889	4.0255409
0.9	4.3450678	4.3561818	4.3596031
1.0	4.7017999	4.7140808	4.7182818

27.2. Use the modified Euler's method to solve $y' = y^2 + 1;\ y(0) = 0$ on the interval $[0, 1]$ with $h = 0.1$.

Here $f(x, y) = y^2 + 1$, $x_0 = 0$, and $y_0 = 0$. From (*27.2*) we have $y_0' = f(0, 0) = (0)^2 + 1 = 1$. Then using (*27.4*) and (*27.3*), we compute

$$
\begin{aligned}
\mathbf{n = 0:}\ & x_1 = 0.1 \\
& py_1 = y_0 + hy_0' = 0 + 0.1(1) = 0.1 \\
& py_1' = f(x_1, py_1) = f(0.1, 0.1) = (0.1)^2 + 1 = 1.01 \\
& y_1 = y_0 + (h/2)(py_1' + y_0') = 0 + 0.05(1.01 + 1) = 0.1005 \\
& y_1' = f(x_1, y_1) = f(0.1, 0.1005) = (0.1005)^2 + 1 = 1.0101003 \\
\mathbf{n = 1:}\ & x_2 = 0.2 \\
& py_2 = y_1 + hy_1' = 0.1005 + 0.1(1.0101003) = 0.2015100 \\
& py_2' = f(x_2, py_2) = f(0.2, 0.2015100) = (0.2015100)^2 + 1 = 1.0406063 \\
& y_2 = y_1 + (h/2)(py_2' + y_1') = 0.1005 + 0.05(1.0406063) + 1.0101002) = 0.2030353 \\
& y_2' = f(x_2, y_2) = f(0.2, 0.2030353) = (0.2030353)^2 + 1 = 1.0412233 \\
\mathbf{n = 2:}\ & x_3 = 0.3 \\
& py_3 = y_2 + hy_2' = 0.2030353 + 0.1(1.0412233) = 0.3071577 \\
& py_3' = f(x_3, py_3) = f(0.3, 0.3071577) = (0.3071577)^2 + 1 = 1.0943458 \\
& y_3 + y_2 + (h/2)(py_3' + y_2') = 0.2030353 + 0.05(1.0943458 + 1.0412233) = 0.3098138 \\
& y_3' = f(x_3, y_3) = f(0.3, 0.3098138) = (0.3098138)^2 + 1 = 1.0959846
\end{aligned}
$$

Continuing in this manner, we generate Table 27-2. Compare it to Table 26-3.

Table 27-2

Method: MODIFIED EULER'S METHOD			
Problem: $y' = y^2 + 1;\ y(0) = 0$			
x_n	$h = 0.1$		True solution $Y(x) = \tan x$
	py_n	y_n	
0.0	—	0.0000000	0.0000000
0.1	0.1000000	0.1005000	0.1003347
0.2	0.2015100	0.2030353	0.2027100
0.3	0.3071577	0.3098138	0.3093363
0.4	0.4194122	0.4234083	0.4227932
0.5	0.5413358	0.5470243	0.5463025
0.6	0.6769479	0.6848990	0.6841368
0.7	0.8318077	0.8429485	0.8422884
0.8	1.0140048	1.0298869	1.0296386
0.9	1.2359536	1·2592993	1.2601582
1.0	1.5178828	1.5537895	1.5574077

27.3. Find $y(1.6)$ for $y' = 2x$; $y(1) = 1$ using the modified Euler's method with $h = 0.2$.

Here $f(x, y) = 2x$, $x_0 = 1$, and $y_0 = 2$. From Eq. (*27.2*) we have $y'_0 = f(1, 2) = 2(1) = 2$. Then using (*27.4*) and (*27.3*), we compute

$$\begin{aligned}
\mathbf{n = 0:}\ \ & x_1 = x_0 + h = 1 + 0.2 = 1.2 \\
& py_1 = y_0 + hy'_0 = 1 + 0.2(2) = 1.4 \\
& py'_1 = f(x_1, py_1) = f(1.2, 1.4) = 2(1.2) = 2.4 \\
& y_1 = y_0 + (h/2)(py'_1 + y'_0) = 1 + 0.1(2.4 + 2) = 1.44 \\
& y'_1 = f(x_1, y_1) = f(1.2, 1.44) = 2(1.2) = 2.4 \\
\mathbf{n = 1:}\ \ & x_2 = x_1 + h = 1.2 + 0.2 = 1.4 \\
& py_2 = y_1 + hy'_1 = 1.44 + 0.2(2.4) = 1.92 \\
& py'_2 = f(x_2, py_2) = f(1.4, 1.92) = 2(1.4) = 2.8 \\
& y_2 = y_1 + (h/2)(py'_2 + y'_1) = 1.44 + 0.1(2.8 + 2.4) = 1.96 \\
& y'_2 = f(x_2, y_2) = f(1.4, 1.96) = 2(1.4) = 2.8 \\
\mathbf{n = 2:}\ \ & x_3 = x_2 + h = 1.4 + 0.2 = 1.6 \\
& py_3 = y_2 + hy'_2 = 1.96 + 0.2(2.8) = 2.52 \\
& py'_3 = f(x_3, py_3) = f(1.6, 2.52) = 2(1.6) = 3.2 \\
& y_3 = y_2 + (h/2)(py'_3 + y'_2) = 1.96 + 0.1(3.2 + 2.8) = 2.56
\end{aligned}$$

The true solution is $Y(x) = x^2$; hence $Y(1.6) = y(1.6) = (1.6)^2 = 2.56$. Since the true solution is a second-degree polynomial and the modified Euler's method is a second-order method, this agreement is expected.

27.4. Use the Runge–Kutta method to solve $y' = y - x$; $y(0) = 2$ on the interval $[0, 1]$ with $h = 0.1$.

Here $f(x, y) = y - x$. Using Eq. (*27.5*) with $n = 0, 1, \ldots, 9$, we compute

$$\begin{aligned}
\mathbf{n = 0:}\ \ & x_0 = 0, \qquad y_0 = 2 \\
& k_1 = hf(x_0, y_0) = hf(0, 2) = (0.1)(2 - 0) = 0.2 \\
& k_2 = hf(x_0 + \tfrac{1}{2}h, y_0 + \tfrac{1}{2}k_1) = hf[0 + \tfrac{1}{2}(0.1), 2 + \tfrac{1}{2}(0.2)] \\
& \quad = hf(0.05, 2.1) = (0.1)(2.1 - 0.05) = 0.205 \\
& k_3 = hf(x_0 + \tfrac{1}{2}h, y_0 + \tfrac{1}{2}k_2) = hf[0 + \tfrac{1}{2}(0.1), 2 + \tfrac{1}{2}(0.205)] \\
& \quad = hf(0.05, 2.103) = (0.1)(2.103 - 0.05) = 0.205 \\
& k_4 = hf(x_0 + h, y_0 + k_3) = hf(0 + 0.1, 2 + 0.205) \\
& \quad = hf(0.1, 2.205) = (0.1)(2.205 - 0.1) = 0.211 \\
& y_1 = y_0 + \tfrac{1}{6}(k_1 + 2k_2 + 2k_3 + k_4) \\
& \quad = 2 + \tfrac{1}{6}[0.2 + 2(0.205) + 2(0.205) + 0.211] = 2.205 \\
\mathbf{n = 1:}\ \ & x_1 = 0.1, \qquad y_1 = 2.205 \\
& k_1 = hf(x_1, y_1) = hf(0.1, 2.205) = (0.1)(2.205 - 0.1) = 0.211 \\
& k_2 = hf(x_1 + \tfrac{1}{2}h, y_1 + \tfrac{1}{2}k_1) = hf[0.1 + \tfrac{1}{2}(0.1), 2.205 + \tfrac{1}{2}(0.211)] \\
& \quad = hf(0.15, 2.311) = (0.1)(2.311 - 0.15) = 0.216 \\
& k_3 = hf(x_1 + \tfrac{1}{2}h, y_1 + \tfrac{1}{2}k_2) = hf[0.1 + \tfrac{1}{2}(0.1), 2.205 + \tfrac{1}{2}(0.216)]
\end{aligned}$$

$= hf(0.15, 2.313) = (0.1)(2.313 - 0.15) = 0.216$

$k_4 = hf(x_1 + h, y_1 + k_3) = hf(0.1 + 0.1, 2.205 + 0.216)$

$= hf(0.2, 2.421) = (0.1)(2.421 - 0.2) = 0.222$

$y_2 = y_1 + \frac{1}{6}(k_1 + 2k_2 + 2k_3 + k_4)$

$= 2.205 + \frac{1}{6}[0.211 + 2(0.216) + 2(0.216) + 0.222] = 2.421$

$n = 2$: $x_2 = 0.2, \quad y_2 = 2.421$

$k_1 = hf(x_2, y_2) = hf(0.2, 2.421) = (0.1)(2.421 - 0.2) = 0.222$

$k_2 = hf(x_2 + \frac{1}{2}h, y_2 + \frac{1}{2}k_1) = hf[0.2 + \frac{1}{2}(0.1), 2.421 + \frac{1}{2}(0.222)]$

$= hf(0.25, 2.532) = (0.1)(2.532 - 0.25) = 0.228$

$k_3 = hf(x_2 + \frac{1}{2}h, y_2 + \frac{1}{2}k_2) = hf[0.2 + \frac{1}{2}(0.1), 2.421 + \frac{1}{2}(0.228)]$

$= hf(0.25, 2.535) = (0.1)(2.535) - 0.25) = 0.229$

$k_4 = hf(x_2 + h, y_2 + k_3) = hf(0.2 + 0.1, 2.421 + 0.229)$

$= hf(0.3, 2.650) = (0.1)(2.650 - 0.3) = 0.235$

$y_3 = y_2 + \frac{1}{6}(k_1 + 2k_2 + 2k_3 + k_4)$

$= 2.421 + \frac{1}{6}[0.222 + 2(0.228) + 2(0.229) + 0.235] = 2.650$

Continuing in this manner, we generate Table 27-3. Compare it with Table 27-1.

Table 27-3

Method: RUNGE–KUTTA METHOD		
Problem: $y' = y - x$; $y(0) = 2$		
x_n	$h = 0.1$ y_n	True solution $Y(x) = e^x + x + 1$
0.0	2.0000000	2.0000000
0.1	2.2051708	2.2051709
0.2	2.4214026	2.4214028
0.3	2.6498585	2.6498588
0.4	2.8918242	2.8918247
0.5	3.1487206	3.1487213
0.6	3.4221180	3.4221188
0.7	3.7137516	3.7137527
0.8	4.0255396	4.0255409
0.9	4.3596014	4.3596031
1.0	4.7182797	4.7182818

27.5. Use the Runge–Kutta method to solve $y' = y$; $y(0) = 1$ on the interval $[0, 1]$ with $h = 0.1$.

Here $f(x, y) = y$. Using Eq. (*27.5*) with $n = 0, 1, \ldots, 9$, we compute

$n = 0$: $x_0 = 0, \quad y_0 = 1$

$k_1 = hf(x_0, y_0) = hf(0, 1) = (0.1)(1) = 0.1$

$k_2 = hf(x_0 + \frac{1}{2}h, y_0 + \frac{1}{2}k_1) = hf[0 + \frac{1}{2}(0.1), 1 + \frac{1}{2}(0.1)]$

$$= hf(0.05, 1.05) = (0.1)(1.05) = 0.105$$

$$k_3 = hf(x_0 + \tfrac{1}{2}h, y_0 + \tfrac{1}{2}k_2) = hf[0 + \tfrac{1}{2}(0.1), 1 + \tfrac{1}{2}(0.105)]$$

$$= hf(0.05, 1.053) = (0.1)(1.053) = 0.105$$

$$k_4 = hf(x_0 + h, y_0 + k_3) = hf(0 + 0.1, 1 + 0.105)$$

$$= hf(0.1, 1.105) = (0.1)(1.105) = 0.111$$

$$y_1 = y_0 + \tfrac{1}{6}(k_1 + 2k_2 + 2k_3 + k_4)$$

$$= 1 + \tfrac{1}{6}[0.1 + 2(0.105) + 2(0.105) + 0.111] = 1.105$$

$n = 1$: $x_1 = 0.1, \quad y_1 = 1.105$

$$k_1 = hf(x_1, y_1) = hf(0.1, 1.105) = (0.1)(1.105) = 0.111$$

$$k_2 = hf(x_1 + \tfrac{1}{2}h, y_1 + \tfrac{1}{2}k_1) = hf[0.1 + \tfrac{1}{2}(0.1), 1.105 + \tfrac{1}{2}(0.111)]$$

$$= hf(0.15, 1.161) = (0.1)(1.161) = 0.116$$

$$k_3 = hf(x_1 + \tfrac{1}{2}h, y_1 + \tfrac{1}{2}k_2) = hf[0.1 + \tfrac{1}{2}(0.1), 1.105 + \tfrac{1}{2}(0.116)]$$

$$= hf(0.15, 1.163) = (0.1)(1.163) = 0.116$$

$$k_4 = hf(x_1 + h, y_1 + k_3) = hf(0.1 + 0.1, 1.105 + 0.116)$$

$$= hf(0.2, 1.221) = (0.1)(1.221) = 0.122$$

$$y_2 = y_1 + \tfrac{1}{6}(k_1 + 2k_2 + 2k_3 + k_4)$$

$$= 1.105 + \tfrac{1}{6}[0.111 + 2(0.116) + 2(0.116) + 0.122] = 1.221$$

$n = 2$: $x_2 = 0.2, \quad y_2 = 1.221$

$$k_1 = hf(x_2, y_2) = hf(0.2, 1.221) = (0.1)(1.221) = 0.122$$

$$k_2 = hf(x_2 + \tfrac{1}{2}h, y_2 + \tfrac{1}{2}k_1) = hf[0.2 + \tfrac{1}{2}(0.1), 1.221 + \tfrac{1}{2}(0.122)]$$

$$= hf(0.25, 1.282) = (0.1)(1.282) = 0.128$$

$$k_3 = hf(x_2 + \tfrac{1}{2}h, y_2 + \tfrac{1}{2}k_2) = hf[0.2 + \tfrac{1}{2}(0.1), 1.221 + \tfrac{1}{2}(0.128)]$$

$$= hf(0.25, 1.285) = (0.1)(1.285) = 0.129$$

$$k_4 = hf(x_2 + h, y_2 + k_3) = hf(0.2 + 0.1, 1.221 + 0.129)$$

$$= hf(0.3, 1.350) = (0.1)(1.350) = 0.135$$

$$y_3 = y_2 + \tfrac{1}{6}(k_1 + 2k_2 + 2k_3 + k_4)$$

$$= 1.221 + \tfrac{1}{6}[0.122 + 2(0.128) + 2(0.129) + 0.135] = 1.350$$

Continuing in this manner, we generate Table 27-4.

27.6. Use the Runge–Kutta method to solve $y' = y^2 + 1$; $y(0) = 0$ on the interval $[0, 1]$ with $h = 0.1$.

Here $f(x, y) = y^2 + 1$. Using Eq. (*27.5*), we compute

$n = 0$: $x_0 = 0, \quad y_0 = 0$

$$k_1 = hf(x_0, y_0) = hf(0, 0) = (0.1)[(0)^2 + 1] = 0.1$$

$$k_2 = hf(x_0 + \tfrac{1}{2}h, y_0 + \tfrac{1}{2}k_1) + hf[0 + \tfrac{1}{2}(0.1), 0 + \tfrac{1}{2}(0.1)]$$

$$= hf(0.05, 0.05) = (0.1)[(0.05)^2 + 1] = 0.1$$

$$k_3 = hf(x_0 + \tfrac{1}{2}h, y_0 + \tfrac{1}{2}k_2) = hf[0 + \tfrac{1}{2}(0.1), 0 + \tfrac{1}{2}(0.1)]$$

$$= hf(0.05, 0.05) = (0.1)[(0.05)^2 + 1] = 0.1$$

$$k_4 = hf(x_0 + h, y_0 + k_3) = hf[0 + 0.1, 0 + 0.1]$$

$$= hf(0.1, 0.1) = (0.1)[(0.1)^2 + 1] = 0.101$$

$$y_1 = y_0 + \tfrac{1}{6}(k_1 + 2k_2 + 2k_3 + k_4)$$

$$= 0 + \tfrac{1}{6}[0.1 + 2(0.1) + 2(0.1) + 0.101] = 0.1$$

Table 27-4

Method: RUNGE–KUTTA METHOD		
Problem: $y' = y;\ y(0) = 1$		
x_n	$h = 0.1$ y_n	True solution $Y(x) = e^x$
0.0	1.0000000	1.0000000
0.1	1.1051708	1.1051709
0.2	1.2214026	1.2214028
0.3	1.3498585	1.3498588
0.4	1.4918242	1.4918247
0.5	1.6487206	1.6487213
0.6	1.8221180	1.8221188
0.7	2.0137516	2.0137527
0.8	2.2255396	2.2255409
0.9	2.4596014	2.4596031
1.0	2.7182797	2.7182818

$n = 1$: $x_1 = 0.1, \qquad y_1 = 0.1$

$$\begin{aligned}
k_1 &= hf(x_1, y_1) = hf(0.1, 0.1) = (0.1)[(0.1)^2 + 1] = 0.101 \\
k_2 &= hf(x_1 + \tfrac{1}{2}h, y_1 + \tfrac{1}{2}k_1) = hf[0.1 + \tfrac{1}{2}(0.1), (0.1) + \tfrac{1}{2}(0.101)] \\
&= hf(0.15, 0.151) = (0.1)[(0.151)^2 + 1] = 0.102 \\
k_3 &= hf(x_1 + \tfrac{1}{2}h, y_1 + \tfrac{1}{2}k_2) = hf[0.1 + \tfrac{1}{2}(0.1), (0.1) + \tfrac{1}{2}(0.102)] \\
&= hf(0.15, 0.151) = (0.1)[(0.151)^2 + 1] = 0.102 \\
k_4 &= hf(x_1 + h, y_1 + k_3) = hf(0.1 + 0.1, 0.1 + 0.102) \\
&= hf(0.2, 0.202) = (0.1)[(0.202)^2 + 1] = 0.104 \\
y_2 &= y_1 + \tfrac{1}{6}(k_1 + 2k_2 + 2k_3 + k_4) \\
&= 0.1 + \tfrac{1}{6}[0.101 + 2(0.102) + 2(0.102) + 0.104] = 0.202
\end{aligned}$$

$n = 2$: $x_2 = 0.2, \qquad y_2 = 0.202$

$$\begin{aligned}
k_1 &= hf(x_2, y_2) = hf(0.2, 0.202) = (0.1)[(0.202)^2 + 1] = 0.104 \\
k_2 &= hf(x_2 + \tfrac{1}{2}h, y_2 + \tfrac{1}{2}k_1) = hf[0.2 + \tfrac{1}{2}(0.1), 0.202 + \tfrac{1}{2}(0.104)] \\
&= hf(0.25, 0.254) = (0.1)[(0.254)^2 + 1] = 0.106 \\
k_3 &= hf(x_2 + \tfrac{1}{2}h, y_2 + \tfrac{1}{2}k_2) = hf[0.2 + \tfrac{1}{2}(0.1), 0.202 + \tfrac{1}{2}(0.106)] \\
&= hf(0.25, 0.255) = (0.1)[(0.255)^2 + 1] = 0.107 \\
k_4 &= hf(x_2 + h, y_2 + k_3) = hf(0.2 + 0.1, 0.202 + 0.107) \\
&= hf(0.3, 0.309) = (0.1)[(0.309)^2 + 1] = 0.110 \\
y_3 &= y_2 + \tfrac{1}{6}(k_1 + 2k_2 + 2k_3 + k_4) \\
&= 0.202 + \tfrac{1}{6}[0.104 + 2(0.106) + 2(0.107) + 0.110] = 0.309
\end{aligned}$$

Continuing in this manner, we generate Table 27-5.

Table 27-5

Method: RUNGE–KUTTA METHOD		
Problem: $y' = y^2 + 1;\ y(0) = 0$		
x_n	$h = 0.1$ y_n	True solution $Y(x) = \tan x$
0.0	0.0000000	0.0000000
0.1	0.1003346	0.1003347
0.2	0.2027099	0.2027100
0.3	0.3093360	0.3093363
0.4	0.4227930	0.4227932
0.5	0.5463023	0.5463025
0.6	0.6841368	0.6841368
0.7	0.8422886	0.8422884
0.8	1.0296391	1.0296386
0.9	1.2601588	1.2601582
1.0	1.5574064	1.5574077

27.7. Use the Adams–Bashforth–Moulton method to solve $y' = y - x;\ y(0) = 2$ on the interval $[0, 1]$ with $h = 0.1$.

Here $f(x, y) = y - x$, $x_0 = 0$, and $y_0 = 2$. Using Table 27-3, we find the three additional starting values to be $y_1 = 2.2051708$, $y_2 = 2.4214026$, and $y_3 = 2.6498585$. Thus,

$$y_0' = y_0 - x_0 = 2 - 0 = 2 \qquad y_1' = y_1 - x_1 = 2.1051708$$

$$y_2' = y_2 - x_2 = 2.2214026 \qquad y_3' = y_3 - x_3 = 2.3498585$$

Then, using Eqs. (*27.6*), beginning with $n = 3$, and Eq. (*27.3*), we compute

$\boldsymbol{n = 3}$: $x_4 = 0.4$

$$\begin{aligned} py_4 &= y_3 + (h/24)(55y_3' - 59y_2' + 37y_1' - 9y_0') \\ &= 2.6498585 + (0.1/24)[55(2.349585) - 59(2.2214026) + 37(2.1051708) - 9(2)] \\ &= 2.8918201 \end{aligned}$$

$$py_4' = py_4 - x_4 = 2.8918201 - 0.4 = 2.4918201$$

$$\begin{aligned} y_4 &= y_3 + (h/24)(9py_4' + 19y_3' - 5y_2' + y_1') \\ &= 2.6498585 + (0.1/24)[9(2.4918201) + 19(2.3498585) - 5(2.2214026) + 2.1051708] \\ &= 2.8918245 \end{aligned}$$

$$y_4' = y_4 - x_4 = 2.8918245 - 0.4 = 2.4918245$$

$n = 4$: $x_5 = 0.5$

$$py_5 = y_4 + (h/24)(55y'_4 - 59y'_3 + 37y'_2 - 9y'_1)$$
$$= 2.8918245 + (0.1/24)[55(2.4918245) - 59(2.3498585) + 37(2.2214026) - 9(2.1051708)]$$
$$= 3.1487164$$
$$py'_5 = py_5 - x_5 = 3.1487164 - 0.5 = 2.6487164$$
$$y_5 = y_4 + (h/24)(9py'_5 + 19y'_4 - 5y'_3 + y'_2)$$
$$= 2.8918245 + (0.1/24)[9(2.6487164) + 19(2.4918245) - 5(2.3498585) + 2.2214026]$$
$$= 3.1487213$$
$$y'_5 = y_5 - x_5 = 3.1487213 - 0.5 = 2.6487213$$

$n = 5$: $x_6 = 0.6$

$$py_6 = y_5 + (h/24)(55y'_5 - 59y'_4 + 37y'_3 - 9y'_2)$$
$$= 3.1487213 + (0.1/24)[55(2.6487213) - 59(2.4918245) + 37(2.3498585) - 9(2.2214026)]$$
$$= 3.4221137$$
$$py'_6 = py_6 - x_6 = 3.4221137 - 0.6 = 2.8221137$$
$$y_6 = y_5 + (h/24)(9py'_6 + 19y'_5 - 5y'_4 + y'_3)$$
$$= 3.1487213 + (0.1/24)[9(2.8221137) + 19(2.6487213) - 5(2.4918245) + 2.3498585]$$
$$= 3.4221191$$
$$y'_6 = y_6 - x_6 = 3.4221191 - 0.6 = 2.8221191$$

Continuing in this manner, we generate Table 27-6.

Table 27-6

Method: ADAMS–BASHFORTH–MOULTON METHOD			
Problem: $y' = y - x;\ y(0) = 2$			
x_n	$h = 0.1$		True solution $Y(x) = e^x + x + 1$
	py_n	y_n	
0.0	—	2.0000000	2.0000000
0.1	—	2.2051708	2.2051709
0.2	—	2.4214026	2.4214028
0.3	—	2.6498585	2.6498588
0.4	2.8918201	2.8918245	2.8918247
0.5	3.1487164	3.1487213	3.1487213
0.6	3.4221137	3.4221191	3.4221188
0.7	3.7137473	3.7137533	3.7137527
0.8	4.0255352	4.0255418	4.0255409
0.9	4.3595971	4.3596044	4.3596031
1.0	4.7182756	4.7182836	4.7182818

27.8. Use the Adams–Bashforth–Moulton method to solve $y' = y^2 + 1$; $y(0) = 0$, on the interval $[0, 1]$ with $h = 0.1$.

Here $f(x, y) = y^2 + 1$, $x_0 = 0$, and $y_0 = 0$. Using Table 27-5, we find the three additional starting values to be $y_1 = 0.1003346$, $y_2 = 0.2027099$, and $y_3 = 0.3093360$. Thus,

$$y'_0 = (y_0)^2 + 1 = (0)^2 + 1 = 1$$

$$y'_1 = (y_1)^2 + 1 = (0.1003346)^2 + 1 = 1.0100670$$

$$y'_2 = (y_2)^2 + 1 = (0.2027099)^2 + 1 = 1.0410913$$

$$y'_3 = (y_3)^2 + 1 = (0.3093360)^2 + 1 = 1.0956888$$

Then, using Eqs. (*27.6*), beginning with $n = 3$, and Eq. (*27.3*), we compute

$\mathbf{n = 3:}$ $x_4 = 0.4$

$$\begin{aligned} py_4 &= y_3 + (h/24)(55y'_3 - 59y'_2 + 37y'_1 - 9y'_0) \\ &= 0.3093360 + (0.1/24)[55(1.0956888) - 59(1.0410913) + 37(1.0100670) - 9(1)] \\ &= 0.4227151 \end{aligned}$$

$$py'_4 = (py_4)^2 + 1 = (0.4227151)^2 + 1 = 1.1786881$$

$$\begin{aligned} y_4 &= y_3 + (h/24)(9py'_4 + 19y'_3 - 5y'_2 + y'_1) \\ &= 0.3093360 + (0.1/24)[9(1.1786881) + 19(1.0956888) - 5(1.0410913) + 1.0100670] \\ &= 0.4227981 \end{aligned}$$

$$y'_4 = (y_4)^2 + 1 = (0.4227981)^2 + 1 = 1.1787582$$

$\mathbf{n = 4:}$ $x_5 = 0.5$

$$\begin{aligned} py_5 &= y_4 + (h/24)(55y'_4 - 59y'_3 + 37y'_2 - 9y'_1) \\ &= 0.4227981 + (0.1/24)[55(1.1787582) - 59(1.0956888) + 37(1.0410913) - 9(1.0100670)] \\ &= 0.5461974 \end{aligned}$$

$$py'_5 = (py_5)^2 + 1 = (0.5461974)^2 + 1 = 1.2983316$$

$$\begin{aligned} y_5 &= y_4 + (h/24)(9py'_5 + 19y'_4 - 5y'_3 + y'_2) \\ &= 0.4227981 + (0.1/24)[9(1.2983316) + 19(1.1787582) - 5(1.0956888) + 1.0410913] \\ &= 0.5463149 \end{aligned}$$

$$y'_5 = (y_5)^2 + 1 = (0.5463149)^2 + 1 = 1.2984600$$

$\mathbf{n = 5:}$ $x_6 = 0.6$

$$\begin{aligned} py_6 &= y_5 + (h/24)(55y'_5 - 59y'_4 + 37y'_3 - 9y'_2) \\ &= 0.5463149 + (0.1/24)[55(1.2984600) - 59(1.1787582) + 37(1.0956888) - 9(1.0410913)] \\ &= 0.6839784 \end{aligned}$$

$$py'_6 = (py_6)^2 + 1 = (0.6839784)^2 + 1 = 1.4678265$$

$$\begin{aligned} y_6 &= y_5 + (h/24)(9py'_6 + 19y'_5 - 5y'_4 + y'_3) \\ &= 0.5463149 + (0.1/24)[9(1.4678265) + 19(1.2984600) - 5(1.1787582) + 1.0956888] \\ &= 0.6841611 \end{aligned}$$

$$y'_6 = (y_6)^2 + 1 = (0.6841611)^2 + 1 = 1.4680764$$

Continuing in this manner, we generate Table 27-7.

27.9. Use the Adams–Bashforth–Moulton method to solve $y' = 2xy/(x^2 - y^2)$; $y(1) = 3$ on the interval $[1, 2]$ with $h = 0.2$.

Here $f(x, y) = 2xy/(x^2 - y^2)$, $x_0 = 1$, and $y_0 = 3$. With $h = 0.2$, $x_1 = x_0 + h = 1.2$, $x_2 = x_1 + h = 1.4$, and $x_3 = x_2 + h = 1.6$. Using the Runge–Kutta method to obtain the corresponding y-values needed to

Table 27-7

Method: ADAMS–BASHFORTH–MOULTON METHOD			
Problem: $y' = y^2 + 1;\ y(0) = 0$			
x_n	$h = 0.1$		True solution $Y(x) = \tan x$
	py_n	y_n	
0.0	—	0.0000000	0.0000000
0.1	—	0.1003346	0.1003347
0.2	—	0.2027099	0.2027100
0.3	—	0.3093360	0.3093363
0.4	0.4227151	0.4227981	0.4227932
0.5	0.5461974	0.5463149	0.5463025
0.6	0.6839784	0.6841611	0.6841368
0.7	0.8420274	0.8423319	0.8422884
0.8	1.0291713	1.0297142	1.0296386
0.9	1.2592473	1.2602880	1.2601582
1.0	1.5554514	1.5576256	1.5574077

start the Adams–Bashforth–Moulton method, we find $y_1 = 2.8232844$, $y_2 = 2.5709342$, and $y_3 = 2.1321698$. It then follows from Eq. (*27.3*) that

$$y_0' = \frac{2x_0 y_0}{(x_0)^2 - (y_0)^2} = \frac{2(1)(3)}{(1)^2 - (3)^2} = -0.75$$

$$y_1' = \frac{2x_1 y_1}{(x_1)^2 - (y_1)^2} = \frac{2(1.2)(2.8232844)}{(1.2)^2 - (2.8232844)^2} = -1.0375058$$

$$y_2' = \frac{2x_2 y_2}{(x_2)^2 - (y_2)^2} = \frac{2(1.4)(2.5709342)}{(1.4)^2 - (2.5709342)^2} = -1.5481884$$

$$y_3' = \frac{2x_3 y_3}{(x_3)^2 - (y_3)^2} = \frac{2(1.6)(2.1321698)}{(1.6)^2 - (2.1321698)^2} = -3.4352644$$

Then, using Eqs. (*27.6*), beginning with $n = 3$, and Eq. (*27.3*), we compute

$\boldsymbol{n = 3}$**:** $x_4 = 1.8$

$$\begin{aligned} py_4 &= y_3 + (h/24)(55y_3' - 59y_2' + 37y_1' - 9y_0') \\ &= 2.1321698 + (0.1/24)[55(-3.4352644) - 59(-1.5481884) + 37(-1.0375058) - 9(-0.75)] \\ &= 1.0552186 \end{aligned}$$

$$py_4' = \frac{2x_4 py_4}{(x_4)^2 - (py_4)^2} = \frac{2(1.8)(1.0552186)}{(1.8)^2 - (1.0552186)^2} = 1.7863919$$

$$\begin{aligned} y_4 &= y_3 + (h/24)(9py_4' + 19y_3' - 5y_2' + y_1') \\ &= 2.1321698 + (0.1/24)[9(1.7863919) + 19(-3.4352644) - 5(-1.5481884) + (-1.0375058)] \\ &= 1.7780943 \end{aligned}$$

$$y'_4 = \frac{2x_4 y_4}{(x_4)^2 - (y_4)^2} = \frac{2(1.8)(1.7780943)}{(1.8)^2 - (1.7780943)^2} = 81.6671689$$

$\boldsymbol{n = 4}$: $x_5 = 2.0$

$$py_5 = y_4 + (h/24)(55y'_4 - 59y'_3 + 37y'_2 - 9y'_1)$$
$$= 1.7780943 + (0.1/24)[55(81.6671689) - 59(-3.4352644) + 37(-1.5481884) - 9(-1.0375058)]$$
$$= 40.4983398$$

$$py'_5 = \frac{2x_5 py_5}{(x_5)^2 - (py_5)^2} = \frac{2(2.0)(40.4983398)}{(2.0)^2 - (40.4983398)^2} = -0.0990110$$

$$y_5 = y_4 + (h/24)(9py'_5 + 19y'_4 - 5y'_3 + y'_2)$$
$$= 1.7780943 + (0.1/24)[9(-0.0990110) + 19(81.6671689) - 5(-3.4352644) + (-1.5481884)]$$
$$= 14.8315380$$

$$y'_5 = \frac{2x_5 y_5}{(x_5)^2 - (y_5)^2} = \frac{2(2.0)(14.8315380)}{(2.0)^2 - (14.8315380)^2} = -0.2746905$$

These results are troubling because the corrected values are not close to the predicted values as they should be. Note that y_5 is significantly different from py_5 and y'_4 is significantly different from py'_4. In any predictor-corrector method, the corrected values of y and y' represent a fine-tuning of the predicted values, and not a major change. When significant changes occur, they are often the result of numerical instability, which can be remedied by a smaller step-size. Sometimes, however, significant differences arise because of a singularity in the solution.

In the computations above, note that the derivative at $x = 1.8$, namely 81.667, generates a nearly vertical slope and suggests a possible singularity near 1.8. Figure 27-1 is a direction field for this differential equation. On this direction field we have plotted the points (x_0, y_0) through (x_4, y_4) as determined by the Adams–Bashforth–Moulton method and then sketched the solution curve through these points consistent with the direction field. The cusp between 1.6 and 1.8 is a clear indicator of a problem.

The analytic solution to the differential equation is given in Problem 3.14 as $x^2 + y^2 = ky$. Applying the initial condition, we find $k = 10/3$, and then using the quadratic formula to solve

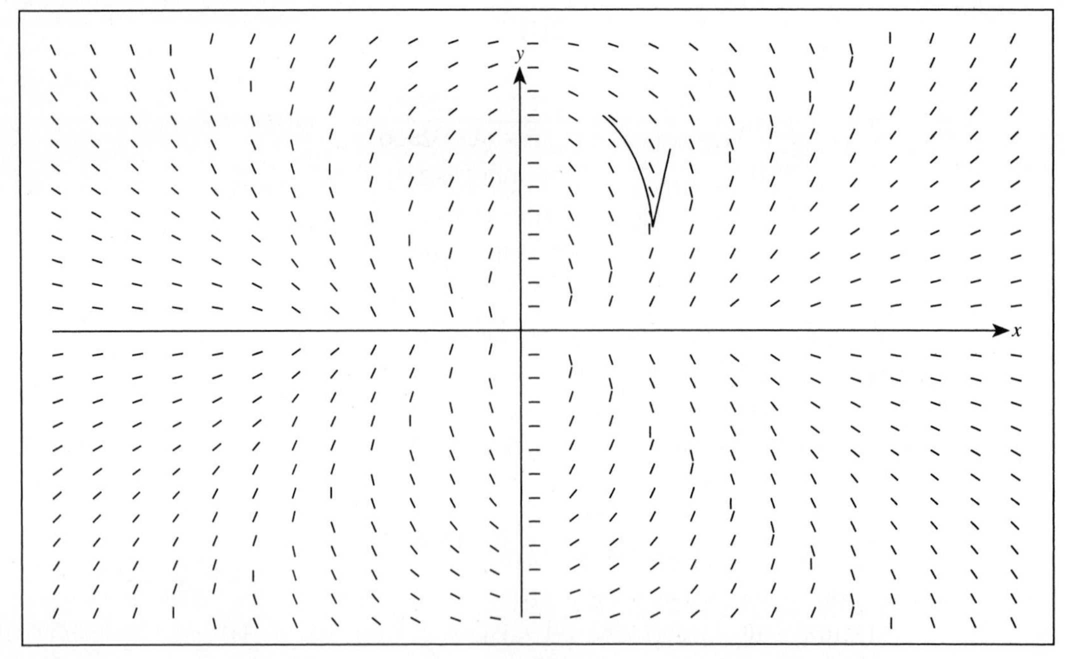

Fig. 27-1

explicitly for y, we obtain the solution

$$y = \frac{5 + \sqrt{25 - 9x^2}}{3}$$

This solution is only defined through $x = 5/3$ and is undefined after that.

27.10. Redo Problem 27.7 using Milne's method.

The values of y_0, y_1, y_2, y_3, and their derivatives are exactly as given in Problem 27.7. Using Eqs. (*27.7*) and (*27.3*), we compute

$$\mathbf{n = 3:}\quad py_4 = y_0 + \frac{4h}{3}(2y_3' - y_2' + 2y_1')$$

$$= 2 + \frac{4(0.1)}{3}[2(2.3498585) - 2.2214026 + 2(2.1051708)]$$

$$= 2.8918208$$

$$py_4' = py_4 - x_4 = 2.4918208$$

$$y_4 = y_2 + \frac{h}{3}(py_4' + 4y_3' + y_2')$$

$$= 2.4214026 + \frac{0.1}{3}[2.4918208 + 4(2.3498585) + 2.2214026]$$

$$= 2.8918245$$

$$\mathbf{n = 4:}\quad x_4 = 0.4, \qquad y_4' = y_4 - x_4 = 2.4918245$$

$$py_5 = y_1 + \frac{4h}{3}(2y_4' - y_3' + 2y_2')$$

$$= 2.2051708 + \frac{4(0.1)}{3}[2(2.4918245) - 2.3498585 + 2(2.2214026)]$$

$$= 3.1487169$$

$$py_5' = py_5 - x_5 = 2.6487169$$

$$y_5 = y_3 + \frac{h}{3}(py_5' + 4y_4' + y_3')$$

$$= 2.6498585 + \frac{0.1}{3}[2.6487169 + 4(2.4918245) + 2.3498585]$$

$$= 3.1487209$$

$$\mathbf{n = 5:}\quad x_5 = 0.5, \qquad y_5' = y_5 - x_5 = 2.6487209$$

$$py_6 = y_2 + \frac{4h}{3}(2y_5' - y_4' + 2y_3')$$

$$= 2.4214026 + \frac{4(0.1)}{3}[2(2.6487209) - 2.4918245 + 2(2.3498585)]$$

$$= 3.4221138$$

$$py_6' = py_6 - x_6 = 2.8221138$$

$$y_6 = y_4 + \frac{h}{3}(py_6' + 4y_5' + y_4')$$

$$= 2.8918245 + \frac{0.1}{3}[2.8221138 + 4(2.6487209) + 2.4918245]$$

$$= 3.4221186$$

Continuing in this manner, we generate Table 27-8.

Table 27-8

Method: MILNE'S METHOD			
Problem: $y' = y - x;\ y(0) = 2$			
x_n	$h = 0.1$		True solution
	py_n	y_n	$Y(x) = e^x + x + 1$
0.0	—	2.0000000	2.0000000
0.1	—	2.2051708	2.2051709
0.2	—	2.4214026	2.4214028
0.3	—	2.6498585	2.6498588
0.4	2.8918208	2.8918245	2.8918247
0.5	3.1487169	3.1487209	3.1487213
0.6	3.4221138	3.4221186	3.4221188
0.7	3.7137472	3.7137524	3.7137527
0.8	4.0255349	4.0255407	4.0255409
0.9	4.3595964	4.3596027	4.3596031
1.0	4.7182745	4.7182815	4.7182818

27.11. Redo Problem 27.8 using Milne's method.

The values of y_0, y_1, y_2, y_3, and their derivatives are exactly as given in Problem 27.8. Using Eqs. (*27.7*) and (*27.3*), we compute

$$\boldsymbol{n=3:}\quad py_4 = y_0 + \frac{4h}{3}(2y_3' - y_2' + 2y_1')$$

$$= 0 + \frac{4(0.1)}{3}[2(1.0956888) - 1.0410913 + 2(1.0100670)]$$

$$= 0.4227227$$

$$py_4' = (py_4)^2 + 1 = (0.4227227)^2 + 1 = 1.1786945$$

$$y_4 = y_2 + \frac{h}{3}(py_4' + 4y_3' + y_2')$$

$$= 0.2027099 + \frac{0.1}{3}[1.1786945 + 4(1.0956888) + 1.0410913]$$

$$= 0.4227946$$

$$\boldsymbol{n=4:}\quad x_4 = 0.4, \qquad y_4' = (y_4)^2 + 1 = (0.4227946)^2 + 1 = 1.1787553$$

$$py_5 = y_1 + \frac{4h}{3}(2y_4' - y_3' + 2y_2')$$

$$= 0.1003346 + \frac{4(0.1)}{3}[2(1.1787553) - 1.0956888 + 2(1.0410913)]$$

$$= 0.5462019$$

$$py'_5 = (py_5)^2 + 1 = (0.5462019)^2 + 1 = 1.2983365$$

$$y_5 = y_3 + \frac{h}{3}(py'_5 + 4y'_4 + y'_3)$$

$$= 0.3093360 + \frac{0.1}{3}[1.2983365 + 4(1.1787553) + 1.0956888]$$

$$= 0.5463042$$

$n = 5$: $x_5 = 0.5, \quad y'_5 = (y_5)^2 + 1 = (0.5463042)^2 + 1 = 1.2984483$

$$py_6 = y_2 + \frac{4h}{3}(2y'_5 - y'_4 + 2y'_3)$$

$$= 0.2027099 + \frac{4(0.1)}{3}[2(1.2984483) - 1.1787553 + 2(1.0956888)]$$

$$= 0.6839791$$

$$py'_6 = (py_6)^2 + 1 = (0.6839791)^2 + 1 = 1.4678274$$

$$y_6 = y_4 + \frac{h}{3}(py'_6 + 4y'_5 + y'_4)$$

$$= 0.4227946 + \frac{0.1}{3}[1.4678274 + 4(1.2984483) + 1.1787553]$$

$$= 0.6841405$$

Continuing in this manner, we generate Table 27-9.

Table 27-9

Method: MILNE'S METHOD			
Problem: $y' = y^2 + 1;\ y(0) = 0$			
x_n	$h = 0.1$		True solution $Y(x) = \tan x$
	py_n	y_n	
0.0	—	0.0000000	0.0000000
0.1	—	0.1003346	0.1003347
0.2	—	0.2027099	0.2027100
0.3	—	0.3093360	0.3093363
0.4	0.4227227	0.4227946	0.4227932
0.5	0.5462019	0.5463042	0.5463025
0.6	0.6839791	0.6841405	0.6841368
0.7	0.8420238	0.8422924	0.8422884
0.8	1.0291628	1.0296421	1.0296386
0.9	1.2592330	1.2601516	1.2601582
1.0	1.5554357	1.5573578	1.5574077

27.12. Use Milne's method to solve $y' = y$; $y(0) = 1$ on the interval $[0, 1]$ with $h = 0.1$.

Here $f(x, y) = y$, $x_0 = 0$, and $y_0 = 1$. From Table 27-4, we find as the three additional starting values $y_1 = 1.1051708$, $y_2 = 1.2214026$, and $y_3 = 1.3498585$. Note that $y_1' = y_1$, $y_2' = y_2$, and $y_3' = y_3$. Then, using Eqs. (*27.7*) and (*27.3*), we compute

$$\begin{aligned}
\boldsymbol{n = 3:}\quad py_4 &= y_0 + \frac{4h}{3}(2y_3' - y_2' + 2y_1') \\
&= 1 + \frac{4(0.1)}{3}[2(1.3498585) - 1.2214026 + 2(1.1051708)] \\
&= 1.4918208 \\
py_4' &= py_4 = 1.4918208 \\
y_4 &= y_2 + \frac{h}{3}(py_4' + 4y_3' + y_2') \\
&= 1.2214026 + \frac{0.1}{3}[1.4918208 + 4(1.3498585) + 1.2214026] \\
&= 1.4918245
\end{aligned}$$

$$\begin{aligned}
\boldsymbol{n = 4:}\quad x_4 &= 0.4, \qquad y_4' = y_4 = 1.4918245 \\
py_5 &= y_1 + \frac{4h}{3}(2y_4' - y_3' + 2y_2') \\
&= 1.1051708 + \frac{4(0.1)}{3}[2(1.4918245) - 1.3498585 + 2(1.2214026)] \\
&= 1.6487169 \\
py_5' &= py_5 = 1.6487169 \\
y_5 &= y_3 + \frac{h}{3}(py_5' + 4y_4' + y_3') \\
&= 1.3498585 + \frac{0.1}{3}[1.6487169 + 4(1.4918245) + 1.3498585] \\
&= 1.6487209
\end{aligned}$$

$$\begin{aligned}
\boldsymbol{n = 5:}\quad x_5 &= 0.5, \qquad y_5' = y_5 = 1.6487209 \\
py_6 &= y_2 + \frac{4h}{3}(2y_5' - y_4' + 2y_3') \\
&= 1.2214026 + \frac{4(0.1)}{3}[2(1.6487209) - 1.4918245 + 2(1.3498585)] \\
&= 1.8221138 \\
py_6' &= py_6 = 1.8221138 \\
y_6 &= y_4 + \frac{h}{3}(py_6' + 4y_5' + y_4') \\
&= 1.4918245 + \frac{0.1}{3}[1.8221138 + 4(1.6487209) + 1.4918245] \\
&= 1.8221186
\end{aligned}$$

Continuing in this manner, we generate Table 27-10.

Supplementary Problems

Carry all computations to three decimal places.

27.13. Use the modified Euler's method to solve $y' = -y + x + 2$; $y(0) = 2$ on the interval $[0, 1]$ with $h = 0.1$.

27.14. Use the modified Euler's method to solve $y' = -y$; $y(0) = 1$ on the interval $[0, 1]$ with $h = 0.1$.

27.15. Use the modified Euler's method to solve $y' = \dfrac{x^2 + y^2}{xy}$; $y(1) = 3$ on the interval $[1, 2]$ with $h = 0.2$.

Table 27-10

Method: MILNE'S METHOD			
Problem: $y' = y;\ y(0) = 1$			
x_n	$h = 0.1$		True solution $Y(x) = e^x$
	py_n	y_n	
0.0	—	1.0000000	1.0000000
0.1	—	1.1051708	1.1051709
0.2	—	1.2214026	1.2214028
0.3	—	1.3498585	1.3498588
0.4	1.4918208	1.4918245	1.4918247
0.5	1.6487169	1.6487209	1.6487213
0.6	1.8221138	1.8221186	1.8221188
0.7	2.0137472	2.0137524	2.0137527
0.8	2.2255349	2.2255407	2.2255409
0.9	2.4595964	2.4596027	2.4596031
1.0	2.7182745	2.7182815	2.7182818

27.16. Use the modified Euler's method to solve $y' = x;\ y(2) = 1$ on the interval $[2, 3]$ with $h = 0.25$.

27.17. Use the modified Euler's method to solve $y' = 4x^3;\ y(2) = 6$ on the interval $[2, 3]$ with $h = 0.2$.

27.18. Redo Problem 27.13 using the Runge–Kutta method.

27.19. Redo Problem 27.14 using the Runge–Kutta method.

27.20. Redo Problem 27.15 using the Runge–Kutta method.

27.21. Redo Problem 27.17 using the Runge–Kutta method.

27.22. Use the Runge–Kutta method to solve $y' = 5x^4;\ y(0) = 0$ on the interval $[0, 1]$ with $h = 0.1$.

27.23. Use the Adams–Bashforth–Moulton method to solve $y' = y;\ y(0) = 1$ on the interval $[0, 1]$ with $h = 0.1$.

27.24. Redo Problem 27.13 using the Adams–Bashforth–Moulton method.

27.25. Redo Problem 27.14 using the Adams–Bashforth–Moulton method.

27.26. Redo Problem 27.15 using the Adams–Bashforth–Moulton method.

27.27. Redo Problem 27.13 using Milne's method.

27.28. Redo Problem 27.14 using Milne's method.

Chapter 28

Numerical Methods for Systems

FIRST-ORDER SYSTEMS

Numerical methods for solving first-order initial-value problems, including Euler's method (see Chapter 26) and all the methods given in Chapter 27, are easily extended to a *system* of first-order initial-value problems. These methods are also applicable to most *higher-order* initial-value problems, in particular those that can be transformed to a system of first-order differential equations by the reduction process described in Chapter 21.

Standard form for a system of two equations is

$$\begin{aligned} y' &= f(x, y, z) \\ z' &= g(x, y, z); \\ y(x_0) &= y_0,\ z(x_0) = z_0 \end{aligned} \tag{28.1}$$

We note that, with $y' = f(x, y, z) \equiv z$, system (*28.1*) represents the second-order initial-value problem

$$y'' = g(x, y, y'); \qquad y(x_0) = y_0, \qquad y'(x_0) = z_0$$

Standard form for a system of three equations is

$$\begin{aligned} y' &= f(x, y, z, w) \\ z' &= g(x, y, z, w) \\ w' &= r(x, y, z, w); \\ y(x_0) &= y_0,\ z(x_0) = z_0,\ w(x_0) = w_0 \end{aligned} \tag{28.2}$$

If, in such a system, $f(x, y, z, w) = z$, and $g(x, y, z, w) = w$, then system (*28.2*) represents the third-order initial-value problem

$$y''' = r(x, y, z, w); \qquad y(x_0) = y_0, \qquad y'(x_0) = z_0, \qquad y''(x_0) = w_0$$

The formulas that follow are for systems of two equations in standard form (*28.1*). Generalizations to systems of three equations in standard form (*28.2*) or systems with four or more equations are straightforward.

EULER'S METHOD

$$\begin{aligned} y_{n+1} &= y_n + hy'_n \\ z_{n+1} &= z_n + hz'_n \end{aligned} \tag{28.3}$$

RUNGE–KUTTA METHOD

$$\begin{aligned} y_{n+1} &= y_n + \frac{1}{6}(k_1 + 2k_2 + 2k_3 + k_4) \\ z_{n+1} &= z_n + \frac{1}{6}(l_1 + 2l_2 + 2l_3 + l_4) \end{aligned} \tag{28.4}$$

where
$$\begin{aligned}
k_1 &= hf(x_n, y_n, z_n)\\
l_1 &= hg(x_n, y_n, z_n)\\
k_2 &= hf(x_n + \tfrac{1}{2}h,\ y_n + \tfrac{1}{2}k_1,\ z_n + \tfrac{1}{2}l_1)\\
l_2 &= hg(x_n + \tfrac{1}{2}h,\ y_n + \tfrac{1}{2}k_1,\ z_n + \tfrac{1}{2}l_1)\\
k_3 &= hf(x_n + \tfrac{1}{2}h,\ y_n + \tfrac{1}{2}k_2,\ z_n + \tfrac{1}{2}l_2)\\
l_3 &= hg(x_n + \tfrac{1}{2}h,\ y_n + \tfrac{1}{2}k_2,\ z_n + \tfrac{1}{2}l_2)\\
k_4 &= hf(x_n + h,\ y_n + k_3,\ z_n + l_3)\\
l_4 &= hg(x_n + h,\ y_n + k_3,\ z_n + l_3)
\end{aligned}$$

ADAMS–BASHFORTH–MOULTON METHOD

$$\begin{aligned}
\text{predictors:}\quad & py_{n+1} = y_n + \frac{h}{24}(55y'_n - 59y'_{n-1} + 37y'_{n-2} - 9y'_{n-3})\\
& pz_{n+1} = z_n + \frac{h}{24}(55z'_n - 59z'_{n-1} + 37z'_{n-2} - 9z'_{n-3})\\
\text{correctors:}\quad & y_{n+1} = y_n + \frac{h}{24}(9py'_{n+1} + 19y'_n - 5y'_{n-1} + y'_{n-2})\\
& z_{n+1} = z_n + \frac{h}{24}(9pz'_{n+1} + 19z'_n - 5z'_{n-1} + z'_{n-2})
\end{aligned} \tag{28.5}$$

Corresponding derivatives are calculated from system (*28.1*). In particular,

$$\begin{aligned}
y'_{n+1} &= f(x_{n+1}, y_{n+1}, z_{n+1})\\
z'_{n+1} &= g(x_{n+1}, y_{n+1}, z_{n+1})
\end{aligned} \tag{28.6}$$

The derivatives associated with the predicted values are obtained similarly, by replacing y and z in Eq. (*28.6*) with py and pz, respectively. As in Chapter 27, four sets of starting values are required for the Adams–Bashforth–Moulton method. The first set comes directly from the initial conditions; the other three sets are obtained from the Runge–Kutta method.

Solved Problems

28.1. Reduce the initial-value problem $y'' - y = x$; $y(0) = 0$, $y'(0) = 1$ to system (*28.1*).

Defining $z = y'$, we have $z(0) = y'(0) = 1$ and $z' = y''$. The given differential equation can be rewritten as $y'' = y + x$, or $z' = y + x$. We thus obtain the first-order system

$$\begin{aligned}
y' &= z\\
z' &= y + x;\\
y(0) &= 0,\ z(0) = 1
\end{aligned}$$

28.2. Reduce the initial-value problem $y'' - 3y' + 2y = 0$; $y(0) = -1$, $y'(0) = 0$ to system (*28.1*).

Defining $z = y'$, we have $z(0) = y'(0) = 0$ and $z' = y''$. The given differential equation can be rewritten as $y'' = 3y' - 2y$, or $z' = 3z - 2y$. We thus obtain the first-order system

$$y' = z$$
$$z' = 3z - 2y;$$
$$y(0) = -1, \; z(0) = 0$$

28.3. Reduce the initial-value problem $3x^2y'' - xy' + y = 0$; $y(1) = 4$, $y'(1) = 2$ to system (*28.1*).

Defining $z = y'$, we have $z(1) = y'(1) = 2$, and $z' = y''$. The given differential equation can be rewritten as

$$y'' = \frac{xy' - y}{3x^2}$$

or

$$z' = \frac{xz - y}{3x^2}$$

We thus obtain the first-order system

$$y' = z$$
$$z' = \frac{xz - y}{3x^2}$$
$$y(1) = 4, \; z(1) = 2$$

28.4. Reduce the initial-value problem $y''' - 2xy'' + 4y' - x^2y = 1$; $y(0) = 1$, $y'(0) = 2$, $y''(0) = 3$ to system (*28.2*).

Following Steps 1 through 3 of Chapter 21, we obtain the system

$$y_1' = y_2$$
$$y_2' = y_3$$
$$y_3' = x^2y_1 - 4y_2 + 2xy_3 + 1;$$
$$y_1(0) = 1, \; y_2(0) = 2, \; y_3(0) = 3$$

To eliminate subscripting, we define $y = y_1$, $z = y_2$, and $w = y_3$. The system then becomes

$$y' = z$$
$$z' = w$$
$$w' = x^2y - 4z + 2xw + 1;$$
$$y(0) = 1, \; z(0) = 2, \; w(0) = 3$$

28.5. Use Euler's method to solve $y'' - y = x$; $y(0) = 0$, $y'(0) = 1$ on the interval $[0, 1]$ with $h = 0.1$.

Using the results of Problem 28.1, we have $f(x, y, z) = z$, $g(x, y, z) = y + x$, $x_0 = 0$, $y_0 = 0$, and $z_0 = 1$. Then, using (*28.3*), we compute

$n = 0$: $y_0' = f(x_0, y_0, z_0) = z_0 = 1$

$$z_0' = g(x_0, y_0, z_0) = y_0 + x_0 = 0 + 0 = 0$$
$$y_1 = y_0 + hy_0' = 0 + (0.1)(1) = 0.1$$
$$z_1 = z_0 + hz_0' = 1 + (0.1)(0) = 1$$

$n = 1$: $y_1' = f(x_1, y_1, z_1) = z_1 = 1$

$$z_1' = g(x_1, y_1, z_1) = y_1 + x_1 = 0.1 + 0.1 = 0.2$$
$$y_2 = y_1 + hy_1' = 0.1 + (0.1)(1) = 0.2$$
$$z_2 = z_1 + hz_1' = 1 + (0.1)(0.2) = 1.02$$

$n = 2$: $y_2' = f(x_2, y_2, z_2) = z_2 = 1.02$

$$z_2' = g(x_2, y_2, z_2) = y_2 + x_2 = 0.2 + 0.2 = 0.4$$
$$y_3 = y_2 + hy_2' = 0.2 + (0.1)(1.02) = 0.302$$
$$z_3 = z_2 + hz_2' = 1.02 + (0.1)(0.4) = 1.06$$

Continuing in this manner, we generate Table 28-1.

Table 28-1

Method: EULER'S METHOD			
Problem: $y'' - y = x;\ y(0) = 0,\ y'(0) = 1$			
x_n	$h = 0.1$		True solution
	y_n	z_n	$Y(x) = e^x - e^{-x} - x$
0.0	0.0000	1.0000	0.0000
0.1	0.1000	1.0000	0.1003
0.2	0.2000	1.0200	0.2027
0.3	0.3020	1.0600	0.3090
0.4	0.4080	1.1202	0.4215
0.5	0.5200	1.2010	0.5422
0.6	0.6401	1.3030	0.6733
0.7	0.7704	1.4270	0.8172
0.8	0.9131	1.5741	0.9762
0.9	1.0705	1.7454	1.1530
1.0	1.2451	1.9424	1.3504

28.6. Use Euler's method to solve $y'' - 3y' + 2y = 0;\ y(0) = -1,\ y'(0) = 0$ on the interval $[0, 1]$ with $h = 0.1$.

Using the results of Problem 28.2, we have $f(x, y, z) = z$, $g(x, y, z) = 3z - 2y$, $x_0 = 0$, $y_0 = -1$, and $z_0 = 0$. Then, using *(28.3)*, we compute

$n = 0$: $y_0' = f(x_0, y_0, z_0) = z_0 = 0$

$$z_0' = g(x_0, y_0, z_0) = 3z_0 - 2y_0 = 3(0) - 2(-1) = 2$$
$$y_1 = y_0 + hy_0' = -1 + (0.1)(0) = -1$$
$$z_1 = z_0 + hz_0' = 0 + (0.1)(2) = 0.2$$

$$\boldsymbol{n = 1:}\quad y_1' = f(x_1, y_1, z_1) = z_1 = 0.2$$

$$z_1' = g(x_1, y_1, z_1) = 3z_1 - 2y_1 = 3(0.2) - 2(-1) = 2.6$$

$$y_2 = y_1 + hy_1' = -1 + (0.1)(0.2) = -0.98$$

$$z_2 = z_1 + hz_1' = 0.2 + (0.1)(2.6) = 0.46$$

Continuing in this manner, we generate Table 28-2.

Table 28-2

Method: EULER'S METHOD			
Problem: $y'' - 3y' + 2y = 0;\ y(0) = -1,\ y'(0) = 0$			
x_n	$h = 0.1$		True solution $Y(x) = e^{2x} - 2e^x$
	y_n	z_n	
0.0	−1.0000	0.0000	−1.0000
0.1	−1.0000	0.2000	−0.9889
0.2	−0.9800	0.4600	−0.9510
0.3	−0.9340	0.7940	−0.8776
0.4	−0.8546	1.2190	−0.7581
0.5	−0.7327	1.7556	−0.5792
0.6	−0.5571	2.4288	−0.3241
0.7	−0.3143	3.2689	0.0277
0.8	0.0126	4.3125	0.5020
0.9	0.4439	5.6037	1.1304
1.0	1.0043	7.1960	1.9525

28.7. Use the Runge–Kutta method to solve $y'' - y = x;\ y(0) = 0,\ y'(0) = 1$ on the interval $[0, 1]$ with $h = 0.1$.

Using the results of Problem 28.1, we have $f(x, y, z) = z$, $g(x, y, z) = y + x$, $x_0 = 0$, $y_0 = 0$, and $z_0 = 1$. Then, using *(28.4)* and rounding all calculations to three decimal places, we compute:

$$\boldsymbol{n = 0:}\quad k_1 = hf(x_0, y_0, z_0) = hf(0, 0, 1) = (0.1)(1) = 0.1$$

$$l_1 = hg(x_0, y_0, z_0) = hg(0, 0, 1) = (0.1)(0 + 0) = 0$$

$$k_2 = hf(x_0 + \tfrac{1}{2}h, y_0 + \tfrac{1}{2}k_1, z_0 + \tfrac{1}{2}l_1)$$

$$= hf[0 + \tfrac{1}{2}(0.1), 0 + \tfrac{1}{2}(0.1), 1 + \tfrac{1}{2}(0)]$$

$$= hf(0.05, 0.05, 1) = (0.1)(1) = 0.1$$

$$l_2 = hg(x_0 + \tfrac{1}{2}h, y_0 + \tfrac{1}{2}k_1, z_0 + \tfrac{1}{2}l_1)$$

$$= hg(0.05, 0.05, 1) = (0.1)(0.05 + 0.05) = 0.01$$

$$k_3 = hf(x_0 + \tfrac{1}{2}h, y_0 + \tfrac{1}{2}k_2, z_0 + \tfrac{1}{2}l_2)$$
$$= hf[0 + \tfrac{1}{2}(0.1), 0 + \tfrac{1}{2}(0.1), 1 + \tfrac{1}{2}(0.01)]$$
$$= hf(0.05, 0.05, 1.005) = (0.1)(1.005) = 0.101$$
$$l_3 = hg(x_0 + \tfrac{1}{2}h, y_0 + \tfrac{1}{2}k_2, z_0 + \tfrac{1}{2}l_2)$$
$$= hg(0.05, 0.05, 1.005) = (0.1)(0.05 + 0.05) = 0.01$$
$$k_4 = hf(x_0 + h, y_0 + k_3, z_0 + l_3)$$
$$= hf(0 + 0.1, 0 + 0.101, 1 + 0.01)$$
$$= hf(0.1, 0.101, 1.01) = (0.1)(1.01) = 0.101$$
$$l_4 = hg(x_0 + h, y_0 + k_3, z_0 + l_3)$$
$$= hg(0.1, 0.101, 1.01) = (0.1)(0.101 + 0.1) = 0.02$$
$$y_1 = y_0 + \tfrac{1}{6}(k_1 + 2k_2 + 2k_3 + k_4)$$
$$= 0 + \tfrac{1}{6}[0.1 + 2(0.1) + 2(0.101) + (0.101)] = 0.101$$
$$z_1 = z_0 + \tfrac{1}{6}(l_1 + 2l_2 + 2l_3 + l_4)$$
$$= 1 + \tfrac{1}{6}[0 + 2(0.01) + 2(0.01) + (0.02)] = 1.01$$

$n = 1$: $k_1 = hf(x_1, y_1, z_1) = hf(0.1, 0.101, 1.01)$

$$= (0.1)(1.01) = 0.101$$
$$l_1 = hg(x_1, y_1, z_1) = hg(0.1, 0.101, 1.01)$$
$$= (0.1)(0.101 + 0.1) = 0.02$$
$$k_2 = hf(x_1 + \tfrac{1}{2}h, y_1 + \tfrac{1}{2}k_1, z_1 + \tfrac{1}{2}l_1)$$
$$= hf[0.1 + \tfrac{1}{2}(0.1), 0.101 + \tfrac{1}{2}(0.101), 1.01 + \tfrac{1}{2}(0.02)]$$
$$= hf(0.15, 0.152, 1.02) = (0.1)(1.02) = 0.102$$
$$l_2 = hg(x_1 + \tfrac{1}{2}h, y_1 + \tfrac{1}{2}k_1, z_1 + \tfrac{1}{2}l_1)$$
$$= hg(0.15, 0.152, 1.02) = (0.1)(0.152 + 0.15) = 0.03$$
$$k_3 = hf(x_1 + \tfrac{1}{2}h, y_1 + \tfrac{1}{2}k_2, z_1 + \tfrac{1}{2}l_2)$$
$$= hf[0.1 + \tfrac{1}{2}(0.1), 0.101 + \tfrac{1}{2}(0.102), 1.01 + \tfrac{1}{2}(0.03)]$$
$$= hf(0.15, 0.152, 1.025) = (0.1)(1.025) = 0.103$$
$$l_3 = hg(x_1 + \tfrac{1}{2}h, y_1 + \tfrac{1}{2}k_2, z_1 + \tfrac{1}{2}l_2)$$
$$= hg(0.15, 0.152, 1.025) = (0.1)(0.152 + 0.15) = 0.03$$
$$k_4 = hf(x_1 + h, y_1 + k_3, z_1 + l_3)$$
$$= hf(0.1 + 0.1, 0.101 + 0.103, 1.01 + 0.03)$$
$$= hf(0.2, 0.204, 1.04) = (0.1)(1.04) = 0.104$$
$$l_4 = hg(x_1 + h, y_1 + k_3, z_1 + l_3)$$
$$= hg(0.2, 0.204, 1.04) = (0.1)(0.204 + 0.2) = 0.04$$
$$y_2 = y_1 + \tfrac{1}{6}(k_1 + 2k_2 + 2k_3 + k_4)$$
$$= 0.101 + \tfrac{1}{6}[0.101 + 2(0.102) + 2(0.103) + (0.104)]$$
$$= 0.204$$
$$z_2 = z_1 + \tfrac{1}{6}(l_1 + 2l_2 + 2l_3 + l_4)$$
$$= 1.01 + \tfrac{1}{6}[0.02 + 2(0.03) + 2(0.03) + 0.04] = 1.04$$

Continuing in this manner, but rounding to seven decimal places, we generate Table 28-3.

Table 28-3

Method: RUNGE–KUTTA METHOD			
Problem: $y'' - y = x$; $y(0) = 0$, $y'(0) = 1$			
x_n	$h = 0.1$		True solution $Y(x) = e^x - e^{-x} - x$
	y_n	z_n	
0.0	0.0000000	1.0000000	0.0000000
0.1	1.1003333	1.0100083	0.1003335
0.2	0.2026717	1.0401335	0.2026720
0.3	0.3090401	1.0906769	0.3090406
0.4	0.4215040	1.1621445	0.4215047
0.5	0.5421897	1.2552516	0.5421906
0.6	0.6733060	1.3709300	0.6733072
0.7	0.8171660	1.5103373	0.8171674
0.8	0.9762103	1.6748689	0.9762120
0.9	1.1530314	1.8661714	1.1530335
1.0	1.3504000	2.0861595	1.3504024

28.8. Use the Runge–Kutta method to solve $y'' - 3y' + 2y = 0$; $y(0) = -1$, $y'(0) = 0$ on the interval $[0, 1]$ with $h = 0.1$.

Using the results of Problem 28.2, we have $f(x, y, z) = z$, $g(x, y, z) = 3z - 2y$, $x_0 = 0$, $y_0 = -1$, and $z_0 = 0$. Then, using *(28.4)*, we compute:

$$\boldsymbol{n = 0}\text{:}\quad k_1 = hf(x_0, y_0, z_0) = hf(0, -1, 0) = (0.1)(0) = 0$$

$$l_1 = hg(x_0, y_0, z_0) = hg(0, -1, 0) = (0.1)[3(0) - 2(-1)] = 0.2$$

$$k_2 = hf(x_0 + \tfrac{1}{2}h, y_0 + \tfrac{1}{2}k_1, z_0 + \tfrac{1}{2}l_1)$$

$$= hf[0 + \tfrac{1}{2}(0.1), -1 + \tfrac{1}{2}(0), 0 + \tfrac{1}{2}(0.2)]$$

$$= hf(0.05, -1, 0.1) = (0.1)(0.1) = 0.01$$

$$l_2 = hg(x_0 + \tfrac{1}{2}h, y_0 + \tfrac{1}{2}k_1, z_0 + \tfrac{1}{2}l_1)$$

$$= hg(0.05, -1, 0.1) = (0.1)[3(0.1) - 2(-1)] = 0.23$$

$$k_3 = hf(x_0 + \tfrac{1}{2}h, y_0 + \tfrac{1}{2}k_2, z_0 + \tfrac{1}{2}l_2)$$

$$= hf[0 + \tfrac{1}{2}(0.1), -1 + \tfrac{1}{2}(0.01), 0 + \tfrac{1}{2}(0.23)]$$

$$= hf(0.05, -0.995, 0.115) = (0.1)(0.115) = 0.012$$

$$l_3 = hg(x_0 + \tfrac{1}{2}h, y_0 + \tfrac{1}{2}k_2, z_0 + \tfrac{1}{2}l_2)$$
$$= hg(0.05, -0.995, 0.115) = (0.1)[3(0.115) - 2(-0.995)]$$
$$= 0.234$$
$$k_4 = hf(x_0 + h, y_0 + k_3, z_0 + l_3)$$
$$= hf(0 + 0.1, -1 + 0.012, 0 + 0.234)$$
$$= hf(0.1, -0.988, 0.234) = (0.1)(0.234) = 0.023$$
$$l_4 = hg(x_0 + h, y_0 + k_3, z_0 + l_3)$$
$$= hg(0.1, -0.988, 0.234) = (0.1)[3(0.234) - 2(-0.988)]$$
$$= 0.268$$
$$y_1 = y_0 + \tfrac{1}{6}(k_1 + 2k_2 + 2k_3 + k_4)$$
$$= -1 + \tfrac{1}{6}[0 + 2(0.01) + 2(0.012) + 0.023] = -0.989$$
$$z_1 = z_0 + \tfrac{1}{6}(l_1 + 2l_2 + 2l_3 + l_4)$$
$$= 0 + \tfrac{1}{6}[0.2 + 2(0.23) + 2(0.234) + 0.268] = 0.233$$

Continuing in this manner, we generate Table 28-4.

Table 28-4

Method: RUNGE-KUTTA METHOD			
Problem: $y'' - 3y' + 2y = 0$; $y(0) = -1$, $y'(0) = 0$			
x_n	$h = 0.1$		True solution $Y(x) = e^{2x} - 2e^x$
	y_n	z_n	
0.0	−1.0000000	0.0000000	−1.0000000
0.1	−0.9889417	0.2324583	−0.9889391
0.2	−0.9509872	0.5408308	−0.9509808
0.3	−0.8776105	0.9444959	−0.8775988
0.4	−0.7581277	1.4673932	−0.7581085
0.5	−0.5791901	2.1390610	−0.5791607
0.6	−0.3241640	2.9959080	−0.3241207
0.7	0.0276326	4.0827685	0.0276946
0.8	0.5018638	5.4548068	0.5019506
0.9	1.1303217	7.1798462	1.1304412
1.0	1.9523298	9.3412190	1.9524924

28.9. Use the Runge–Kutta method to solve $3x^2y'' - xy' + y = 0$; $y(1) = 4$, $y'(1) = 2$ on the interval $[1, 2]$ with $h = 0.2$.

It follows from Problem 28.3 that $f(x, y, z) = z$, $g(x, y, z) = (xz - y)/(3x^2)$, $x_0 = 1$, $y_0 = 4$, and $z_0 = 2$. Using (*28.4*), we compute

$$\textbf{n = 0:}\quad k_1 = hf(x_0, y_0, z_0) = hf(1, 4, 2) = 0.2(2) = 0.4$$

$$l_1 = hg(x_0, y_0, z_0) = hg(1, 4, 2) = 0.2\left[\frac{1(2) - 4}{3(1)^2}\right] = -0.1333333$$

$$k_2 = hf(x_0 + \tfrac{1}{2}h, y_0 + \tfrac{1}{2}k_1, z_0 + \tfrac{1}{2}l_1)$$

$$= hf(1.1, 4.2, 1.9333333) = 0.2(1.9333333) = 0.3866666$$

$$l_2 = hg(x_0 + \tfrac{1}{2}h, y_0 + \tfrac{1}{2}k_1, z_0 + \tfrac{1}{2}l_1) = hg(1.1, 4.2, 1.9333333)$$

$$= 0.2\left[\frac{1.1(1.9333333) - 4.2}{3(1.1)^2}\right] = -0.1142332$$

$$k_3 = hf(x_0 + \tfrac{1}{2}h, y_0 + \tfrac{1}{2}k_2, z_0 + \tfrac{1}{2}l_2)$$

$$= hf(1.1, 4.1933333, 1.9428834) = 0.2(1.9428834) = 0.3885766$$

$$l_3 = hg(x_0 + \tfrac{1}{2}h, y_1 + \tfrac{1}{2}k_2, z_1 + \tfrac{1}{2}l_2) = hg(1.1, 4.1933333, 1.9428834)$$

$$= 0.2\left[\frac{1.1(1.9428834) - 4.1933333}{3(1.1)^2}\right] = -0.1132871$$

$$k_4 = hf(x_0 + h, y_0 + k_3, z_0 + l_3)$$

$$= hf(1.2, 4.3885766, 1.8867129) = 0.2(1.8867129) = 0.3773425$$

$$l_4 = hgx_0 + h, y_0 + k_3, z_0 + l_3) = hg(1.2, 4.3885766, 1.8867129)$$

$$= 0.2\left[\frac{1.2(1.8867129) - 4.3885766}{3(1.2)^2}\right] = -0.0983574$$

$$y_1 = y_0 + \tfrac{1}{6}(k_1 + 2k_2 + 2k_3 + k_4)$$

$$= 4 + \tfrac{1}{6}[0.4 + 2(0.3866666) + 2(0.3885766) + 0.3773425] = 4.3879715$$

$$z_1 = z_0 + \tfrac{1}{6}(l_1 + 2l_2 + 2l_3 + l_4)$$

$$= 2 + \tfrac{1}{6}[-0.1333333 + 2(-0.1142332) + 2(-0.1132871) + (-0.0983574)] = 1.8855447$$

Continuing in this manner, we generate Table 28-5.

Table 28-5

Method: RUNGE–KUTTA METHOD			
Problem: $3x^2y'' - xy' + y = 0$; $y(1) = 4$, $y'(1) = 2$			
x_n	$h = 0.2$		True solution $Y(x) = x + 3x^{1/3}$
	y_n	z_n	
1.0	4.0000000	2.0000000	4.0000000
1.2	4.3879715	1.8855447	4.3879757
1.4	4.7560600	1.7990579	4.7560668
1.6	5.1088123	1.7309980	5.1088213
1.8	5.4493105	1.6757935	5.4493212
2.0	5.7797507	1.6299535	5.7797632

28.10. Use the Adams–Bashforth–Moulton method to solve $3x^2y'' - xy' + y = 0$; $y(1) = 4$, $y'(1) = 2$ on the interval $[1, 2]$ with $h = 0.2$.

It follows from Problem 28.3 that $f(x, y, z) = z$, $g(x, y, z) = (xz - y)/(3x^2)$, $x_0 = 1$, $y_0 = 4$, and $z_0 = 2$. From Table 28-5, we have

$$x_1 = 1.2 \qquad y_1 = 4.3879715 \qquad z_1 = 1.8855447$$

$$x_2 = 1.4 \qquad y_2 = 4.7560600 \qquad z_2 = 1.7990579$$

$$x_3 = 1.6 \qquad y_3 = 5.1088123 \qquad z_3 = 1.7309980$$

Using (*28.6*), we compute

$$y_0' = z_0 = 2 \qquad y_1' = z_1 = 1.8855447$$

$$y_2' = z_2 = 1.7990579 \qquad y_3' = z_3 = 1.7309980$$

$$z_0' = \frac{x_0 z_0 - y_0}{3x_0^2} = \frac{1(2) - 4}{3(1)^2} = -0.6666667$$

$$z_1' = \frac{x_1 z_1 - y_1}{3x_1^2} = \frac{1.2(1.8855447) - 4.3879715}{3(1.2)^2} = -0.4919717$$

$$z_2' = \frac{x_2 z_2 - y_2}{3x_2^2} = \frac{1.4(1.7990579) - 4.7560600}{3(1.4)^2} = -0.3805066$$

$$z_3' = \frac{x_3 z_3 - y_3}{3x_3^2} = \frac{1.6(1.7309980) - 5.1088123}{3(1.6)^2} = -0.3045854$$

Then using (*26.5*), we compute

***n* = 3:** $x_4 = 1.8$

$$py_4 = y_3 + \frac{h}{24}(55y_3' - 59y_2' + 37y_1' - 9y_0')$$

$$= 5.1088123 + (0.2/24)[55(1.7309980) - 59(1.7990579) + 37(1.8855447) - 9(2)] = 5.4490260$$

$$pz_4 = z_3 + \frac{h}{24}(55z_3' - 59z_2' + 37z_1' - 9z_0')$$

$$= 1.7309980 + (0.2/24)[55(-0.3045854) - 59(-0.3805066) + 37(-0.4919717) - 9(-0.6666667)] = 1.6767876$$

$$py_4' = pz_4 = 1.6767876$$

$$pz_4' = \frac{x_4 pz_4 - py_4}{3x_4^2} = \frac{1.8(1.6767876) - 5.4490260}{3(1.8)^2} = -0.2500832$$

$$y_4 = y_3 + \frac{h}{24}(9py_4' + 19y_3' - 5y_2' + y_1')$$

$$= 5.1088123 + (0.2/24)[9(1.6767876) + 19(1.7309980) - 5(1.7990579) + 1.8855447]$$

$$= 5.4493982$$

$$z_4 = z_3 + \frac{h}{24}(9pz_4' + 19z_3' - 5z_2' + z_1')$$

$$= 1.7309980 + (0.2/24)[9(-0.2500832) + 19(-0.3045854) - 5(-0.3805066) + (-0.4919717)]$$

$$= 1.6757705$$

$$y_4' = z_4 = 1.6757705$$

$$z_4' = \frac{x_4 z_4 - y_4}{3x_4^2} = \frac{1.8(1.6757705) - 5.4493982}{3(1.8)^2} = -0.2503098$$

$n = 4$: $x_5 = 2.0$

$$py_5 = y_4 + \frac{h}{24}(55y_4' - 59y_3' + 37y_2' - 9y_1')$$

$$= 5.4493982 + (0.2/24)[55(1.6757705) - 59(1.7309980) + 37(1.7990579) - 9(1.8855447)]$$

$$= 5.7796793$$

$$pz_5 = z_4 + \frac{h}{24}(55z_4' - 59z_3' + 37z_2' - 9z_1')$$

$$= 1.6757705 + (0.2/24)[55(-0.2503098) - 59(-0.3045854) + 37(-0.3805066) - 9(-0.4919717)]$$

$$= 1.6303746$$

$$py_5' = pz_5 = 1.6303746$$

$$pz_5' = \frac{x_5 pz_5 - py_5}{3x_5^2} = \frac{2.0(1.6303746) - 5.7796793}{3(2.0)^2} = -0.2099108$$

$$y_5 = y_4 + \frac{h}{24}(9py_5' + 19y_4' - 5y_3' + y_2')$$

$$= 5.4493982 + (0.2/24)[9(1.6303746) + 19(1.6757705) - 5(1.7309980) + 1.7990579]$$

$$= 5.7798739$$

$$z_5 = z_4 + \frac{h}{24}(9pz_5' + 19z_4' - 5z_3' + z_2')$$

$$= 1.6757705 + (0.2/24)[9(-0.2099108) + 19(-0.2503098) - 5(-0.3045854) + (-0.3805066)]$$

$$= 1.6299149$$

$$y_5' = z_5 = 1.6299149$$

$$z_5' = \frac{x_5 z_5 - y_5}{3x_5^2} = \frac{2.0(1.6299149) - 5.7798739}{3(2.0)^2} = -0.2100037$$

See Table 28-6.

Table 28-6

Method: ADAMS–BASHFORTH–MOULTON METHOD					
Problem: $3x^2y'' - xy' + y = 0$; $y(1) = 4$, $y'(1) = 2$					
x_n	$h = 0.2$				True solution $Y(x) = x + 3x^{1/3}$
	py_n	pz_n	y_n	z_n	
1.0	—	—	4.0000000	2.0000000	4.0000000
1.2	—	—	4.3879715	1.8855447	4.3879757
1.4	—	—	4.7560600	1.7990579	4.7560668
1.6	—	—	5.1088123	1.7309980	5.1088213
1.8	5.4490260	1.6767876	5.4493982	1.6757705	5.4493212
2.0	5.7796793	1.6303746	5.7798739	1.6299149	5.7797632

28.11. Use the Adams–Bashforth–Moulton method to solve $y'' - y = x;\ y(0) = 0,\ y'(0) = 1$ on the interval $[0, 1]$ with $h = 0.1$.

It follows from Problem 28.1 that $f(x, y, z) = z$ and $g(x, y, z) = y + x$ and from Table 28-3 that

$$\begin{array}{lll} x_0 = 0 & y_0 = 0 & z_0 = 1 \\ x_1 = 0.1 & y_1 = 0.1003333 & z_1 = 1.0100083 \\ x_2 = 0.2 & y_2 = 0.2026717 & z_2 = 1.0401335 \\ x_3 = 0.3 & y_3 = 0.3090401 & z_3 = 1.0906769 \end{array}$$

Using (*28.6*), we compute

$$y_0' = z_0 = 1 \qquad y_1' = z_1 = 1.0100083$$

$$y_2' = z_2 = 1.0401335 \qquad y_3' = z_3 = 1.0906769$$

$$z_0' = y_0 + x_0 = 0 + 0 = 0$$

$$z_1' = y_1 + x_1 = 0.1003333 + 0.1 = 0.2003333$$

$$z_2' = y_2 + x_2 = 0.2026717 + 0.2 = 0.4026717$$

$$z_3' = y_3 + x_3 = 0.3090401 + 0.3 = 0.6090401$$

Then using (*26.5*), we compute

$n = 3$: $x_4 = 0.4$

$$py_4 = y_3 + \frac{h}{24}(55y_3' - 59y_2' + 37y_1' - 9y_0')$$

$$= 0.3090401 + (0.1/24)[55(1.0906769) - 59(1.0401335) + 37(1.0100083) - 9(1)]$$

$$= 0.4214970$$

$$pz_4 = z_3 + \frac{h}{24}(55z_3' - 59z_2' + 37z_1' - 9z_0')$$

$$= 1.0906769 + (0.1/24)[55(0.6090401) - 59(0.4026717) + 37(0.2003333) - 9(0)]$$

$$= 1.1621432$$

$$py_4' = pz_4 = 1.1621432$$

$$pz_4' = py_4 + x_4 = 0.4214970 + 0.4 = 0.8214970$$

$$y_4 = y_3 + \frac{h}{24}(9py_4' + 19y_3' - 5y_2' + y_1')$$

$$= 0.3090401 + (0.1/24)[9(1.1621432) + 19(1.0906769) - 5(1.0401335) + 1.0100083]$$

$$= 0.4215046$$

$$z_4 = z_3 + \frac{h}{24}(9pz_4' + 19z_3' - 5z_2' - z_1')$$

$$= 1.0906769 + (0.1/24)[9(0.8214970) + 19(0.6090401) - 5(0.4026717) + (0.2003333)]$$

$$= 1.1621445$$

$$y_4' = z_4 = 1.1621445$$

$$z_4' = y_4 + x_4 = 0.4215046 + 0.4 = 0.8215046$$

Continuing in this manner, we generate Table 28-7.

Table 28-7

Method: ADAMS–BASHFORTH–MOULTON METHOD					
Problem: $y'' - y = x;\ y(0) = 0,\ y'(0) = 1$					
x_n	$h = 0.1$				True solution
	py_n	pz_n	y_n	z_n	$Y(x) = e^x - e^{-x} - x$
0.0	—	—	0.0000000	1.0000000	0.0000000
0.1	—	—	0.1003333	1.0100083	0.1003335
0.2	—	—	0.2026717	1.0401335	0.2026720
0.3	—	—	0.3090401	1.0906769	0.3090406
0.4	0.4214970	1.1621432	0.4215046	1.1621445	0.4215047
0.5	0.5421832	1.2552496	0.5421910	1.2552516	0.5421906
0.6	0.6733000	1.3709273	0.6733080	1.3709301	0.6733072
0.7	0.8171604	1.5103342	0.8171687	1.5103378	0.8171674
0.8	0.9762050	1.6748654	0.9762138	1.6748699	0.9762120
0.9	1.1530265	1.8661677	1.1530358	1.8661731	1.1530335
1.0	1.3503954	2.0861557	1.3504053	2.0861620	1.3504024

28.12. Formulate the Adams–Bashforth–Moulton method for system (*28.2*).

predictors:

$$py_{n+1} = y_n + \frac{h}{24}(55y'_n - 59y'_{n-1} + 37y'_{n-2} - 9y'_{n-3})$$

$$pz_{n+1} = z_n + \frac{h}{24}(55z'_n - 59z'_{n-1} + 37z'_{n-2} - 9z'_{n-3})$$

$$pw_{n+1} = w_n + \frac{h}{24}(55w'_n - 59w'_{n-1} + 37w'_{n-2} - 9w'_{n-3})$$

correctors:

$$y_{n+1} = y_n + \frac{h}{24}(9py'_{n+1} + 19y'_n - 5y'_{n-1} + y'_{n-2})$$

$$z_{n+1} = z_n + \frac{h}{24}(9pz'_{n+1} + 19z'_n - 5z'_{n-1} + z'_{n-2})$$

$$w_{n+1} = w_n + \frac{h}{24}(9pw'_{n+1} + 19w'_n - 5w'_{n-1} + w'_{n-2})$$

28.13. Formulate Milne's method for system (*28.1*).

predictors: $$py_{n+1} = y_{n-3} + \frac{4h}{3}(2y'_n - y'_{n-1} + 2y'_{n-2})$$

$$pz_{n+1} = z_{n-3} + \frac{4h}{3}(2z'_n - z'_{n-1} + 2z'_{n-2})$$

correctors: $$y_{n+1} = y_{n-1} + \frac{h}{3}(py'_{n+1} + 4y'_n + y'_{n-1})$$

$$z_{n+1} = z_{n-1} + \frac{h}{3}(pz'_{n+1} + 4z'_n + z'_{n-1})$$

28.14. Use Milne's method to solve $y'' - y = x$; $y(0) = 0$, $y'(0) = 1$ on the interval $[0, 1]$ with $h = 0.1$.

All starting values and their derivatives are identical to those given in Problem 28.11. Using the formulas given in Problem 28.13, we compute

n = 3: $$py_4 = y_0 + \frac{4h}{3}(2y'_3 - y'_2 + 2y'_1)$$

$$= 0 + \frac{4(0.1)}{3}[2(1.0906769) - 1.0401335 + 2(1.0100083)]$$

$$= 0.4214983$$

$$pz_4 = z_0 + \frac{4h}{3}(2z'_3 - z'_2 + 2z'_1)$$

$$= 1 + \frac{4(0.1)}{3}[2(0.6090401) - 0.4026717 + 2(0.2003333)]$$

$$= 1.1621433$$

$$py'_4 = pz_4 = 1.1621433$$

$$pz'_4 = py_4 + x_4 = 0.4214983 + 0.4 = 0.8214983$$

$$y_4 = y_2 + \frac{h}{3}(py'_4 + 4y'_3 + y'_2)$$

$$= 0.2026717 + \frac{0.1}{3}[1.1621433 + 4(1.0906769) + 1.0401335]$$

$$= 0.4215045$$

$$z_4 = z_2 + \frac{h}{3}(pz'_4 + 4z'_3 + z'_2)$$

$$= 1.0401335 + \frac{0.1}{3}[0.8214983 + 4(0.6090401) + 0.4026717]$$

$$= 1.1621445$$

n = 4: $$y'_4 = z_4 = 1.1621445$$

$$z'_4 = y_4 + x_4 = 0.4215045 + 0.4 = 0.8215045$$

$$py_5 = y_1 + \frac{4h}{3}(2y'_4 - y'_3 + 2y'_2)$$

$$= 0.1003333 + \frac{4(0.1)}{3}[2(1.1621445) - 1.0906769 + 2(1.0401335)]$$

$$= 0.5421838$$

$$pz_5 = z_1 + \frac{4h}{3}(2z_4' - z_3' + 2z_2')$$

$$= 1.0100083 + \frac{4(0.1)}{3}[2(0.8215045) - 0.6090401 + 2(0.4026717)]$$

$$= 1.2552500$$

$$py_5' = pz_5 = 1.2552500$$

$$pz_5' = py_5 + x_5 = 0.5421838 + 0.5 = 1.0421838$$

$$y_5 = y_3 + \frac{h}{3}(py_5' + 4y_4' + y_3')$$

$$= 0.3090401 + \frac{0.1}{3}[1.2552500 + 4(1.1621445) + 1.0906769]$$

$$= 0.5421903$$

$$z_5 = z_3 + \frac{h}{3}(pz_5' + 4z_4' + z_3')$$

$$= 1.0906769 + \frac{0.1}{3}[1.0421838 + 4(0.8215045) + 0.6090401]$$

$$= 1.2552517$$

Continuing in this manner, we generate Table 28-8.

Table 28-8

Method: MILNE'S METHOD					
Problem: $y'' - y = x$; $y(0) = 0$, $y'(0) = 1$					
x_n	$h = 0.1$				True solution $Y(x) = e^x - e^{-x} - x$
	py_n	pz_n	y_n	z_n	
0.0	—	—	0.0000000	1.0000000	0.0000000
0.1	—	—	0.1003333	1.0100083	0.1003335
0.2	—	—	0.2026717	1.0401335	0.2026720
0.3	—	—	0.3090401	1.0906769	0.3090406
0.4	0.4214983	1.1621433	0.4215045	1.1621445	0.4215047
0.5	0.5421838	1.2552500	0.5421903	1.2552517	0.5421906
0.6	0.6733000	1.3709276	0.6733071	1.3709300	0.6733072
0.7	0.8171597	1.5103347	0.8171671	1.5103376	0.8171674
0.8	0.9762043	1.6748655	0.9762120	1.6748693	0.9762120
0.9	1.1530250	1.8661678	1.1530332	1.8661723	1.1530335
1.0	1.3503938	2.0861552	1.3504024	2.0861606	1.3504024

Supplementary Problems

25.15. Reduce the initial-value problem $y'' + y = 0;\ y(0) = 1,\ y'(0) = 0$ to system (*28.1*).

28.16. Reduce the initial-value problem $y'' - y = x;\ y(0) = 0,\ y'(0) = -1$ to system (*28.1*).

28.17. Reduce the initial-value problem $2yy'' - 4xy^2y' + 2(\sin x)y^4 = 6;\ y(1) = 0,\ y'(1) = 15$ to system (*28.1*).

28.18. Reduce the initial-value problem $xy''' - x^2y'' + (y')^2y = 0;\ y(0) = 1,\ y'(0) = 2,\ y''(0) = 3$ to system (*28.2*).

28.19. Use Euler's method with $h = 0.1$ to solve the initial-value problem given in Problem 28.15 on the interval $[0, 1]$.

28.20. Use Euler's method with $h = 0.1$ to solve the initial-value problem given in Problem 28.16 on the interval $[0, 1]$.

28.21. Use the Runge–Kutta method with $h = 0.1$ to solve the initial-value problem given in Problem 28.15 on the interval $[0, 1]$.

28.22. Use the Runge–Kutta method with $h = 0.1$ to solve the initial-value problem given in Problem 28.16 on the interval $[0, 1]$.

28.23. Use the Adams–Bashforth–Moulton method with $h = 0.1$ to solve the initial-value problem given in Problem 28.2 on the interval $[0, 1]$. Obtain appropriate starting values from Table 28-4.

28.24. Use the Adams–Bashforth–Moulton method with $h = 0.1$ to solve the initial-value problem given in Problem 28.15 on the interval $[0, 1]$.

28.25. Use the Adams–Bashforth–Moulton method with $h = 0.1$ to solve the initial-value problem given in Problem 28.16 on the interval $[0, 1]$.

28.26. Use Milne's method with $h = 0.1$ to solve the initial-value problem given in Problem 28.2 on the interval $[0, 1]$. Obtain appropriate starting values from Table 28-4.

28.27. Use Milne's method with $h = 0.1$ to solve the initial-value problem given in Problem 28.15 on the interval $[0, 1]$.

28.28. Formulate the modified Euler's method for system (*28.1*).

28.29. Formulate the Runge–Kutta method for system (*28.2*).

28.30. Formulate Milne's method for system (*28.2*).

Chapter 29

Second-Order Boundary-Value Problems

STANDARD FORM

A boundary-value problem in standard form consists of the second-order linear differential equation

$$y'' + P(x)y' + Q(x)y = \phi(x) \tag{29.1}$$

and the boundary conditions

$$\begin{aligned} \alpha_1 y(a) + \beta_1 y'(a) &= \gamma_1 \\ \alpha_2 y(b) + \beta_2 y'(b) &= \gamma_2 \end{aligned} \tag{29.2}$$

where $P(x)$, $Q(x)$, and $\phi(x)$ are continuous in $[a, b]$ and α_1, α_2, β_1, β_2, γ_1, and γ_2 are all real constants. Furthermore, it is assumed that α_1 and β_1 are not both zero, and also that α_2 and β_2 are not both zero.

The boundary-value problem is said to be *homogeneous* if both the differential equation and the boundary conditions are homogeneous (i.e., $\phi(x) \equiv 0$ and $\gamma_1 = \gamma_2 = 0$). Otherwise the problem is *nonhomogeneous.* Thus a homogeneous boundary-value problem has the form

$$\begin{aligned} y'' + P(x)y' + Q(x)y &= 0; \\ \alpha_1 y(a) + \beta_1 y'(a) &= 0 \\ \alpha_2 y(b) + \beta_2 y'(b) &= 0 \end{aligned} \tag{29.3}$$

A somewhat more general homogeneous boundary-value problem than (29.3) is one where the coefficients $P(x)$ and $Q(x)$ also depend on an arbitrary constant λ. Such a problem has the form

$$\begin{aligned} y'' + P(x, \lambda)y' + Q(x, \lambda)y &= 0; \\ \alpha_1 y(a) + \beta_1 y'(a) &= 0 \\ \alpha_2 y(b) + \beta_2 y'(b) &= 0 \end{aligned} \tag{29.4}$$

Both (29.3) and (29.4) always admit the trivial solution $y(x) \equiv 0$.

SOLUTIONS

A boundary-value problem is solved by first obtaining the general solution to the differential equation, using any of the appropriate methods presented heretofore, and then applying the boundary conditions to evaluate the arbitrary constants.

Theorem 29.1. Let $y_1(x)$ and $y_2(x)$ be two linearly independent solutions of

$$y'' + P(x)y' + Q(x)y = 0$$

Nontrivial solutions (i.e., solutions not identically equal to zero) to the homogeneous boundary-value problem (29.3) exist if and only if the determinant

$$\begin{vmatrix} \alpha_1 y_1(a) + \beta_1 y_1'(a) & \alpha_1 y_2(a) + \beta_1 y_2'(a) \\ \alpha_2 y_1(b) + \beta_2 y_1'(b) & \alpha_2 y_2(b) + \beta_2 y_2'(b) \end{vmatrix} \tag{29.5}$$

equals zero.

Theorem 29.2. The nonhomogeneous boundary-value problem defined by (*29.1*) and (*29.2*) has a unique solution if and only if the associated homogeneous problem (*29.3*) has only the trivial solution.

In other words, *a nonhomogeneous problem has a unique solution when and only when the associated homogeneous problem has a unique solution.*

EIGENVALUE PROBLEMS

When applied to the boundary-value problem (*29.4*), Theorem 29.1 shows that nontrivial solutions may exist for certain values of λ but not for other values of λ. Those values of λ for which nontrivial solutions do exist are called *eigenvalues*; the corresponding nontrivial solutions are called *eigenfunctions.*

STURM–LIOUVILLE PROBLEMS

A second-order *Sturm–Liouville problem* is a homogeneous boundary-value problem of the form

$$[p(x)y']' + q(x)y + \lambda w(x)y = 0; \tag{29.6}$$

$$\begin{aligned} \alpha_1 y(a) + \beta_1 y'(a) &= 0 \\ \alpha_2 y(b) + \beta_2 y'(b) &= 0 \end{aligned} \tag{29.7}$$

where $p(x)$, $p'(x)$, $q(x)$, and $w(x)$ are continuous on $[a, b]$, and both $p(x)$ and $w(x)$ are positive on $[a, b]$.

Equation (*29.6*) can be written in standard form (*29.4*) by dividing through by $p(x)$. Form (*29.6*), when attainable, is preferred, because Sturm–Liouville problems have desirable features not shared by more general eigenvalue problems. The second-order differential equation

$$a_2(x)y'' + a_1(x)y' + a_0(x)y + \lambda r(x)y = 0 \tag{29.8}$$

where $a_2(x)$ does not vanish on $[a, b]$, is equivalent to Eq. (*29.6*) if and only if $a_2'(x) = a_1(x)$. (See Problem 29.15.) This condition can always be forced by multiplying Eq. (*29.8*) by a suitable factor. (See Problem 29.16.)

PROPERTIES OF STURM–LIOUVILLE PROBLEMS

Property 29.1. The eigenvalues of a Sturm–Liouville problem are all real and nonnegative.

Property 29.2. The eigenvalues of a Sturm–Liouville problem can be arranged to form a strictly increasing infinite sequence; that is, $0 \le \lambda_1 < \lambda_2 < \lambda_3 < \cdots$. Furthermore, $\lambda_n \to \infty$ as $n \to \infty$.

Property 29.3. For each eigenvalue of a Sturm–Liouville problem, there exists one and only one linearly independent eigenfunction.

[By Property 29.3 there corresponds to each eigenvalue λ_n a unique eigenfunction with lead coefficient unity; we denote this eigenfunction by $e_n(x)$.]

Property 29.4. The set of eigenfunctions $\{e_1(x), e_2(x), \ldots\}$ of a Sturm–Liouville problem satisfies the relation

$$\int_a^b w(x)e_n(x)e_m(x)\,dx = 0 \tag{29.9}$$

for $n \ne m$, where $w(x)$ is given in Eq. (*29.6*).

Solved Problems

29.1. Solve $y'' + 2y' - 3y = 0$; $y(0) = 0$, $y'(1) = 0$.

This is a homogeneous boundary-value problem of the form (*29.3*), with $P(x) \equiv 2$, $Q(x) \equiv -3$, $\alpha_1 = 1$, $\beta_1 = 0$, $\alpha_2 = 0$, $\beta_2 = 1$, $a = 0$, and $b = 1$. The general solution to the differential equation is $y = c_1 e^{-3x} + c_2 e^x$. Applying the boundary conditions, we find that $c_1 = c_2 = 0$; hence, the solution is $y \equiv 0$.

The same result follows from Theorem 29.1. Two linearly independent solutions are $y_1(x) = e^{-3x}$ and $y_2(x) = e^x$; hence, the determinant (*29.5*) becomes

$$\begin{vmatrix} 1 & 1 \\ -3e^{-3} & e \end{vmatrix} = e + 3e^{-3}$$

Since this determinant is not zero, the only solution is the trivial solution $y(x) \equiv 0$.

29.2. Solve $y'' = 0$; $y(-1) = 0$, $y(1) - 2y'(1) = 0$.

This is a homogeneous boundary-value problem of form (*29.3*), where $P(x) = Q(x) \equiv 0$, $\alpha_1 = 1$, $\beta_1 = 0$, $\alpha_2 = 1$, $\beta_2 = -2$, $a = -1$, and $b = 1$. The general solution to the differential equation is $y = c_1 + c_2 x$. Applying the boundary conditions, we obtain the equations $c_1 - c_2 = 0$ and $c_1 - c_2 = 0$, which have the solution $c_1 = c_2$, c_2 arbitrary. Thus, the solution to the boundary-value problem is $y = c_2(1 + x)$, c_2 arbitrary. As a different solution is obtained for each value of c_2, the problem has infinitely many nontrivial solutions.

The existence of nontrivial solutions is also immediate from Theorem 29.1. Here $y_1(x) = 1$, $y_2(x) = x$, and determinant (*29.5*) becomes

$$\begin{vmatrix} 1 & -1 \\ 1 & -1 \end{vmatrix} = 0$$

29.3. Solve $y'' + 2y' - 3y = 9x$; $y(0) = 1$, $y'(1) = 2$.

This is a nonhomogeneous boundary-value problem of forms (*29.1*) and (*29.2*), where $\phi(x) = x$, $\gamma_1 = 1$, and $\gamma_2 = 2$. Since the associated homogeneous problem has only the trivial solution (Problem 29.1), it follows from Theorem 29.2 that the given problem has a unique solution. Solving the differential equation by the method of Chapter 10, we obtain

$$y = c_1 e^{-3x} + c_2 e^x - 3x - 2$$

Applying the boundary conditions, we find

$$c_1 + c_2 - 2 = 1 \qquad -3c_1 e^{-3} + c_2 e - 3 = 2$$

whence

$$c_1 = \frac{3e - 5}{e + 3e^{-3}} \qquad c_2 = \frac{5 + 9e^{-3}}{e + 3e^{-3}}$$

Finally,

$$y = \frac{(3e - 5)e^{-3x} + (5 + 9e^{-3})e^x}{e + 3e^{-3}} - 3x - 2$$

29.4. Solve $y'' = 2$; $y(-1) = 5$, $y(1) - 2y'(1) = 1$.

This is a nonhomogeneous boundary-value problem of forms (*29.1*) and (*29.2*), where $\phi(x) \equiv 2$, $\gamma_1 = 5$, and $\gamma_2 = 1$. Since the associated homogeneous problem has nontrivial solutions (Problem 29.2), this problem does not have a unique solution. There are, therefore, either no solutions or more than one solution. Solving the differential equation, we find that $y = c_1 + c_2 x + x^2$. Then, applying the boundary conditions, we obtain the equations $c_1 - c_2 = 4$ and $c_1 - c_2 = 4$; thus, $c_1 = 4 + c_2$, c_2 arbitrary. Finally, $y = c_2(1 + x) + 4 + x^2$; and this problem has infinitely many solutions, one for each value of the arbitrary constant c_2.

29.5. Solve $y'' = 2$; $y(-1) = 0$, $y(1) - 2y'(1) = 0$.

This is a nonhomogeneous boundary-value problem of forms *(29.1)* and *(29.2)*, where $\phi(x) \equiv 2$ and $\gamma_1 = \gamma_2 = 0$. As in Problem 29.4, there are either no solutions or more than one solution. The solution to the differential equation is $y = c_1 + c_2 x + x^2$. Applying the boundary conditions, we obtain the equations $c_1 - c_2 = -1$ and $c_1 - c_2 = 3$. Since these equations have no solution, the boundary-value problem has no solution.

29.6. Find the eigenvalues and eigenfunctions of

$$y'' - 4\lambda y' + 4\lambda^2 y = 0; \qquad y(0) = 0, \qquad y(1) + y'(1) = 0$$

The coefficients of the given differential equation are constants (with respect to x); hence, the general solution can be found by use of the characteristic equation. We write the characteristic equation in terms of the variable m, since λ now has another meaning. Thus we have $m^2 - 4\lambda m + 4\lambda^2 = 0$, which has the double root $m = 2\lambda$; the solution to the differential equation is $y = c_1 e^{2\lambda x} + c_2 x e^{2\lambda x}$. Applying the boundary conditions and simplifying, we obtain

$$c_1 = 0 \qquad c_1(1 + 2\lambda) + c_2(2 + 2\lambda) = 0$$

It now follows that $c_1 = 0$ and either $c_2 = 0$ or $\lambda = -1$. The choice $c_2 = 0$ results in the trivial solution $y \equiv 0$; the choice $\lambda = -1$ results in the nontrivial solution $y = c_2 x e^{-2x}$, c_2 arbitrary. Thus, the boundary-value problem has the eigenvalue $\lambda = -1$ and the eigenfunction $y = c_2 x e^{-2x}$.

29.7. Find the eigenvalues and eigenfunctions of

$$y'' - 4\lambda y' + 4\lambda^2 y = 0; \qquad y'(1) = 0, \qquad y(2) + 2y'(2) = 0$$

As in Problem 29.6, the solution to the differential equation is $y = c_1 e^{2\lambda x} + c_2 x e^{2\lambda x}$. Applying the boundary conditions and simplifying, we obtain the equations

$$\begin{aligned} (2\lambda)c_1 + (1 + 2\lambda)c_2 &= 0 \\ (1 + 4\lambda)c_1 + (4 + 8\lambda)c_2 &= 0 \end{aligned} \tag{1}$$

This system of equations has a nontrivial solution for c_1 and c_2 if and only if the determinant

$$\begin{vmatrix} 2\lambda & 1 + 2\lambda \\ 1 + 4\lambda & 4 + 8\lambda \end{vmatrix} = (1 + 2\lambda)(4\lambda - 1)$$

is zero; that is, if and only if either $\lambda = -\frac{1}{2}$ or $\lambda = \frac{1}{4}$. When $\lambda = -\frac{1}{2}$, *(1)* has the solution $c_1 = 0$, c_2 arbitrary; when $\lambda = \frac{1}{4}$, *(1)* has the solution $c_1 = -3c_2$, c_2 arbitrary. It follows that the eigenvalues are $\lambda_1 = -\frac{1}{2}$ and $\lambda_2 = \frac{1}{4}$ and the corresponding eigenfunctions are $y_1 = c_2 x e^{-x}$ and $y_2 = c_2(-3 + x)e^{x/2}$.

29.8. Find the eigenvalues and eigenfunctions of

$$y'' + \lambda y' = 0; \qquad y(0) + y'(0) = 0, \qquad y'(1) = 0$$

In terms of the variable m, the characteristic equation is $m^2 + \lambda m = 0$. We consider the cases $\lambda = 0$ and $\lambda \neq 0$ separately, since they result in different solutions.

$\lambda = 0$: The solution to the differential equation is $y = c_1 + c_2 x$. Applying the boundary conditions, we obtain the equations $c_1 + c_2 = 0$ and $c_2 = 0$. It follows that $c_1 = c_2 = 0$, and $y \equiv 0$. Therefore, $\lambda = 0$ is not an eigenvalue.

$\lambda \neq 0$: The solution to the differential equation is $y = c_1 + c_2 e^{-\lambda x}$. Applying the boundary conditions, we obtain

$$c_1 + (1 - \lambda)c_2 = 0$$
$$(-\lambda e^{-\lambda})c_2 = 0$$

These equations have a nontrivial solution for c_1 and c_2 if and only if

$$\begin{vmatrix} 1 & 1-\lambda \\ 0 & -\lambda e^{-\lambda} \end{vmatrix} = -\lambda e^{-\lambda} = 0$$

which is an impossibility, since $\lambda \neq 0$.

Since we obtain only the trivial solution for $\lambda = 0$ and $\lambda \neq 0$, can conclude that the problem does not have any eigenvalues.

29.9. Find the eigenvalues and eigenfunctions of

$$y'' - 4\lambda y' + 4\lambda^2 y = 0; \qquad y(0) + y'(0) = 0, \qquad y(1) - y'(1) = 0$$

As in Problem 29.6, the solution to the differential equation is $y = c_1 e^{2\lambda x} + c_2 x e^{2\lambda x}$. Applying the boundary conditions and simplifying, we obtain the equations

$$\begin{aligned} (1 + 2\lambda)c_1 + \qquad c_2 &= 0 \\ (1 - 2\lambda)c_1 + (-2\lambda)c_2 &= 0 \end{aligned} \tag{1}$$

Equations (*1*) have a nontrivial solution for c_1 and c_2 if and only if the determinant

$$\begin{vmatrix} 1+2\lambda & 1 \\ 1-2\lambda & -2\lambda \end{vmatrix} = -4\lambda^2 - 1$$

is zero; that is, if and only if $\lambda = \pm\frac{1}{2}i$. These eigenvalues are complex. In order to keep the differential equation under consideration real, we require that λ be real. Therefore this problem has no (real) eigenvalues and the only (real) solution is the trivial one: $y(x) \equiv 0$.

29.10. Find the eigenvalues and eigenfunctions of

$$y'' + \lambda y = 0; \qquad y(0) = 0, \qquad y(1) = 0$$

The characteristic equation is $m^2 + \lambda = 0$. We consider the cases $\lambda = 0$, $\lambda < 0$, and $\lambda > 0$ separately, since they lead to different solutions.

$\lambda = 0$: The solution is $y = c_1 + c_2 x$. Applying the boundary conditions, we obtain $c_1 = c_2 = 0$, which results in the trivial solution.

$\lambda < 0$: The solution is $y = c_1 e^{\sqrt{-\lambda}x} + c_2 e^{-\sqrt{-\lambda}x}$, where $-\lambda$ and $\sqrt{-\lambda}$ are positive. Applying the boundary conditions, we obtain

$$c_1 + c_2 = 0 \qquad c_1 e^{\sqrt{-\lambda}} - c_2 e^{-\sqrt{-\lambda}} = 0$$

Here

$$\begin{vmatrix} 1 & 1 \\ e^{\sqrt{-\lambda}} & e^{-\sqrt{-\lambda}} \end{vmatrix} = e^{-\sqrt{-\lambda}} - e^{\sqrt{-\lambda}}$$

which is never zero for any value of $\lambda < 0$. Hence, $c_1 = c_2 = 0$ and $y \equiv 0$.

$\lambda > 0$: The solution is $A \sin \sqrt{\lambda}\, x + B \cos \sqrt{\lambda}\, x$. Applying the boundary conditions, we obtain $B = 0$ and $A \sin \sqrt{\lambda} = 0$. Note that $\sin \theta = 0$ if and only if $\theta = n\pi$, where $n = 0, \pm 1, \pm 2, \ldots$. Furthermore, if $\theta > 0$, then n must be positive. To satisfy the boundary conditions, $B = 0$ and either $A = 0$ or $\sin \sqrt{\lambda} = 0$. This last equation is equivalent to $\sqrt{\lambda} = n\pi$ where $n = 1, 2, 3, \ldots$. The choice $A = 0$

results in the trivial solution; the choice $\sqrt{\lambda} = n\pi$ results in the nontrivial solution $y_n = A_n \sin n\pi x$. Here the notation A_n signifies that the arbitrary constant A_n can be different for different values of n.

Collecting the results of all three cases, we conclude that the eigenvalues are $\lambda_n = n^2\pi^2$ and the corresponding eigenfunctions are $y_n = A_n \sin n\pi x$, for $n = 1, 2, 3, \ldots$.

29.11. Find the eigenvalues and eigenfunctions of

$$y'' + \lambda y = 0; \qquad y(0) = 0, \qquad y'(\pi) = 0$$

As in Problem 29.10, the cases $\lambda = 0$, $\lambda < 0$, and $\lambda > 0$ must be considered separately.

$\lambda = 0$: The solution is $y = c_1 + c_2 x$. Applying the boundary conditions, we obtain $c_1 = c_2 = 0$; hence $y \equiv 0$.

$\lambda < 0$: The solution is $y = c_1 e^{\sqrt{-\lambda}x} + c_2 e^{-\sqrt{-\lambda}x}$, where $-\lambda$ and $\sqrt{-\lambda}$ are positive. Applying the boundary conditions, we obtain

$$c_1 + c_2 = 0 \qquad c_1\sqrt{-\lambda}\, e^{\sqrt{-\lambda}\pi} - c_2\sqrt{-\lambda}\, e^{-\sqrt{-\lambda}\pi} = 0$$

which admits only the solution $c_1 = c_2 = 0$; hence $y \equiv 0$.

$\lambda > 0$: The solution is $y = A \sin\sqrt{\lambda}\,x + B\cos\sqrt{\lambda}\,x$. Applying the boundary conditions, we obtain $B = 0$
and $A\sqrt{\lambda}\cos\sqrt{\lambda}\,\pi = 0$. For $\theta > 0$, $\cos\theta = 0$ if and only if θ is a positive odd multiple of $\pi/2$; that is, when $\theta = (2n-1)(\pi/2) = (n - \frac{1}{2})\pi$, where $n = 1, 2, 3, \ldots$. Therefore, to satisfy the boundary conditions, we must have $B = 0$ and either $A = 0$ or $\cos\sqrt{\lambda}\,\pi = 0$. This last equation is equivalent to $\sqrt{\lambda} = n - \frac{1}{2}$. The choice $A = 0$ results in the trivial solution; the choice $\sqrt{\lambda} = n - \frac{1}{2}$ results in the nontrivial solution $y_n = A_n \sin(n - \frac{1}{2})x$.

Collecting all three cases, we conclude that the eigenvalues are $\lambda_n = (n - \frac{1}{2})^2$ and the corresponding eigenfunctions are $y_n = A_n \sin(n - \frac{1}{2})x$, where $n = 1, 2, 3, \ldots$.

29.12. Show that the boundary-value problem given in Problem 29.10 is a Sturm–Liouville problem.

It has form *(29.6)* with $p(x) \equiv 1$, $q(x) \equiv 0$, and $w(x) \equiv 1$. Here both $p(x)$ and $w(x)$ are positive and continuous everywhere, in particular on $[0, 1]$.

29.13. Determine whether the boundary-value problem

$$(xy')' + [x^2 + 1 + \lambda e^x]y = 0; \qquad y(1) + 2y'(1) = 0, \qquad y(2) - 3y'(2) = 0$$

is a Sturm–Liouville problem.

Here $p(x) = x$, $q(x) = x^2 + 1$, and $w(x) = e^x$. Since both $p(x)$ and $q(x)$ are continuous and positive on $[1, 2]$, the interval of interest, the boundary problem is a Sturm–Liouville problem.

29.14. Determine which of the following differential equations with the boundary conditions $y(0) = 0$, $y'(1) = 0$ form Sturm–Liouville problems:

(*a*) $e^x y'' + e^x y' + \lambda y = 0$ (*b*) $xy'' + y' + (x^2 + 1 + \lambda)y = 0$

(*c*) $\left(\dfrac{1}{x}y'\right)' + (x + \lambda)y = 0$ (*d*) $y'' + \lambda(1 + x)y = 0$

(*e*) $e^{2x}y'' + e^{2x}y' + \lambda y = 0$

(*a*) The equation can be rewritten as $(e^x y')' + \lambda y = 0$; hence $p(x) = e^x$, $q(x) \equiv 0$, and $w(x) \equiv 1$. This is a Sturm–Liouville problem.

(*b*) The equation is equivalent to $(xy')' + (x^2 + 1)y + \lambda y = 0$; hence $p(x) = x$, $q(x) = x^2 + 1$, and $w(x) \equiv 1$. Since $p(x)$ is zero at a point in the interval $[0, 1]$, this is not a Sturm–Liouville problem.

(c) Here $p(x) = 1/x$, $q(x) = x$, and $w(x) \equiv 1$. Since $p(x)$ is not continuous in $[0, 1]$, in particular at $x = 0$, this is not a Sturm–Liouville problem.

(d) The equation can be rewritten as $(y')' + \lambda(1 + x)y = 0$; hence $p(x) \equiv 1$, $q(x) \equiv 0$, and $w(x) = 1 + x$. This is a Sturm–Liouville problem.

(e) The equation, in its present form, is not equivalent to Eq. (29.6); this is not a Sturm–Liouville problem. However, if we first multiply the equation by e^{-x}, we obtain $(e^x y')' + \lambda e^{-x} y = 0$; this is a Sturm–Liouville problem with $p(x) = e^x$, $q(x) \equiv 0$, and $w(x) = e^{-x}$.

29.15. Prove that Eq. (29.6) is equivalent to Eq. (29.8) if and only if $a_2'(x) = a_1(x)$.

Applying the product rule of differentiation to (29.6), we find that

$$p(x)y'' + p'(x)y' + q(x)y + \lambda w(x)y = 0 \tag{1}$$

Setting $a_2(x) = p(x)$, $a_1(x) = p'(x)$, $a_0(x) = q(x)$, and $r(x) = w(x)$, it follows that (1), which is (29.6) rewritten, is precisely (29.8) with $a_2'(x) = p'(x) = a_1(x)$.

Conversely, if $a_2'(x) = a_1(x)$, then (29.8) has the form

$$a_2(x)y'' + a_2'(x)y' + a_0(x)y + \lambda r(x)y = 0$$

which is equivalent to $[a_2(x)y']' + a_0(x)y + \lambda r(x)y = 0$. This last equation is precisely (29.6) with $p(x) = a_2(x)$, $q(x) = a_0(x)$, and $w(x) = r(x)$.

29.16. Show that if Eq. (29.8) is multiplied by $I(x) = e^{\int [a_1(x)/a_2(x)]\,dx}$, the resulting equation is equivalent to Eq. (29.6).

Multiplying (29.8) by $I(x)$, we obtain

$$I(x)a_2(x)y'' + I(x)a_1(x)y' + I(x)a_0(x)y + \lambda I(x)r(x)y = 0$$

which can be rewritten as

$$a_2(x)[I(x)y']' + I(x)a_0(x)y + \lambda I(x)r(x)y = 0 \tag{1}$$

Divide (1) by $a_2(x)$ and then set $p(x) = I(x)$, $q(x) = I(x)a_0(x)/a_2(x)$ and $w(x) = I(x)r(x)/a_2(x)$; the resulting equation is precisely (29.6). Note that since $I(x)$ is an exponential and since $a_2(x)$ does not vanish, $I(x)$ is positive.

29.17. Transform $y'' + 2xy' + (x + \lambda)y = 0$ into Eq. (29.6) by means of the procedure outlined in Problem 29.16.

Here $a_2(x) \equiv 1$ and $a_1(x) = 2x$; hence $a_1(x)/a_2(x) = 2x$ and $I(x) = e^{\int 2x\,dx} = e^{x^2}$. Multiplying the given differential equation by $I(x)$, we obtain

$$e^{x^2}y'' + 2xe^{x^2}y' + xe^{x^2}y + \lambda e^{x^2}y = 0$$

which can be rewritten as

$$(e^{x^2}y')' + xe^{x^2}y + \lambda e^{x^2}y = 0$$

This last equation is precisely Eq. (29.6) with $p(x) = e^{x^2}$, $q(x) = xe^{x^2}$, and $w(x) = e^{x^2}$.

29.18. Transform $(x + 2)y'' + 4y' + xy + \lambda e^x y = 0$ into Eq. (29.6) by means of the procedure outlined in Problem 29.16.

Here $a_2(x) = x + 2$ and $a_1(x) \equiv 4$; hence $a_1(x)/a_2(x) = 4/(x + 2)$ and

$$I(x) = e^{\int [4/(x+2)]\,dx} = e^{4 \ln |x+2|} = e^{\ln (x+2)^4} = (x + 2)^4$$

Multiplying the given differential equation by $I(x)$, we obtain

$$(x + 2)^5 y'' + 4(x + 2)^4 y' + (x + 2)^4 xy + \lambda(x + 2)^4 e^x y = 0$$

which can be rewritten as

$$(x+2)[(x+2)^4y']' + (x+2)^4xy + \lambda(x+2)^4e^xy = 0$$

or

$$[(x+2)^4y']' + (x+2)^3xy + \lambda(x+2)^3e^xy = 0$$

This last equation is precisely (*29.6*) with $p(x)=(x+2)^4$, $q(x)=(x+2)^3x$, and $w(x)=(x+2)^3e^x$. Note that since we divided by $a_2(x)$, it is necessary to restrict $x \neq -2$. Furthermore, in order that both $p(x)$ and $w(x)$ be positive, we must require $x > -2$.

29.19. Verify Properties 29.1 through 29.4 for the Sturm–Liouville problem

$$y'' + \lambda y = 0; \qquad y(0) = 0, \qquad y(1) = 0$$

Using the results of Problem 29.10, we have that the eigenvalues are $\lambda_n = n^2\pi^2$ and the corresponding eigenfunctions are $y_n(x) = A_n \sin n\pi x$, for $n = 1, 2, 3, \ldots$. The eigenvalues are obviously real and nonnegative, and they can be ordered as $\lambda_1 = \pi^2 < \lambda_2 = 4\pi^2 < \lambda_3 = 9\pi^2 < \cdots$. Each eigenvalue has a single linearly independent eigenfunction $e_n(x) = \sin n\pi x$ associated with it. Finally, since

$$\sin n\pi x \sin m\pi x = \frac{1}{2}\cos(n-m)\pi x - \frac{1}{2}\cos(n+m)\pi x$$

we have for $n \neq m$ and $w(x) \equiv 1$:

$$\begin{aligned}\int_a^b w(x)e_n(x)e_m(x)\,dx &= \int_0^1 \left[\frac{1}{2}\cos(n-m)\pi x - \frac{1}{2}\cos(n+m)\pi x\right]dx \\ &= \left[\frac{1}{2(n-m)\pi}\sin(n-m)\pi x - \frac{1}{2(n+m)\pi}\sin(n+m)\pi x\right]_{x=0}^{x=1} \\ &= 0\end{aligned}$$

29.20. Verify Properties 29.1 through 29.4 for the Sturm–Liouville problem

$$y'' + \lambda y = 0; \qquad y'(0) = 0, \qquad y(\pi) = 0$$

For this problem, we calculate the eigenvalues $\lambda_n = (n - \frac{1}{2})^2$ and the corresponding eigenfunctions $y_n(x) = A_n \cos(n - \frac{1}{2})x$, for $n = 1, 2, \ldots$. The eigenvalues are real and positive, and can be ordered as

$$\lambda_1 = \frac{1}{4} < \lambda_2 = \frac{9}{4} < \lambda_3 = \frac{25}{4} < \cdots$$

Each eigenvalue has only one linearly independent eigenfunction $e_n(x) = \cos(n - \frac{1}{2})x$ associated with it. Also, for $n \neq m$ and $w(x) \equiv 1$,

$$\begin{aligned}\int_a^b w(x)e_n(x)e_m(x)\,dx &= \int_0^\pi \cos\left(n-\frac{1}{2}\right)x\cos\left(m-\frac{1}{2}\right)x\,dx \\ &= \int_0^\pi\left[\frac{1}{2}\cos(n+m-1)x + \frac{1}{2}\cos(n-m)x\right]dx \\ &= \left[\frac{1}{2(n+m-1)}\sin(n+m-1)x + \frac{1}{2(n-m)}\sin(n-m)x\right]_{x=0}^{x=\pi} \\ &= 0\end{aligned}$$

29.21. Prove that if the set of nonzero functions $\{y_1(x), y_2(x), \ldots, y_p(x)\}$ satisfies (*29.9*), then the set is linearly independent on $[a, b]$.

From (*7.7*), we consider the equation

$$c_1y_1(x) + c_2y_2(x) + \cdots + c_ky_k(x) + \cdots + c_py_p(x) \equiv 0 \tag{1}$$

Multiplying this equation by $w(x)y_k(x)$ and then integrating from a to b, we obtain

$$c_1 \int_a^b w(x)y_k(x)y_1(x)\,dx + c_2 \int_a^b w(x)y_k(x)y_2(x)\,dx + \cdots$$

$$+ c_k \int_a^b w(x)y_k(x)y_k(x)\,dx + \cdots + c_p \int_a^b w(x)y_k(x)y_p(x)\,dx = 0$$

From Eq. (*29.9*) we conclude that for $i \neq k$,

$$c_k \int_a^b w(x)y_k(x)y_i(x)\,dx = 0$$

But since $y_k(x)$ is a nonzero function and $w(x)$ is positive on $[a, b]$, it follows that

$$\int_a^b w(x)[y_k(x)]^2\,dx \neq 0$$

hence, $c_k = 0$. Since $c_k = 0$, $k = 1, 2, \ldots, p$, is the only solution to (*1*), the given set of functions is linearly independent on $[a, b]$.

Supplementary Problems

In Problems 29.22 through 29.29, find all solutions, if solutions exist, to the given boundary-value problems.

29.22. $y'' + y = 0;\ y(0) = 0,\ y(\pi/2) = 0$

29.23. $y'' + y = x;\ y(0) = 0,\ y(\pi/2) = 0$

29.24. $y'' + y = 0;\ y(0) = 0,\ y(\pi/2) = 1$

29.25. $y'' + y = x;\ y(0) = -1,\ y(\pi/2) = 1$

29.26. $y'' + y = 0;\ y'(0) = 0,\ y(\pi/2) = 0$

29.27. $y'' + y = 0;\ y'(0) = 1,\ y(\pi/2) = 0$

29.28. $y'' + y = x;\ y'(0) = 1,\ y(\pi/2) = 0$

29.29. $y'' + y = x;\ y'(0) = 1,\ y(\pi/2) = \pi/2$

In Problems 29.30 through 29.36, find the eigenvalues and eigenfunctions, if any, of the given boundary-value problems.

29.30. $y'' + 2\lambda y' + \lambda^2 y = 0;\ y(0) + y'(0) = 0,\ y(1) + y'(1) = 0$

29.31. $y'' + 2\lambda y' + \lambda^2 y = 0;\ y(0) = 0,\ y(1) = 0$

29.32. $y'' + 2\lambda y' + \lambda^2 y = 0;\ y(1) + y'(1) = 0,\ 3y(2) + 2y'(2) = 0$

29.33. $y'' + \lambda y' = 0;\ y(0) + y'(0) = 0;\ y(2) + y'(2) = 0$

29.34. $y'' - \lambda y = 0;\ y(0) = 0,\ y(1) = 0$

29.35. $y'' + \lambda y = 0;\ y'(0) = 0,\ y(5) = 0$

29.36. $y'' + \lambda y = 0;\ y'(0) = 0,\ y'(\pi) = 0$

In Problems 29.37 through 29.43, determine whether each of the given differential equations with the boundary conditions $y(-1) + 2y'(-1) = 0$, $y(1) + 2y'(1) = 0$ is a Sturm–Liouville problem.

29.37. $(2 + \sin x)y'' + (\cos x)y' + (1 + \lambda)y = 0$

29.38. $(\sin \pi x)y'' + (\pi \cos \pi x)y' + (x + \lambda)y = 0$

29.39. $(\sin x)y'' + (\cos x)y' + (1 + \lambda)y = 0$

29.40. $(x + 2)^2 y'' + 2(x + 2)y' + (e^x + \lambda e^{2x})y = 0$

29.41. $(x + 2)^2 y'' + (x + 2)y' + (e^x + \lambda e^{2x})y = 0$

29.42. $y'' + \dfrac{3}{x^2}\lambda y = 0$

29.43. $y'' + \dfrac{3}{(x - 4)^2}\lambda y = 0$

29.44. Transform $e^{2x}y'' + e^{2x}y' + (x + \lambda)y = 0$ into Eq. *(29.6)* by means of the procedure outlined in Problem 29.16.

29.45. Transform $x^2y'' + xy' + \lambda xy = 0$ into Eq. *(29.6)* by means of the procedure outlined in Problem 29.16.

29.46. Verify Properties 29.1 through 29.4 for the Sturm–Liouville problem

$$y'' + \lambda y = 0; \qquad y'(0) = 0, \qquad y'(\pi) = 0$$

29.47. Verify Properties 29.1 through 29.4 for the Sturm–Liouville problem

$$y'' + \lambda y = 0; \qquad y(0) = 0, \qquad y(2\pi) = 0$$

Chapter 30

Eigenfunction Expansions

PIECEWISE SMOOTH FUNCTIONS

A wide class of functions can be represented by infinite series of eigenfunctions of a Sturm–Liouville problem (see Chapter 29).

Definition: A function $f(x)$ is *piecewise continuous on the open interval* $a<x<b$ if (1) $f(x)$ is continuous everywhere in $a<x<b$ with the possible exception of at most a *finite* number of points $x_1, x_2, \ldots, x_n$, and (2) at these points of discontinuity, the right- and left-hand limits of $f(x)$, respectively $\lim\limits_{\substack{x \to x_j \\ x > x_j}} f(x)$ and $\lim\limits_{\substack{x \to x_j \\ x < x_j}} f(x)$, exist ($j = 1, 2, \ldots, n$).

(Note that a continuous function is piecewise continuous.)

Definition: A function $f(x)$ is *piecewise continuous on the closed interval* $a \le x \le b$ if (1) it is piecewise continuous on the open interval $a<x<b$, (2) the right-hand limit of $f(x)$ exists at $x = a$, and (3) the left-hand limit of $f(x)$ exists at $x = b$.

Definition: A function $f(x)$ is *piecewise smooth* on $[a, b]$ if both $f(x)$ and $f'(x)$ are piecewise continuous on $[a, b]$.

Theorem 30.1. If $f(x)$ is piecewise smooth on $[a, b]$ and if $\{e_n(x)\}$ is the set of all eigenfunctions of a Sturm–Liouville problem (see Property 29.3), then

$$f(x) = \sum_{n=1}^{\infty} c_n e_n(x) \tag{30.1}$$

where

$$c_n = \frac{\int_a^b w(x) f(x) e_n(x)\, dx}{\int_a^b w(x) e_n^2(x)\, dx} \tag{30.2}$$

The representation (30.1) is valid at all points in the open interval (a, b) where $f(x)$ is continuous. The function $w(x)$ in (30.2) is given by Eq. (29.6).

Because different Sturm–Liouville problems usually generate different sets of eigenfunctions, a given piecewise smooth function will have many expansions of the form (30.1). The basic features of all such expansions are exhibited by the trigonometric series discussed below.

FOURIER SINE SERIES

The eigenfunctions of the Sturm–Liouville problem $y'' + \lambda y = 0$; $y(0) = 0$, $y(L) = 0$, where L is a real positive number, are $e_n(x) = \sin(n\pi x/L)$ ($n = 1, 2, 3, \ldots$). Substituting these functions into (30.1), we obtain

$$f(x) = \sum_{n=1}^{\infty} c_n \sin\frac{n\pi x}{L} \tag{30.3}$$

For this Sturm–Liouville problem, $w(x) \equiv 1$, $a = 0$, and $b = L$; so that

$$\int_a^b w(x) e_n^2(x)\, dx = \int_0^L \sin^2 \frac{n\pi x}{L}\, dx = \frac{L}{2}$$

and (30.2) becomes

$$c_n = \frac{2}{L}\int_0^L f(x) \sin\frac{n\pi x}{L}\, dx \tag{30.4}$$

The expansion (30.3) with coefficients given by (30.4) is the *Fourier sine series* for $f(x)$ on $(0, L)$.

FOURIER COSINE SERIES

The eigenfunctions of the Sturm–Liouville problem $y'' + \lambda y = 0$; $y'(0) = 0$, $y'(L) = 0$, where L is a real positive number, are $e_0(x) = 1$ and $e_n(x) = \cos(n\pi x/L)$ $(n = 1, 2, 3, \ldots)$. Here $\lambda = 0$ is an eigenvalue with corresponding eigenfunction $e_0(x) = 1$. Substituting these functions into *(30.1)*, where because of the additional eigenfunction $e_0(x)$ the summation now begins at $n = 0$, we obtain

$$f(x) = c_0 + \sum_{n=1}^{\infty} c_n \cos\frac{n\pi x}{L} \tag{30.5}$$

For this Sturm–Liouville problem, $w(x) \equiv 1$, $a = 0$, and $b = L$; so that

$$\int_a^b w(x)e_0^2(x)\,dx = \int_0^L dx = L \qquad \int_a^b w(x)e_n^2(x)\,dx = \int_0^L \cos^2\frac{n\pi x}{L}\,dx = \frac{L}{2}$$

Thus *(30.2)* becomes

$$c_0 = \frac{1}{L}\int_0^L f(x)\,dx \qquad c_n = \frac{2}{L}\int_0^L f(x)\cos\frac{n\pi x}{L}\,dx \qquad (n = 1, 2, \ldots) \tag{30.6}$$

The expansion *(30.5)* with coefficients given by *(30.6)* is the *Fourier cosine series* for $f(x)$ on $(0, L)$.

Solved Problems

30.1. Determine whether $f(x) = \begin{cases} x^2+1 & x \geq 0 \\ 1/x & x < 0 \end{cases}$ is piecewise continuous on $[-1, 1]$.

The given function is continuous everywhere on $[-1, 1]$ except at $x = 0$. Therefore, if the right- and left-hand limits exist at $x = 0$, $f(x)$ will be piecewise continuous on $[-1, 1]$. We have

$$\lim_{\substack{x\to 0\\ x>0}} f(x) = \lim_{\substack{x\to 0\\ x>0}} (x^2 + 1) = 1 \qquad \lim_{\substack{x\to 0\\ x<0}} f(x) = \lim_{\substack{x\to 0\\ x<0}} \frac{1}{x} = -\infty$$

Since the left-hand limit does not exist, $f(x)$ is not piecewise continuous on $[-1, 1]$.

30.2. Is $f(x) = \begin{cases} \sin \pi x & x > 1 \\ 0 & 0 \leq x \leq 1 \\ e^x & -1 < x < 0 \\ x^3 & x \leq -1 \end{cases}$ piecewise continuous on $[-2, 5]$?

The given function is continuous on $[-2, 5]$ except at the two points $x_1 = 0$ and $x_2 = -1$. (Note that $f(x)$ is continuous at $x = 1$.) At the two points of discontinuity, we find that

$$\lim_{\substack{x\to 0\\ x>0}} f(x) = \lim_{x\to 0} 0 = 0 \qquad \lim_{\substack{x\to 0\\ x<0}} f(x) = \lim_{x\to 0} e^x = e^0 = 1$$

and

$$\lim_{\substack{x\to -1\\ x>-1}} f(x) = \lim_{x\to -1} e^x = e^{-1} \qquad \lim_{\substack{x\to -1\\ x<-1}} f(x) = \lim_{x\to -1} x^3 = -1$$

Since all required limits exist, $f(x)$ is piecewise continuous on $[-2, 5]$.

30.3. Is the function

$$f(x) = \begin{cases} x^2 + 1 & x < 0 \\ 1 & 0 \le x \le 1 \\ 2x + 1 & x > 1 \end{cases}$$

piecewise smooth on $[-2, 2]$?

The function is continuous everywhere on $[-2, 2]$ except at $x_1 = 1$. Since the required limits exist at x_1, $f(x)$ is piecewise continuous. Differentiating $f(x)$, we obtain

$$f'(x) = \begin{cases} 2x & x < 0 \\ 0 & 0 \le x < 1 \\ 2 & x > 1 \end{cases}$$

The derivative does not exist at $x_1 = 1$ but is continuous at all other points in $[-2, 2]$. At x_1 the required limits exist; hence $f'(x)$ is piecewise continuous. It follows that $f(x)$ is piecewise smooth on $[-2, 2]$.

30.4. Is the function

$$f(x) = \begin{cases} 1 & x < 0 \\ \sqrt{x} & 0 \le x \le 1 \\ x^3 & x > 1 \end{cases}$$

piecewise smooth on $[-1, 3]$?

The function $f(x)$ is continuous everywhere on $[-1, 3]$ except at $x_1 = 0$. Since the required limits exist at x_1, $f(x)$ is piecewise continuous. Differentiating $f(x)$, we obtain

$$f'(x) = \begin{cases} 0 & x < 0 \\ \dfrac{1}{2\sqrt{x}} & 0 < x < 1 \\ 3x^2 & x > 1 \end{cases}$$

which is continuous everywhere on $[-1, 3]$ except at the two points $x_1 = 0$ and $x_2 = 1$ where the derivative does not exist. At x_1,

$$\lim_{\substack{x \to x_1 \\ x > x_1}} f'(x) = \lim_{\substack{x \to 0 \\ x > 0}} \frac{1}{2\sqrt{x}} = \infty$$

Hence, one of the required limits does not exist. It follows that $f'(x)$ is not piecewise continuous, and therefore that $f(x)$ is not piecewise smooth, on $[-1, 3]$.

30.5. Find a Fourier sine series for $f(x) = 1$ on $(0, 5)$.

Using Eq. (*30.4*) with $L = 5$, we have

$$c_n = \frac{2}{L}\int_0^L f(x) \sin\frac{n\pi x}{L}\, dx = \frac{2}{5}\int_0^5 (1) \sin\frac{n\pi x}{5}\, dx$$

$$= \frac{2}{5}\left[-\frac{5}{n\pi}\cos\frac{n\pi x}{5}\right]_{x=0}^{x=5} = \frac{2}{n\pi}[1 - \cos n\pi] = \frac{2}{n\pi}[1 - (-1)^n]$$

Thus Eq. (*30.3*) becomes

$$1 = \sum_{n=1}^{\infty} \frac{2}{n\pi}[1 - (-1)^n] \sin\frac{n\pi x}{5}$$

$$= \frac{4}{\pi}\left(\sin\frac{\pi x}{5} + \frac{1}{3}\sin\frac{3\pi x}{5} + \frac{1}{5}\sin\frac{5\pi x}{5} + \cdots\right) \tag{1}$$

Since $f(x) = 1$ is piecewise smooth on $[0, 5]$ and continuous everywhere in the open interval $(0, 5)$, it follows from Theorem 30.1 that (*1*) is valid for all x in $(0, 5)$.

30.6. Find a Fourier cosine series for $f(x) = x$ on $(0, 3)$.

Using Eq. (*30.6*) with $L = 3$, we have

$$c_0 = \frac{1}{L}\int_0^L f(x)\,dx = \frac{1}{3}\int_0^3 x\,dx = \frac{3}{2}$$

$$\begin{aligned} c_n &= \frac{2}{L}\int_0^L f(x)\cos\frac{n\pi x}{L}\,dx = \frac{2}{3}\int_0^3 x\cos\frac{n\pi x}{3}\,dx \\ &= \frac{2}{3}\left[\frac{3x}{n\pi}\sin\frac{n\pi x}{3} + \frac{9}{n^2\pi^2}\cos\frac{n\pi x}{3}\right]_{x=0}^{x=3} \\ &= \frac{2}{3}\left(\frac{9}{n^2\pi^2}\cos n\pi - \frac{9}{n^2\pi^2}\right) = \frac{6}{n^2\pi^2}[(-1)^n - 1] \end{aligned}$$

Thus Eq. (*30.5*) becomes

$$\begin{aligned} x &= \frac{3}{2} + \sum_{n=1}^{\infty}\frac{6}{n^2\pi^2}[(-1)^n - 1]\cos\frac{n\pi x}{3} \\ &= \frac{3}{2} - \frac{12}{\pi^2}\left(\cos\frac{\pi x}{3} + \frac{1}{9}\cos\frac{3\pi x}{3} + \frac{1}{25}\cos\frac{5\pi x}{3} + \cdots\right) \end{aligned} \tag{1}$$

Since $f(x) = x$ is piecewise smooth on $[0, 3]$ and continuous everywhere in the open interval $(0, 3)$, it follows from Theorem 30.1 that (*1*) is valid for all x in $(0, 3)$.

30.7. Find a Fourier sine series for $f(x) = \begin{cases} 0 & x \le 2 \\ 2 & x > 2 \end{cases}$ on $(0, 3)$.

Using Eq. (*30.4*) with $L = 3$, we obtain

$$\begin{aligned} c_n &= \frac{2}{3}\int_0^3 f(x)\sin\frac{n\pi x}{3}\,dx \\ &= \frac{2}{3}\int_0^2 (0)\sin\frac{n\pi x}{3}\,dx + \frac{2}{3}\int_2^3 (2)\sin\frac{n\pi x}{3}\,dx \\ &= 0 + \frac{4}{3}\left[-\frac{3}{n\pi}\cos\frac{n\pi x}{3}\right]_{x=2}^{x=3} = \frac{4}{n\pi}\left[\cos\frac{2n\pi}{3} - \cos n\pi\right] \end{aligned}$$

Thus Eq. (*30.3*) becomes

$$f(x) = \sum_{n=1}^{\infty}\frac{4}{n\pi}\left[\cos\frac{2n\pi}{3} - (-1)^n\right]\sin\frac{n\pi x}{3}$$

Furthermore, $\cos\frac{2\pi}{3} = -\frac{1}{2}$, $\cos\frac{4\pi}{3} = -\frac{1}{2}$, $\cos\frac{6\pi}{3} = 1, \ldots$

Hence,
$$f(x) = \frac{4}{\pi}\left(\frac{1}{2}\sin\frac{\pi x}{3} - \frac{3}{4}\sin\frac{2\pi x}{3} + \frac{2}{3}\sin\frac{3\pi x}{3} - \cdots\right) \tag{1}$$

Since $f(x)$ is piecewise smooth on $[0, 3]$ and continuous everywhere in $(0, 3)$ except at $x = 2$, it follows from Theorem 30.1 that (*1*) is valid everywhere in $(0, 3)$ except at $x = 2$.

30.8. Find a Fourier sine series for $f(x) = e^x$ on $(0, \pi)$.

Using Eq. (*30.4*) with $L = \pi$, we obtain

$$c_n = \frac{2}{\pi}\int_0^\pi e^x \sin\frac{n\pi x}{\pi}\,dx = \frac{2}{\pi}\left[\frac{e^x}{1+n^2}(\sin nx - n\cos nx)\right]_{x=0}^{x=\pi}$$

$$= \frac{2}{\pi}\left(\frac{n}{1+n^2}\right)(1 - e^\pi \cos n\pi)$$

Thus Eq. (*30.3*) becomes

$$e^x = \frac{2}{\pi}\sum_{n=1}^{\infty}\frac{n}{1+n^2}[1 - e^\pi(-1)^n]\sin nx$$

It follows from Theorem 30.1 that this last equation is valid for all x in $(0, \pi)$.

30.9. Find a Fourier cosine series for $f(x) = e^x$ on $(0, \pi)$.

Using Eq. (*30.6*) with $L = \pi$, we have

$$c_0 = \frac{1}{\pi}\int_0^\pi e^x\,dx = \frac{1}{\pi}(e^\pi - 1)$$

$$c_n = \frac{2}{\pi}\int_0^\pi e^x \cos\frac{n\pi x}{\pi}\,dx = \frac{2}{\pi}\left[\frac{e^x}{1+n^2}(\cos nx + n\sin nx)\right]_{x=0}^{x=\pi}$$

$$= \frac{2}{\pi}\left(\frac{1}{1+n^2}\right)(e^\pi \cos n\pi - 1)$$

Thus Eq. (*30.5*) becomes

$$e^x = \frac{1}{\pi}(e^\pi - 1) + \frac{2}{\pi}\sum_{n=1}^{\infty}\frac{1}{1+n^2}[(-1)^n e^\pi - 1]\cos nx$$

As in Problem 30.8, this last equation is valid for all x in $(0, \pi)$.

30.10. Find an expansion for $f(x) = e^x$ in terms of the eigenfunctions of the Sturm–Liouville problem $y'' + \lambda y = 0$; $y'(0) = 0$, $y(\pi) = 0$.

From Problem 29.20, we have $e_n(x) = \cos\left(n - \frac{1}{2}\right)x$ for $n = 1, 2, \ldots$. Substituting these functions and $w(x) \equiv 1$, $a = 0$, and $b = \pi$ into Eq. (*30.2*), we obtain for the numerator:

$$\int_a^b w(x)f(x)e_n(x)\,dx = \int_0^\pi e^x \cos\left(n - \frac{1}{2}\right)x\,dx$$

$$= \frac{e^x}{1 + (n-\frac{1}{2})^2}\left[\cos\left(n - \frac{1}{2}\right)x + \left(n - \frac{1}{2}\right)\sin\left(n - \frac{1}{2}\right)x\right]\Bigg|_{x=0}^{x=\pi}$$

$$= \frac{-1}{1 + (n-\frac{1}{2})^2}\left[e^\pi\left(n - \frac{1}{2}\right)(-1)^n + 1\right]$$

and for the denominator:

$$\int_a^b w(x)e_n^2(x)\,dx = \int_0^\pi \cos^2\left(n - \frac{1}{2}\right)x\,dx$$

$$= \left[\frac{x}{2} + \frac{\sin(2n-1)x}{4(n-\frac{1}{2})}\right]_{x=0}^{x=\pi} = \frac{\pi}{2}$$

Thus

$$c_n = \frac{2}{\pi}\left[\frac{-1}{1 + (n-\frac{1}{2})^2}\right]\left[e^\pi\left(n - \frac{1}{2}\right)(-1)^n + 1\right]$$

and Eq. (*30.1*) becomes

$$e^x = \frac{-2}{\pi}\sum_{n=1}^{\infty}\frac{1 + (-1)^n e^\pi(n - \frac{1}{2})}{1 + (n-\frac{1}{2})^2}\cos\left(n - \frac{1}{2}\right)x$$

By Theorem 30.1 this last equation is valid for all x in $(0, \pi)$.

30.11. Find an expansion for $f(x) = 1$ in terms of the eigenfunctions of the Sturm–Liouville problem $y'' + \lambda y = 0;\ y(0) = 0,\ y'(1) = 0$.

We can show that the eigenfunctions are $e_n(x) = \sin\left(n - \frac{1}{2}\right)\pi x$ $(n = 1, 2, \ldots)$. Substituting these functions and $w(x) \equiv 1$, $a = 0$, $b = 1$ into Eq. (*30.2*), we obtain for the numerator:

$$\int_a^b w(x)f(x)e_n(x)\,dx = \int_0^1 \sin\left(n - \frac{1}{2}\right)\pi x\,dx$$

$$= \frac{-1}{\left(n - \frac{1}{2}\right)\pi}\cos\left(n - \frac{1}{2}\right)\pi x\bigg|_0^1 = \frac{1}{\left(n - \frac{1}{2}\right)\pi}$$

and for the denominator:

$$\int_a^b w(x)e_n^2(x)\,dx = \int_0^1 \sin^2\left(n - \frac{1}{2}\right)\pi x\,dx$$

$$= \left[\frac{x}{2} - \frac{\sin(2n-1)\pi x}{4\left(n - \frac{1}{2}\right)}\right]_{x=0}^{x=1} = \frac{1}{2}$$

Thus

$$c_n = \frac{2}{\left(n - \frac{1}{2}\right)\pi}$$

and Eq. (*30.1*) becomes

$$1 = \frac{2}{\pi}\sum_{n=1}^{\infty}\frac{\sin\left(n - \frac{1}{2}\right)\pi x}{n - \frac{1}{2}}$$

By Theorem 30.1 this last equation is valid for all x in $(0, 1)$.

Supplementary Problems

30.12. Find a Fourier sine series for $f(x) = 1$ on $(0, 1)$.

30.13. Find a Fourier sine series for $f(x) = x$ on $(0, 3)$.

30.14. Find a Fourier cosine series for $f(x) = x^2$ on $(0, \pi)$.

30.15. Find a Fourier cosine series for $f(x) = \begin{cases} 0 & x \le 2 \\ 2 & x > 2 \end{cases}$ on $(0, 3)$.

30.16. Find a Fourier cosine series for $f(x) = 1$ on $(0, 7)$.

30.17. Find a Fourier sine series for $f(x) = \begin{cases} x & x \le 1 \\ 2 & x > 1 \end{cases}$ on $(0, 2)$.

30.18. Find an expansion for $f(x) = 1$ in terms of the eigenfunctions of the Sturm–Liouville problem $y'' + \lambda y = 0;\ y'(0) = 0,\ y(\pi) = 0$.

30.19. Find an expansion for $f(x) = x$ in terms of the eigenfunctions of the Sturm–Liouville problem $y'' + \lambda y = 0;\ y(0) = 0,\ y'(\pi) = 0$.

30.20. Determine whether the following functions are piecewise continuous on $[-1, 5]$:

(*a*) $f(x) = \begin{cases} x^2 & x \geq 2 \\ 4 & 0 < x < 2 \\ x & x \leq 0 \end{cases}$ (*b*) $f(x) = \begin{cases} 1/(x-2)^2 & x > 2 \\ 5x^2 - 1 & x \leq 2 \end{cases}$

(*c*) $f(x) = \dfrac{1}{(x-2)^2}$ (*d*) $f(x) = \dfrac{1}{(x+2)^2}$

30.21. Which of the following functions are piecewise smooth on $[-2, 3]$?

(*a*) $f(x) = \begin{cases} x^3 & x < 0 \\ \sin \pi x & 0 \leq x \leq 1 \\ x^2 - 5x & x > 1 \end{cases}$ (*b*) $f(x) = \begin{cases} e^x & x < 1 \\ \sqrt{x} & x \geq 1 \end{cases}$

(*c*) $f(x) = \ln |x|$ (*d*) $f(x) = \begin{cases} (x-1)^2 & x \leq 1 \\ (x-1)^{1/3} & x > 1 \end{cases}$

Appendix A

Laplace Transforms

	$f(x)$	$F(s) = \mathscr{L}\{f(x)\}$
1.	1	$\dfrac{1}{s} \quad (s > 0)$
2.	x	$\dfrac{1}{s^2} \quad (s > 0)$
3.	$x^{n-1} \quad (n = 1, 2, \ldots)$	$\dfrac{(n-1)!}{s^n} \quad (s > 0)$
4.	$\sqrt{x}$	$\dfrac{1}{2}\sqrt{\pi}\, s^{-3/2} \quad (s > 0)$
5.	$1/\sqrt{x}$	$\sqrt{\pi}\, s^{-1/2} \quad (s > 0)$
6.	$x^{n-1/2} \quad (n = 1, 2, \ldots)$	$\dfrac{(1)(3)(5)\cdots(2n-1)\sqrt{\pi}}{2^n} s^{-n-1/2} \quad (s > 0)$
7.	e^{ax}	$\dfrac{1}{s-a} \quad (s > a)$
8.	$\sin ax$	$\dfrac{a}{s^2 + a^2} \quad (s > 0)$
9.	$\cos ax$	$\dfrac{s}{s^2 + a^2} \quad (s > 0)$
10.	$\sinh ax$	$\dfrac{a}{s^2 - a^2} \quad (s > \lvert a \rvert)$
11.	$\cosh ax$	$\dfrac{s}{s^2 - a^2} \quad (s > \lvert a \rvert)$
12.	$x \sin ax$	$\dfrac{2as}{(s^2 + a^2)^2} \quad (s > 0)$
13.	$x \cos ax$	$\dfrac{s^2 - a^2}{(s^2 + a^2)^2} \quad (s > 0)$
14.	$x^{n-1}e^{ax} \quad (n = 1, 2, \ldots)$	$\dfrac{(n-1)!}{(s-a)^n} \quad (s > a)$
15.	$e^{bx} \sin ax$	$\dfrac{a}{(s-b)^2 + a^2} \quad (s > b)$
16.	$e^{bx} \cos ax$	$\dfrac{s-b}{(s-b)^2 + a^2} \quad (s > b)$

Laplace Transforms (*cont.*)

	$f(x)$	$F(s) = \mathscr{L}\{f(x)\}$
17.	$\sin ax - ax\cos ax$	$\dfrac{2a^3}{(s^2+a^2)^2} \quad (s>0)$
18.	$\dfrac{1}{a}e^{-x/a}$	$\dfrac{1}{1+as}$
19.	$\dfrac{1}{a}(e^{ax}-1)$	$\dfrac{1}{s(s-a)}$
20.	$1-e^{-x/a}$	$\dfrac{1}{s(1+as)}$
21.	$\dfrac{1}{a^2}xe^{-x/a}$	$\dfrac{1}{(1+as)^2}$
22.	$\dfrac{e^{ax}-e^{bx}}{a-b}$	$\dfrac{1}{(s-a)(s-b)}$
23.	$\dfrac{e^{-x/a}-e^{-x/b}}{a-b}$	$\dfrac{1}{(1+as)(1+bs)}$
24.	$(1+ax)e^{ax}$	$\dfrac{s}{(s-a)^2}$
25.	$\dfrac{1}{a^3}(a-x)e^{-x/a}$	$\dfrac{s}{(1+as)^2}$
26.	$\dfrac{ae^{ax}-be^{bx}}{a-b}$	$\dfrac{s}{(s-a)(s-b)}$
27.	$\dfrac{ae^{-x/b}-be^{-x/a}}{ab(a-b)}$	$\dfrac{s}{(1+as)(1+bs)}$
28.	$\dfrac{1}{a^2}(e^{ax}-1-ax)$	$\dfrac{1}{s^2(s-a)}$
29.	$\sin^2 ax$	$\dfrac{2a^2}{s(s^2+4a^2)}$
30.	$\sinh^2 ax$	$\dfrac{2a^2}{s(s^2-4a^2)}$
31.	$\dfrac{1}{\sqrt{2}}\left(\cosh\dfrac{ax}{\sqrt{2}}\sin\dfrac{ax}{\sqrt{2}} - \sinh\dfrac{ax}{\sqrt{2}}\cos\dfrac{ax}{\sqrt{2}}\right)$	$\dfrac{a^3}{s^4+a^4}$
32.	$\sin\dfrac{ax}{\sqrt{2}}\sinh\dfrac{ax}{\sqrt{2}}$	$\dfrac{a^2s}{s^4+a^4}$

Laplace Transforms (*cont.*)

	$f(x)$	$F(s) = \mathscr{L}\{f(x)\}$
33.	$\frac{1}{\sqrt{2}}\left(\cos\frac{ax}{\sqrt{2}}\sinh\frac{ax}{\sqrt{2}} + \sin\frac{ax}{\sqrt{2}}\cosh\frac{ax}{\sqrt{2}}\right)$	$\frac{as^2}{s^4+a^4}$
34.	$\cos\frac{ax}{\sqrt{2}}\cosh\frac{ax}{\sqrt{2}}$	$\frac{s^3}{s^4+a^4}$
35.	$\frac{1}{2}(\sinh ax - \sin ax)$	$\frac{a^3}{s^4-a^4}$
36.	$\frac{1}{2}(\cosh ax - \cos ax)$	$\frac{a^2s}{s^4-a^4}$
37.	$\frac{1}{2}(\sinh ax + \sin ax)$	$\frac{as^2}{s^4-a^4}$
38.	$\frac{1}{2}(\cosh ax + \cos ax)$	$\frac{s^3}{s^4-a^4}$
39.	$\sin ax \sinh ax$	$\frac{2a^2s}{s^4+4a^4}$
40.	$\cos ax \sinh ax$	$\frac{a(s^2-2a^2)}{s^4+4a^4}$
41.	$\sin ax \cosh ax$	$\frac{a(s^2+2a^2)}{s^4+4a^4}$
42.	$\cos ax \cosh ax$	$\frac{s^3}{s^4+4a^4}$
43.	$\frac{1}{2}(\sin ax + ax\cos ax)$	$\frac{as^2}{(s^2+a^2)^2}$
44.	$\cos ax - \frac{ax}{2}\sin ax$	$\frac{s^3}{(s^2+a^2)^2}$
45.	$\frac{1}{2}(ax\cosh ax - \sinh ax)$	$\frac{a^3}{(s^2-a^2)^2}$
46.	$\frac{x}{2}\sinh ax$	$\frac{as}{(s^2-a^2)^2}$
47.	$\frac{1}{2}(\sinh ax + ax\cosh ax)$	$\frac{as^2}{(s^2-a^2)^2}$
48.	$\cosh ax + \frac{ax}{2}\sinh ax$	$\frac{s^3}{(s^2-a^2)^2}$

Laplace Transforms (*cont.*)

	$f(x)$	$F(s) = \mathscr{L}\{f(x)\}$
49.	$\dfrac{a \sin bx - b \sin ax}{a^2 - b^2}$	$\dfrac{ab}{(s^2 + a^2)(s^2 + b^2)}$
50.	$\dfrac{\cos bx - \cos ax}{a^2 - b^2}$	$\dfrac{s}{(s^2 + a^2)(s^2 + b^2)}$
51.	$\dfrac{a \sin ax - b \sin bx}{a^2 - b^2}$	$\dfrac{s^2}{(s^2 + a^2)(s^2 + b^2)}$
52.	$\dfrac{a^2 \cos ax - b^2 \cos bx}{a^2 - b^2}$	$\dfrac{s^3}{(s^2 + a^2)(s^2 + b^2)}$
53.	$\dfrac{b \sinh ax - a \sinh bx}{a^2 - b^2}$	$\dfrac{ab}{(s^2 - a^2)(s^2 - b^2)}$
54.	$\dfrac{\cosh ax - \cosh bx}{a^2 - b^2}$	$\dfrac{s}{(s^2 - a^2)(s^2 - b^2)}$
55.	$\dfrac{a \sinh ax - b \sinh bx}{a^2 - b^2}$	$\dfrac{s^2}{(s^2 - a^2)(s^2 - b^2)}$
56.	$\dfrac{a^2 \cosh ax - b^2 \cosh bx}{a^2 - b^2}$	$\dfrac{s^3}{(s^2 - a^2)(s^2 - b^2)}$
57.	$x - \dfrac{1}{a} \sin ax$	$\dfrac{a^2}{s^2(s^2 + a^2)}$
58.	$\dfrac{1}{a} \sinh ax - x$	$\dfrac{a^2}{s^2(s^2 - a^2)}$
59.	$1 - \cos ax - \dfrac{ax}{2} \sin ax$	$\dfrac{a^4}{s(s^2 + a^2)^2}$
60.	$1 - \cosh ax + \dfrac{ax}{2} \sinh ax$	$\dfrac{a^4}{s(s^2 - a^2)^2}$
61.	$1 + \dfrac{b^2 \cos ax - a^2 \cos bx}{a^2 - b^2}$	$\dfrac{a^2 b^2}{s(s^2 + a^2)(s^2 + b^2)}$
62.	$1 + \dfrac{b^2 \cosh ax - a^2 \cosh bx}{a^2 - b^2}$	$\dfrac{a^2 b^2}{s(s^2 - a^2)(s^2 - b^2)}$
63.	$\dfrac{1}{8}[(3 - a^2x^2) \sin ax - 3ax \cos ax]$	$\dfrac{a^5}{(s^2 + a^2)^3}$
64.	$\dfrac{x}{8}[\sin ax - ax \cos ax]$	$\dfrac{a^3 s}{(s^2 + a^2)^3}$

Laplace Transforms (*cont.*)

	$f(x)$	$F(s) = \mathscr{L}\{f(x)\}$
65.	$\frac{1}{8}[(1+a^2x^2)\sin ax - ax\cos ax]$	$\frac{a^3s^2}{(s^2+a^2)^3}$
66.	$\frac{1}{8}[(3+a^2x^2)\sinh ax - 3ax\cosh ax]$	$\frac{a^5}{(s^2-a^2)^3}$
67.	$\frac{x}{8}(ax\cosh ax - \sinh ax)$	$\frac{a^3s}{(s^2-a^2)^3}$
68.	$\frac{1}{8}[ax\cosh ax - (1-a^2x^2)\sinh ax]$	$\frac{a^3s^2}{(s^2-a^2)^3}$
69.	$\frac{1}{n!}(1-e^{-x/a})^n$	$\frac{1}{s(as+1)(as+2)\cdots(as+n)}$
70.	$\sin(ax+b)$	$\frac{s\sin b + a\cos b}{s^2+a^2}$
71.	$\cos(ax+b)$	$\frac{s\cos b - a\sin b}{s^2+a^2}$
72.	$e^{-ax} - e^{ax/2}\left(\cos\frac{ax\sqrt{3}}{2} - \sqrt{3}\sin\frac{ax\sqrt{3}}{2}\right)$	$\frac{3a^2}{s^3+a^3}$
73.	$\frac{1+2ax}{\sqrt{\pi x}}$	$\frac{s+a}{s\sqrt{s}}$
74.	$e^{-ax}/\sqrt{\pi x}$	$\frac{1}{\sqrt{s+a}}$
75.	$\frac{1}{2x\sqrt{\pi x}}(e^{bx} - e^{ax})$	$\sqrt{s-a} - \sqrt{s-b}$
76.	$\frac{1}{\sqrt{\pi x}}\cos 2\sqrt{ax}$	$\frac{1}{\sqrt{s}}e^{-a/s}$
77.	$\frac{1}{\sqrt{\pi x}}\cosh 2\sqrt{ax}$	$\frac{1}{\sqrt{s}}e^{a/s}$
78.	$\frac{1}{\sqrt{a\pi}}\sin 2\sqrt{ax}$	$s^{-3/2}e^{-a/s}$
79.	$\frac{1}{\sqrt{a\pi}}\sinh 2\sqrt{ax}$	$s^{-3/2}e^{a/s}$
80.	$J_0(2\sqrt{ax})$	$\frac{1}{s}e^{-a/s}$

Laplace Transforms (*cont.*)

	$f(x)$	$F(s) = \mathscr{L}\{f(x)\}$
81.	$\sqrt{x/a}\, J_1(2\sqrt{ax})$	$\frac{1}{s^2} e^{-a/s}$
82.	$(x/a)^{(p-1)/2} J_{p-1}(2\sqrt{ax}) \quad (p > 0)$	$s^{-p} e^{-a/s}$
83.	$J_0(x)$	$\frac{1}{\sqrt{s^2+1}}$
84.	$J_1(x)$	$\frac{\sqrt{s^2+1} - s}{\sqrt{s^2+1}}$
85.	$J_p(x) \quad (p > -1)$	$\frac{(\sqrt{s^2+1} - s)^p}{\sqrt{s^2+1}}$
86.	$x^p J_p(ax) \quad \left(p > -\frac{1}{2}\right)$	$\frac{(2a)^p \Gamma(p + \frac{1}{2})}{\sqrt{\pi}(s^2 + a^2)^{p+(1/2)}}$
87.	$\frac{x^{p-1}}{\Gamma(p)} \quad (p > 0)$	$\frac{1}{s^p}$
88.	$\frac{4^n n!}{(2n)!\sqrt{\pi}} x^{n-(1/2)}$	$\frac{1}{s^n \sqrt{s}}$
89.	$\frac{x^{p-1}}{\Gamma(p)} e^{-ax} \quad (p > 0)$	$\frac{1}{(s+a)^p}$
90.	$\frac{1 - e^{ax}}{x}$	$\ln \frac{s-a}{s}$
91.	$\frac{e^{bx} - e^{ax}}{x}$	$\ln \frac{s-a}{s-b}$
92.	$\frac{2}{x} \sinh ax$	$\ln \frac{s+a}{s-a}$
93.	$\frac{2}{x}(1 - \cos ax)$	$\ln \frac{s^2 + a^2}{s^2}$
94.	$\frac{2}{x}(\cos bx - \cos ax)$	$\ln \frac{s^2 + a^2}{s^2 + b^2}$
95.	$\frac{\sin ax}{x}$	$\arctan \frac{a}{s}$
96.	$\frac{2}{x} \sin ax \cos bx$	$\arctan \frac{2as}{s^2 - a^2 + b^2}$
97.	$\sin \lvert ax \rvert$	$\left(\frac{a}{s^2 + a^2}\right)\left(\frac{1 + e^{-(\pi/a)s}}{1 - e^{-(\pi/a)s}}\right)$

Answers to Supplementary Problems

CHAPTER 1

1.14. (*a*) 2; (*b*) y; (*c*) x

1.15. (*a*) 4; (*b*) y; (*c*) x

1.16. (*a*) 2; (*b*) s; (*c*) t

1.17. (*a*) 4; (*b*) y; (*c*) x

1.18. (*a*) n; (*b*) x; (*c*) y

1.19. (*a*) 2; (*b*) r; (*c*) y

1.20. (*a*) 2; (*b*) y; (*c*) x

1.21. (*a*) 7; (*b*) b; (*c*) p

1.22. (*a*) 1; (*b*) b; (*c*) p

1.23. (*a*) 6; (*b*) y; (*c*) x

1.24. (*d*) and (*e*)

1.25. (*a*), (*c*), and (*e*)

1.26. (*b*), (*d*), and (*e*)

1.27. (*a*), (*c*), and (*d*)

1.28. (*d*)

1.29. (*a*), (*c*), and (*d*)

1.30. (*b*) and (*e*)

1.31. (*a*), (*c*), and (*d*)

1.32. $c = 0$

1.33. $c = 1$

1.34. $c = e^{-2}$

1.35. $c = -3e^{-4}$

1.36. $c = 1$

1.37. c can be any real number

1.38. $c = -1/3$

1.39. No solution

1.40. $c_1 = 2$, $c_2 = 1$; initial conditions

1.41. $c_1 = 1$, $c_2 = 2$; initial conditions

1.42. $c_1 = 1$, $c_2 = -2$; initial conditions

1.43. $c_1 = c_2 = 1$; boundary conditions

1.44. $c_1 = 1$, $c_2 = -1$; boundary conditions

1.45. $c_1 = -1$, $c_2 = 1$; boundary conditions

1.46. No values; boundary conditions

1.47. $c_1 = c_2 = 0$; initial conditions

1.48. $c_1 = \dfrac{-2}{\sqrt{3} - 1}$, $c_2 = \dfrac{2}{\sqrt{3} - 1}$; boundary conditions

1.49. No values; boundary conditions

1.50. $c_1 = -2$, $c_2 = 3$

1.51. $c_1 = 0$, $c_2 = 1$

1.52. $c_1 = 3$, $c_2 = -6$

1.53. $c_1 = 0$, $c_2 = 1$

1.54. $c_1 = 1 + \dfrac{3}{e}$, $c_2 = -2 - \dfrac{2}{e}$

CHAPTER 2

2.15. $y' = -y^2/x$

2.16. $y' = x/(e^x - 1)$

2.17. $y' = (\sin x - y^2 - y)^{1/3}$

2.18. Cannot reduce to standard form

2.19. $y' = -y + \ln x$

2.20. $y' = 2$ and $y' = x + y + 3$

2.21. $y' = \dfrac{y - x}{y^2}$

2.22. $y' = \dfrac{x + y}{x - y}$

2.23. $y' = \dfrac{y - x}{x + y}$

2.24 $y' = ye^{-x} - e^x$

2.25. $y' = -1$

2.26. Linear

2.27. Linear, separable, and exact

2.28. Linear

2.29. Homogeneous, Bernoulli

2.30. Homogeneous, Bernoulli, separable, and exact

2.31. Linear, homogeneous, and exact

2.32. Homogeneous

2.33. Exact

2.34. Bernoulli

2.35. Linear and exact

CHAPTER 3

3.23. $y = \pm\sqrt{k - x^2},\ k = 2c$

3.24. $y = \pm(k + 2x^2)^{1/4},\ k = -4c$

3.25. $y = (k + 3x)^{-1/3},\ k = -3c$

3.26. $y = -\left(\frac{1}{2}t^2 + t - c\right)^{-1}$

3.27. $y = kx,\ k = \pm e^{-c}$

3.28. $y = \ln\left|\frac{k}{x}\right|,\ c = \ln|k|$

3.29. $y = ke^{-x^2/2},\ k = \pm e^c$

3.30. $2t^3 + 6t + 2y^3 + 3y^2 = k,\ k = 6c$

3.31. $y^3t^4 = ke^y,\ k = \pm e^c$

3.32. $y = \tan(x - c)$

3.33. $y = 3 + 2\tan(2x + k),\ k = -2c$

3.34. $\frac{1}{x^2}dx - \frac{1}{y}dy = 0;\ y = ke^{-1/x},\ k = \pm e^{-c}$

3.35. $xe^x\,dx - 2y\,dy = 0;\ y = \pm\sqrt{xe^x - e^x - c}$

3.36. $y = \pm\sqrt{x^2 + 2x + k},\ k = -2c$

3.37. $y = -1/(x - c)$

3.38. $x = -3/(t^3 + k),\ k = 3c$

3.39. $x = kt,\ k = \pm e^c$

3.40. $y = -\frac{3}{5} + ke^{5t},\ k = \pm\frac{1}{5}e^{5c}$

3.41. $y = -\sqrt{2 + 2\cos x}$

3.42. $y = e^{-1/3(x^3+3x+4)}$

3.43. $\frac{1}{2}e^{x^2} + \frac{1}{6}y^6 - y = \frac{1}{2}$

3.44. $\frac{x^3}{3} - x - y - \ln|y| = 7$

3.45. $x = \frac{8}{3} + \frac{4}{3}e^{-3t}$

3.46. $y = x \ln|k/x|$

3.47. $y = kx^2 - x$

3.48. $y^2 = kx^4 - x^2$

3.49. Not homogeneous

3.50. $y^2 = x^2 - kx$

3.51. $3yx^2 - y^3 = k$

3.52. $-2\sqrt{x/y} + \ln|y| = c$

3.53. Not homogeneous

3.54. $y^2 = -x^2\left(1 + \frac{1}{\ln|kx^2|}\right)$

CHAPTER 4

4.24. $xy + x^2y^3 + y = c_2$

4.25. Not exact

4.26. $y = c_2e^{-x^3} + \frac{1}{3}$

4.27. $x^3y^2 + y^4 = c_2$

4.28. $xy = c_2$

4.29. Not exact

4.30. $xy \sin x + y = c_2$

4.31. $y^2 = c_2t$

4.32. $y = c_2t^2$

4.33. Not exact

4.34. $t^4y^3 - t^2y = c_2$

4.35. $y = \frac{-1}{t \ln|kt|}$

4.36. $x = \frac{1}{3}t^2 - \frac{c_2}{t}$

4.37. $2t^3 + 6tx^2 - 3x^2 = c_2$ or $x = \pm\sqrt{\frac{c_2 - 2t^3}{6t - 3}}$

4.38. $x = \frac{k}{1 + e^{2t}}$

4.39. Not exact

4.40. $t \cos x + x \sin t = c_2$

4.41. $I(x, y) = \frac{-1}{x^2}$; $y = cx - 1$

4.42. $I(x, y) = \frac{1}{y^2}$; $cy = x - 1$

4.43. $I(x, y) = -\frac{1}{x^2 + y^2}$; $y = x \tan(x + c)$

4.44. $I(x, y) = \frac{1}{(xy)^3}$; $\frac{1}{y^2} = 2x^2(x - c)$

4.45. $I(x, y) = \frac{1}{(xy)^2}$; $\frac{1}{y} = \frac{1}{3}x^4 - cx$

4.46. $I(x, y) = e^{x^3}$; $y = ce^{-x^3} + \frac{1}{3}$

4.47. $I(x, y) = e^{-y^2}$; $y^2 = \ln|kx|$

4.48. $I(x, y) = \frac{1}{y}$; $y^2 = 2(c - x^2)$

4.49. $I(x, y) = y^2$; $y^3 = \frac{c}{x}$

4.50. $I(x, y) = y^2$; $x^2y^4 + \frac{x^2}{2} = c$

4.51. $I(x, y) = \frac{1}{(xy)^2}$; $\ln|xy| = c - y$

4.52. $I(x, y) = 1$ (the equation is exact); $\frac{1}{2}x^2y^2 = c$

4.53. $I(x, y) = -\frac{1}{x^2 + y^2}$; $y = x \tan\left(\frac{1}{2}x^2 + c\right)$

4.54. $I(x, y) = \frac{1}{(xy)^2}$; $3x^3y + 2xy^4 + kxy = -6$

4.55. $I(x, y) = e^{y^3/3}$; $x^3y^2e^{y^3/3} = c$

4.56. $x(t) = \frac{t + \sqrt{t^2 + 16}}{2}$

4.57. $x(t) \equiv 0$

4.58. $x(t) = \frac{t - \sqrt{t^2 + 120}}{2}$

4.59. $xy + x^2y^3 + y = -135$

4.60. $y(x) = -\frac{4}{3}e^{-x^3} + \frac{1}{3}$

4.61. No solution

4.62. $y(t) = -\sqrt{2t}$

4.63. $y(t) = -\frac{1}{2}t^2$

4.64. $x(t) = \frac{1}{3}t^2 + \frac{14}{3}\left(\frac{1}{t}\right)$

4.65. $x(t) = \frac{-2(1 + e^2)}{1 + e^{2t}}$

CHAPTER 5

5.20. $y = ce^{-5x}$

5.21. $y = ce^{5x}$

5.22. $y = ce^{0.01x}$

5.23. $y = ce^{-x^2}$

5.24. $y = ce^{-x^3}$

5.25. $y = ce^{x^3/3}$

5.26. $y = ce^{3x^5/5}$

5.27. $y = c/x$

5.28. $y = c/x^2$

5.29. $y = cx^2$

5.30. $y = ce^{-2/x}$

5.31. $y = ce^{7x} - \frac{1}{6}e^x$

5.32. $y = ce^{7x} - 2x - \frac{2}{7}$

5.33. $y = ce^{7x} - \frac{2}{53}\cos 2x - \frac{7}{53}\sin 2x$

5.34. $y = ce^{-x^3/3} + 1$

5.35. $y = ce^{-3/x} - \frac{1}{3}$

5.36. $y = c + \sin x$

5.37. $\frac{1}{y} = ce^x + 1$

5.38. $y^2 = 1/(2x + cx^2)$

5.39. $y = (6 + ce^{-x^2/4})^2$

5.40. $y = 1/(1 + ce^x)$

5.41. $y = (1 + ce^{-3x})^{1/3}$

5.42. $y = e^{-x}/(c - x)$

5.43. $y = ce^{-50t}$

5.44. $z = c\sqrt{t}$

5.45. $N = ce^{kt}$

5.46. $p = \frac{1}{2}t^3 + 3t^2 - 2t \ln |t| + ct$

5.47. $Q = 4(20 - t) + c(20 - t)^2$

5.48. $T = (3.2t + c)e^{-0.04t}$

5.49. $p = \frac{4}{3}z + cz^{-2}$

5.50. $y = \frac{1}{4}(-x^{-2} + x^2)$

5.51. $y = 5e^{-3(x^2-\pi^2)}$

5.52. $y = 2e^{-x^2} + x^2 - 1$

5.53. $\frac{1}{y^4} = -\frac{31}{16}x^8 + 2x^{10}$

5.54. $v = -16e^{-2t} + 16$

5.55. $q = \frac{1}{5}e^{-t} + \frac{8}{5}\sin 2t + \frac{4}{5}\cos 2t$

5.56. $N = \frac{1}{3}\left(t^2 + \frac{40}{t}\right)$

5.57. $T = -60e^{-0.069t} + 30$

CHAPTER 6

6.26. (*a*) $N = 250e^{0.166t}$; (*b*) 11.2 hr

6.27. (*a*) $N = 300e^{0.0912t}$; (*b*) 7.6 hr

6.28. (*a*) 2.45 oz; (*b*) 15.19 oz

6.29. 32 fold increase

6.30. 3.17 hr

6.31. (*a*) $N = 80e^{0.0134t}$ (in millions);
(*b*) 91.5 million

6.32. $N = 16{,}620e^{0.11t}$; $N_0 = 16{,}620$

6.33. (*a*) $N = 100e^{-0.026t}$; (*b*) 4.05 yr

6.34. $N = N_0e^{-0.105t}$; $t_{1/2} = 6.6$ hr

6.35. $N = \dfrac{500}{1 + 99e^{-500kt}}$

6.36. $15,219.62

6.37. $16,904.59

6.38. $14,288.26

6.39. 8.67 percent

6.40. 10.99 percent

6.41. 20.93 yr

6.42. 7.93 yr

6.43. 12.78 percent

6.44. 8.38 yr

6.45. $T = -100e^{-0.029t} + 100$;
(*a*) 23.9 min; (*b*) 44° F

6.46. $T = 80e^{-0.035t}$; $T_0 = 80°$ F

6.47. $T = -100e^{-0.029t} + 150$; $t_{100} = 23.9$ min

6.48. (*a*) 138.6° F; (*b*) 3.12 min

6.49. (*a*) 113.9° F; (*b*) 6.95 min

6.50. An additional 1.24 min

6.51. (*a*) $v = 32t$; (*b*) $16t^2$

6.52. (*a*) 5.59 sec; (*b*) 5.59 sec

6.53. (*a*) $32t + 30$; (*b*) 3.49 sec

6.54. (*a*) $32t + 10$; (*b*) 5 sec

6.55. 31.25 sec

6.56. 976.6 ft

6.57. (*a*) $\dfrac{dv}{dt} = -g$; (*b*) $v = -gt + v_0$;
(*c*) $t_m = \dfrac{v_0}{g}$; (*d*) $x = -\dfrac{1}{2}gt^2 + v_0 t$;
(*e*) $x_m = \dfrac{v_0^2}{2g}$

6.58. (*a*) $v = 48 - 48e^{-2t/3}$;
(*b*) $x = 72e^{-2t/3} + 48t - 72$

6.59. (*a*) $v = 128 - 118e^{-t/4}$;
(*b*) 6.472 sec

6.60. 320 ft/sec

6.61. 0.392 m/sec with $g = 9.8\ \text{m/sec}^2$

6.62. (*a*) $v = -320e^{-0.1t} + 320$;
(*b*) $x = 3200e^{-0.1t} + 320t - 3200$;
(*c*) 6.9 sec

6.63. (*a*) $v = 4 - 4e^{-8t}$;
(*b*) $x = \dfrac{1}{2}e^{-8t} + 4t - \dfrac{1}{2}$

6.64. (*a*) $v = 320 - 320e^{-t/10}$;
(*b*) $x = 3200e^{-t/10} + 320t - 3200$

6.65. (*a*) $Q = -5e^{-0.2t} + 5$;
(*b*) $\dfrac{Q}{V} = \dfrac{1}{2}(-e^{-0.2t} + 1)$

6.66. $Q = -\dfrac{7}{40}(20 - t)^2 + 4(20 - t)$;
at $t = 10$, $Q = 22.5$ lb
(Note that $a = 80(1/8) = 10$ lb.)

6.67. (*a*) $Q = 80e^{-0.04t}$; (*b*) 17.3 min

6.68. 56.3 lb

6.69 111.1 g

6.70. 80 g

6.71. (*a*) $-\dfrac{99}{2}e^{-10t}$; (*b*) 0 amp

6.72. (*a*) $q = 2 + 3e^{-10t}$; (*b*) $I = -30e^{-10t}$

6.73. (*a*) $q = 10e^{-2.5t}$; (*b*) $I = -25e^{-2.5t}$

6.74. (*a*) $q = \dfrac{1}{50}(2\sin t - \cos t + e^{-2t})$;
(*b*) $I_s = \dfrac{1}{50}(2\cos t + \sin t)$

6.75. (*a*) $q = \dfrac{1}{5}(\sin 2t + 2\cos 2t + 23e^{-4t})$;
(*b*) $I_s = \dfrac{1}{5}(2\cos 2t - 4\sin 2t)$

6.76. (*a*) $I = \dfrac{1}{10}(1 - e^{-50t})$; (*b*) $I_s = \dfrac{1}{10}$

6.77. (*a*) $I = 10e^{-25t}$; (*b*) $I_t = 10e^{-25t}$

6.78. (*a*) $I = \dfrac{1}{10}(9 + 51e^{-20t/3})$;
(*b*) $I_t = \dfrac{51}{10}e^{-20t/3}$

6.79. $I = \dfrac{1}{626}(e^{-25t} + 25\sin t - \cos t)$

6.80. $A = \dfrac{2}{\sqrt{34}}$ $\phi = \arctan\dfrac{3}{5}$

6.81. $A = -\dfrac{3}{\sqrt{101}}$ $\phi = \arctan 10$

6.82. $xy = k$

6.83. $y^2 = -2x + k$

6.84. $x^2y + \frac{1}{3}y^3 = k$

6.85. $x^2 + y^2 = kx$

6.86. $x^2 + \frac{1}{2}y^2 = k \ (k > 0)$

6.87. $N = \dfrac{1000}{1 + 9e^{-0.1158t}}$

6.88. $N = \dfrac{1{,}000{,}000}{1 + 999e^{-0.275t}}$

6.89. $\dfrac{2 + v}{2 - v} = 3e^{32t}$ or $v = 2(3e^{32t} - 1)/(3e^{32t} + 1)$

CHAPTER 7

7.33. (e), (g), (j), and (k) are nonlinear; all the rest are linear. Note that (f) has the form $y' - (2 + x)y = 0$.

7.34. (a), (c), and (f) are homogeneous. Note that (l) has the form $y'' = -e^x$.

7.35. (b), (c), and (l) have constant coefficients.

7.36. $W = 0$

7.37. $W = -x^2$; the set is linearly independent.

7.38. $W = -x^4$; the set is linearly independent.

7.39. $W = -2x^3$; the set is linearly independent.

7.40. $W = -10x$; the set is linearly independent.

7.41. $W = 0$

7.42. $W = -4$; the set is linearly independent.

7.43. $W = e^{5x}$; the set is linearly independent.

7.44. $W = 0$

7.45. $W = 0$

7.46. $W = 0$

7.47. $W = 2x^6$; the set is linearly independent.

7.48. $W = 6e^{2x}$; the set is linearly independent.

7.49. $W = 0$

7.50. $[4]3x + [-3]4x \equiv 0$

7.51. $[1]x^2 + [1](-x^2) \equiv 0$

7.52. $[5](3e^{2x}) + [-3](5e^{2x}) \equiv 0$

7.53. $[-2]x + [7](1) + [1](2x - 7) \equiv 0$

7.54. $[3](x + 1) + [-2](x^2 + x) + [1](2x^2 - x - 3) \equiv 0$

7.55 $[-6]\sin x + [-1](2\cos x) + [2](3\sin x + \cos x) \equiv 0$

7.56. $y = c_1e^{2x} + c_2e^{-2x}$

7.57. $y = c_1e^{2x} + c_2e^{3x}$

7.58. $y = c_1 \sin 4x + c_2 \cos 4x$

7.59. $y = c_1e^{8x} + c_2$

7.60. Since y_1 and y_2 are linearly dependent, there is not enough information provided to exhibit the general solution.

7.61. $y = c_1x + c_2e^x + c_3y_3$ where y_3 is a third particular solution, linearly independent from the other two.

7.62. Since the given set is linearly dependent, not enough information is provided to exhibit the general solution.

7.63. $y = c_1e^{-x} + c_2e^{x} + c_3e^{2x}$

7.64. $y = c_1x^2 + c_2x^3 + c_3x^4 + c_4y_4 + c_5y_5$, where y_4 and y_5 are two other solutions having the property that the set $\{x^2, x^3, x^4, y_4, y_5\}$ is linearly independent.

7.65. $y = c_1 \sin x + c_2 \cos x + x^2 - 2$

7.66. Since e^x and $3e^x$ are linearly dependent, there is not enough information given to find the general solution.

7.67. $y = c_1e^{x} + c_2e^{-x} + c_3xe^{x} + 5$

7.68. Theorem 7.1 does not apply, since $a_0(x) = -(2/x)$ is not continuous about $x_0 = 0$.

7.69. Yes; $a_0(x)$ is continuous about $x_0 = 1$.

7.70. Theorem 7.1 does not apply, since $b_1(x)$ is zero at the origin.

CHAPTER 8

8.17. $y = c_1e^{x} + c_2e^{-x}$

8.18. $y = c_1e^{-5x} + c_2e^{6x}$

8.19. $y = c_1e^{x} + c_2xe^{x}$

8.20. $y = c_1 \cos x + c_2 \sin x$

8.21. $y = c_1e^{-x}\cos x + c_2e^{-x}\sin x$

8.22. $y = c_1e^{\sqrt{7}x} + c_2e^{-\sqrt{7}x}$

8.23. $y = c_1e^{-3x} + c_2xe^{-3x}$

8.24. $y = c_1e^{-x}\cos\sqrt{2}\,x + c_2e^{-x}\sin\sqrt{2}\,x$

8.25. $y = c_1e^{[(3+\sqrt{29})/2]x} + c_2e^{[(3-\sqrt{29})/2]x}$
$= e^{(3/2)x}\left(k_1\cosh\frac{\sqrt{29}}{2}x + k_2\sinh\frac{\sqrt{29}}{2}x\right)$

8.26. $y = c_1e^{-(1/2)x} + c_2xe^{-(1/2)x}$

8.27. $x = c_1e^{4t} + c_2e^{16t}$

8.28. $x = c_1e^{-50t} + c_2e^{-10t}$

8.29. $x = c_1e^{(3+\sqrt{5})t/2} + c_2e^{(3-\sqrt{5})t/2}$

8.30. $x = c_1e^{5t} + c_2te^{5t}$

8.31. $x = c_1\cos 5t + c_2\sin 5t$

8.32. $x = c_1 + c_2e^{-25t}$

8.33. $x = c_1e^{-t/2}\cos\frac{\sqrt{7}}{2}t + c_2e^{-t/2}\sin\frac{\sqrt{7}}{2}t$

8.34. $u = c_1e^{t}\cos\sqrt{3}\,t + c_2e^{t}\sin\sqrt{3}\,t$

8.35. $u = c_1e^{(2+\sqrt{2})t} + c_2e^{(2-\sqrt{2})t}$

8.36. $u = c_1 + c_2e^{36t}$

8.37. $u = c_1e^{6t} + c_2e^{-6t} = k_1\cosh 6t + k_2\sinh 6t$

8.38. $Q = c_1e^{5t/2}\cos\frac{\sqrt{3}}{2}t + c_2e^{5t/2}\sin\frac{\sqrt{3}}{2}t$

8.39. $Q = c_1e^{(7+\sqrt{29})t/2} + c_2e^{(7-\sqrt{29})t/2}$

8.40. $P = c_1e^{9t} + c_2te^{9t}$

8.41. $P = c_1e^{-x}\cos 2\sqrt{2}\,x + c_2e^{-x}\sin 2\sqrt{2}\,x$

8.42. $N = c_1e^{3x} + c_2e^{-8x}$

8.43. $N = c_1e^{-5x/2}\cos\frac{\sqrt{71}}{2}x + c_2e^{-5x/2}\sin\frac{\sqrt{71}}{2}x$

8.44. $T = c_1e^{-15\theta} + c_2\theta e^{-15\theta}$

8.45. $R = c_1 + c_2e^{-5\theta}$

CHAPTER 9

9.16. $y = c_1e^{-x} + c_2e^{x} + c_3e^{2x}$

9.17. $y = c_1e^{x} + c_2xe^{x} + c_3e^{-x}$

9.18. $y = c_1e^{x} + c_2xe^{x} + c_3x^2e^{x}$

9.19. $y = c_1e^{x} + c_2\cos x + c_3\sin x$

9.20. $y = (c_1 + c_2x)\cos x + (c_3 + c_4x)\sin x$

9.21. $y = c_1e^{x} + c_2e^{-x} + c_3\cos x + c_4\sin x$

9.22. $y = c_1e^{x} + c_2e^{-x} + c_3xe^{-x} + c_4x^2e^{-x}$

9.23. $y = c_1e^{-2x} + c_2xe^{-2x} + c_3e^{2x}\cos 2x + c_4e^{2x}\sin 2x$

9.24. $y = c_1 + c_2x + c_3x^2 + c_4e^{-5x}$

9.25. $y = (c_1 + c_3x)e^{-(1/2)x}\cos\frac{\sqrt{3}}{2}x + (c_2 + c_4x)e^{-(1/2)x}\sin\frac{\sqrt{3}}{2}x$

9.26. $y = c_1e^{2x}\cos 2x + c_2e^{2x}\sin 2x + c_3e^{-2x} + c_4xe^{-2x} + c_5e^{x} + c_6e^{-x}$

9.27. $x = c_1e^{-t} + c_2te^{-t} + c_3t^2e^{-t} + c_4t^3e^{-t}$

9.28. $x = c_1 + c_2t + c_3t^2$

9.29. $x = c_1\cos t + c_2\sin t + c_3\cos 3t + c_4\sin 3t$

9.30. $x = c_1e^{5t} + c_2\cos 5t + c_3\sin 5t$

9.31. $q = c_1e^{x} + c_2e^{-x} + c_3\cos\sqrt{2}\,x + c_4\sin\sqrt{2}\,x$

9.32. $q = c_1e^{x} + c_2e^{-x} + c_3e^{\sqrt{2}x} + c_4e^{-\sqrt{2}x}$

9.33. $N = c_1e^{-6x} + c_2e^{8x} + c_3e^{10x}$

9.34. $r = c_1e^{-\theta} + c_2\theta e^{-\theta} + c_3\theta^2e^{-\theta} + c_4\theta^3e^{-\theta} + c_5\theta^4e^{-\theta}$

9.35. $y = c_1e^{2x} + c_2e^{8x} + c_3e^{-14x}$

9.36. $y = c_1 + c_2\cos 19x + c_3\sin 19x$

9.37. $y = c_1 + c_2x + c_3e^{2x}\cos 9x + c_4e^{2x}\sin 9x$

9.38. $y = c_1e^{2x}\cos 9x + c_2e^{2x}\sin 9x + c_3xe^{2x}\cos 9x + c_4xe^{2x}\sin 9x$

9.39. $y = c_1e^{5x} + c_2xe^{5x} + c_3x^2e^{5x} + c_4e^{-5x} + c_5xe^{-5x}$

9.40. $y = c_1\cos 6x + c_2\sin 6x + c_3x\cos 6x + c_4x\sin 6x + c_5x^2\cos 6x + c_6x^2\sin 6x$

9.41. $y = e^{-3x}(c_1\cos x + c_2\sin x + c_3x\cos x + c_4x\sin x) + e^{3x}(c_5\cos x + c_6\sin x + c_7x\cos x + c_8x\sin x)$

9.42. $y''' + 4y'' - 124y' + 224y = 0$

9.43. $y''' + 361y' = 0$

9.44. $y^{(4)} - 4y''' + 85y'' = 0$

9.45. $y^{(4)} - 8y''' + 186y'' - 680y' + 7225y = 0$

9.46. $y^{(5)} - 5y^{(4)} - 50y^{(3)} + 250y'' + 625y' - 3125y = 0$

9.47. $y = c_1e^{-x} + c_2xe^{-x} + c_3x^2e^{-x} + c_4x^3e^{-x}$

9.48. $y = c_1\cos 4x + c_2\sin 4x + c_3\cos 3x + c_4\sin 3x$

9.49. $y = c_1\cos 4x + c_2\sin 4x + c_3x\cos 4x + c_4x\sin 4x$

9.50. $y = c_1e^{2x} + c_2xe^{2x} + c_3e^{5x} + c_4xe^{5x}$

CHAPTER 10

10.15. $y_p = A_1x + A_0$

10.16. $y_p = A_2x^2 + A_1x + A_0$

10.17. $y_p = A_2x^2 + A_1x + A_0$

10.18. $y_p = Ae^{-2x}$

10.19. $y_p = Ae^{5x}$

10.20. $y_p = Axe^{2x}$

10.21. $y_p = A\sin 3x + B\cos 3x$

10.22. $y_p = A\sin 3x + B\cos 3x$

10.23. $y_p = (A_1x + A_0)\sin 3x + (B_1x + B_0)\cos 3x$

10.24. $y_p = A_1x + A_0 + Be^{8x}$

10.25. $y_p = (A_1x + A_0)e^{5x}$

10.26. $y_p = x(A_1x + A_0)e^{3x}$

10.27. $y_p = Ae^{3x}$

10.28. $y_p = (A_1x + A_0)e^{3x}$

10.29. $y_p = Ae^{5x}$

10.30. $y_p = (A_2x^2 + A_1x + A_0)e^{5x}$

10.31. $y_p = A\sin\sqrt{2}\,x + B\cos\sqrt{2}\,x$

10.32. $y_p = (A_2x^2 + A_1x + A_0)\sin\sqrt{2}\,x + (B_2x^2 + B_1x + B_0)\cos\sqrt{2}\,x$

10.33. $y_p = A\sin 3x + B\cos 3x$

10.34. $y_p = A\sin 4x + B\cos 4x + C\sin 7x + D\cos 7x$

10.35. $y_p = Ae^{-x}\sin 3x + Be^{-x}\cos 3x$

10.36. $y_p = x(Ae^{5x}\sin 3x + Be^{5x}\cos 3x)$

10.37. $x_p = t(A_1t + A_0)$

10.38. $x_p = t(A_2t^2 + A_1t + A_0)$

10.39. $x_p = (A_1t + A_0)e^{-2t} + Bt$

10.40. $x_p = t^2(Ae^t)$

10.41. $x_p = t^2(A_1t + A_0)e^t$

10.42. $x_p = At + (B_1t + B_0)\sin t + (C_1t + C_0)\cos t$

10.43. $x_p = (A_1t + A_0)e^{2t}\sin 3t + (B_1t + B_0)e^{2t}\cos 3t$

10.44. $y = c_1e^x + c_2xe^x + x^2 + 4x + 5$

10.45. $y = c_1e^x + c_2xe^x + 3e^{2x}$

10.46. $y = c_1e^x + c_2xe^x - 2\sin x$

10.47. $y = c_1e^x + c_2xe^x + \frac{3}{2}x^2e^x$

10.48. $y = c_1e^x + c_2xe^x + \frac{1}{6}x^3e^x$

10.49. $y = c_1e^x + xe^x$

10.50. $y = c_1e^x + xe^{2x} - e^{2x} - 1$

10.51. $y = c_1e^x - \frac{1}{2}\sin x - \frac{1}{2}\cos x + \frac{2}{5}\sin 2x - \frac{1}{5}\cos 2x$

10.52. $y = c_1e^x + c_2xe^x + c_3x^2e^x + \frac{1}{6}x^3e^x - 1$

CHAPTER 11

11.9. $y = c_1e^x + c_2xe^x + \frac{1}{12}x^{-3}e^x$

11.10. $y = c_1\cos x + c_2\sin x + (\cos x)\ln|\cos x| + x\sin x$

11.11. $y = c_1e^{-x} + c_2e^{2x} + \frac{1}{4}e^{3x}$

11.12. $y = c_1e^{30x} + c_2xe^{30x} + \frac{1}{80}e^{10x}$

11.13. $y = c_3 + c_2e^{7x} + \frac{3}{7}x$ $\left(\text{with } c_3 = c_1 + \frac{3}{49}\right)$

11.14. $y = c_1x + \frac{c_2}{x} + \frac{x^2}{3}\ln|x| - \frac{4}{9}x^2$

11.15. $y = c_1 + c_2x^2 + xe^x - e^x$

11.16. $y = c_1x + \frac{1}{2}x^3$

11.17. $y = c_1e^{-x^2} + \frac{1}{2}$

11.18. $y = c_1 + c_2x + c_3x^2 + 2x^3$

11.19. $x = c_1e^t + c_2te^t + \frac{e^t}{2t}$

11.20. $x = c_3e^{3t} + c_2te^{3t} - e^{3t}\ln|t|$ (with $c_3 = c_1 - 1$)

11.21. $x = c_1\cos 2t + c_2\sin 2t - 1 + (\sin 2t)\ln|\sec 2t + \tan 2t|$

11.22. $x = c_3e^t + c_4e^{3t} + \frac{e^t}{2}\ln(1+e^{-t}) - \frac{e^{3t}}{2}\ln(1+e^{-t}) + \frac{e^{2t}}{2}$ $\left(\text{with } c_3 = c_1 - \frac{1}{4};\ c_4 = c_2 + \frac{3}{4}\right)$

11.23. $x = c_1t + c_2(t^2+1) + \frac{t^4}{6} - \frac{t^2}{2}$

11.24. $x = c_1e^t + \frac{c_2}{t} - \frac{t^2}{3} - t - 1$

11.25. $r = c_1e^t + c_2te^t + c_3t^2e^t + \frac{t^2}{2}e^t\ln|t|$

11.26. $r = c_1e^{-2t} + c_2te^{-2t} + c_3t^2e^{-2t} + 2t^3e^{-2t}$

11.27. $r = c_1e^{5t} + c_2\cos 5t + c_3\sin 5t - 8$

11.28. $z = c_1 + c_2e^{\theta} + c_3e^{2\theta} + \frac{1}{4}(1+e^{\theta})^2[-3 + 2\ln(1+e^{\theta})]$

11.29. $y = \frac{c_1}{t} + c_2 + c_3t - \ln|t|$

11.30. $y = c_1 + c_2x + c_3x^2 + c_6e^{2x} + c_5e^{-2x} + xe^{2x}$ $\left(\text{with } c_6 = c_4 - \frac{7}{4}\right)$

CHAPTER 12

12.7. $y = \frac{1}{12}e^{-x} + \frac{2}{3}e^{2x} + \frac{1}{4}e^{3x}$

12.8. $y = \frac{13}{12}e^{-x} + \frac{2}{3}e^{2x} + \frac{1}{4}e^{3x}$

12.9. $y = e^{-x} + e^{2x}$

12.10. $y = \left(1 + \frac{1}{12}e^3\right)e^{-(x-1)} + \left(1 - \frac{1}{3}e^3\right)e^{2(x-1)} + \frac{1}{4}e^{3x}$

12.11. $y = -\cos 1\cos x - \sin 1\sin x + x = -\cos(x-1) + x$

12.12. $y = -\frac{1}{6}\cos 2x + \frac{1}{4}\cos^2 2x - \frac{1}{12}\cos^4 2x + \frac{1}{12}\sin^4 2x = \frac{1}{12}(1 + \cos^2 2x - 2\cos 2x)$

12.13. $y \equiv 0$

12.14. $y = -2 + 6x - 6x^2 + 2x^3$

12.15. $y = e^{-t}\left(\frac{3}{10}\cos t + \frac{11}{10}\sin t\right) + \frac{1}{10}\sin 2t - \frac{3}{10}\cos 2t$

CHAPTER 13

13.26. 60 lb/ft

13.27. 17.07 lb/ft

13.28. 130.7 dynes/cm

13.29. 19.6 N/m

13.30. $x = \frac{1}{6}\cos 8t$

13.31. $x = -\frac{1}{6}\cos 8t + \frac{1}{4}\sin 8t$

13.32. $x = 3\cos 12t + \frac{5}{6}\sin 12t$

13.33. $x = \sin 2t - \cos 2t$

13.34. (*a*) $\omega = 8$ Hz; (*b*) $f = 4/\pi$ Hz; (*c*) $T = \pi/4$ sec

13.35. (*a*) $\omega = 12$ Hz; (*b*) $f = 6/\pi$ Hz; (*c*) $T = \pi/6$ sec

13.36. (*a*) $\omega = 2$ Hz; (*b*) $f = 1/\pi$ Hz; (*c*) $T = \pi$ sec

13.37. $x = x_0 \cos\sqrt{k/m}\,t + v_0\sqrt{m/k}\sin\sqrt{k/m}\,t$

13.38. $x = -\frac{1}{3}\sqrt{3}\,e^{-4t}\sin 4\sqrt{3}\,t$

13.39. $x = -\frac{1}{2}e^{-4t} - 2te^{-4t}$

13.40. $x = \frac{3}{4}e^{-2t} - \frac{1}{4}e^{-6t}$

13.41. $x = -0.1e^{-4t} - 2.4te^{-4t}$

13.42. $x = -0.1e^{-4t}\cos\sqrt{0.02}t - \frac{2.4}{\sqrt{0.02}}e^{-4t}\sin\sqrt{0.02}t$

13.43. $x = -8.62e^{-3.86t} + 8.52e^{-4.14t}$

13.44. $x = e^{-2t}\left(\frac{2}{5}\cos 2t - \frac{6}{5}\sin 2t\right) + \frac{4}{5}\sin 4t - \frac{2}{5}\cos 4t$

13.45. $x = -\frac{4}{105}\sin 5t + \frac{2}{21}\sin 2t$

13.46. $x = \frac{1}{16}\sin 4t - \frac{1}{2}\cos 4t - \frac{t}{4}\cos 4t$

13.47. $x = e^{-4t}\cos 4\sqrt{3}\,t - \cos 8t$

13.48. $x = -\frac{5}{4}e^{-2t}\cos 4t - \frac{3}{4}e^{-2t}\sin 4t + \frac{1}{2}\cos 2t + \frac{1}{4}\sin 2t$

13.49. $x_s = \frac{1}{2}\cos 2t + \frac{1}{4}\sin 2t = \frac{\sqrt{5}}{4}\cos(2t - 0.46)$

13.50. $x = -4e^{-3t}\cos\sqrt{3}\,t - 6\sqrt{3}\,e^{-3t}\sin\sqrt{3}\,t + 4\cos 3t + 2\sin 3t$

13.51. $x_s = 4\cos 3t + 2\sin 3t = \sqrt{20}\cos(3t - 0.46)$

13.52. $q = \frac{1}{100}(3e^{-50t} - 15e^{-10t} + 12);$
$I = \frac{3}{2}(e^{-10t} - e^{-50t})$

13.53. $I = 10.09e^{-50t} \sin 50\sqrt{19}\,t;\ q = \frac{11}{250}\left(1 - e^{-50t} \cos 50\sqrt{19}\,t - \frac{1}{\sqrt{19}} e^{-50t} \sin 50\sqrt{19}\,t\right)$

13.54. $I = \frac{5}{4}(e^{-50t} - e^{-10t})$

13.55. $I = 24te^{-500t}$

13.56. $I = -\frac{2}{5}e^{-4t} \cos 6t + \frac{82}{15}e^{-4t} \sin 6t + \frac{2}{5}\cos 2t - \frac{6}{5}\sin 2t$

13.57. $I_s = \frac{2}{5}\cos 2t - \frac{6}{5}\sin 2t = \frac{\sqrt{40}}{5}\cos(2t + 1.25)$

13.58. $I = -\frac{150}{52}e^{-4t} \cos 3t - \frac{425}{52}e^{-4t} \sin 3t + \frac{150}{52}\cos 3t + \frac{225}{52}\sin 3t$

13.59. $I_s = \frac{150}{52}\cos 3t + \frac{225}{52}\sin 3t = 5.2\cos(3t - 0.983)$

13.60. $I = -e^{-200t} \cos 400t + \frac{11}{2}e^{-200t} \sin 400t + \cos 200t - 2\sin 200t$

13.61. $I_s = \cos 200t - 2\sin 200t = \sqrt{5}\cos(200t + 1.11)$

13.62. $q = \frac{30}{61}e^{-10t} \cos 50t + \frac{36}{61}e^{-10t} \sin 50t - \frac{30}{61}\cos 60t - \frac{25}{61}\sin 60t$

13.63. $q_s = -\frac{30}{61}\cos 60t - \frac{25}{61}\sin 60t = -0.64\cos(60t - 0.69)$

13.64. 0

13.65. $\frac{1}{640{,}001}(6392\cos t + 320\sin t)$

13.66. 1.28 ft = 15.36 in submerged

13.67. $x = -\frac{1}{6}\cos 5t - \frac{1}{5}\sin 5t$

13.68. $x = -0.260\cos(5t - 0.876)$

13.69. 0.764 ft submerged

13.70. $x = -0.236\cos 6.47t$

13.71. (*a*) $\omega = 6.47$ Hz; (*b*) $f = 1.03$ Hz; (*c*) $T = 0.97$ sec

13.72. (*a*) $\omega = 5$ Hz; (*b*) $f = 5/(2\pi)$ Hz; (*c*) $T = 2\pi/5$ sec

13.73. No equilibrium position; it sinks.

13.74. No equilibrium position; it sinks.

13.75. 9.02 cm submerged

13.76. $x = -4.80\sin 10.42t$

13.77. $x = c_1 \cos\sqrt{\frac{\pi\rho r^2}{m}}\,t + c_2 \sin\sqrt{\frac{\pi\rho r^2}{m}}\,t;$
$T = \frac{2}{r}\sqrt{\frac{\pi m}{\rho}}$

13.78. 0.236 ft = 2.84 in

13.79. 159.15 lb

13.80. $\ddot{x} + \frac{wl\rho}{m}x = 0$

13.81. (*a*) $T = 2\pi\sqrt{\frac{m}{wl\rho}}$; (*b*) period is reduced by $1/\sqrt{2}$

CHAPTER 14

14.27. $\dfrac{3}{s}$

14.28. $\dfrac{\sqrt{5}}{s}$

14.29. $\dfrac{1}{s-2}$

14.30. $\dfrac{1}{s+6}$

14.31. $\dfrac{1}{s^2}$

14.32. $-\dfrac{8}{s^2}$

14.33. $\dfrac{s}{s^2+9}$

14.34. $\dfrac{s}{s^2+16}$

14.35. $\dfrac{s}{s^2+b^2}$

14.36. $\dfrac{1}{(s+8)^2}$

14.37. $\dfrac{1}{(s-b)^2}$

14.38. $\dfrac{6}{s^4}$

14.39. $\dfrac{1-e^{-2s}}{s^2}$

14.40. $\dfrac{1-e^{-s}}{s}+\dfrac{e^{-(s-1)}-e^{-4(s-1)}}{s-1}$

14.41. $\dfrac{2}{s}(1-e^{-3s})$

14.42. $\dfrac{2(1-e^{-2s})}{s^2}$

14.43. $\dfrac{7!}{s^8}$

14.44. $\dfrac{s^2-9}{(s^2+9)^2}$

14.45. $\dfrac{120}{(s+1)^6}$

14.46. $\sqrt{\dfrac{\pi}{s}}$

14.47. $\dfrac{1}{1+3s}$

14.48. $\dfrac{15}{1+3s}$

14.49. $2\left[\dfrac{6}{s(s^2+12)}\right]=\dfrac{12}{s^3+12s}$

14.50. $\dfrac{8}{s+5}$

14.51. $3\dfrac{1/2}{s^2+1/4}=\dfrac{6}{4s^2+1}$

14.52. $\dfrac{-s}{s^2+19}$

14.53. $-0.9\sqrt{\pi}\,s^{-3/2}$

14.54. $\dfrac{2}{(s+1)^2+4}$

14.55. $\dfrac{2}{(s-1)^2+4}$

14.56. $\dfrac{s-1}{(s-1)^2+4}$

14.57. $\dfrac{s-3}{(s-3)^2+4}$

14.58. $\dfrac{s-3}{(s-3)^2+25}$

14.59. $\dfrac{\sqrt{\pi}}{2}(s-5)^{-3/2}$

14.60. $\dfrac{\sqrt{\pi}}{2}(s+5)^{-3/2}$

14.61. $\dfrac{2}{(s+2)[(s+2)^2+4]}$

14.62. $\dfrac{6}{s^4}+\dfrac{3s}{s^2+4}$

14.63. $\dfrac{5}{s-2}+\dfrac{7}{s+1}$

14.64. $\dfrac{2}{s}+\dfrac{3}{s^2}$

14.65. $\dfrac{3}{s}-\dfrac{8}{s^3}$

14.66. $\dfrac{2}{s^2}+\dfrac{15}{s^2+9}$

14.67. $\dfrac{2s-3}{s^2+9}$

14.68. $\dfrac{4s(s^2+3)}{(s^2-1)^3}$

14.69. $\dfrac{4(s+1)[(s+1)^2+3]}{[(s+1)^2-1]^3}$

14.70. $\dfrac{8(3s^2-16)}{(s^2+16)^3}$

14.71. $\dfrac{1}{2}\sqrt{\pi}\,(s-2)^{-3/2}$

14.72. $\dfrac{2}{(s^2-1)^2}$

14.73. $\dfrac{1}{s}\left[\dfrac{s-3}{(s-3)^2+1}\right]$

14.74. $\dfrac{1}{s(1+e^{-s})}$

14.75. $\dfrac{1-e^{-s}-se^{-2s}}{s^2(1-e^{-2s})}$

14.76. $\dfrac{(s+1)e^{-2s}+s-1}{s^2(1-e^{-2s})}$

CHAPTER 15

15.20. x

15.21. $2x$

15.22. x^2

15.23. $x^2/2$

15.24. $x^3/6$

15.25. e^{-2x}

15.26. $-2e^{2x}$

15.27. $4e^{-3x}$

15.28. $\dfrac{1}{2}e^{3x/2}$

15.29. $\dfrac{1}{2}x^2e^{2x}$

15.30. $2x^3e^{-5x}$

15.31. $\dfrac{3}{2}(\sin x + x\cos x)$

15.32. $\dfrac{\sqrt{3}}{6}(\sin\sqrt{3}\,x+\sqrt{3}\,x\cos\sqrt{3}\,x)$

15.33. $\dfrac{1}{2}\sin 2x$

15.34. $\dfrac{2}{3}e^{2x}\sin 3x$

15.35. $e^{-x}\cos\sqrt{5}\,x-\dfrac{1}{\sqrt{5}}e^{-x}\sin\sqrt{5}\,x$

15.36. $2e^{x}\cos\sqrt{7}\,x+\dfrac{3}{\sqrt{7}}e^{x}\sin\sqrt{7}\,x$

15.37. $\dfrac{1}{\sqrt{2}}\sin\dfrac{1}{\sqrt{2}}x$

15.38. $e^{x}\sin x$

15.39. $e^{-x}\cos 2x+e^{-x}\sin 2x$

15.40. $e^{(1/2)x}\cos 2x + \frac{1}{4}e^{(1/2)x}\sin 2x$

15.41. $e^{-(3/2)x}\cos\frac{\sqrt{11}}{2}x - \frac{1}{\sqrt{11}}e^{-(3/2)x}\sin\frac{\sqrt{11}}{2}x$

15.42. $e^x + \cos x + \sin x$

15.43. $\frac{1}{2}e^x - \frac{1}{2}e^{-x}$

15.44. $\cos x - e^x + xe^x$

15.45. $x + x^2$

15.46. $-x + 3x^2$

15.47. $x^2/2 + x^4/8$

15.48. $2x^3 + \frac{8}{\sqrt{\pi}}x^{5/2}$

15.49. $-1 + e^{2x}\cos 3x$

15.50. $2e^{(1/2)x}\cos\frac{\sqrt{3}}{2}x - \frac{2}{\sqrt{3}}e^{(1/2)x}\sin\frac{\sqrt{3}}{2}x$

15.51. $\frac{1}{6}x\sin 3x$

15.52. $-\frac{1}{2}e^x + \frac{1}{2}e^{(1/2)x}\cosh\frac{\sqrt{5}}{2}x + \frac{1}{2\sqrt{5}}e^{(1/2)x}\sinh\frac{\sqrt{5}}{2}x$

15.53. $\frac{1}{2}e^{-x}\cos\frac{1}{2}x - e^{-x}\sin\frac{1}{2}x$

CHAPTER 16

16.17. $x^3/6$

16.18. x^2

16.19. $e^{2x} - (2x + 1)$

16.20. $\frac{1}{6}(e^{4x} - e^{-2x})$

16.21. $e^x - x - 1$

16.22. $xe^{-x} + 2e^{-x} + x - 2$

16.23. $\frac{3}{2}(1 - \cos 2x)$

16.24. $1 - \cos x$

16.25. $e^{2x} - e^x$

16.26. x

16.27. $2(1 - e^{-x})$

16.28. $\frac{1}{13}(e^{5x} - e^{-8x})$

16.29. $x - \frac{1}{\sqrt{3}}\sin\sqrt{3}\,x$

16.30. $\frac{1}{4}(1 - \cos 2x)$

16.31. $1 - \cos 3x$

16.32. $x - \frac{1}{3}\sin 3x$

16.33. See Fig. 16-9.

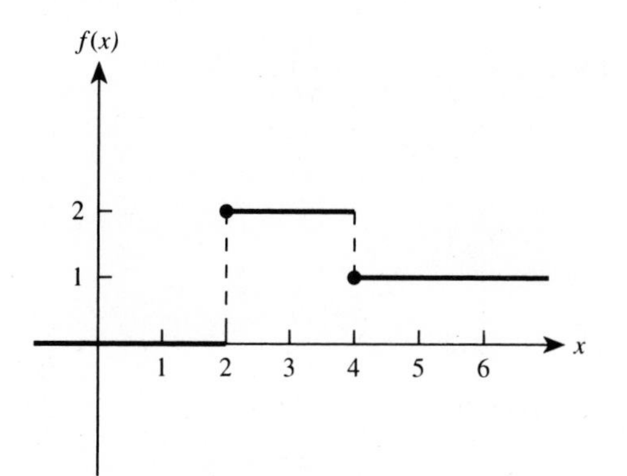

Fig. 16-9

16.34. See Fig. 16-10.

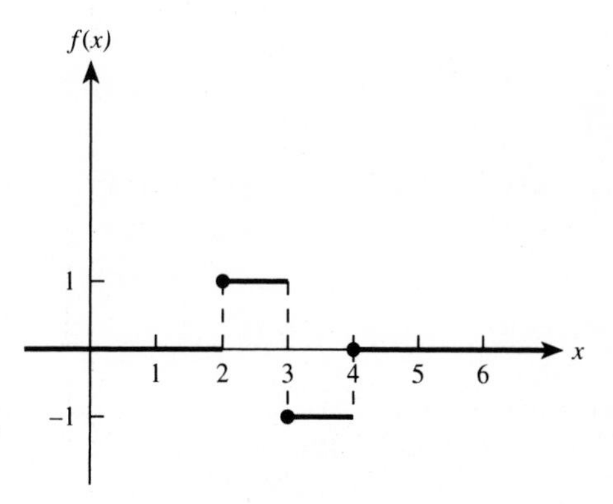

Fig. 16-10

16.35. $u(x) - u(x - c)$

16.36. See Fig. 16-11.

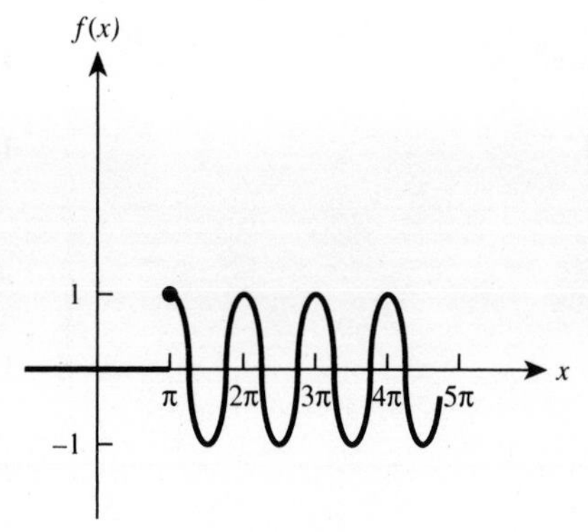

Fig. 16-11

16.37. See Fig. 16-12.

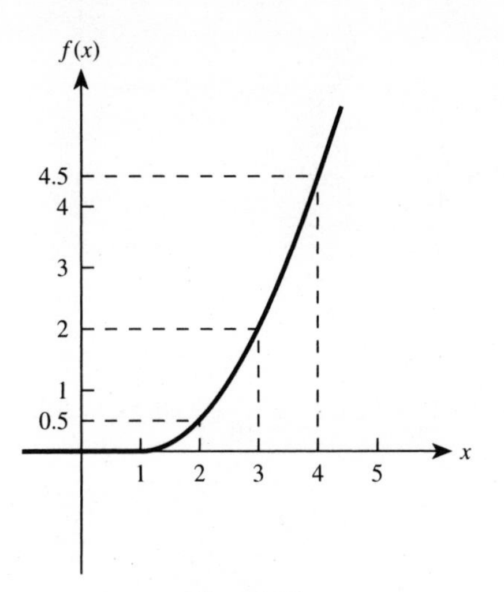

Fig. 16-12

16.38. $\dfrac{e^{-s}}{s^2 + 1}$

16.39. $\dfrac{e^{-3s}}{s^2}$

16.40. $g(x) = u(x-3)f(x-3)$ if $f(x) = x+3$

Then $G(s) = e^{-3s}\left(\dfrac{1}{s^2} + \dfrac{3}{s}\right)$

16.41. $g(x) = u(x-3)f(x-3)$ if $f(x) = x+4$.

16.42. $\dfrac{e^{-5s}}{s-1}$

16.43. $\dfrac{e^{-5(s-1)}}{s-1}$

16.44. $\dfrac{e^{-2s-3}}{s-1}$

16.45. $g(x) = u(x-2)f(x-2)$
if $f(x) = x^3 + 6x^2 + 12x + 9$.
Then $G(s) = e^{-2s}\left(\dfrac{6}{s^4} + \dfrac{12}{s^3} + \dfrac{12}{s^2} + \dfrac{9}{s}\right)$

16.46. $u(x-3)\cos 2(x-3)$

16.47. $\dfrac{1}{2}u(x-5)\sin 2(x-5)$

16.48. $\dfrac{1}{2}u(x-\pi)\sin 2(x-\pi)$

16.49. $2u(x-2)e^{3(x-2)}$

16.50. $8u(x-1)e^{-3(x-1)}$

16.51. $u(x-2)$

16.52. $(x-\pi)u(x-\pi)$

CHAPTER 17

17.17. $y = e^{-2x}$

17.18. $y = 1$

17.19. $y = \dfrac{2}{3}e^{-2x} + \dfrac{1}{3}e^{x}$

17.20. $y = e^{-2(x-1)}$

17.21. $y = 0$

17.22. $y = 2e^{5x} + xe^{5x}$

17.23. $y = -2e^{-x} + \dfrac{x^2}{2}e^{-x}$

17.24. $y = \dfrac{1}{2}\sin x - \dfrac{1}{2}\cos x + d_0e^{-x} \quad \left(d_0 = c_0 + \dfrac{1}{2}\right)$

17.25. $y = \dfrac{1}{101}(609e^{-20x} + 30\sin 2x - 3\cos 2x)$

17.26. $y = e^{x}$

17.27. $y = \dfrac{3}{4}e^{x} - \dfrac{3}{4}e^{-x} - \dfrac{1}{2}\sin x$

17.28. $y = \dfrac{1}{4}e^{x} + \dfrac{3}{4}e^{-x} + \dfrac{1}{2}xe^{x}$

17.29. $y = \dfrac{1}{10}e^{x} - \dfrac{1}{26}e^{-3x} - \dfrac{4}{65}\cos 2x - \dfrac{7}{65}\sin 2x$

17.30. $y = \dfrac{5}{2}\sin x - \dfrac{1}{2}x\cos x$

17.31. $y = 4e^{-(1/2)x}\cos\dfrac{\sqrt{3}}{2}x - \dfrac{2}{\sqrt{3}}e^{-(1/2)x}\sin\dfrac{\sqrt{3}}{2}x$

17.32. $y = \dfrac{3}{5}e^{-2x} + \dfrac{2}{5}e^{-x}\cos 2x + \dfrac{13}{10}e^{-x}\sin 2x$

17.33. $$y = \left[-\frac{1}{3} + \frac{1}{3}e^{-(5/2)(x-4)}\cosh\frac{\sqrt{37}}{2}(x-4) + \frac{5}{3\sqrt{37}}e^{-(5/2)(x-4)}\sinh\frac{\sqrt{37}}{2}(x-4)\right]u(x-4)$$

17.34. $y = \sin x$

17.35. $y = -5 + \frac{5}{3}e^{x} + \frac{10}{3}e^{-(1/2)x}\cos\frac{\sqrt{3}}{2}x$

17.36. $y = \frac{1}{4}e^{x} + \frac{1}{4}e^{-x} + \frac{1}{2}\cos x$

17.37. $y = e^{x}\left(1 + x + \frac{x^5}{60}\right)$

17.38. $N = 5000e^{0.085t}$

17.39. $T = 100e^{3t}$

17.40. $T = 70e^{-3t} + 30$

17.41. $v = d_0e^{-2t} + 16 \qquad (d_0 = c_0 - 16)$

17.42. $q = -\frac{4}{5}e^{-t} + \frac{8}{5}\sin 2t + \frac{4}{5}\cos 2t$

17.43. $x = \frac{1}{5}e^{-7t} - \frac{1}{5}e^{-2t}$

17.44. $x = 2(1 + t)e^{-2t}$

17.45. $x = -2e^{-4(t-\pi)}\sin 3t$

17.46. $q = \frac{1}{500}(110e^{-2t} - 101e^{-7t} + 13\sin t - 9\cos t)$

CHAPTER 18

18.7. $u(x) = x^2 + x \qquad v(x) = x - 1$

18.8. $u(x) = e^{2x} + 2e^{-x} \qquad v(x) = e^{2x} + e^{-x}$

18.9. $u(x) = 2e^{x} + 6e^{-x} \qquad v(x) = e^{x} + 2e^{-x}$

18.10. $y(x) = 1 \qquad z(x) = x$

18.11. $y(x) = e^{x} \qquad z(x) = e^{x}$

18.12. $w(x) = e^{5x} - e^{-x} + 1$
$y(x) = 2e^{5x} + e^{-x} - 1$

18.13. $w(x) = \cos x + \sin x$
$y(x) = \cos x - \sin x \qquad z(x) = 1$

18.14. $u(x) = -e^{x} + e^{-x} \qquad v(x) = e^{x} - e^{-x}$

18.15. $u(x) = e^{2x} + 1 \qquad v(x) = 2e^{2x} - 1$

18.16. $w(x) = x^2 \qquad y(x) = x \qquad z(x) = 1$

18.17. $w(x) = \sin x \qquad y(x) = -1 + \cos x$
$z(x) = \sin x - \cos x$

CHAPTER 19

19.18. $\begin{bmatrix} 3 & -1 \\ 2 & -1 \end{bmatrix}$

19.19. $\begin{bmatrix} 4 & 17 \\ -9 & -8 \end{bmatrix}$

19.20. $\begin{bmatrix} 2 & 5 & -2 \\ -3 & -3 & -1 \\ -1 & 1 & -3 \end{bmatrix}$

19.21. $\begin{bmatrix} 11 & 10 & 10 \\ 1 & -6 & 5 \\ 12 & 2 & 22 \end{bmatrix}$

19.22. Not defined

9.23. Not defined

19.24. (*a*) $\begin{bmatrix} 11 & -5 \\ -7 & 2 \end{bmatrix}$ (*b*) $\begin{bmatrix} 6 & 11 \\ 5 & 7 \end{bmatrix}$

19.25. $\begin{bmatrix} 1 & 0 \\ 0 & 1 \end{bmatrix} = \mathbf{I}$

19.26. $\begin{bmatrix} 2 & 3 \\ -1 & -2 \end{bmatrix}$

19.27. $\begin{bmatrix} -11 & -8 \\ 6 & -11 \end{bmatrix}$

19.28. (*a*) $\begin{bmatrix} 8 & 0 & 11 \\ -5 & 0 & -7 \\ 4 & 0 & 7 \end{bmatrix}$ (*b*) $\begin{bmatrix} 5 & 7 & 2 \\ 4 & 6 & 1 \\ 10 & 14 & 4 \end{bmatrix}$

19.29. (*a*) $\begin{bmatrix} -4 \\ 3 \end{bmatrix}$ (*b*) Not defined

19.30. Not defined

19.31. $\begin{bmatrix} 13 \\ -2 \\ 14 \end{bmatrix}$

19.32. $\lambda^2 - 1 = 0;\ \lambda_1 = 1,\ \lambda_2 = -1$

19.33. $\lambda^2 - 2\lambda + 13 = 0;\ \lambda_1 = 1 + 2\sqrt{3}\,i,$ $\lambda_2 = 1 - 2\sqrt{3}\,i$

19.34. $\lambda^2 - 2\lambda - 1 = 0;\ \lambda_1 = 1 + \sqrt{2},\ \lambda_2 = 1 - \sqrt{2}$

19.35. $\lambda^2 - 9 = 0;\ \lambda_1 = 3,\ \lambda_2 = -3$

19.36. $\lambda^2 - 10\lambda + 24 = 0;\ \lambda_1 = 4,\ \lambda_2 = 6$

19.37. $(1 - \lambda)(\lambda^2 + 1) = 0;\ \lambda_1 = 1,\ \lambda_2 = i,\ \lambda_3 = -i$

Each eigenvalue has multiplicity one.

19.38. $(-\lambda)(\lambda^2 - 5\lambda) = 0;\ \lambda_1 = 0,\ \lambda_2 = 0,\ \lambda_3 = 5$

The eigenvalue $\lambda = 0$ has multiplicity two, while $\lambda = 5$ has multiplicity one.

19.39. $\lambda^2 - 3t\lambda + t^2 = 0;\ \lambda_1 = \left(\frac{3}{2} + \frac{1}{2}\sqrt{5}\right)t,$ $\lambda_2 = \left(\frac{3}{2} - \frac{1}{2}\sqrt{5}\right)t$

19.40. $(5t - \lambda)(\lambda^2 - 25t^2) = 0;\ \lambda_1 = 5t,$ $\lambda_2 = 5t,\ \lambda_3 = -5t$

19.41. $\begin{bmatrix} 1 & 2t \\ 0 & 2 \end{bmatrix}$

19.42. $\begin{bmatrix} -2\sin 2t \\ (1 + 6t^2)e^{3t^2} \end{bmatrix}$

19.43. $\begin{bmatrix} \frac{1}{2}\sin 2 \\ \frac{1}{6}(e^3 - 1) \end{bmatrix}$

CHAPTER 20

20.13. $\lambda_1 = 2t,\ \lambda_2 = -3t;\ \begin{bmatrix} e^{2t} & 0 \\ 0 & e^{-3t} \end{bmatrix}$

20.14. $\lambda_1 = -t,\ \lambda_2 = 5t;$

$$\frac{1}{6}\begin{bmatrix} 4e^{5t} + 2e^{-t} & 2e^{5t} - 2e^{-t} \\ 4e^{5t} - 4e^{-t} & 2e^{5t} + 4e^{-t} \end{bmatrix}$$

20.15. $\lambda_1 = t,\ \lambda_2 = -t;\ \begin{bmatrix} 3e^t - 2e^{-t} & 3e^t - 3e^{-t} \\ -2e^t + 2e^{-t} & -2e^t + 3e^{-t} \end{bmatrix}$

20.16. $\lambda_1 = 2t,\ \lambda_2 = -4t;$

$$\frac{1}{6}\begin{bmatrix} 4e^{2t} + 2e^{-4t} & e^{2t} - e^{-4t} \\ 8e^{2t} - 8e^{-4t} & 2e^{2t} + 4e^{-4t} \end{bmatrix}$$

20.17. $\lambda_1 = -2t,\ \lambda_2 = -7t;$

$$\frac{1}{5}\begin{bmatrix} 7e^{-2t} - 2e^{-7t} & e^{-2t} - e^{-7t} \\ -14e^{-2t} + 14e^{-7t} & -2e^{-2t} + 7e^{-7t} \end{bmatrix}$$

20.18. $\lambda_1 = \lambda_2 = 2t;\ e^{2t}\begin{bmatrix} 1 & 0 \\ 0 & 1 \end{bmatrix}$

20.19. $\lambda_1 = \lambda_2 = 2t;\ e^{2t}\begin{bmatrix} 1 & t \\ 0 & 1 \end{bmatrix}$

20.20. $\lambda_1 = 2ti,\ \lambda_2 = -2ti;$

$$\begin{bmatrix} \cos 2t + 2\sin 2t & (5/2)\sin 2t \\ -2\sin 2t & \cos 2t - 2\sin 2t \end{bmatrix}$$

20.21. $\lambda_1 = 4it,\ \lambda_2 = -4it;\ \frac{1}{4}\begin{bmatrix} 4\cos 4t & \sin 4t \\ -16\sin 4t & 4\cos 4t \end{bmatrix}$

20.22. $\lambda_1 = \lambda_2 = -8t;\ e^{-8t}\begin{bmatrix} 1+8t & t \\ -64t & 1-8t \end{bmatrix}$

20.23. $\lambda_1 = \lambda_2 = -2t;\ e^{-2t}\begin{bmatrix} 1+2t & t \\ -4t & 1-2t \end{bmatrix}$

20.24. $\lambda_1 = 6it,\ \lambda_2 = -6it;\ \frac{1}{6}\begin{bmatrix} 6\cos 6t & \sin 6t \\ -36\sin 6t & 6\cos 6t \end{bmatrix}$

20.25. $\lambda_1 = (-4+3i)t,\ \lambda_2 = (-4-3i)t;\ \frac{e^{-4t}}{3}\begin{bmatrix} 3\cos 3t + 4\sin 3t & \sin 3t \\ -25\sin 3t & 3\cos 3t - 4\sin 3t \end{bmatrix}$

20.26. $\lambda_1 = (3+\sqrt{15}\,i)t,\ \lambda_2 = (3-\sqrt{15}\,i)t;\ \frac{e^{3t}}{\sqrt{15}}\begin{bmatrix} \sqrt{15}\cos\sqrt{15}\,t + \sin\sqrt{15}\,t & -2\sin\sqrt{15}\,t \\ 8\sin\sqrt{15}\,t & \sqrt{15}\cos\sqrt{15}\,t - \sin\sqrt{15}\,t \end{bmatrix}$

20.27. $\lambda_1 = \lambda_2 = \lambda_3 = 2t;\ e^{2t}\begin{bmatrix} 1 & t & t^2/2 \\ 0 & 1 & t \\ 0 & 0 & 1 \end{bmatrix}$

20.28. $\lambda_1 = \lambda_2 = \lambda_3 = 2t;\ e^{2t}\begin{bmatrix} 1 & 0 & 0 \\ 0 & 1 & t \\ 0 & 0 & 1 \end{bmatrix}$

20.29. $\lambda_1 = -t,\ \lambda_2 = \lambda_3 = 2t;$

$$\frac{1}{9}\begin{bmatrix} 9e^{-t} & -3e^{-t}+3e^{2t} & e^{-t}-e^{2t}+3te^{2t} \\ 0 & 9e^{2t} & 9te^{2t} \\ 0 & 0 & 9e^{2t} \end{bmatrix}$$

20.30. $\lambda_1 = \lambda_2 = \lambda_3 = 0;\ \begin{bmatrix} 1 & 0 & 0 \\ 0 & 1 & 0 \\ 0 & 0 & 1 \end{bmatrix}$

(see Problem 20.12)

20.31. $\lambda_1 = \lambda_2 = 0,\ \lambda_3 = t;\ \begin{bmatrix} 1 & t & 0 \\ 0 & 1 & 0 \\ 0 & 0 & e^t \end{bmatrix}$

20.32. $\lambda_1 = \lambda_2 = 0,\ \lambda_3 = t;\ \begin{bmatrix} 1 & 0 & 0 \\ t & 1 & 0 \\ e^t - 1 & 0 & e^t \end{bmatrix}$

CHAPTER 21

21.10. $\mathbf{x}(t) = \begin{bmatrix} x_1(t) \\ x_2(t) \end{bmatrix} \quad \mathbf{A}(t) = \begin{bmatrix} 0 & 1 \\ -1 & 2 \end{bmatrix} \quad \mathbf{f}(t) = \begin{bmatrix} 0 \\ t+1 \end{bmatrix} \quad \mathbf{c} = \begin{bmatrix} 1 \\ 2 \end{bmatrix} \quad t_0 = 1$

21.11. $\mathbf{x}(t) = \begin{bmatrix} x_1(t) \\ x_2(t) \end{bmatrix} \quad \mathbf{A}(t) = \begin{bmatrix} 0 & 1 \\ -\frac{1}{2} & 0 \end{bmatrix} \quad \mathbf{f}(t) = \begin{bmatrix} 0 \\ 2e^t \end{bmatrix} \quad \mathbf{c} = \begin{bmatrix} 1 \\ 1 \end{bmatrix} \quad t_0 = 0$

21.12. $\mathbf{x}(t) = \begin{bmatrix} x_1(t) \\ x_2(t) \end{bmatrix} \quad \mathbf{A}(t) = \begin{bmatrix} 0 & 1 \\ t & 3/t \end{bmatrix} \quad \mathbf{f}(t) = \begin{bmatrix} 0 \\ \dfrac{\sin t}{t} \end{bmatrix} \quad \mathbf{c} = \begin{bmatrix} 3 \\ 4 \end{bmatrix} \quad t_0 = 2$

21.13. $\mathbf{x}(t) = \begin{bmatrix} y_1(t) \\ y_2(t) \end{bmatrix} \quad \mathbf{A}(t) = \begin{bmatrix} 0 & 1 \\ 2t & -5 \end{bmatrix} \quad \mathbf{f}(t) = \begin{bmatrix} 0 \\ t^2+1 \end{bmatrix} \quad \mathbf{c} = \begin{bmatrix} 11 \\ 12 \end{bmatrix} \quad t_0 = 0$

21.14. $\mathbf{x}(t) = \begin{bmatrix} y_1(t) \\ y_2(t) \end{bmatrix} \quad \mathbf{A}(t) = \begin{bmatrix} 0 & 1 \\ 6 & 5 \end{bmatrix} \quad \mathbf{f}(t) = \begin{bmatrix} 0 \\ 0 \end{bmatrix}$ **c** and t_0 not specified

21.15. $\mathbf{x}(t) = \begin{bmatrix} x_1(t) \\ x_2(t) \\ x_3(t) \end{bmatrix} \quad \mathbf{A}(t) = \begin{bmatrix} 0 & 1 & 0 \\ 0 & 0 & 1 \\ 1 & -e^{-t} & te^{-t} \end{bmatrix} \quad \mathbf{f}(t) = \begin{bmatrix} 0 \\ 0 \\ 0 \end{bmatrix} \quad \mathbf{c} = \begin{bmatrix} 1 \\ 0 \\ 1 \end{bmatrix} \quad t_0 = -1$

21.16. $\mathbf{x}(t) = \begin{bmatrix} y_1(t) \\ y_2(t) \\ y_3(t) \end{bmatrix}$ $\mathbf{A}(t) = \begin{bmatrix} 0 & 1 & 0 \\ 0 & 0 & 1 \\ -2.5 & 2 & -1.5 \end{bmatrix}$ $\mathbf{f}(t) = \begin{bmatrix} 0 \\ 0 \\ 0.5t^2 + 8t + 10 \end{bmatrix}$ $\mathbf{c} = \begin{bmatrix} -1 \\ -2 \\ -3 \end{bmatrix}$ $t_0 = \pi$

21.17. $\mathbf{x}(t) = \begin{bmatrix} x_1(t) \\ x_2(t) \\ x_3(t) \end{bmatrix}$ $\mathbf{A}(t) = \begin{bmatrix} 0 & 1 & 0 \\ 0 & 0 & 1 \\ 0 & 0 & 0 \end{bmatrix}$ $\mathbf{f}(t) = \begin{bmatrix} 0 \\ 0 \\ t \end{bmatrix}$ $\mathbf{c} = \begin{bmatrix} 0 \\ 0 \\ 0 \end{bmatrix}$ $t_0 = 0$

21.18. $x(t) = \begin{bmatrix} x_1(t) \\ x_2(t) \\ y_1(t) \\ y_2(t) \\ z_1(t) \end{bmatrix}$ $\mathbf{A}(t) = \begin{bmatrix} 0 & 1 & 0 & 0 & 0 \\ 0 & 1 & 0 & 1 & -1 \\ 0 & 0 & 0 & 1 & 0 \\ t & 0 & -2 & 1 & 0 \\ 1 & 0 & -1 & 1 & 1 \end{bmatrix}$ $\mathbf{f}(t) = \begin{bmatrix} 0 \\ t \\ 0 \\ t^2 + 1 \\ 0 \end{bmatrix}$ $\mathbf{c} = \begin{bmatrix} 1 \\ 15 \\ 0 \\ -7 \\ 4 \end{bmatrix}$ $t_0 = 1$

21.19. $\mathbf{x}(t) = \begin{bmatrix} x_1(t) \\ x_2(t) \\ y_1(t) \end{bmatrix}$ $\mathbf{A}(t) = \begin{bmatrix} 0 & 1 & 0 \\ 0 & 2 & 5 \\ 0 & -1 & -2 \end{bmatrix}$ $\mathbf{f}(t) = \begin{bmatrix} 0 \\ 3 \\ 0 \end{bmatrix}$ $\mathbf{c} = \begin{bmatrix} 0 \\ 0 \\ 1 \end{bmatrix}$ $t_0 = 0$

21.20. $\mathbf{x}(t) = \begin{bmatrix} x_1(t) \\ y_1(t) \end{bmatrix}$ $\mathbf{A}(t) = \begin{bmatrix} 1 & 2 \\ 4 & 3 \end{bmatrix}$ $\mathbf{f}(t) = \begin{bmatrix} 0 \\ 0 \end{bmatrix}$ $\mathbf{c} = \begin{bmatrix} 2 \\ -3 \end{bmatrix}$ $t_0 = 7$

CHAPTER 22

22.9. $x = \frac{1}{3}e^{-4(t-1)} + \frac{2}{3}e^{2(t-1)}$

22.10. $x = \frac{1}{6}e^{-4t} + \frac{1}{3}e^{2t} - \frac{1}{2}$

22.11. $x = \frac{1}{6}e^{-4(t-1)} + \frac{1}{3}e^{2(t-1)} - \frac{1}{2}$

22.12. $x = \frac{1}{6}e^{-4t} + \frac{4}{3}e^{2t} - \frac{1}{2}$

22.13. $x = \frac{1}{2}e^{-4t} + \frac{1}{2}e^{2t} - e^{-t}$

22.15. $x = k_1 \cos t + k_2 \sin t$

22.16. $x = 0$

22.17. $x = -\cos(t - 1) + t$

22.18. $y = k_3 e^{-t} + k_4 e^{2t}$

22.19. $y = e^{-t} + e^{2t}$

22.20. $y = \frac{13}{12}e^{-t} + \frac{2}{3}e^{2t} + \frac{1}{4}e^{3t}$

22.21. $y = \frac{1}{12}e^{-t} + \frac{2}{3}e^{2t} + \frac{1}{4}e^{3t}$

22.22. $z = \frac{1}{500}(13 \sin t - 9 \cos t - 90e^{-2t} + 99e^{-7t})$

22.23. $x = e^{2t} + 2e^{-t}$ $y = e^{2t} + e^{-t}$

22.24. $x = 2e^t + 6e^{-t}$ $y = e^t + 2e^{-t}$

22.25. $x = t^2 + t$ $y = t - 1$

22.26. $x = k_3 e^{5t} + k_4 e^{-t}$ $y = 2k_3 e^{5t} - k_4 e^{-t}$

22.27. $x = \frac{1}{4}t^4 + 6t^2$

22.28. $x = -e^t + e^{-t}$ $y = e^t - e^{-t}$

22.29. $x = -8 \cos t - 6 \sin t + 8 + 6t$ $y = 4 \cos t - 2 \sin t - 3$

CHAPTER 23

23.26. Ordinary point

23.27. Ordinary point

23.28. Singular point

23.29. Singular point

23.30. Singular point

23.31. Singular point

23.32. Ordinary point

23.33. Singular point

23.34. Singular point

23.35. $y = a_0 + a_1\left(x + \frac{x^2}{2} + \frac{x^3}{6} + \cdots\right) = c_1 + c_2e^x$, where $c_1 = a_0 - a_1$ and $c_2 = a_1$

23.36. RF (recurrence formula): $a_{n+2} = \frac{-1}{(n+2)(n+1)}a_{n-1}$

$$y = a_0\left(1 - \frac{1}{6}x^3 + \frac{1}{180}x^6 + \cdots\right) + a_1\left(x - \frac{1}{12}x^4 + \frac{1}{504}x^7 + \cdots\right)$$

23.37. RF: $a_{n+2} = \frac{2}{n+2}a_n$

$$y = a_0\left(1 + x^2 + \frac{1}{2}x^4 + \frac{1}{6}x^6 + \cdots\right) + a_1\left(x + \frac{2}{3}x^3 + \frac{4}{15}x^5 + \frac{8}{105}x^7 + \cdots\right)$$

23.38. RF: $a_{n+2} = \frac{-1}{n+2}a_{n-1}$

$$y = a_0\left(1 - \frac{1}{3}x^3 + \frac{1}{18}x^6 + \cdots\right) + a_1\left(x - \frac{1}{4}x^4 + \frac{1}{28}x^7 + \cdots\right)$$

23.39. RF: $a_{n+2} = \frac{n-1}{(n+2)(n+1)}a_{n-1} + \frac{1}{(n+2)(n+1)}a_n$

$$y = a_0\left(1 + \frac{1}{2}x^2 + \frac{1}{24}x^4 + \frac{1}{20}x^5 + \cdots\right) + a_1\left(x + \frac{1}{6}x^3 + \frac{1}{12}x^4 + \frac{1}{120}x^5 + \cdots\right)$$

23.40. RF: $a_{n+2} = \frac{-2}{(n+2)(n+1)}a_{n-2}$

$$y = a_0\left(1 - \frac{1}{6}x^4 + \frac{1}{168}x^8 + \cdots\right) + a_1\left(x - \frac{1}{10}x^5 + \frac{1}{360}x^9 + \cdots\right)$$

23.41. RF: $a_{n+2} = \frac{n-1}{n+2}a_n$

$$y = a_0\left(1 - \frac{1}{2}x^2 - \frac{1}{8}x^4 - \frac{1}{16}x^6 - \cdots\right) + a_1x$$

23.42. RF: $a_{n+2} = \frac{1}{(n+2)(n+1)}a_{n-1}$

$$y = a_0\left(1 + \frac{1}{6}x^3 + \frac{1}{180}x^6 + \cdots\right) + a_1\left(x + \frac{1}{12}x^4 + \frac{1}{504}x^7 + \cdots\right)$$

23.43. RF: $a_{n+2} = \dfrac{1}{(n+2)(n+1)}(a_n + a_{n-1})$

$$y = a_0\left[1 + \frac{1}{2}(x-1)^2 + \frac{1}{6}(x-1)^3 + \frac{1}{24}(x-1)^4 + \cdots\right]$$

$$+ a_1\left[(x-1) + \frac{1}{6}(x-1)^3 + \frac{1}{12}(x-1)^4 + \cdots\right]$$

23.44. RF: $a_{n+2} = \dfrac{n-2}{(n+2)(n+1)}a_{n-1} - \dfrac{4n}{(n+2)(n+1)}a_n + \dfrac{4}{n+2}a_{n+1}$

$$y = a_0\left[1 - \frac{1}{6}(x+2)^3 - \frac{1}{6}(x+2)^4 + \cdots\right]$$

$$+ a_1\left[(x+2) + 2(x+2)^2 + 2(x+2)^3 + \frac{2}{3}(x+2)^4 + \cdots\right]$$

23.45. RF: $a_{n+2} = -\dfrac{n^2 - n + 1}{4(n+2)(n+1)}a_n, \qquad n > 1$

$$y = \left(\frac{1}{24}x^3 - \frac{7}{1920}x^5 + \cdots\right) + a_0\left(1 - \frac{1}{8}x^2 + \frac{1}{128}x^4 + \cdots\right) + a_1\left(x - \frac{1}{24}x^3 + \frac{7}{1920}x^5 + \cdots\right)$$

23.46. RF: $a_{n+2} = \dfrac{n}{(n+2)(n+1)}a_n, \qquad n > 2$

$$y = -\frac{1}{2}(x-1)^2 + a_0 + a_1\left[(x-1) + \frac{1}{6}(x-1)^3 + \frac{1}{40}(x-1)^5 + \cdots\right]$$

23.47. RF: $a_{n+2} = \dfrac{n}{(n+2)(n+1)}a_n + \dfrac{(-1)^n}{n!\,(n+2)(n+1)}$

$$y = \left(\frac{1}{2}x^2 - \frac{1}{6}x^3 + \frac{1}{8}x^4 - \frac{1}{30}x^5 + \cdots\right) + a_0 + a_1\left(x + \frac{1}{6}x^3 + \frac{1}{40}x^5 + \cdots\right)$$

23.48. $y = 1 - x - \dfrac{1}{3}x^3 - \dfrac{1}{12}x^4 - \cdots$

23.49. $y = 2(x-1) + \dfrac{1}{2}(x-1)^2 + (x-1)^3 + \cdots$

CHAPTER 24

24.25. RF (recurrence formula): $a_n = \dfrac{1}{[2(\lambda + n) - 1][(\lambda + n) - 1]}a_{n-1}$

$$y_1(x) = a_0 x\left(1 + \frac{1}{3}x + \frac{1}{30}x^2 + \frac{1}{630}x^3 + \cdots\right)$$

$$y_2(x) = a_0\sqrt{x}\left(1 + x + \frac{1}{6}x^2 + \frac{1}{90}x^3 + \cdots\right)$$

24.26. RF: $a_n = \dfrac{-1}{2(\lambda + n) - 1} a_{n-1}$

$$y_1(x) = a_0 x\left(1 - \frac{1}{3}x + \frac{1}{15}x^2 - \frac{1}{105}x^3 + \cdots\right)$$

$$y_2(x) = a_0\sqrt{x}\left(1 - \frac{1}{2}x + \frac{1}{8}x^2 - \frac{1}{48}x^3 + \cdots\right)$$

24.27. RF: $a_n = \dfrac{1}{[3(\lambda + n) + 1][(\lambda + n) - 2]} a_{n-2}$

$$y_1(x) = a_0 x^2\left(1 + \frac{1}{26}x^2 + \frac{1}{1976}x^4 + \cdots\right)$$

$$y_2(x) = a_0 x^{-1/3}\left(1 - \frac{1}{2}x^2 - \frac{1}{40}x^4 - \frac{1}{2640}x^6 - \cdots\right)$$

24.28. For convenience, first multiply the differential equation by x. Then

RF: $a_n = \dfrac{1}{(\lambda + n)^2} a_{n-1}$

$$y_1(x) = a_0\left(1 + x + \frac{1}{4}x^2 + \frac{1}{36}x^3 + \cdots\right)$$

$$y_2(x) = y_1(x)\ln x + a_0\left(-2x - \frac{3}{4}x^2 + \cdots\right)$$

24.29. RF: $a_n = \dfrac{-1}{(\lambda + n)^2} a_{n-3}$

$$y_1(x) = a_0\left(1 - \frac{1}{9}x^3 + \frac{1}{324}x^6 + \cdots\right)$$

$$y_2(x) = y_1(x)\ln x + a_0\left(\frac{2}{27}x^3 - \frac{1}{324}x^6 + \cdots\right)$$

24.30. RF: $a_n = \dfrac{1}{(\lambda + n) + 1} a_{n-1}$

$$y_1(x) = a_0 x\left(1 + \frac{1}{3}x + \frac{1}{12}x^2 + \frac{1}{60}x^3 + \cdots\right) = \frac{2}{x}a_0(e^x - 1 - x)$$

$$y_2(x) = a_0 x^{-1}\left(1 + x + \frac{1}{2!}x^2 + \frac{1}{3!}x^3 + \cdots\right) = a_0 x^{-1}e^x$$

24.31. For convenience, first multiply the differential equation by x. Then

RF: $a_n = \dfrac{1}{(\lambda + n) - 2} a_{n-1}$

$$y_1(x) = a_0 x^2\left(1 + x + \frac{1}{2!}x^2 + \frac{1}{3!}x^3 + \cdots\right) = a_0 x^2 e^x$$

$$y_2(x) = -y_1(x)\ln x + a_0(1 - x - x^2 + 0x^3 + \cdots)$$

24.32. RF: $a_n = \dfrac{-1}{2(\lambda + n) - 1} a_{n-1}$

$$y_1(x) = a_0\sqrt{x}\left(1 - \frac{1}{2}x + \frac{1}{8}x^2 - \frac{1}{48}x^3 + \cdots\right)$$

$$y_2(x) = -\frac{1}{2}y_1(x)\ln x + a_0x^{-1/2}\left(1 - \frac{1}{8}x^2 + \frac{3}{32}x^3 + \cdots\right)$$

24.33. RF: $a_n = \dfrac{-1}{(\lambda + n) - 2} a_{n-1}$

$$y_1(x) = a_0x^2\left(1 - x + \frac{1}{2!}x^2 - \frac{1}{3!}x^3 + \cdots\right) = a_0x^2e^{-x}$$

$$y_2(x) = y_1(x)\ln x + a_0x^2\left(x - \frac{3}{4}x^2 + \frac{11}{36}x^3 + \cdots\right)$$

24.34. $y = c_1x^{1/2} + c_2x^{-1/2}$

24.35. $y = c_1x^2 + c_2x^2\ln x$

24.36. $y = c_1x^{-1/2} + c_2x^{-4}$

24.37. $y = c_1x^{-1} + c_2x^2$

24.38. $y = c_1 + c_2x^7$

CHAPTER 25

25.19. 1.4296

25.20. 2.6593

25.21. 7.1733

25.22. −0.8887

25.23. 3.0718

25.24. $\frac{1}{3}\Gamma\left(\frac{1}{3}\right)$

25.25. $\frac{1}{2}\Gamma(2) = \frac{1}{2}$

25.26. First separate the $k = 0$ term from the series, then make the change of variables $j = k - 1$, and finally change the dummy index from j to k.

25.29. $\frac{1}{2}[J_0^2(1) + J_1^2(1)]$

CHAPTER 26

26.17. See Fig. 26-20.

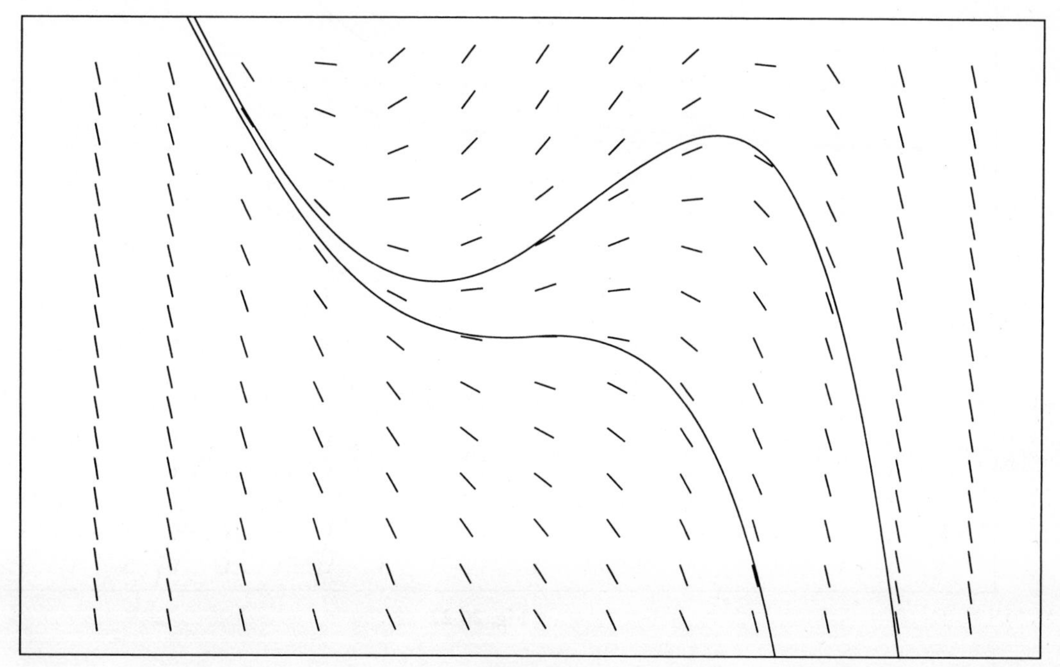

Fig. 26-20

26.18. See Fig. 26-21.

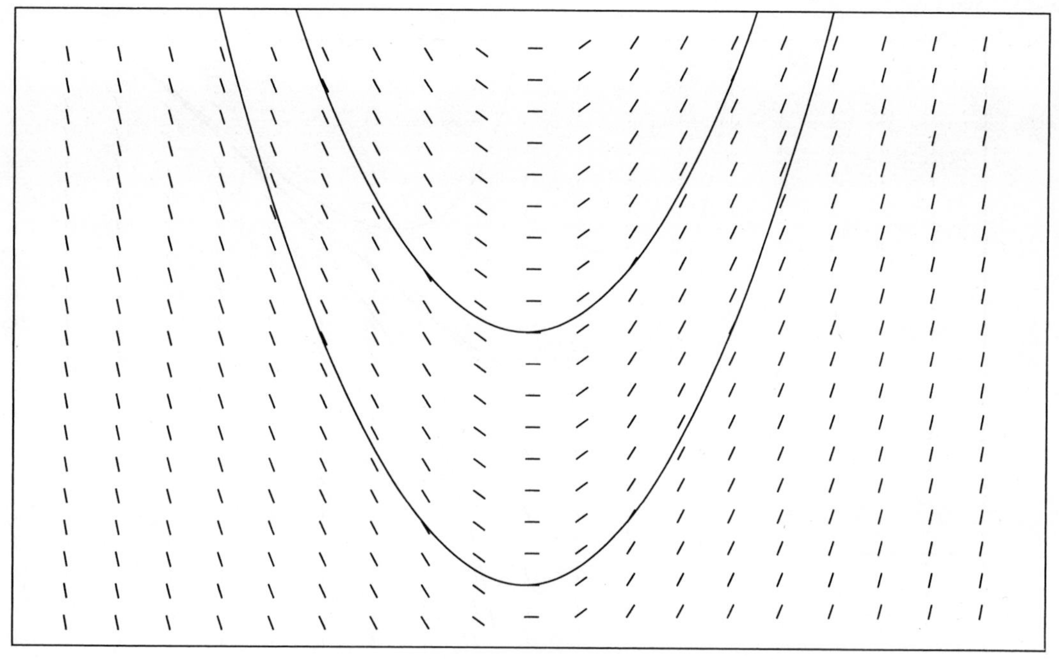

Fig. 26-21

26.19. See Fig. 26-22.

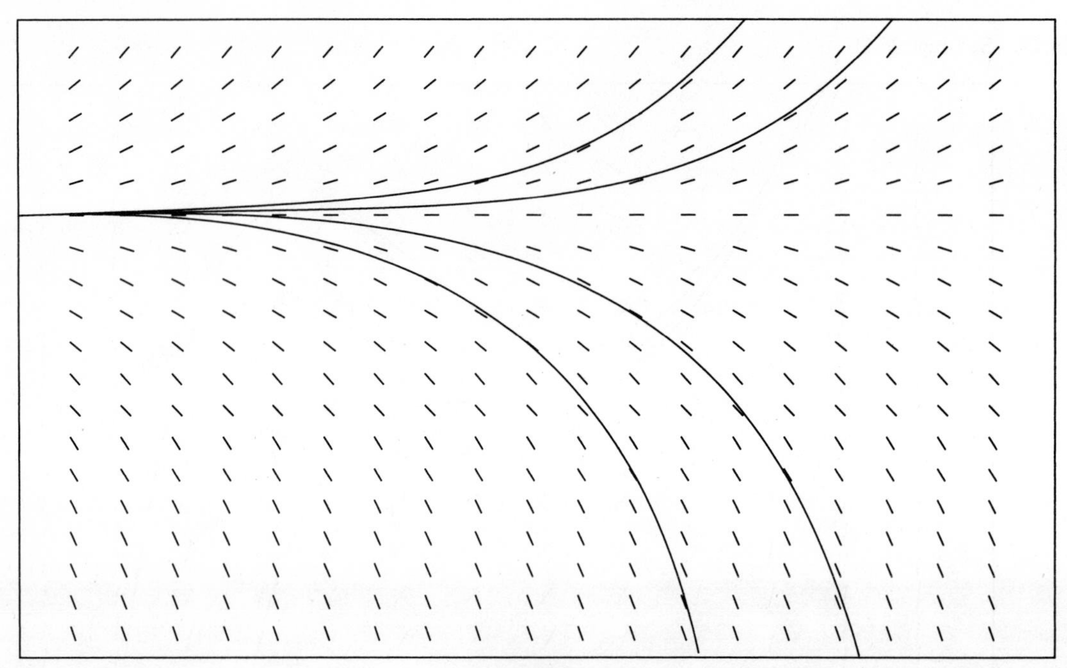

Fig. 26-22

26.20. See Fig. 26-23.

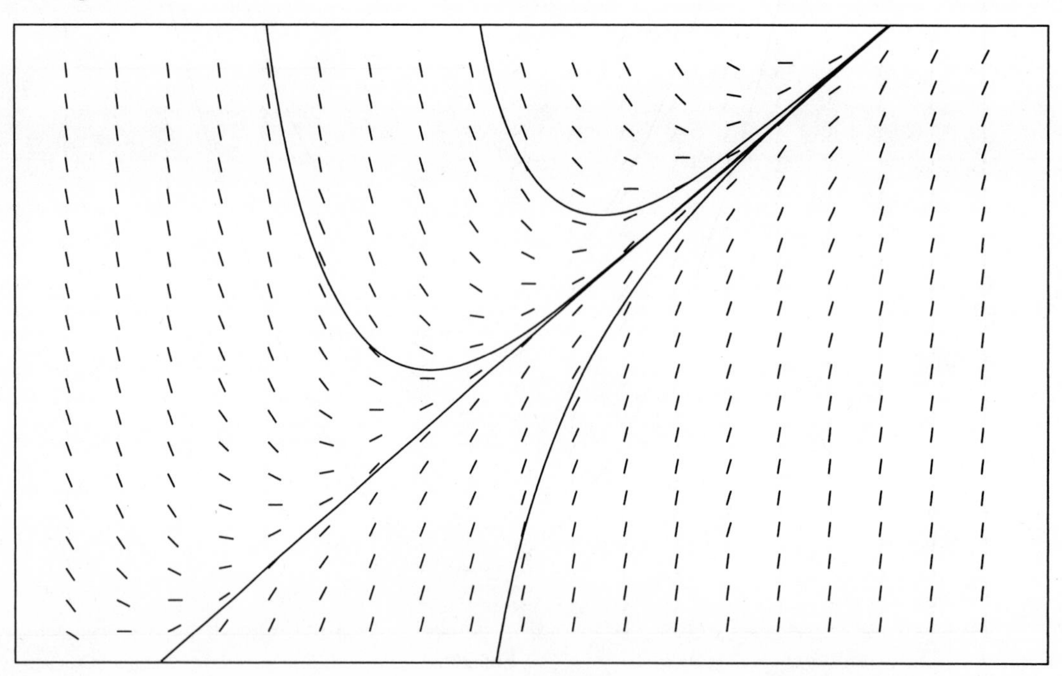

Fig. 26-23

26.21. See Fig. 26-24.

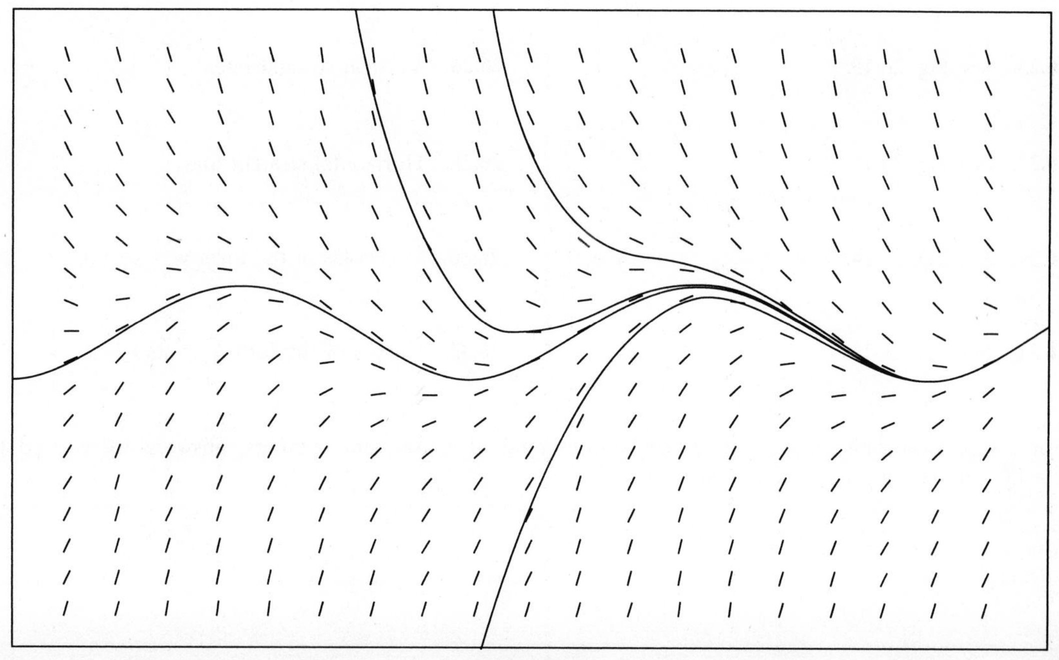

Fig. 26-24

26.22. Four solution curves are drawn, beginning at the points $(1,3)$, $(1,-3)$, $(-1,-3)$, and $(-1,3)$, respectively, and continuing in the positive x-direction. See Fig. 26-25.

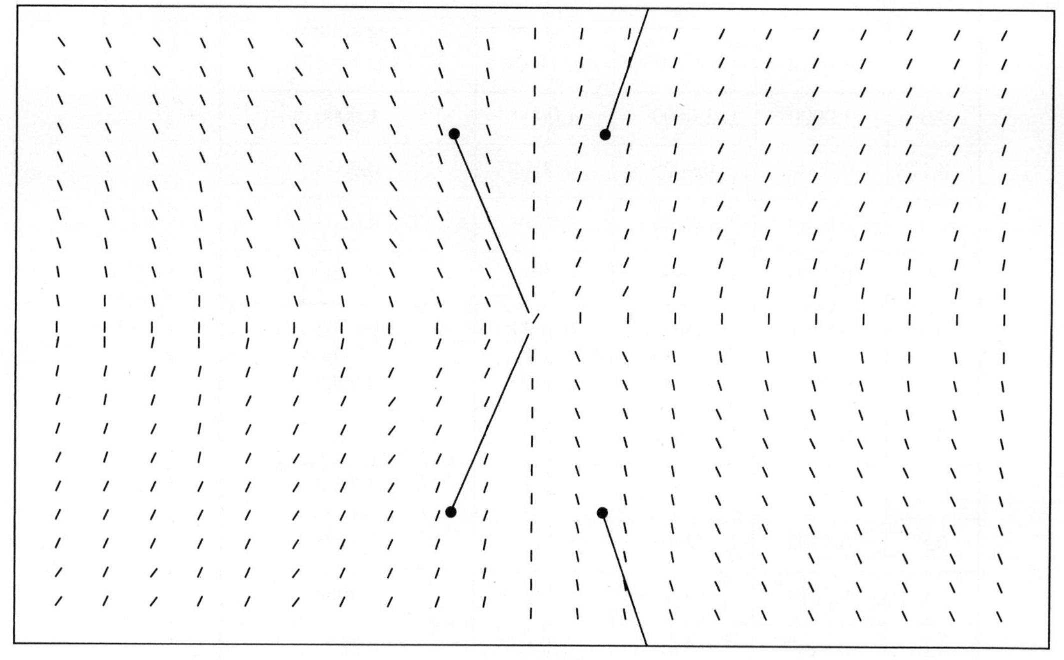

Fig. 26-25

26.23. See Fig. 26-17.

26.24. Straight lines of the form $y = x + (1 - c)$

26.25. See Fig. 26-15.

26.26. Vertical straight lines

26.27. See Fig. 26-16.

26.28. Horizontal straight lines

26.29. See Fig. 26-14.

26.30. Parabolas of the form $y = x^2 + c$

26.31. See Fig. 26-18.

26.32. Curves of the form $y = \sin x - c$

For comparison with other methods to be presented in subsequent chapters, answers are carried through $x = 1.0$, and are given for additional values of h.

26.33.

Method: EULER'S METHOD				
Problem: $y' = -y;\ y(0) = 1$				
x_n	y_n			True solution
	$h = 0.1$	$h = 0.05$	$h = 0.01$	$Y(x) = e^{-x}$
0.0	1.0000	1.0000	1.0000	1.0000
0.1	0.9000	0.9025	0.9044	0.9048
0.2	0.8100	0.8145	0.8179	0.8187
0.3	0.7290	0.7351	0.7397	0.7408
0.4	0.6561	0.6634	0.6690	0.6703
0.5	0.5905	0.5987	0.6050	0.6065
0.6	0.5314	0.5404	0.5472	0.5488
0.7	0.4783	0.4877	0.4948	0.4966
0.8	0.4305	0.4401	0.4475	0.4493
0.9	0.3874	0.3972	0.4047	0.4066
1.0	0.3487	0.3585	0.3660	0.3679

26.34.

Method: EULER'S METHOD				
Problem: $y' = 2x;\ y(0) = 0$				
x_n	y_n			True solution $Y(x) = x^2$
	$h = 0.1$	$h = 0.05$	$h = 0.01$	
0.0	0.0000	0.0000	0.0000	0.0000
0.1	0.0000	0.0050	0.0090	0.0100
0.2	0.0200	0.0300	0.0380	0.0400
0.3	0.0600	0.0750	0.0870	0.0900
0.4	0.1200	0.1400	0.1560	0.1600
0.5	0.2000	0.2250	0.2450	0.2500
0.6	0.3000	0.3300	0.3540	0.3600
0.7	0.4200	0.4550	0.4830	0.4900
0.8	0.5600	0.6000	0.6320	0.6400
0.9	0.7200	0.7650	0.8010	0.8100
1.0	0.9000	0.9500	0.9900	1.0000

26.35.

Method: EULER'S METHOD				
Problem: $y' = -y + x + 2;\ y(0) = 2$				
x_n	y_n			True solution $Y(x) = e^{-x} + x + 1$
	$h = 0.1$	$h = 0.05$	$h = 0.01$	
0.0	2.0000	2.0000	2.0000	2.0000
0.1	2.0000	2.0025	2.0044	2.0048
0.2	2.0100	2.0145	2.0179	2.0187
0.3	2.0290	2.0351	2.0397	2.0408
0.4	2.0561	2.0634	2.0690	2.0703
0.5	2.0905	2.0987	2.1050	2.1065
0.6	2.1314	2.1404	2.1472	2.1488
0.7	2.1783	2.1877	2.1948	2.1966
0.8	2.2305	2.2401	2.2475	2.2493
0.9	2.2874	2.2972	2.3047	2.3066
1.0	2.3487	2.3585	2.3660	2.3679

26.36.

Method: EULER'S METHOD

Problem: $y' = 4x^3$; $y(0) = 0$

x_n	y_n			True solution $Y(x) = x^4$
	$h = 0.1$	$h = 0.05$	$h = 0.01$	
0.0	0.0000	0.0000	0.0000	0.0000
0.1	0.0000	0.0000	0.0001	0.0001
0.2	0.0004	0.0009	0.0014	0.0016
0.3	0.0036	0.0056	0.0076	0.0081
0.4	0.0144	0.0196	0.0243	0.0256
0.5	0.0400	0.0506	0.0600	0.0625
0.6	0.0900	0.1089	0.1253	0.1296
0.7	0.1764	0.2070	0.2333	0.2401
0.8	0.3136	0.3600	0.3994	0.4096
0.9	0.5184	0.5852	0.6416	0.6561
1.0	0.8100	0.9025	0.9801	1.0000

CHAPTER 27

27.13.

Method: MODIFIED EULER'S METHOD

Problem: $y' = -y + x + 2$; $y(0) = 2$

x_n	$h = 0.1$		True solution $Y(x) = e^{-x} + x + 1$
	py_n	y_n	
0.0	—	2.000000	2.000000
0.1	2.000000	2.005000	2.004837
0.2	2.014500	2.019025	2.018731
0.3	2.037123	2.041218	2.040818
0.4	2.067096	2.070802	2.070320
0.5	2.103722	2.107076	2.106531
0.6	2.146368	2.149404	2.148812
0.7	2.194463	2.197210	2.196585
0.8	2.247489	2.249975	2.249329
0.9	2.304978	2.307228	2.306570
1.0	2.366505	2.368541	2.367879

27.14.

Method: MODIFIED EULER'S METHOD			
Problem: $y' = -y;\ y(0) = 1$			
x_n	$h = 0.1$		True solution $Y(x) = e^{-x}$
	py_n	y_n	
0.0	—	1.0000000	1.0000000
0.1	0.9000000	0.9050000	0.9048374
0.2	0.8145000	0.8190250	0.8187308
0.3	0.7371225	0.7412176	0.7408182
0.4	0.6670959	0.6708020	0.6703201
0.5	0.6037218	0.6070758	0.6065307
0.6	0.5463682	0.5494036	0.5488116
0.7	0.4944632	0.4972102	0.4965853
0.8	0.4474892	0.4499753	0.4493290
0.9	0.4049777	0.4072276	0.4065697
1.0	0.3665048	0.3685410	0.3678794

27.15.

Method: MODIFIED EULER'S METHOD			
Problem: $y' = \dfrac{x^2 + y^2}{xy};\ y(1) = 3$			
x_n	$h = 0.2$		True solution $Y(x) = x\sqrt{9 + \ln x^2}$
	py_n	y_n	
1.0	—	3.0000	3.0000
1.2	3.6667	3.6716	3.6722
1.4	4.3489	4.3530	4.3542
1.6	5.0393	5.0429	5.0444
1.8	5.7367	5.7399	5.7419
2.0	6.4404	6.4432	6.4456

27.16. The true solution is $Y(x) = x^2/2 - 1$, a second-degree polynomial. Since the modified Euler's method is a second-order method, it will generate the exact solution.

27.17.

Method: MODIFIED EULER'S METHOD			
Problem: $y' = 4x^3$; $y(2) = 6$			
x_n	$h = 0.2$		True solution $Y(x) = x^4 - 10$
	py_n	y_n	
2.0	—	6.0000	6.0000
2.2	12.4000	13.4592	13.4256
2.4	21.9776	23.2480	23.1776
2.6	34.3072	35.8080	35.6976
2.8	49.8688	51.6192	51.4656
3.0	69.1808	71.2000	71.0000

27.18.

Method: RUNGE–KUTTA METHOD		
Problem: $y' = -y + x + 2$; $y(0) = 2$		
x_n	$h = 0.1$ y_n	True solution $Y(x) = e^{-x} + x + 1$
0.0	2.000000	2.000000
0.1	2.004838	2.004837
0.2	2.018731	2.018731
0.3	2.040818	2.040818
0.4	2.070320	2.070320
0.5	2.106531	2.106531
0.6	2.148812	2.148812
0.7	2.196586	2.196585
0.8	2.249329	2.249329
0.9	2.306570	2.306570
1.0	2.367880	2.367879

27.19.

Method: RUNGE–KUTTA METHOD		
Problem: $y' = -y$; $y(0) = 1$		
x_n	$h = 0.1$ y_n	True solution $Y(x) = e^{-x}$
0.0	1.0000000	1.0000000
0.1	0.9048375	0.9048374
0.2	0.8187309	0.8187308
0.3	0.7408184	0.7408182
0.4	0.6703203	0.6703201
0.5	0.6065309	0.6065307
0.6	0.5488119	0.5488116
0.7	0.4965856	0.4965853
0.8	0.4493293	0.4493290
0.9	0.4065700	0.4065697
1.0	0.3678798	0.3678794

27.20.

Method: RUNGE–KUTTA METHOD		
Problem: $y' = \dfrac{x^2 + y^2}{xy}$; $y(1) = 3$		
x_n	$h = 0.2$ y_n	True solution $Y(x) = x\sqrt{9 + \ln x^2}$
1.0	3.0000000	3.0000000
1.2	3.6722028	3.6722045
1.4	4.3541872	4.3541901
1.6	5.0444406	5.0444443
1.8	5.7418469	5.7418514
2.0	6.4455497	6.4455549

27.21. Since the true solution $Y(x) = x^4 - 10$ is a fourth-degree polynomial, the Runge–Kutta method, which is a fourth-order numerical method, generates an exact solution.

27.22.

Method: RUNGE–KUTTA METHOD		
Problem: $y' = 5x^4$; $y(0) = 0$		
x_n	$h = 0.1$ y_n	True solution $Y(x) = x^5$
0.0	0.0000000	0.0000000
0.1	0.0000104	0.0000100
0.2	0.0003208	0.0003200
0.3	0.0024313	0.0024300
0.4	0.0102417	0.0102400
0.5	0.0312521	0.0312500
0.6	0.0777625	0.0777600
0.7	0.1680729	0.1680700
0.8	0.3276833	0.3276800
0.9	0.5904938	0.5904900
1.0	1.0000042	1.0000000

27.23.

Method: ADAMS–BASHFORTH–MOULTON METHOD			
Problem: $y' = y$; $y(0) = 1$			
x_n	$h = 0.1$		True solution $Y(x) = e^x$
	py_n	y_n	
0.0	—	1.0000000	1.0000000
0.1	—	1.1051708	1.1051709
0.2	—	1.2214026	1.2214028
0.3	—	1.3498585	1.3498588
0.4	1.4918201	1.4918245	1.4918247
0.5	1.6487164	1.6487213	1.6487213
0.6	1.8221137	1.8221191	1.8221188
0.7	2.0137473	2.0137533	2.0137527
0.8	2.2255352	2.2255418	2.2255409
0.9	2.4595971	2.4596044	2.4596031
1.0	2.7182756	2.7182836	2.7182818

27.24.

Method: ADAMS–BASHFORTH–MOULTON METHOD			
Problem: $y' = -y + x + 2;\ y(0) = 2$			
x_n	$h = 0.1$		True solution $Y(x) = e^{-x} + x + 1$
	py_n	y_n	
0.0	—	2.000000	2.000000
0.1	—	2.004838	2.004837
0.2	—	2.018731	2.018731
0.3	—	2.040818	2.040818
0.4	2.070323	2.070320	2.070320
0.5	2.106533	2.106530	2.106531
0.6	2.148814	2.148811	2.148812
0.7	2.196587	2.196585	2.196585
0.8	2.249330	2.249328	2.249329
0.9	2.306571	2.306569	2.306570
1.0	2.367880	2.367878	2.367879

27.25.

Method: ADAMS–BASHFORTH–MOULTON METHOD			
Problem: $y' = -y;\ y(0) = 1$			
x_n	$h = 0.1$		True solution $Y(x) = e^{-x}$
	py_n	y_n	
0.0	—	1.0000000	1.0000000
0.1	—	0.9048375	0.9048374
0.2	—	0.8187309	0.8187308
0.3	—	0.7408184	0.7408182
0.4	0.6703231	0.6703199	0.6703201
0.5	0.6065332	0.6065303	0.6065307
0.6	0.5488136	0.5488110	0.5488116
0.7	0.4965869	0.4965845	0.4965853
0.8	0.4493302	0.4493281	0.4493290
0.9	0.4065706	0.4065687	0.4065697
1.0	0.3678801	0.3678784	0.3678794

27.26.

Method: ADAMS–BASHFORTH–MOULTON METHOD

Problem: $y' = \dfrac{x^2 + y^2}{xy}$; $y(1) = 3$

x_n	$h = 0.2$		True solution $Y(x) = x\sqrt{9 + \ln x^2}$
	py_n	y_n	
1.0	—	3.0000000	3.0000000
1.2	—	3.6722028	3.6722045
1.4	—	4.3541872	4.3541901
1.6	—	5.0444406	5.0444443
1.8	5.7419118	5.7418465	5.7418514
2.0	6.4455861	6.4455489	6.4455549

27.27.

Method: MILNE'S METHOD

Problem: $y' = -y + x + 2$; $y(0) = 2$

x_n	$h = 0.1$		True solution $Y(x) = e^{-x} + x + 1$
	py_n	y_n	
0.0	—	2.000000	2.000000
0.1	—	2.004838	2.004837
0.2	—	2.018731	2.018731
0.3	—	2.040818	2.040818
0.4	2.070323	2.070320	2.070320
0.5	2.106533	2.106531	2.106531
0.6	2.148814	2.148811	2.148812
0.7	2.196588	2.196585	2.196585
0.8	2.249331	2.249329	2.249329
0.9	2.306571	2.306570	2.306570
1.0	2.367881	2.367879	2.367879

27.28.

Method: MILNE'S METHOD			
Problem: $y' = -y;\ y(0) = 1$			
x_n	$h = 0.1$		True solution $Y(x) = e^{-x}$
	py_n	y_n	
0.0	—	1.0000000	1.0000000
0.1	—	0.9048375	0.9048374
0.2	—	0.8187309	0.8187308
0.3	—	0.7408184	0.7408182
0.4	0.6703225	0.6703200	0.6703201
0.5	0.6065331	0.6065307	0.6065307
0.6	0.5488138	0.5488114	0.5488116
0.7	0.4965875	0.4965852	0.4965853
0.8	0.4493306	0.4493287	0.4493290
0.9	0.4065714	0.4065695	0.4065697
1.0	0.3678807	0.3678791	0.3678794

CHAPTER 28

28.15. $y' = z,\ z' = -y;\ y(0) = 1,\ z(0) = 0$

28.16. $y' = z,\ z' = y + x;\ y(0) = 0,\ z(0) = -1$

28.17. $y' = z,\ z' = 2xyz - (\sin x)y^3 + \dfrac{3}{y};\ y(1) = 0,\ z(1) = 15$

28.18. $y' = z,\ z' = w,\ w' = xw - \dfrac{z^2 y}{x};\ y(0) = 1,\ z(0) = 2,\ w(0) = 3$

28.19.

Method: EULER'S METHOD			
Problem: $y'' + y = 0$; $y(0) = 1$, $y'(0) = 0$			
x_n	$h = 0.1$		True solution
	y_n	z_n	$Y(x) = \cos x$
0.0	1.0000	0.0000	1.0000
0.1	1.0000	−0.1000	0.9950
0.2	0.9900	−0.2000	0.9801
0.3	0.9700	−0.2990	0.9553
0.4	0.9401	−0.3960	0.9211
0.5	0.9005	−0.4900	0.8776
0.6	0.8515	−0.5801	0.8253
0.7	0.7935	−0.6652	0.7648
0.8	0.7270	−0.7446	0.6967
0.9	0.6525	−0.8173	0.6216
1.0	0.5708	−0.8825	0.5403

28.20. Since the true solution is $Y(x) = -x$, a first-degree polynomial, Euler's method is exact and generates the true solution $y_n = -x_n$ at each x_n.

28.21.

Method: RUNGE–KUTTA METHOD			
Problem: $y'' + y = 0$; $y(0) = 1$, $y'(0) = 0$			
x_n	$h = 0.1$		True solution
	y_n	z_n	$Y(x) = \cos x$
0.0	1.0000000	0.0000000	1.0000000
0.1	0.9950042	−0.0998333	0.9950042
0.2	0.9800666	−0.1986692	0.9800666
0.3	0.9553365	−0.2955200	0.9553365
0.4	0.9210611	−0.3894180	0.9210610
0.5	0.8775827	−0.4794252	0.8775826
0.6	0.8253359	−0.5646420	0.8253356
0.7	0.7648425	−0.6442172	0.7648422
0.8	0.6967071	−0.7173556	0.6967067
0.9	0.6216105	−0.7833264	0.6216100
1.0	0.5403030	−0.8414705	0.5403023

28.22. Since the true solution is $Y(x) = -x$, a first-degree polynomial, the Runge–Kutta method is exact and generates the true solution $y_n = -x_n$ at each x_n.

28.23.

Method: ADAMS–BASHFORTH–MOULTON METHOD

Problem: $y'' - 3y' + 2y = 0$; $y(0) = -1$, $y'(0) = 0$

x_n	$h = 0.1$				True solution
	py_n	pz_n	y_n	z_n	$Y(x) = e^{2x} - 2e^x$
0.0	—	—	−1.0000000	0.0000000	−1.0000000
0.1	—	—	−0.9889417	0.2324583	−0.9889391
0.2	—	—	−0.9509872	0.5408308	−0.9509808
0.3	—	—	−0.8776105	0.9444959	−0.8775988
0.4	−0.7582805	1.4670793	−0.7581212	1.4674067	−0.7581085
0.5	−0.5793682	2.1386965	−0.5791739	2.1390948	−0.5791607
0.6	−0.3243735	2.9954802	−0.3241340	2.9959702	−0.3241207
0.7	0.0273883	4.0822712	0.0276819	4.0828703	0.0276946
0.8	0.5015797	5.4542298	0.5019396	5.4549628	0.5019506
0.9	1.1299923	7.1791788	1.1304334	7.1800757	1.1304412
1.0	1.9519493	9.3404498	1.9524898	9.3415469	1.9524924

28.24.

Method: ADAMS–BASHFORTH–MOULTON METHOD

Problem: $y'' + y = 0$; $y(0) = 1$, $y'(0) = 0$

x_n	$h = 0.1$				True solution
	py_n	pz_n	y_n	z_n	$Y(x) = \cos x$
0.0	—	—	1.0000000	0.0000000	1.0000000
0.1	—	—	0.9950042	−0.0998333	0.9950042
0.2	—	—	0.9800666	−0.1986692	0.9800666
0.3	—	—	0.9553365	−0.2955200	0.9553365
0.4	0.9210617	−0.3894147	0.9210611	−0.3894184	0.9210610
0.5	0.8775837	−0.4794223	0.8775827	−0.4794259	0.8775826
0.6	0.8253371	−0.5646396	0.8253357	−0.5646431	0.8253356
0.7	0.7648439	−0.6442153	0.7648422	−0.6442186	0.7648422
0.8	0.6967086	−0.7173541	0.6967066	−0.7173573	0.6967067
0.9	0.6216119	−0.7833254	0.6216096	−0.7833284	0.6216100
1.0	0.5403043	−0.8414700	0.5403017	−0.8414727	0.5403023

28.25. Since the true solution is $Y(x) = -x$, a first-degree polynomial, the Adams–Bashforth–Moulton method is exact and generates the true solution $y_n = -x_n$ at each x_n.

28.26.

Method: MILNE'S METHOD

Problem: $y'' - 3y' + 2y = 0$; $y(0) = -1$, $y'(0) = 0$

x_n	$h = 0.1$				True solution
	py_n	pz_n	y_n	z_n	$Y(x) = e^{2x} - 2e^x$
0.0	—	—	−1.0000000	0.0000000	−1.0000000
0.1	—	—	−0.9889417	0.2324583	−0.9889391
0.2	—	—	−0.9509872	0.5408308	−0.9509808
0.3	—	—	−0.8776105	0.9444959	−0.8775988
0.4	−0.7582563	1.4671290	−0.7581224	1.4674042	−0.7581085
0.5	−0.5793451	2.1387436	−0.5791820	2.1390779	−0.5791607
0.6	−0.3243547	2.9955182	−0.3241479	2.9959412	−0.3241207
0.7	0.0274045	4.0823034	0.0276562	4.0828171	0.0276946
0.8	0.5015908	5.4542513	0.5019008	5.4548828	0.5019506
0.9	1.1299955	7.1791838	1.1303739	7.1799534	1.1304412
1.0	1.9519398	9.3404286	1.9524049	9.3413729	1.9524924

28.27.

Method: MILNE'S METHOD

Problem: $y'' + y = 0$; $y(0) = 1$, $y'(0) = 0$

x_n	$h = 0.1$				True solution
	py_n	pz_n	y_n	z_n	$Y(x) = \cos x$
0.0	—	—	1.0000000	0.0000000	1.0000000
0.1	—	—	0.9950042	−0.0998333	0.9950042
0.2	—	—	0.9800666	−0.1986692	0.9800666
0.3	—	—	0.9553365	−0.2955200	0.9553365
0.4	0.9210617	−0.3894153	0.9210611	−0.3894183	0.9210610
0.5	0.8775835	−0.4794225	0.8775827	−0.4794254	0.8775826
0.6	0.8253369	−0.5646395	0.8253358	−0.5646426	0.8253356
0.7	0.7648437	−0.6442148	0.7648423	−0.6442178	0.7648422
0.8	0.6967086	−0.7173535	0.6967069	−0.7173564	0.6967067
0.9	0.6216120	−0.7833245	0.6216101	−0.7833272	0.6216100
1.0	0.5403047	−0.8414690	0.5403024	−0.8414715	0.5403023

28.28. predictors: $py_{n+1} = y_n + hy'_n$

$pz_{n+1} = z_n + hz'_n$

correctors: $y_{n+1} = y_n + \frac{h}{2}(py'_{n+1} + y'_n)$

28.29. $y_{n+1} = y_n + \frac{1}{6}(k_1 + 2k_2 + 2k_3 + k_4)$

$$z_{n+1} = z_n + \frac{1}{6}(l_1 + 2l_2 + 2l_3 + l_4)$$

$$w_{n+1} = w_n + \frac{1}{6}(m_1 + 2m_2 + 2m_3 + m_4)$$

where $k_1 = hf(x_n, y_n, z_n, w_n)$

$$l_1 = hg(x_n, y_n, z_n, w_n)$$

$$m_1 = hr(x_n, y_n, z_n, w_n)$$

$$k_2 = hf\left(x_n + \frac{1}{2}h, y_n + \frac{1}{2}k_1, z_n + \frac{1}{2}l_1, w_n + \frac{1}{2}m_1\right)$$

$$l_2 = hg\left(x_n + \frac{1}{2}h, y_n + \frac{1}{2}k_1, z_n + \frac{1}{2}l_1, w_n + \frac{1}{2}m_1\right)$$

$$m_2 = hr\left(x_n + \frac{1}{2}h, y_n + \frac{1}{2}k_1, z_n + \frac{1}{2}l_1, w_n + \frac{1}{2}m_1\right)$$

$$k_3 = hf\left(x_n + \frac{1}{2}h, y_n + \frac{1}{2}k_2, z_n + \frac{1}{2}l_2, w_n + \frac{1}{2}m_2\right)$$

$$l_3 = hg\left(x_n + \frac{1}{2}h, y_n + \frac{1}{2}k_2, z_n + \frac{1}{2}l_2, w_n + \frac{1}{2}m_2\right)$$

$$m_3 = hr\left(x_n + \frac{1}{2}h, y_n + \frac{1}{2}k_2, z_n + \frac{1}{2}l_2, w_n + \frac{1}{2}m_2\right)$$

$$k_4 = hf(x_n + h, y_n + k_3, z_n + l_3, w_n + m_3)$$

$$l_4 = hg(x_n + h, y_n + k_3, z_n + l_3, w_n + m_3)$$

$$m_4 = hr(x_n + h, y_n + k_3, z_n + l_3, w_n + m_3)$$

28.30. Same equations as given in Problem 28.13 with the addition of

$$pw_{n+1} = w_{n-3} + \frac{4h}{3}(2w'_n - w'_{n-1} + 2w'_{n-2})$$

$$w_{n+1} = w_{n-1} + \frac{h}{3}(pw'_{n+1} + 4w'_n + w'_{n-1})$$

CHAPTER 29

29.22. $y \equiv 0$

29.23. $y = x - \frac{\pi}{2}\sin x$

29.24. $y = \sin x$

29.25. $y = x + \left(1 - \frac{1}{2}\pi\right)\sin x - \cos x$

29.26. $y = B\cos x$, B arbitrary

29.27. No solution

29.28. No solution

29.29. $y = x + B\cos x$, B arbitrary

29.30. $\lambda = 1$, $y = c_1 e^{-x}$

29.31. No eigenvalues or eigenfunctions

29.32. $\lambda = 2$, $y = c_2 x e^{-2x}$ and $\lambda = \frac{1}{2}$, $y = c_2(-3 + x)e^{-x/2}$

29.33. $\lambda = 1$, $y = c_2 e^{-x}$ (c_2 arbitrary)

29.34. $\lambda_n = -n^2\pi^2$, $y_n = A_n \sin n\pi x \quad (n = 1, 2, \ldots)$ (A_n arbitrary)

29.35. $\lambda_n = \left(\frac{1}{5}n - \frac{1}{10}\right)^2 \pi^2$, $y_n = B_n \cos\left(\frac{1}{5}n - \frac{1}{10}\right)\pi x \quad (n = 1, 2, \ldots)$ (B_n arbitrary)

29.36. $\lambda_n = n^2$, $y_n = B_n \cos nx \quad (n = 0, 1, 2, \ldots)$ (B_n arbitrary)

29.37. Yes

29.38. No, $p(x) = \sin \pi x$ is zero at $x = \pm 1, 0$.

29.39. No, $p(x) = \sin x$ is zero at $x = 0$.

29.40. Yes

29.41. No, the equation is not equivalent to (*29.6*).

29.42. No, $w(x) = \frac{3}{x^2}$ is not continuous at $x = 0$.

29.43. Yes

29.44. $I(x) = e^x$; $(e^x y')' + xe^{-x}y + \lambda e^{-x}y = 0$

29.45. $I(x) = x$; $(xy')' + \lambda y = 0$

29.46. $\lambda_n = n^2$, $e_n(x) = \cos nx \quad (n = 0, 1, 2, \ldots)$

29.47. $\lambda_n = \frac{n^2}{4}$, $e_n(x) = \sin\frac{nx}{2} \quad (n = 1, 2, \ldots)$

CHAPTER 30

30.12. $\frac{2}{\pi}\sum_{n=1}^{\infty}\frac{1}{n}[1 - (-1)^n]\sin n\pi x$

30.13. $-\frac{6}{\pi}\sum_{n=1}^{\infty}\frac{(-1)^n}{n}\sin\frac{n\pi x}{3}$

30.14. $\frac{1}{3}\pi^2 + 4\sum_{n=1}^{\infty}\frac{(-1)^n}{n^2}\cos nx$

30.15. $\frac{2}{3} - \frac{4}{\pi}\sum_{n=1}^{\infty}\frac{1}{n}\sin\frac{2n\pi}{3}\cos\frac{n\pi x}{3}$

30.16. 1

30.17. $\sum_{n=1}^{\infty}\left(\frac{4}{n^2\pi^2}\sin\frac{n\pi}{2} + \frac{2}{n\pi}\cos\frac{n\pi}{2} - \frac{4}{n\pi}\cos n\pi\right)\sin\frac{n\pi x}{2}$

30.18. $-\frac{2}{\pi}\sum_{n=1}^{\infty}\frac{(-1)^n}{n - \frac{1}{2}}\cos\left(n - \frac{1}{2}\right)x$

30.19. $-\frac{2}{\pi}\sum_{n=1}^{\infty}\frac{(-1)^n}{\left(n - \frac{1}{2}\right)^2}\sin\left(n - \frac{1}{2}\right)x$

30.20. (*a*) yes; (*b*) no, $\lim_{\substack{x\to 2\\ x>2}} f(x) = \infty$; (*c*) no, $\lim_{\substack{x\to 2\\ x>2}} f(x) = \infty$;
(*d*) yes, $f(x)$ is continuous on $[-1, 5]$

30.21. (*a*) yes; (*b*) yes; (*c*) no, since $\lim_{\substack{x\to 0\\ x>0}} \ln|x| = -\infty$; (*d*) no, since $\lim_{\substack{x\to 1\\ x>1}} \frac{1}{3(x - 1)^{2/3}} = \infty$

Index

Adams–Bashforth–Moulton method, 255
  for systems, 273, 284
Addition of matrices, 166
Amplitude, 111
Analytic functions, 199
Applications:
  to buoyancy problems, 110
  to cooling problems, 43
  to dilution problems, 44
  to electrical circuits, 45, 109
  to falling-body problems, 43
  of first-order equations, 43
  to growth and decay problems, 43
  to orthogonal trajectories, 45
  of second-order equations, 108
  to spring problems, 108
  to temperature problems, 43
Archimedes principle, 110

Bernoulli equation, 8, 35
Bessel functions, 228
Bessel's equation, of order *p*, 228
  of order zero, 232
Boundary conditions, 2, 288
Boundary-value problems:
  definition, 2, 288
  Sturm–Liouville problems, 289
Buoyancy problems, 110

Cayley–Hamilton theorem, 168
Characteristic equation:
  for a linear differential equation, 77, 83
  of a matrix, 167
Characteristic value (*see* Eigenvalue)
Circular frequency, 111
Complementary solution, 68
Completing the square, method of, 138
Constant coefficients, 67, 77, 83, 88, 191
Constant matrix, 166
Convolution, 146
Cooling problems, 43
Critically damped motion, 111

Damped motion, 111
Decay problems, 43
Derivative:
  of a Laplace transform, 125
  of a matrix, 167
Differential equation, 1
  Bernoulli, 35
  with boundary conditions, 2, 288
  exact, 9, 24
  homogeneous, 8, 14, 67 (*See also* Homogeneous linear differential equations)
  with initial conditions, 2, 104
  linear, 8, 35, 67 (*See also* Linear differential equations)
  order of, 1
  ordinary or partial, 1
  separable, 8, 14
  solution of (*see* Solutions of ordinary differential equations)
  systems of (*see* Systems of differential equations)
Differential form, 8
Dilution problems, 44
Direction field, 236

$e^{\mathbf{A}t}$, 175, 191
Eigenfunctions, 289, 298
Eigenvalues:
  for a boundary-value problem, 289
  of a matrix, 168
  for a Sturm–Liouville problem, 289
Electrical circuits, 45, 109
Equilibrium point:
  for a buoyant body, 110
  for a spring, 108
Euler's constant, 232
Euler's equation, 225
Euler's method, 236
  modified, 254
  for systems, 272
Euler's relations, 81
Exact differential equation, 9, 24
Existence of solutions:
  of first-order equations, 12
  of linear initial-value problems, 67
  near an ordinary point, 199
  near a regular singular point, 213
Exponential of a matrix, 175

Factorial, 203, 230
Falling-body problem, 43
First-order differential equations:
  applications of, 43
  Bernoulli, 8
  differential form, 8

First-order differential equations (*Cont.*):
exact, 9, 24
existence and uniqueness theorem, 12
graphical methods, 236
homogeneous, 8, 14, 22
integrating factors, 24
linear, 8, 35
numerical solutions of (*see* Numerical methods)
separable, 8, 14
standard form, 8
systems of (*see* Systems of differential equations)
Fourier cosine series, 299
Fourier sine series, 298
Free motion, 111
Frequency, circular, 111
natural, 111
Frobenius, method of, 213

Gamma function, 228
table of, 229
General solution, 68 (*See also* Solutions of ordinary differential equations)
Graphical methods for solutions, 236
Growth problems, 43

Half-life, 49
Harmonic motion, simple, 111
Homogeneous boundary conditions, 288
Homogeneous boundary-value problem, 288
Sturm–Liouville problem, 289
Homogeneous linear differential equation, 67
characteristic equation for, 77, 83
with constant coefficients, 77, 83, 191
solution of (*see* Solutions of ordinary differential equations)
with variable coefficients, 199, 213
Homogeneous first-order equations, 8, 14, 22
Homogeneous function of degree n, 22
Hooke's law, 108
Hypergeometric equation, 225
Hypergeometric series, 226

Identity matrix, 167
Indicial equation, 213
Initial conditions, 2, 184
Initial-value problems, 2
solutions of, 2, 14, 104, 154, 191, 201
Instability, numerical, 237
Integral of a matrix, 167
Integrating factors, 24
Inverse Laplace transform, 138
Isocline, 236

$J_p(x)$ (*see* Bessel functions)

Kirchhoff's loop law, 109

$\mathbf{L}(y)$, 67
Laplace transforms, 125
applications to differential equations, 154
of convolution, 146
of derivatives, 154
derivatives of, 125
of integrals, 125
inverse of, 138
of periodic functions, 125
for systems, 161
table of, 305
of the unit step function, 147
Legendre polynomials, 206
Legendre's equation, 206
Limiting velocity, 44
Line element, 236
Linear dependence of functions, 67
Linear differential equations:
applications of, 43, 108
characteristic equation for, 77, 83
with constant coefficients, 67, 77, 83, 88, 191
existence and uniqueness of solution of, 67
first-order, 8, 35
general solution of, 68 (*See also* Solutions of ordinary differential equations)
homogeneous, 67, 199
nth-order, 83
nonhomogeneous, 67, 88, 97
ordinary point of, 199
regular singular point of, 213
second-order, 77, 199, 213
series solution of (*see* Series solutions)
singular point, 199
solutions of, 67 (*See also* Solutions of ordinary differential equations)
superposition of solutions of, 74
systems of (*see* Systems of differential equations)
with variable coefficients, 67, 199, 213
Linear independence:
of functions, 68
of solutions of a linear differential equation, 68

Matrices, 166
$e^{\mathbf{A}t}$, 175, 191
Method of Frobenius, 213
general solutions of, 214
Milne's method, 255
for systems, 284

Modified Euler's method, 254
Multiplication of matrices, 166
Multiplicity of an eigenvalue, 168

$n!$, 203, 230
Natural frequency, 111
Natural length of a spring, 109
Newton's law of cooling, 43
Newton's second law of motion, 43, 109
Nonhomogeneous boundary conditions, 288
Nonhomogeneous boundary-value problem, 288
Nonhomogeneous linear differential equations, 67
  existence of solutions, 68
  matrix solutions, 191
  power series solutions, 200
  undetermined coefficients, 88
  variation of parameters, 97
Nontrivial solutions, 288
Numerical instability, 237
Numerical methods, 254
  Adams–Bashforth–Moulton method, 255, 273, 284
  Euler's method, 236, 272
  Milne's method, 255, 284
  Modified Euler's method, 254
  order of, 255
  Runge–Kutta method, 255, 272
  stability of, 237
  starting values, 255
  for systems, 272

Order, of a differential equation, 1
  of a numerical method, 255
Ordinary differential equation, 1 (*see* Differential equation)
Ordinary point, 199
Orthogonal trajectories, 45
Oscillatory damped motion, 111
Overdamped motion, 111

Partial differential equation, 1
Partial fractions, method of, 138
Particular solution, 68
Period, 111
Periodic function, 125
Phase angle, 59, 111
Piecewise continuous function, 298
Piecewise smooth function, 298
Power series method, 200
Powers of a matrix, 167
Predictor-corrector methods, 254
Pure resonance, 115

RC circuits, 45
RCL circuits, 109
Recurrence formula, 200
Reduction to a system of differential equations, 183
Regular singular point, 213
Resonance, 115
RL circuit, 45
Runge–Kutta method, 255
  for systems, 272

Scalar multiplication, 166
Second-order linear equations, 77, 199, 213
  (*See also* Linear differential equations)
Separable equations, 8, 14
Series solutions:
  existence theorems for, 199
  indicial equation, 213
  method of Frobenius, 213
  near an ordinary point, 199
  recurrence formula, 200
  near a regular singular point, 213
  Taylor series method, 210
Simple harmonic motion, 111
Singular point, 199
Solutions of ordinary differential equations, 2, 67
  boundary-value problems, 2, 288
  from the characteristic equation, 77, 83
  complementary, 68
  for exact, 24
  existence of (*see* Existence of solutions)
  general, 68, 214
  by graphical methods, 236
  homogeneous, 14, 68, 77, 83
  by infinite series, (*see* Series solutions)
  for initial-value problem, 2, 67, 104
  by integrating factors, 24
  by Laplace transforms, 154
  for linear first order, 35
  linearly independent, 68
  by matrix methods, 191
  by the method of Frobenius, 213
  by numerical methods (*see* Numerical methods)
  near an ordinary point, 199
  particular, 68
  by power series, 200
  near a regular singular point, 213
  for separable equations, 14
  by superposition, 74
  of systems, 161, 191, 272
  by undetermined coefficients, 88
  uniqueness of (*see* Uniqueness of solutions)
  by variation of parameters, 97
Spring constant, 108
Spring problems, 108
Square matrix, 166

Standard form, 8
Starting values, 255
Steady-state current, 58, 111
Steady-state motion, 111
Step size, 237
Sturm–Liouville problems, 289, 298
Superposition, 74
Systems of differential equations, 161
  homogeneous, 191
  in matrix notation, 183
  solutions of, 161, 191, 272

Taylor series, 199, 210
Temperature problems, 43
Transient current, 58, 111
Transient motion, 111
Trivial solution, 288

Underdamped motion, 111
Undetermined coefficients, method of, 88
Uniqueness of solutions:
  of boundary-value problems, 289
  of first-order equations, 12
  of linear equations, 67
Unit step function, 146

Variable coefficients, 67, 199, 213
Variable separated, 8
Variation of parameters, method of, 97
Vectors, 166
Vibrating springs, 108

Wronskian, 68

Zero factorial, 231